Walker's

Carnivores of the World

Walker's

Carnivores of the World

Ronald M. Nowak

Introduction by
David W. Macdonald and
Roland W. Kays

The Johns Hopkins University Press
Baltimore and London

Printed in the United States of America on acid-free paper
9 8 7 6 5 4 3 2 1

Portions of this book have been adapted from *Walker's Mammals of the World,* 6th edition, by Ronald M. Nowak, © 1999 by The Johns Hopkins University Press.

The Johns Hopkins University Press
2715 North Charles Street
Baltimore, Maryland 21218-4363
www.press.jhu.edu

Library of Congress Cataloging-in-Publication Data

Nowak, Ronald M.
Walker's carnivores of the world / Ronald M. Nowak ; introduction by David W. Macdonald and Roland W. Kays.
p. cm.
Portions of this book have been adapted from Walker's mammals of the world, 6th ed., by Ronald M. Nowak.
Includes bibliographical references and index.
ISBN 0-8018-8033-5 (hardcover : alk. paper)—ISBN 0-8018-8032-7 (pbk. : alk. paper)
1. Carnivora. 2. Carnivora—Classification. I. Nowak, Ronald M. Walker's mammals of the world. II. Title.
QL737.C2N69 2005
599.7—dc22

2004012073

A catalog record for this book is available from the British Library.

Contents

Walker's

Carnivores of the World

Carnivores of the World: An Introduction

David W. Macdonald and Roland W. Kays

Carnivores are predacious, intelligent, bloodthirsty, and rare. Their predatory habits can limit populations of their prey, leading to a trickle-down effect on their prey's prey, and so on. For example, predation by gray wolves *(Canis lupus)* on Isle Royale regulates the local moose *(Alces)* population, which otherwise would overgraze the vegetation (McLaren and Peterson 1994). Hunting for a living requires keen senses, opportunism, and, in some cases, the ability to outwit prey. These traits have been widely recognized throughout human culture (from Aesop's fables to tales of the medieval bestiaries) and even co-opted for our own purposes in domestic cats and dogs (Clutton-Brock 1996). However, the opportunism, weaponry, and killer instincts that have ensured the carnivores' evolutionary success all too often led them into conflict with humans: not only do carnivores tend to compete with people for prey—either domestic or game—but occasionally we are their prey. Our consequent animosity and fear are deeply embedded in human culture (e.g., the brothers Grimm and their Little Red Riding Hood have a lot to answer for in stigmatizing the Big Bad Wolf and his kin). Indeed, Kruuk (2002) argues that for just these reasons human behavior toward carnivores (including persecution) is an innate heirloom of our evolutionary past.

Unfortunately, the onslaught of humans has eliminated many predators from their historic range and driven them toward, and in some cases over, the edge of extinction. Carnivores living at the top of the food pyramid are by definition never common. Even relatively few such individuals can significantly affect those on the lower strata of the pyramid and thus the structure and functioning of the ecosystem of which they are a part. From their precarious perch atop the food pyramid, these individuals can in turn be toppled by the actions of just a few humans. In the evolutionary marketplace of professional carnivores, there are many ways to make a living—some have gone extinct and some of those have been reinvented more than once in the 65 million years of the order's history. Today, 271 species of carnivores exist (including the pinnipeds; Wozencraft 1993), many of them threatened by one or another form of human activity and together exhibiting the amazing diversity described in this book.

Following this introduction, which aims first to summarize the salient features of the Carnivora, the bulk of this book comprises detailed accounts of the biology of this intriguing mammalian order. The accounts are grouped taxonomically, with a brief introduction preceding each family. To make sense of the detail packed into this book, we use this opening chapter to draw on a broader canvas, identifying evolutionary themes that weave through the natural history of the Carnivora, highlighting the adaptive significance of their diversity, and fi-

Female gray wolf (*Canus lupus*). Photo by Gerry Ellis/Minden Pictures.

nally considering their future in terms of the main issues in carnivore conservation. A general synthesis of carnivores—their behavior and evolution—is given by Macdonald (1992), while Gittleman et al. (2001a) detail the challenges posed by their conservation and Macdonald (2001a) illustrates their natural history.

What Is a Carnivore: Morphology

As members of the class Mammalia, the Carnivora are united in having mammary glands, which produce milk to nourish offspring; a covering of hair; and a muscular diaphragm separating the guts from the thoracic cavity. Traits shared with a few other taxonomic classes include endothermy (warm-bloodedness), a four-chambered heart, a relatively large brain, and live birth. Although all mammals have a relatively keen sense of smell, the evolution of most carnivore species has taken this to the extreme, with corresponding large olfactory lobes in the brain and elaborate turbinal skull bones to provide a large surface area to support the nasal mucosa. Most carnivores are easily recognized as predators by their specialized teeth, including long, sharp canines for killing and sharp scissorlike cheek teeth for cutting. Indeed, these scissor teeth—the carnassials—are the sole unique evolutionary hallmark of the Carnivora (they are not apparent in a few dietary specialists, although they were in their ancestors). This "carnassial connection" unites all major strands of carnivore evolution back through more than 60 million years, when the very first carnivores became distinguishable from their shrewlike ancestors. Although modern Carnivora are united by the heirloom of carnassial teeth, not much else in their anatomy unites them. For many important biological traits, carnivores exhibit a greater range than is seen in any other mammalian order. Body sizes span four orders of magnitude, ranging from the 80 g least weasel *(Mustela nivalis)* to the gigantic 800,000 g (800 kg) polar bear *(Ursus maritimus)* (Kays and Wilson 2002). In addition to this interspecific variation there is considerable intraspecific variation and flexibility. For example, within a population of gray wolves one can spot color variations from all white to all black, and across their range adult weight can vary between 15 and 80 kg.

TABLE 1. CONSERVATION STATUS OF THE 38 MOST ENDANGERED CARNIVORE SPECIES

Latin Name	Common Name	Risk Category
Dusicyon australis	Falkland Island wolf	Extinct
Monachus tropicalis	Caribbean monk seal	Extinct
Mustela macrodon	sea mink	Extinct
Procyon gloveralleni	Barbados raccoon	Extinct
Zalophus japonicus	Japanese sea lion	Extinct
Mustela nigripes	black-footed ferret	Formerly extinct in the the wild, recently reintroduced
Canis rufus	red wolf	Critically endangered
Canis simensis	Abyssian wolf	Critically endangered
Felis pardina	Iberian lynx	Critically endangered
Monachus monachus	Mediterranean monk seal	Critically endangered
Viverra civettina	Malabar civet	Critically endangered
Ailuropoda melanoleuca	giant panda	Endangered
Ailurua fulgens	lesser panda	Endangered
Bassaricyon lasius	Harris's olingo	Endangered
Bassaricyon pauli	Chiriqui olingo	Endangered
Catopuma badia	bay cat	Endangered
Crytoprocta ferox	fossa	Endangered
Cynogale bennettii	otter civet	Endangered
Enhydra lutris	sea otter	Endangered
Eumetopias jubatus	northern sea lion	Endangered
Eupleres goudotti	falanouc	Endangered
Galidictis grandidieri	giant-striped mongoose	Endangered
Genetta cristata	crested genet	Endangered
Herpestes palustris	Bengal mongoose	Endangered
Liberiictis kuhni	Liberian mongoose	Endangered
Lontra felina	marine otter	Endangered
Lontra provocax	huillin	Endangered
Lycaon pictus	African hunting dog	Endangered
Monachus schauinslandi	Hawaiian monk seal	Endangered
Mungotictus decemlineata	Malagasy narrow-striped mongoose	Endangered
Mustela felipei	Colombian weasel	Endangered
Mustela lutreola	European mink	Endangered
Mustela lutreolina	Indonesian mountain weasel	Endangered
Nasua nelsoni	Cozumel Island coati	Endangered
Oreailurus jacobita	Andean cat	Endangered
Panthera tigris	tiger	Endangered
Panthera uncia	snow leopard	Endangered
Procyon insularis	Tres Marias Islands raccoon	Endangered
Procyon maynardi	Bahaman raccoon	Endangered
Procyon pygmaeus	Cozumel Island raccoon	Endangered
Pteronura brasiliensis	giant Brazilian otter	Endangered

Source: IUCN 2003.

Note: Five species are extinct and gone forever. One is extinct in the wild but presently being reintroduced at a few sites. Five are critically endangered and facing an extremely high risk of extinction in the wild in the immediate future. The 28 endangered species are thought to face a high risk of extinction in the wild in the near future. In addition to those listed, there are an additional 58 carnivore species considered threatened or vulnerable and 21 for which we do not have enough data to evaluate their conservation status. More information can be found at www.redlist.org.

Within the extant members of the order, the felid, canid, and ursid families have instantly recognizable body designs. The mustelids, herpestids, and viverrids have, in general, converged on a "weasel-like" form, while the hyenas have their own distinctive sloping profile. The procyonids are a small yet diverse group of long-tailed mammals.

The cat family, Felidae, has one of the most conservative appearances of the Carnivora and is therefore one of the most easily recognized. All members have a certain "cat-ness," which is obvious at a glance. This is the result of morphological specializations for hunting which have allowed them to focus more on meat-eating than have the members of any other family. Their faces are blunt and flattened for more powerful bites (see below). Their large eyes and ears provide keen senses for stalking nocturnal prey, and their strong forelegs and retractable claws pack a powerful first punch.

The dog family, Canidae, is another group of carnivores united by a similar morphology familiar to most of us. This "dog-ness" is the result of an early evolutionary adaptation for fast pursuit of prey in open habitats. Compared with cats, dogs generally have lithe builds and long legs, giving them a slower initial strike but more stamina for the long chase. Canids do not typically swipe at prey with their nonretractile claws but lead with their sharp canine teeth.

The bear family, Ursidae, is a third unmistakable group of carnivores. Bears rely on strength more than speed to make a living, and they're built for it. Bulky and muscular, they can easily tear up a rotten log while hunting grubs or dispatch any large prey they can get their paws on. Their large size also allows them to pack on the fat during the seasons of plenty, in preparation for the lean season (sometimes including hibernation).

As they star in more nature documentaries, the Hyaenidae, with their unusual profile, are becoming increasingly familiar to Western cultures. Only hyenas have shorter hind legs than forelegs, giving them an unmistakable sloping profile. This posture enables them to cover long distances (up to 80 km in a night) using an energy-efficient "loping gallop" (Eloff 1964; Hofer and East 1993).

The Carnivora abounds with long, skinny predators adapted to chasing prey into burrows or other refuges. This body form has been so successful that it evolved multiple times in the order, and it characterizes most of the species of Mustelidae, Mephitidae, Viverridae, and Herpestidae. This morphological strategy usually includes short legs, a long and flexible torso, small ears and eyes, and a long, bushy tail. The trade-off to this body plan is that a large surface area per body mass results in high heat losses. To compensate, many of these species have very high metabolic rates and must eat constantly (e.g., small weasels must eat one-third to two-thirds of their body weight per day) (King 1989).

The Procyonidae are a small, varied group, typically mid-sized with long tails. Raccoons *(Procyon)* are stout creatures with short legs and bushy tails. Coatis (*Nasua* and *Nasuella*) stretch this plan a bit with pointy noses and longer tails. Olingos *(Bassaricyon)* and ringtails *(Bassariscus)* both have slender bodies with long, bushy tails, while the highly arboreal kinkajou *(Potos flavus)* is a bit stouter, with a muscular, prehensile tail.

THE BUSINESS END: CRANIODENTAL ADAPTATIONS FOR KILLING

A skilled anatomist can read the story of an animal's life in its teeth, from which much can be deduced about not just its diet but its lifestyle, ancestry, and society. This dental perspective reveals that, while the teeth of herbivores reflect their evolutionary battle with cellulose and are typically designed to maximize

Laughing spotted hyena (*Crocuta crocuta*) showing its canine teeth. Photo by Mitsuaki Iwago/Minden Pictures.

grinding ability, predators need strong, sharp teeth to kill and dismember their prey. But there is more than one way to acquire dinner, and the techniques used for killing vary across the order, with corresponding dental adaptations.

The stalk, chase, and pounce of a hunt are all designed to bring the predator's mouth in contact with the prey. Once there, the teeth do their work. The effectiveness of this first bite is critical for solitary hunters (e.g., most felids and mustelids), enabling them to avoid damage from a prey's counterattack (Ewer 1973). For some group hunters of large prey (including several examples from the canids, together with spotted hyenas, *Crocuta crocuta*), the first bite is merely one of many to follow; often a few individuals will hold or distract the prey with nonlethal bites and nips as preliminaries to disemboweling the exhausted quarry (Estes and Goddard 1967; Mech 1970; Schaller 1972). While these group attacks can be impressive in their collective power, it is the solitary killers that have more specialized bites. These are generally focused on the face, head, or neck of the prey, designed for a quick penetration of the brain case or spinal cord or slower, more methodical suffocation (Kruuk and Turner 1967; Rowe-Rowe 1976; Leyhausen 1979). Certain species of mustelids, viverrids, herpestids, and small felids are almost surgical killers in aiming for a specific bull's-eye on the lower neck of their prey. By stabbing their daggerlike upper canine between the last cervical and first thoracic vertebrae, they dislocate the two bones, lacerate the spinal cord, and inflict paralysis or death (Gossow 1970; Heidt 1972; Leyhausen 1979).

Making the kill is only a first step toward securing a meal; predators must also be able to open the carcass, chewing through skin, flesh, and bone. Specialized teeth are needed to do this, especially for larger prey. The importance of this is evident in the carcass of a deer that dies of starvation or cold; scavengers such as vultures, ravens, and small mammals will congregate around the potential meal and peck out the eyes, but they cannot consume it without the assistance of a properly toothed carnivore in opening the hide. Muscle and skin are best cut with sharp, bladelike teeth, while bone-cracking is better achieved with stout, blunt, cone-shaped teeth (Lucas and Luke 1984).

In carnivores, this butchering is completed with their unique adaptation—the carnassials. These specially adapted, bladelike cheek teeth (upper last premolar and lower first molar) come together like a pair of scissors, slicing through soft but tough food such as skin and muscle. Molars behind the carnassials are often modified for crushing bone and other hard food items in the predaceous carnivores that retain them (e.g., canids, mustelids, ursids). Carnivores that lack these postcarnassial molars may process hard foods using special buttressed precarnassial cheek teeth (e.g., the robust premolars in hyenas). Although this carnassial design is a Carnivoran patent, intriguingly, a similar order of mammals evolved the same general machinery at much the same time. These creatures, the creodonts, went extinct 8 million years ago, perhaps because they housed their scissor teeth farther back in the jaw than do carnivores, thereby limiting the opportunities for further adaptations of their postcarnassial cheek teeth.

The differences in killing and feeding behavior among contemporary carnivores are reflected in the morphology and biomechanics of their skulls and teeth (Ewer 1973; Biknevicius and Van Valkenburgh 1996). The most obvious difference is that jaw and snout lengths vary between groups. Mustelids and felids make up the short-snouted end of this variation, whereas canids are typified by longer snouts (Van Valkenburgh and Ruff 1987). The trade-off here is biomechanical; short snouts (brachycephalic) permit increased biting force at the front of the mouth, while long snouts (telocephalic) focus this force toward the back. Skull musculature has evolved in concert, so that the large temporalis muscles of felids and mustelids combine with their short dentition to maximize the power packed behind each canine tooth (Radinsky 1981; Van Valkenburgh and Ruff 1987; Biknevicius and Van Valkenburgh 1996). The snouts and teeth of group-hunting canids have evolved with less focus on a single killing bite and more on chewing the food after the kill is made. Although the bite of their canines is not as powerful, their longer rostrum gives their carnassials much greater shearing power (Kleiman and Eisenberg 1973). Among the canids, African wild dogs *(Lycaon pictus)*—often the butt of fierce competition from more powerful carnivores—have developed a capacity to dismember a carcass and gulp it down at astonishing speed, before it is stolen from them. Hyena dentition is characterized by massive bone-crushing premolars and powerful cranial musculature, both of which are particularly robust in the spotted hyena, the only extant hyena that actively preys on large mammals (Werdelin and Solounias 1991). A further adaptation of spotted hyenas is seen in the strong acid of their stomach, which enables them to digest bone.

Because canine teeth are the primary weapon for most carnivores, breaking these teeth can spell disaster. Thus, it is not surprising that canine teeth have evolved to be resistant to extreme stresses during prey capture and consumption. The different shapes observed in these teeth reflect the different physical stresses imposed by the variety of prey capture techniques observed in the order (Van Valkenburgh and Ruff 1987; Biknevicius and Van Valkenburgh 1996). Canid canines are not specifically adapted to deliver a lethal bite (Ewer 1973), nor are they as strong as the canines of felids (Van Valkenburgh and Ruff 1987). Compared to the somewhat knifelike upper canines of canids, those of felids are relatively round in cross section. A round cross section is more resistant to pressure from all angles, which is interpreted as an adaptation allowing felids to inflict deep, sustained bites while grasping struggling prey, during which stresses across the teeth may come from any direction. In contrast, the elliptical cross section of canid upper canines can withstand the loads created when they are used to inflict rapid, shallow, slashing wounds that apply pressure in one direction—along the long axis of the ellipse. This interpretation might seem to be weakened by the case of spotted hyenas, which kill in the manner of pack-hunting canids (Kruuk 1972) but have canines reminiscent in shape to those of the felids. However, the bone-cracking massive premolars may impose high incidental loads on the hyena's canines from all directions (Mills 1990).

VARIATION IN PELAGE AND ITS ADAPTIVE VALUE

The texture, color, and pattern of carnivore fur have been important to humans since our ancestors wore furs as our first clothes in prehistory; later, in pioneering days, the fur trade had a major influence on the history of colonization and commerce. These furs provided early humans with warmth and camouflage and probably also indicated the individual's social status within the clan. Not surprisingly, these three characteristics of fur are also important to the animals themselves. Some of these adaptations are obvious: thick fur provides more warmth; patterns break up the animals' outline to enhance camouflage. In general, coloration in mammals appears to follow Gloger's rule, with darker colors in riparian habitats and lighter colors in arid regions providing appropriate camouflage and protective coloration as well as thermoregulation to absorb or repel heat, respectively (Cott 1940; Hamilton 1973). Ortolani and Caro (1996) report on a comparative analysis (controlling for phylogenetic relationships) of differences in pelage appearance across the Carnivora. They found that white coats are associated with arctic conditions, pale coats with living in deserts, and dark fur with living in tropical forests. Spotted carnivores in general tend to be arboreal, while spotted felids, specifically, are usually forest dwellers. Intriguingly, this study could detect no significant relationship between striped pelage and grassland habitats or between sociality and contrasting throat patches, tail tips, or ear marks. This negative result challenges the prevailing wisdom, which assumes a functional association between these traits. Much remains to be discovered (or even considered) in the intricate pelage patterns of many carnivores. While it seems obvious that the dramatic black-and-white patterns of skunks (Mephitinae) render them conspicuous in the memory of any predator that has threatened them, it is less obvious what functions the striped faces of Eurasian badgers *(Meles meles)* or the ringed faces of polecats *(Mustela putorius)* may serve—although they too may be aposematic warnings.

A clear example of the adaptive significance of pelage colors is provided by Arctic foxes *(Alopex lagopus)*, which occur as two color morphs that change seasonally. The polar form is pure white in winter and grayish in summer. The so-called blue form is dark, steely blue in winter and a lighter chocolate brown in summer. Populations of Arctic foxes vary in the proportions of these two genetically determined color morphs. However, there is pattern in this variation: smaller landmasses tend to have higher proportions of the blue form, which is adapted to foraging on coastal habitats, whereas the white, or polar, form is adapted to be camouflaged in the snow. Blue foxes predominate on smaller islands because these have a higher ratio of coastline to inland than do larger landmasses.

Evolutionary History and Modern Diversity

The nonscientist might be puzzled to learn that even determining how many carnivore species there are can be tricky. The problem is that taxonomists disagree about what constitutes separate species and what are merely subspecies. Nowadays there is a strategic as well as a purely scientific basis to this issue, which has classically divided the so-called splitters and lumpers among taxonomists. The strategic point is that elevating a subspecies to specific status may create a rare species, and around the world conservation legislation tends to be concerned with rare species and less so with rare subspecies. The most recent revision of the genetic and morphological variation of the Carnivora recognized 271 species (including the pinnipeds; Wozencraft 1993). Wozencraft's total represents 5.8% of the 4,629 described mammalian species (Wilson and Reeder 1993) and is the fourth largest group behind the Rodentia (2,021 species), Chiroptera (925), and Insectivora (428). The modern Carnivora descend from an old lineage, diverging from other mammals perhaps as long as 65 million years ago.

Evolutionary biologists consider the physical adaptations and DNA shared by species to build phylogenetic trees plotting hypotheses about their evolutionary history. At least 177 of these trees have been constructed for members of the Car-

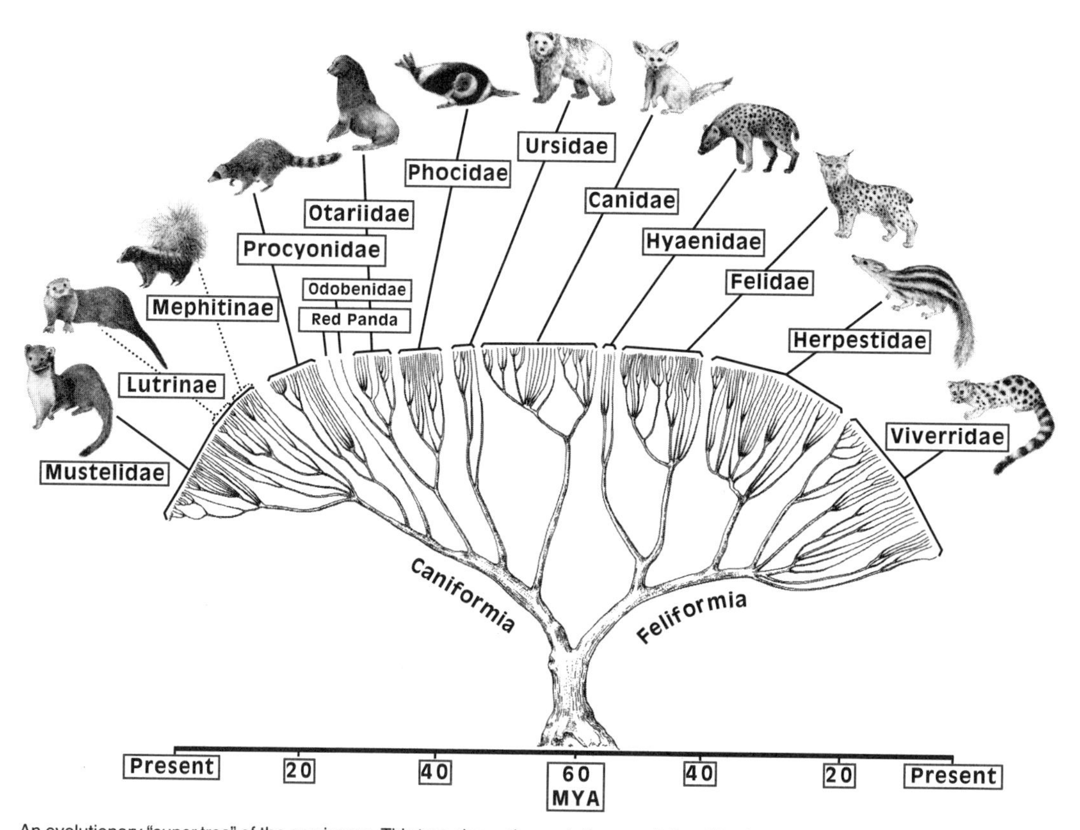

An evolutionary "super tree" of the carnivores. This tree shows the evolutionary relationships between the 271 living species of the order Carnivora and is based on the combination of 177 separate, smaller phylogenetic trees (Bininda-Emonds et al. 1999). The tip of each branch represents one species, and each cluster of branches indicates a presently recognized taxonomic family. The distance of a branching point from the branch tips suggests the estimated age of each evolutionary divergence, as indicated by the scale below the tree. Near the trunk the two major branches of the carnivores are highlighted: suborders Feliformia and Caniformia. Illustration by Patricia Kernan.

nivora, revealing some well-accepted relationships and a few controversial ones. Thankfully, a group of biologists has recently combined these different studies into one "super tree" showing the consensus relationships between all carnivores (Bininda-Emonds et al. 1999). This tree shows that the Carnivora (including the pinnipeds, see below) are monophyletic—that is, all carnivores are more closely related to one another than they are to any species from another order. The deepest split in the tree separates the two major clades: suborders (or superfamilies) Feliformia and Caniformia. Within the catlike carnivores are four well-defined groups presently classified as families: Felidae, Herpestidae, Hyaenidae, and Viverridae. Within the doglike carnivore branch there are seven living families including four terrestrial groups (Canidae, Mustelidae, Ursidae, and Procyonidae) and three aquatic families (Otariidae, Odobenidae, and Phocidae) (Wozencraft 1993). We also highlighted on the super tree three subgroups within the Caniformia that are particularly old and may deserve family-level recognition: the red panda *(Ailurus fulgens)*, the skunks, and the otters (Lutrinae) (Dragoo and Honeycutt 1997; Bininda-Emonds et al. 2002).

The carnivore super tree groups walrus and seals together into a monophyletic clade (Pinnipedia) but places them smack in the middle of the order, between the bears and a group consisting of the Procyonidae and Mustelidae. Few would argue that the 34 species of seal (Phocidae) and walrus (Odobenidae) are each other's closest relatives, but their inclusion within the Carnivora has been more contentious. Although their flippers and streamlined bodies make seals easily recognizable, the overall physical and life-history differences between aquatic and terrestrial carnivores are no greater than those that occur within each group (Bininda-Emonds et al. 2001). While *Walker's Mammals of the World* recognizes their position as an offshoot of the Carnivora, it retains the traditional taxonomy by recognizing Pinnipedia as an order by itself. Sensibly, pinnipeds are covered in *Walker's Marine Mammals of the World* (Nowak 2003) but not in this publication. We will maintain this convention and focus the rest of our introduction on the terrestrial Carnivora.

The Where: Geographic Distribution and Habitat Use

Carnivores live in virtually every habitat on earth, from short grassland (meerkat, *Suricata suricatta*) to sparse woodland (dwarf mongoose, *Helogale parvula*) to desert (fennec fox, *Fennecus zerda*) to thick tropical forest (kinkajou) to oceanic waters (sea otter, *Enhydra lutris*) to the arctic icecap (polar bear). There is remarkable variation in the size of a species' geographic range; foxes offer an excellent example, with ranges spanning five orders of magnitude, from the island fox *(Urocyon littoralis)*, circa 700 km^2, to the intercontinental range of the red fox *(Vulpes vulpes)*, at nearly 70,000,000 km^2.

When considering the distribution of any animal, the magnification on a microscope can be increased to expose finer and finer scale of detail. Thus, we begin this section at the geographic scale of species range maps before using greater magnification to disclose a population's regional distributions and, finally, the individual's home range.

GEOGRAPHIC RANGES

As a group, the Carnivora cover nearly the entire globe, and by the standards of other mammals, carnivore species tend to have unusually large distributions

(Hunt 1996). Modern ranges are the result of a combination of factors, including habitat preferences, evolutionary history, colonization ability, and sensitivity to human-caused environmental change.

Considering the number of species that overlap in any particular location is an important conservation tool. Knowing which areas have the most species is one criterion used to prioritize conservation efforts, and it also has biogeographic and evolutionary consequences. This assessment can be improved by explicitly distinguishing areas that contain species that live nowhere else (i.e., endemic species). Analyses of the worldwide distribution of a large variety of taxa (including carnivores) has revealed 25 hotspots across the globe that contain uncommonly rich biodiversity (Mittermeier et al. 1999). Approximately 22% of carnivore species are endemic to these hotspots, with two of the largest of these endemic taxonomic groups being the Malagasy viverrids and the New World procyonids (Sechrest et al. 2002). The procyonids include the wide-ranging coatis and raccoons and also the more restricted olingos and mountain coati *(Nasuella olivaceai)* (Macdonald 2001a). This radiation represents a group that colonized the New World from Eurasia in the Miocene and then became endemic to the area after the extinction of related Eurasian forms in prehistoric times (Hunt 1996). Similarly, the endemic Malagasy viverrids are the result of a single colonization event from Africa to Madagascar, probably in the late Oligocene (Hunt 1996; Yoder et al. 2003).

Recently, biologists have taken these geographic analyses to the next level by combining the information on evolutionary history found in the phylogeny of the Carnivora with the overlap of geographic ranges to quantify the amount of evolutionary history (not just species diversity) endemic to different areas (Sechrest et al. 2002). Thus, their maps show the conservation importance of a given area not just by counting how many species are found in it but also by giving more importance to species with no close evolutionary cousins (e.g., the red panda) than they do to those within a broad radiation of closely related taxa (e.g., island raccoons). They found that these 25 biodiversity hotspots contain a greater amount of evolutionary history than under a random model, with approximately one-third of the evolutionary history of all carnivores completely encompassed by them, and 70% at least partially included. Thus, these hotspots are not only vital areas of species-level endemism but also highly significant reservoirs of unique and threatened evolutionary history. The most important of these geographic hotspots in terms of all endemic Carnivoran evolutionary history are Sundaland (Borneo, Java, and Sumatra, 53 million years of unique evolutionary history), Madagascar (51), Mesoamerica (35), Western Ghats and Sri Lanka (26), and the Guinean Forests of West Africa (22).

Because they are the most widely distributed family of the Carnivora, the Canidae provide an interesting example. They include species on every continent besides Antarctica (Sillero-Zubiri et al. 2004a). During the last century, the ranges of seven species have increased, whereas eight have contracted; often one species has replaced another, and intraguild competition has emerged as a major force in their behavioral ecology (Macdonald and Sillero-Zubiri 2004). For example, climate change has caused red foxes to expand their range and to displace Arctic foxes (Hersteinsson and Macdonald 1982), whereas the demise of gray wolves and the spread of coyotes *(Canis latrans)* has proved inimical to swift foxes *(Vulpes velox)* (Moehrenschlager et al., in press). In general, canid communities often involve three species, one large (>20 kg) and mainly carnivorous (e.g., gray wolf, African wild dog, dhole, *Cuon alpinus*), one medium-sized (10–20 kg) and omnivorous (e.g., coyote, golden jackal, *Canis aureus*), and another

Red fox *(Vulpes vulpes)* in winter. Photo by Konrad Wothe/Minden Pictures.

small and more omnivorous with flexible food habits (e.g., red fox, Cape fox, *Vulpes chama,* Johnson et al. 1996). These patterns arise from the rules of community assembly, which are often associated with character displacement and do not apply only to canids (e.g., Dayan et al. 1991). These phenomena have provided major complications for conservation, as discussed below.

REGIONAL DISTRIBUTION

By increasing our magnification to focus on details within a species' geographic range we can learn how populations use space in finer detail, thus enabling biologists to quantify how the habitats used compare with those available. Not only do these animal-landscape relationships facilitate understanding of the species—revealing the biological basis of its distribution—but they also provide a powerful tool for conservation. They can be used to predict how the species will be distributed in areas where it has not been surveyed, and how it may react to different future scenarios for regional development or new management strategies (Scott et al. 2002). In the case of carnivores, for example, Macdonald and Rushton (2003) showed how such models might be used in the United Kingdom to predict, on the one hand, how a rare species (the pine marten, *Martes martes*) might be restored and, on the other hand, how an invasive species (the American mink, *Mustela vison*) might be controlled. Predictive models have been revolutionized by the development of computerized mapping programs (i.e., Geographical Information Systems, or GIS), which enable biologists to explore causal relationships between a species and its environment and thus to predict likely reactions to such large-scale changes as global warming or habitat modification (Davison et al. 2002). Conservation action often requires choosing between different options, and models can help ensure that the worst errors are weeded out in the virtual reality of a computer screen (Zielinski and Kucera 1995; Moruzzi et al. 2002; Palomares et al. 2002).

The broad-scale approach to quantifying the habitat use of a species is to monitor a series of sites (e.g., using traps, cameras, or sign surveys) revealing preferences in terms of the use of habitats relative to their availability. For example, the prevailing opinion that gray wolves were habitat generalists was confirmed when researchers found that the only landscape feature significantly predicting their regional distribution was the density of roads (Mladenoff et al. 1999). Similar studies of puma *(Felis concolor)* revealed slightly more specific habitat associations: a model of their distribution in Montana explained 73% of the variation using only indices of terrain ruggedness and forested cover (Riley and Malecki 2001). The habitat preferences of fisher *(Martes pennanti)* and American marten *(M. americana)* may be two of the most studied, and most specific. Both species are sensitive to the physical structure of the forest, preferring the structure typical of old growth forest and avoiding open areas (Buskirk and Powell 1994), although to a lesser extent for eastern fishers (Kays and Bogan 2002; Moruzzi et al. 2002).

A common theme of population-level studies is that habitat fragmentation caused by human activities affects the distribution of carnivore species. When we plough a field, fell a forest, or build a new neighborhood we not only modify that parcel of carnivore habitat but also affect the remaining habitat that it borders through a variety of "edge effects" (Debinski and Holt 2000). This type of landscape change is almost always bad for large carnivores likely to face high human-caused mortality (Woodroffe and Ginsberg 1998, 2000; Crooks 2002). In contrast, some small to medium-sized carnivores may benefit from fragmentation, especially if they have evolved to feed on a variety of plant and animal foods and use both the matrix and the interior of habitat patches. For example, the abundances of gray foxes *(Urocyon cinereoargenteus)*, raccoons, and domestic cats were increased in smaller natural fragments in urbanized Southern California (Crooks 2002), and raccoons were more abundant in forest edges near farms in rural Missouri (Dijak and Thompson 2000).

Paradoxically, and from the standpoint of smaller predators, there may be a second, indirect benefit of fragmentation: the local extinction of larger carnivores. The phenomenon of "mesopredator release" is thought to occur when large carnivores are no longer around to keep smaller carnivore populations in check through competition and direct predation (Courchamp et al. 1999; Crooks and Soulé 1999). Because of the naturally dynamic nature of animal populations, this is a difficult hypothesis to test, but it has found support with the absence of coyotes promoting an abundance of house cats and foxes in suburban California (Crooks and Soulé 1999) and raccoons in Michigan (Rogers and Caro 1998). In both of these studies, the increase in mesopredator abundance trickled down through the ecosystem, creating a decline in the bird species on which they preyed.

Although the foregoing examples highlight the effect of people, larger predators, and landscape features on carnivore distribution, the abundance of food is often the major natural force influencing carnivore population density and viability (Fuller and Sievert 2001; Carbone and Gittleman 2002). Broad comparisons of predator and prey density across 25 carnivore species found that, on average, 10,000 kg of prey is needed to support 90 kg of a given species of carnivore, irrespective of body mass (Carbone and Gittleman 2002). This general predator density–prey density relationship has been borne out repeatedly in fieldwork. Brown bear *(Ursus arctos)* densities, for example, are highest in coastal Alaska, where runs of salmon are seasonally superabundant, and lowest north of the Arctic Circle, where the growing season is short and both plant and animal biomass are low (McLellan 1994). Snow leopard *(Panthera uncia)* densities seem to be highest where blue sheep *(Pseudois nayaur)* densities are highest (Oli 1994).

Comprehensive analyses across geographic ranges indicate that the densities of gray wolf (Fuller 1989; Messier 1995), African lion (*Panthera leo;* Van Orsdol et al. 1985), and European badger (Kruuk and Parrish 1982; Johnson et al. 2002b), and the biomass of cheetah (*Acinonyx jubatus;* Gros et al. 1996) and leopard (*Panthera pardus;* Stander et al. 1997) are strongly correlated with measures of prey biomass or lean season biomass (reviewed by Fuller and Sievert 2001).

HOME RANGE

At the finest level, we can focus our study on the movement of *individual* animals across the landscape. Indeed, the behavior of a population is an emergent property of the behavior of the individuals that compose it. Mammal species do not roam the landscape at random. Instead, each individual repeatedly uses much the same limited area—its home range. A home range is generally considered to be the "area utilized by an individual during its normal activities such as food gathering, mating, and caring for young" (Burt 1943). All or part of the home range of some species may constitute a defended zone, a territory. Considering the huge interspecific variation in their body sizes, it is not surprising that home-range sizes of the Carnivora vary by nearly six orders of magnitude. For example, those of coatis are small (0.45 km^2) and those of polar bears are vast (up to 196,000 km^2; Born et al. 1997; Valenzuela and Macdonald 2002).

Among mammals as a whole, it has long been realized that the basic algebra of body size, metabolic rate, and thus food requirements leads to a relationship in which home-range size increases with body size with the function of approximately $0.71–0.75^2$ (McNab 1989). A variant of this overall relationship is that average home-range size increases with increasing metabolic needs of the species, as indexed by body weight and foraging-group size (Gittleman and Harvey 1982). Once this relationship has been taken into account (and controlled by statistical means), the predominant diet of a species explains some of the remaining variation; for example, meat eaters characteristically use larger areas than do omnivores or frugivores. However, body size is not always a straightforward guide to range size; the 20–30 kg African wild dog is famed for its relatively enormous ranges, which span hundreds of square kilometers and sometimes as much as 1,500–2,000 km^2 (Creel and Macdonald 1995; Valenzuela and Macdonald 2002).

However impressive the variation in home-range size between carnivore species of different size and diet, it is more interesting that several orders of magnitude variation can exist in the home-range sizes of different populations of the same species. For example, Macdonald (1981, 1987) showed that the ranges of red foxes could span two orders of magnitude between different habitats. Similarly, in Poland small family groups of European badgers might occupy 18 km^2, while in Wytham Woods, a social group of 30 crammed into a 0.5 km^2 range (Goszczynski 1999; Kowalczyk et al. 2000; Johnson et al. 2000; Macdonald et al. 2004). Quantitative comparisons of bobcats (*Lynx rufus*), coyotes, and American black bears (*Ursus americanus*) all reveal major intraspecific variations in home-range sizes that were positively associated with latitude (i.e., larger areas in the north). The magnitude of these relationships were related to the diet of the species (Gompper and Gittleman 1991). The carnivorous bobcat had the steepest relationship, the omnivorous coyote had a mid-level relationship, and the frugivorous/folivorous black bear showed the least effect of latitude. Overall, this relationship is probably related to broad-scale variation in habitat productivity with latitude (more food per area nearer the equator), with species feeding higher on the food chain (e.g., highly carnivorous bobcats) being more sensitive to these changes. In general, there have been too few studies of single species under different circumstances to reveal the true extent of variation, with the consequence that impressions in the literature, and average values given for species, are heavily influenced by the chance location of the most thorough studies of particular species. This is particularly important when one considers that it is not only home-range sizes that vary intraspecifically but many other life-history parameters as well, including group size and social system.

This leads to the intriguing question of the extent to which species differ in behavioral flexibility. For example, Macdonald (1979) described a population of golden jackals associated with a clumped food source in Israel, which not only occupied a home range of about one-tenth the area used by conspecifics on the grassy plains of Africa but also lived in large groups that displayed behavior patterns (e.g., the marking of their range with peripheral latrines) completely unknown in lower-density populations. The exploration of the extent of flexibility, and how it differs between species, remains a potentially rich vein for study in carnivore biology.

At the finest scale of resolution, often aided by radio-tracking devices, it is possible to explore habitat preferences within individual home ranges. The recent use of Global Positioning System (GPS) collars collected unprecedented detail (>350 fixes per animal) on the movement of individual brown bears in the mountains of Alberta (Nielsen et al. 2002). This detail revealed habitat features preferred by all bears in the area (e.g., "greenness") but also highlighted the fact that individual animals can have specific, and different, habitat preferences. For example, bear G3 always used springtime herbaceous cover when available, while bear G6 always avoided it. Bears also varied in their habitat specificity, with springtime habitat use of bear G2 described by just 3 habitat variables, whereas bear G20 was significant at 11 different habitat variables. Often human-related features dominate the landscape and have strong effects on the use of that landscape by carnivores. For example, radio-telemetry in Illinois farmland found that coyotes avoided areas of high human traffic, preferring habitats that provided cover, while red foxes did exactly the opposite, probably in their efforts to avoid coyotes (Gosselink et al. 2003). Similarly, in lowland England, American mink preferred using the habitat not characterized by farming activities (Yamaguchi et al. 2003).

Making a Living: Diet and Foraging

To judge by their name, one might assume that all members of the Carnivora are meat eaters. However, the confusion arises because the word *carnivore* can be used as either a proper or a common noun. Carnivores, that is, members of the Carnivora, belong to an order of mammals that shares common descent from an ancestor. Some of these carnivores, along with many other predators from eagles to beetles, are carnivorous, that is, they eat meat. Thus, not all carnivores are carnivorous, and not all carnivorous creatures are Carnivora.

Among the carnivores, some groups are indeed exclusively carnivorous (e.g., all felids and most mustelids); some of these are real specialists—more than 90% of the Ethiopian wolf *(Canis simensis)* diet is made up of rodents (Sillero-Zubiri and Gottelli 1995). Many more Carnivora are at least partly carnivorous and generally predatory, also consuming plenty of invertebrates along with vertebrate meat. Among that majority, which is at least occasionally carnivorous, including many canids, procyonids, and ursids, many eat a great deal of vegetable matter. Thus, from brown bears to side-striped jackals *(Canis adustus)* the lifestyles of many carnivores are heavily influenced by the seasonality of fruit (Rode and Robbins 2000; Atkinson et al. 2002). Likewise, from aardwolves *(Proteles cristatus)* to bat-eared foxes *(Otocyon megalotis)*, others are almost entirely dependent on the abundance of insects (Nel and Mackie 1990; Waser 1981; Richardson 1987) or, from badgers to sea otters *(Enhydra lutris)*, on other invertebrates, such as earthworms and sea urchins (Estes et al. 1978; Kruuk 1978a). However, and perhaps most unexpectedly, there are carnivores that are not carnivorous at all, and these include the giant *(Ailuropoda melanoleuca)* and red pandas (related only in name; Slattery and O'Brien 1995) and the kinkajou (Kays 1999). Nonetheless, even the most avant garde of dietary habits among carnivores cannot disguise their ancestry, as adaptations for killing have driven the evolution of the order and resonate throughout their contemporary appearance and behavior.

KILLING FOR A LIVING

Obviously, the size of a predator is a fundamental factor in determining what type of prey it can overwhelm. No matter how specialized its teeth, a least weasel would face unpromising odds if it sought to kill a moose! The dangers associated with obtaining large prey mean that many carnivores choose prey that are much smaller than themselves. However, larger meals have the advantage of feeding an animal for longer, or of feeding more animals at once. Thus, specializations that increase a predator's effectiveness and reduce the risks of its dangerous profession will be favored by evolution. Among solitary hunters such specializations include protractile claws in felids and locking jaws in mustelids (King 1989; Kitchener 1991). Others hunt cooperatively in groups, and it has been widely assumed (although remarkably hard to demonstrate quantitatively) that by doing so they increase their efficiency and collectively take down larger prey than any individual could by itself (Mech 1970). This assumption remains controversial, and whether cooperative hunting would or would not universally increase food intake per capita (or net energy intake per capita, although difficult to estimate) needs further investigation (Schaller 1972; Mills 1990; Creel and Creel 2002).

There may also be large differences in adaptations to prey size *within* a species; males and females are often dimorphic not only in body size but also in traits specifically related to prey capture (McDonald 2002). In mink, for example, tooth size and several skull variables are much larger in males than females, even after the effects of different body size are taken into account (Thom et al. 2004a). Thus, a female mink scaled up to the body weight of a male would still have smaller canine teeth. These differences are most likely to be related to prey capture—tooth size is thought to be associated with prey size, and larger prey also require larger and more powerful musculature to tear them up, which in turn leads to changes in skull architecture.

Surprisingly, there is also a flipside to the relationship between the body size of predator and prey—large carnivores cannot survive on small prey. In comparing the diet and body mass of 158 carnivore species, Carbone et al. (1999) found the order split into two distinct weight classes, with the transition at a body mass of about 21.5 kg. Predators below this weight (the lightweights) focus on prey less (often much less) than half their own weight, while heavier predators (the heavyweights) primarily eat prey near their own body weight. Insects were the most popular small prey, with 75% of the lightweights feeding at least partially on invertebrate prey; this is probably related to the fact that invertebrates can be a superabundant resource (sometimes 90% of animal biomass; Wilson 1987). Because smaller predators have lower absolute energy requirements, they can survive on small vertebrate and invertebrate prey. This is not sustainable for the heavyweights because of their larger appetites and the physical limitations on how many insects they can eat in one meal; they have no choice but to specialize on larger meals.

The effort required to hunt and kill prey means that a carcass is a valuable commodity and protecting it from competitors is essential. While food theft—known pretentiously as kleptoparasitism—is probably rare in forested habitats, where smaller carnivores can easily hide their kills in the brush, it is a drama that plays out regularly in the African savannah. With open terrain and overhead vultures as scouts, the largest predators, such as lions and hyenas, regularly find and steal kills from the smaller predators (Schaller 1972). Cheetahs, which are particularly at risk, often hunt during the day to minimize their encounters with the primarily nocturnal lions and hyenas (Caro 1994). Leopards attempt to safeguard their food from larger nonclimbing predators by caching carcasses in trees (Schaller 1972). Indeed, throughout the Carnivora, faced as they are by uncertainty in prey capture, food caching is an important adaptation, an insurance policy against an uncertain future (Macdonald 1976). Even the two large predators made mighty by teamwork, lions and hyenas, can lose kills to each other or to conspecifics (Packer 1986). To avoid these potentially dangerous aggressive encounters they often consume their food extremely fast; hyenas may even swallow ribs whole (Marean et al. 1992).

NOT ALL CARNIVORES ARE CARNIVOROUS

Throughout the evolutionary history of the carnivores a number of groups have adapted to a vegetarian diet. Some of these noncarnivorous carnivores have only recently split from their meat-eating ancestry and maintain the appearance of a predator (e.g., the frugivorous/folivorous American black bear). Other species gave up chasing prey long ago and now share few traits with the rest of the order (e.g., the monkeylike, frugivorous kinkajou was first described as a lemur; Ford and Hoffmann 1988).

Logically, the first evolutionary step away from carnivory would be omnivory, and this is seen in a variety of species (Ewer 1973). A flexible, omnivorous diet enabled coyotes to adapt to a variety of habitats and, in the last 50 years, spread from their historical range in prairie heartland to nearly every corner of North and Central America and, recently, into South America (Méndez et al. 1981; Parker 1995; Gompper 2002). In many cases, the use of anthropogenic food sources such as livestock, pets, garbage, and fruit probably helps them flourish in these new areas (Fedriani et al. 2001). While coyotes take advantage of what food is available to them, they can still be discriminating, for example, focusing on hunting hares over mice even when hares are relatively rare (Hernandez et al. 2002). Indeed, carnivores can exhibit strong food preferences between what appear, to human gastronomes, to be rather similar prey; red foxes, for example, favor field voles *(Microtus agrestis)* over bank voles (*Clethrionomys glareolus;* Macdonald 1977).

Progressing along the continuum between carnivory and herbivory, some carnivores have now become such specialized vegetarians that they rarely eat vertebrate or invertebrate foods. Fruit is easy to catch and commonly found in the diet of carnivores (Pigozzi 1992; Rodríguez-Bolaños et al. 2000; Fedriani et al. 2000). However, fruit is often seasonal, so species that specialize on eating fruit must have strategies to deal with annual shortages in food. Most bear species eat fruit, and those in temperate areas fatten up on autumnal fruit crops before hibernating through the snowy and fruit-less winter (Stratman and Pelton 1999; Rode and Robbins 2000; Persson et al. 2001). Tropical plant communities typically produce fruit year-round, allowing tropical frugivores to be more dedicated fruit eaters. In fact, some members of the Carnivora are actually some of the most frugivorous mammalian species on record. For example, fruit makes up approximately 80% of the diet of the African palm civet *(Nandinia binotata)* (Charles-Dominique 1978) and 90–99% of the diet of kinkajous in Panama (Kays 1999).

Compared with flesh, insects, and fruit, plant foliage and stems are a very difficult food to digest and eat; their nutrients are robustly protected by the cellulose walls that encase plant cells. Solving this problem has driven the evolution of the two main mammalian orders of ungulates, the even-toed artiodactyls and the odd-toed perissodactyls, and forced both into alliance with symbiotic bacteria. Worse, plants co-evolve with their predators, evolving a second line of defense in a fearsome array of gut-troubling protective chemicals. Considering their starting point, of simple digestive systems designed for the rather unambitious task of processing meat, it is not surprising that few carnivores have evolved into folivores. Nor is it surprising that the two most folivorous species, giant pandas and red pandas, both specialized on the same superabundant leafy resource—bamboo. Giant pandas eat almost every part of the plant, while red pandas use only leaves and shoots, supplementing their diet with arboreal fruits and mushrooms (Wei et al. 1999).

Social Systems and Mating: Solitary to Group Living

The social systems of the Carnivora range from solitary individuals, which come together briefly during breeding (e.g., weasels), to those species that form prolonged monogamous pair bonds (e.g., golden jackals), to those that live in extended social groups with as many as 80 individuals (e.g., spotted hyenas). These social systems are the product of evolutionary cost-benefit analysis that balances the behaviors that allow individuals to both eat and avoid being eaten, reproducing as successfully as possible along the way. Lifetime reproduction is the ultimate bottom line, and social organizations represent the best behavioral solution for individuals of a particular species (or population), given their morphology, ecology, and evolutionary history. Teasing out the costs and benefits of these adaptations offers a fascinating glimpse into the behavioral diversity of the Carnivora.

SOLITARY HUNTERS

Most (85–90%) of the Carnivora live apparently solitary lives, with individuals coming together only for breeding or territorial confrontations (Bekoff et al. 1984). While future fieldwork will certainly reveal more cryptic sociality in the order (see below), the overall solitary nature of many carnivores needs explanation. As pointed out by Sandell (1989), this explanation, which is probably not complicated, hinges on two generalities: most predators hunt more efficiently by themselves, and most mammalian species do not need biparental care.

Rarely is there a net benefit to predators of working together to make a kill because most potential prey are too small to be worth sharing with companions (Carbone et al. 1999). Indeed, even among African lions, it has proven remarkably difficult to avoid the conclusion that the per capita rewards of hunting collaboratively decline in groups larger than two (Caraco and Wolf 1975). Furthermore, groups of animals are more conspicuous, and the stealth needed to capture small hidden prey would be compromised by other group members.

Among vertebrates, female mammals are unique in being able to feed their young directly through mammary glands (Eisenberg 1981). Most female mammals are sufficiently well equipped to care for their young that no help is needed from males, "releasing" the males to pursue other strategies of maximizing their fitness (Clutton-Brock 1991). In the case of some carnivores (but not canids or social mongooses), males are not only unnecessary for females with young, but females often actually avoid them to minimize competition over food and potential infanticide (Sandell 1989; Kays and Gittleman 2001).

GROUP-LIVING CARNIVORES

If group members are not typically needed to raise young and may compete for food, why do 10–15% of carnivore species live in groups? What tips the "Cost of Group Living vs. Benefits of Group Living" equation to the right? Creel and Macdonald (1995) summarized five types of selective pressure that may select for sociality in carnivores. The first two may reduce the costs of tolerating conspecific group members: (1) *resource dispersion:* most obviously, abundant prey, rich or variable prey patches, or rapid prey renewal may all lead to lower costs of tolerating conspecifics; (2) *high dispersal costs:* constraints on dispersal opportunities such as lack of suitable habitat, low mate availability, or intraspecific competition may encourage young to stay within their natal group past the age of maturity. In short, these two ecological factors may facilitate group formation by making the costs of doing so minimal and the costs of not doing so high.

A calling black-backed jackal *(Canis mesomelas)*. Photo by Mitsuaki Iwago/Minden Pictures.

In contrast, the final three selective pressures may increase the benefits of tolerating group members through behavioral mechanisms: (3) *better resource acquisition:* groups may use strength of numbers in the acquisition and retention of resources—for example, hunting in groups may increase foraging success where prey are large or difficult to kill, and groups may also fare better in territorial defense and intra- and interspecific competition for food, especially at large kills; (4) *protection against predation:* groups may be less vulnerable to predation; and (5) *reproductive advantages:* group membership offers the opportunity for cooperative feeding and protection of young.

It is clear that various different types of society would result depending on the balance of these five categories of selective pressure (Macdonald and Carr 1989). For example, all else being equal, group size is likely to be smaller where dispersal opportunities are greater, whereas cub survival might increase in the presence of alloparents within a group. Below, we will explore how these factors have led to a variety of social groups in the Carnivora.

RESOURCE DISPERSION REDUCING THE COST OF SOCIALITY

While behavioral benefits may be the most conspicuous selective pressures favoring sociality in carnivores (see below), ecological factors create the framework within which these pressures operate and dictate the balance of costs and benefits between group membership and dispersal. Alexander (1974) was the first to suggest that the formation of groups (and cooperative behavior between their members) may be a secondary consequence of some ecological factor. One of the most crucial of such factors is resource dispersion, especially (but not limited to) food. Irrespective of the advantages of group living, resource dispersion may significantly affect the costs of grouping, most obviously where a superabundant resource provides more food per unit area than one pair of animals can eat, resulting in no food-based costs of letting other animals share the resource (Wrangham et al. 1993).

However, the real world is usually more complicated, with food abundance changing over time and space. The idea that these dynamic patterns of resource availability might facilitate group formation by making coexistence feasible grew especially out of observations on Eurasian badgers (Kruuk 1978a, b) that lived in groups without accruing any obvious benefits from it. These observations were later formalized as the Resource Dispersion Hypothesis (RDH; Macdonald 1983; Carr and Macdonald 1986; Johnson et al. 2002a). This hypothesis suggests that groups may develop where resources are dispersed in such a way that the smallest defensible territory for a pair can also sustain additional animals. This happens because, where resource patches are heterogeneous (in space and/or time), the primary breeding pair will have to defend a relatively larger area to include sufficient resource patches to guarantee enough usable patches over time. The RDH offers an explanation for variance in group size regardless of whether individuals gain from each other's presence or not. Not only can it apply to current societies but it can describe the conditions that favored the evolution of sociality. While the RDH has yet to be tested by manipulative experiments (Johnson et al. 2002a), several field studies of carnivores, other mammals, and birds are broadly in line with its predictions (but see Baker et al. 2000, reviewed by Johnson et al. 2002a). Extensive field studies in the site that inspired the hypothesis, Wytham Woods, have raised increasing numbers of questions (Macdonald and Newman 2002) but, perhaps most importantly, have failed over decades to reject the hypothesis (Johnson et al. 2001a, b; Macdonald et al. 2004).

GROUP LIVING AS DISPERSAL AVOIDANCE

Dispersal is poorly understood in carnivores, yet it is crucial to conservation (Macdonald and Johnson 2001) and to understanding life-history processes (Waser 1996), specifically, group living. No matter how low the costs of group living, there are few reasons for breeding-age adults to remain in their natal group and not breed if there are opportunities to do so in the surrounding environment (and no cost to move there). But there are real costs to dispersal, and the dynamic between the benefits of staying at home in the group (called natal philopatry) versus dispersing determines the group size in many species. The balance of advantage will shift with circumstances (MacDonald and Carr 1989). For example, female gray wolves in Alaska may be the least likely to emigrate because hunter-caused mortality improves their chances of attaining breeding status in their natal group; the greater likelihood of more dominant members of their pack being killed creates an opportunity for ascendance to the throne (Ballard et al. 1987). Among coyotes, while larger groups may enjoy some fitness benefits in the efficiency of securing and/or retaining prey, the major selective force favoring larger social groups is thought to be delayed dispersal due to the lack of breeding opportunities in saturated habitats (Bowen 1981; Messier and Barrette 1982).

One extreme of this continuum of dispersal costs is illustrated by Ethiopian wolves in the Bale Mountains. These animals are effectively restricted to "islands" of afro-alpine grassland and heath surrounded by agricultural land, which makes dispersal an unpromising option. However, philopatry brings with it the risk of inbreeding, which may be one reason why breeding females surreptitiously seek liaisons with neighboring males (another reason may be the benefit of confused paternity to reduce the risk of infanticide, Sillero-Zubiri et al. 1996). Within groups of Ethiopian wolves, females effectively accept mating with the dominant male only, but during cross-border liaisons they mate indiscriminately. This must blur the genetic discontinuities between neighbors, as does the movement of crab-eating foxes *(Cerdocyon thous)* (Macdonald and Courtenay 1996). These South American canids live in groups of two to five individuals. Some dispersers settle in territories at the borders of the natal range, and return intermittently (seemingly not accompanied by their new mate) to their original territory in amicable company with their parents (one male tending the next generation of his siblings during his return visits). Examples of both phenomena, neighborhood relatedness and return from dispersal, are increasing and may be widespread (e.g., gray wolves, Wayne et al. 1992; Girman et al. 1997; bat-eared foxes, Maas and Macdonald, in press).

The tendency for males to disperse more often or farther than females is a pattern found across the Mammalia, in sharp contrast to the opposite pattern found in birds (Greenwood 1980). This is thought to stem from strategies of mate defense in most mammals and resource defense in most birds, with male philopatry more useful where sons can inherit their father's territory of resources. Storm et al.'s (1976) monumental tagging study was the first to reveal that more male yearling red foxes dispersed, and dispersed farther, than did females, a result replicated in many studies of other carnivore species (reviewed by Waser 1996). In contrast, tagging studies found no sex bias in dispersal distance or tendency among gray wolves (Mech 1987; Gese and Mech 1991), although genetic evidence suggests either that males engage in more long-range dispersal or suffer greater mortality en route than do females (Wayne et al. 1992).

One exception to the rule of male dispersal is found in the kinkajou, in which genetic studies showed females to be the dominant disperser (Kays et al. 2000). This supports Greenwood's hypothesis insofar as behavioral studies suggest philopatric resource defense (rather than mate defense) by male kinkajous. Female-biased dispersal is also the rule among Ethiopian wolves (Sillero-Zubiri et al. 1996) and is also suspected in African wild dogs (Frame et al. 1979; Fuller et al. 1992; Creel and Creel 2003). Although, mindful as ever of intraspecific variation, we note that other populations of African wild dogs have been documented with male-biased dispersal (McNutt 1996; Girman et al. 1997).

In Eurasian badger populations the liability of individuals to disperse varies with density. In the high-density Wytham Woods population (Macdonald and Newman 2002) dispersal between groups was rare (Macdonald and Johnson 2001) compared to a lower-density suburban population of badgers (Harris and Cresswell 1987; Cheeseman et al. 1988). Woodroffe et al. (1995) established a significant female bias in the dispersal patterns of the Wytham badger population, while by contrast, in a similar habitat and at similar density, Rogers et al. (1998) observed a male bias. More recent studies at Wytham Woods (Macdonald et al., in press) have established neither sex to have a greater dispersal liability under present conditions, though dispersing females move farther than males.

Although most species disperse as individuals, males of social species may benefit from sticking together to compete for new territories, so they disperse as male coalitions (African lions, Schaller 1972; slender mongooses *[Herpestes sanguineus]*, Waser et al. 1994; cheetahs, Caro 1994; meerkats, Doolan and Macdonald 1996). This in turn, however, leads to an "arms race," with resident males forming their own coalitions to fend off would-be invaders (Frame et al. 1979; seen in dwarf mongoose, lions, banded mongoose *[Mungos mungos]*, and African wild dogs, Rood 1986).

All are not equal within a social group, and this also affects its dispersal tendencies. It seems logical that subordinates with no possibilities of breeding at home will be more likely to risk looking for new opportunities by dispersing. However, proof of that prediction is not abundant, and Bekoff and Wells (1982) raised the opposite suggestion, that dominant coyote pups might be the most likely to disperse, being the toughest and most apt to succeed. In cases where the dominants benefit from the presence of subordinates (e.g., in dwarf mongooses) a conflict of interests arises, and subordinates may be "encouraged" to stay by relaxing the reproductive suppression and allowing occasional breeding (Keane et al. 1994; i.e., "power sharing," Creel and Waser 1994). However, while dominance is apparent in many carnivore interactions and hierarchies are sometimes obvious, there has been little detailed study (in the fashion common among primatologists) of the dynamics of their sociality; two of the detailed studies that have been done did not reveal a simply linear hierarchy. In one, among farm cats, interactions (and perhaps status) appeared to flow toward females at the center of successful matrilines; in another, within groups of Eurasian badgers, there was no evidence of a stable hierarchy (Macdonald et al. 2002). In contrast, dominance seems pivotal among some other hierarchies—for example, separately within both sexes of gray wolves and dwarf mongooses (Zimen 1976; Rasa 1986).

Dispersal is not always a sudden and abrupt change, and dispersing gradually with repeated exploratory forays may decrease the risk associated with finding a new home, especially for gregarious species (Baker 1978). For example, predispersal spotted hyenas commonly make forays lasting from a few hours to a few months into other territories before moving (Mills 1990; Holekamp et al. 1993), and they use long-distance calls to assess the breeding opportunities in the region (East and Hofer 1991). In wolves, "extraterritorial" forays usually last for a few weeks and extend over a few tens of kilometers (Messier 1985; Gese and Mech 1991). A young female kinkajou made numerous trips to a vacant neighboring territory more than 50 days before permanently settling with the resident males of that social group (Kays and Gittleman 2001). Some animals also mate while on forays but never actually settle in with the group with whom they mate (female African lions, Packer and Pusey 1987; male wolves, Mech 1987; female badgers, Woodroffe et al. 1995; female Ethiopian wolves, Sillero-Zubiri et al. 1996; male white-nosed coatis *[Nasua narica]*, Gompper et al. 1998).

Dispersal is a dangerous proposition, and the mortality of dispersers is often high. There are two dangerous elements—traveling, generally through unfamiliar terrain, and settlement—which are likely to involve competition. For example, fights between nomadic male African lion coalitions and the resident males they are attempting to depose can result in serious injury or death (Packer and Pusey 1982). Eurasian badger immigrant males have higher rates of scarring than do philopatric males (Woodroffe et al. 1995). Similarly, immigrating male spotted hyenas are the subject of intense aggression and are initially dominated even by the pups of their new group; the process of their assimilation may take months (Henschel and Skinner 1987; Mills 1990; Smale et al. 1993).

Yet most species do disperse eventually—so why? The obvious answer is that they have no choice, with no room at the family inn, so somebody has to leave. Added to this, avoiding inbreeding is ultimately likely to bring evolutionary benefit to dispersers (Girman et al. 1997). Inbreeding can cause minor and severe genetic problems (Maehr and Caddick 1995; Hansson and Westerberg 2002). However, low levels are sometimes observed without obvious problems (Shankaranarayanan et al. 1997; Shivaji et al. 1998; Wisely et al. 2002). For example, breeding may occur locally between related pack founders derived from neighboring packs (Mech 1987; Lehman et al. 1991). Circumstantial evidence suggests that inbreeding can at least sometimes be the norm among bat-eared foxes where, through natal philopatry, 7 out of 54 females were mounted by their father and one by her brother (Maas and Macdonald 2004).

The proximate causes for dispersal have proven hard to confirm. Family ties characteristically weaken as juveniles mature, but evidence of parents triggering dispersal by expelling their offspring aggressively is scant and hard to gather. Harris and White (1992) emphasize instead a general decline in affiliative behavior toward young adults by the breeding pair in red foxes. The first outcome of food shortage among both wolves and African wild dogs seems to be not that young helpers disperse but that pups starve (Malcom and Marten 1982; Harrington et al. 1983)—explicable, perhaps, in terms of parents investing where they are likely to secure the greatest returns.

GROUP LIVING FOR STRENGTH IN NUMBERS

The notion that carnivores hunt together to more effectively overwhelm prey too challenging to be hunted alone is intuitively plausible but has proved extremely difficult to support with hard data. Early data on black-backed jackals *(Canis mesomelas)* (Wyman 1967; Lamprecht 1978) proved inconclusive. Even with the most conspicuous pack hunters, African wild dogs, evidence that individual pack members did better when hunting in larger groups proved equivocal when the measure of success was the quantity of prey eaten (Fanshawe and Fitzgibbon 1993). Later, a review of gray wolf hunting success revealed a general decline in food intake per wolf per day with larger pack sizes (Schmidt and Mech 1997). Vucetich et al. (2003) illustrate that in the absence of scavengers, wolves would do better hunting in pairs, not packs. However, when currency is shifted from the bulk of prey eaten to the energy needed to catch them, it became more convincing that hunting groups of African wild dogs may provide a net benefit to participating individuals (Creel and Creel 1995, 2003).

Interference from other carnivore species at kills can be considerable and, as one would expect, the outcome of such competition appears to be affected by group size. For example, larger groups of coyotes emerged as more successful at defending kills (Camenzind 1978; Bekoff and Wells 1980). Clearly, strength of numbers is important when, for example, wild dogs strive to repel spotted hyenas from a kill (McNutt 1996) or gray wolves seek to keep scavenging corvids at bay (Vucetich et al. 2003). Strength of numbers may also be important in territorial defense, and evidence accumulates that larger groups are more successful at defending territories (Wrangham 1980; Bekoff and Wells 1980; Creel et al. 1998). A large pack of golden jackals habitually stole food from a smaller pack (Macdonald 1979), and among Ethiopian wolves, larger packs invariably prevail in territorial clashes (Sillero-Zubiri and Macdonald 1998). Territorial clashes may account for a substantial proportion of adult mortality among both wild dogs and wolves (Mech 1977; Creel et al. 1998). Among felids, the males of two species form coalitions to compete for females. Coalitions of male cheetahs are almost always litter mates (Caro 1994). Among male African lions, when coalitions number only two individuals they are typically unrelated, whereas larger male groups, requiring more cooperation, are more likely to contain related animals (Gilbert et al. 1992). Even mostly solitary hunters may have loose coalitions, as, for example, between related slender mongooses (Waser et al. 1994).

GROUPING TO AVOID BECOMING PREY

As a group, predators have been broadly implicated in forcing their prey to live in groups in self-defense (van Schaik and van Hooff 1983; Isbell 1994; Hill and Dunbar 1998). Groups of animals are often better able to detect and defend against predators than are singletons, so the costs paid in terms of increased food competition is often balanced toward grouping. But predators, especially smaller species, can themselves become prey. Although not common in the Carnivora, minimizing individual predation risk has been suggested as the cause of group living for a handful of species.

This is probably most important in the case of small social herpestids, which have lower resource-related costs from group living because of their rapidly renewing insect food (Waser 1981). At the same time, their small body size and diurnal habits make them vulnerable to a diversity of predators. Living in groups improves their vigilance and defense against prospective predators (Rasa 1986). For example, groups of meerkats must maintain social group size above nine or fall into the vicious cycle of higher predation rates further reducing their group size, which then further increases their predation rate, until their social group is hunted to extinction (Clutton-Brock et al. 1999). Such advantages of companionship may explain why some social herpestids even form alliances beyond their own species. Rasa (1986), for example, documents a wonderful example of mutualism whereby dwarf mongooses and hornbills (Bucerotidae) respond to each other's alarm calls.

Other carnivores also deal better with would-be predators in groups, although the relative importance of this factor is probably lower than in the herpestids. For example, larger groups of bat-eared foxes appear better able to repel predators approaching their dens (Maas and Macdonald 2004). Also in Africa, young male cheetahs that disperse in pairs are more effective at deterring the much larger spotted hyenas than are singletons (Caro 1994).

Carnivores can also be attacked by another sort of predator—parasites. Alexander (1974) suggested that defense against parasites might have been one of the principle factors favoring group living in animals (and humans). There is a variety of means by which social living can improve one's chances of avoiding infestation (e.g., the dilution effect). However, it has recently been suggested that the evolution of cooperative allogrooming might have been a potent selective pressure on certain species to live together (Johnson et al. 2004). Eurasian badgers, for example, suffer from considerable infestations of a host-specific flea (Cox et al. 1999), and they appear to be in a sort of Prisoner's Dilemma, which favors cooperation in allogrooming. The spread of fleas among individuals in a group means that allogrooming is favored and, at the same time, cheating (not bothering to groom others) is not advantageous. Hence, it represents an interesting "public goods" game problem without the usual situation of cheaters that can destroy cooperation.

GROUPING FOR REPRODUCTIVE ADVANTAGES

Newborn carnivores have a prolonged period of dependency on adults, and many species have evolved sociality to improve their survival (Kleiman and Eisenberg 1973; Wolff and Peterson 1998). Fathers, aunts, uncles, and even "good friends" can help raise a litter in a number of ways. Babysitting the young while their mother finds her own food may be the least costly of these. However, many species take it further. The original list of species known to feed and tend the young of others (e.g., red foxes, black-backed jackals, gray wolves; Moehlman 1979; Macdonald 1979; Fentress and Ryon 1982) has expanded with the number of species studied, revealing alloparental care as a widespread trait of the canids (Woodroffe and Vincent 1994). In addition to guarding, wolflike canids actually bring back a stomach full of food and regurgitate it for the young; an interesting phylogenetic distinction is that foxlike canids do not (Macdonald and Sillero-Zubiri, in press).

Perhaps the most surprising form of alloparental behavior is allosuckling, in which a female other than the mother nurses a pup. Lactation is the most energetically expensive aspect of rearing young for females (Gittleman and Thompson 1988; Clutton-Brock 1991), so nonbreeding lactation represents an extreme form of helping with the young. In dwarf mongooses, for every four subordinates that lactated following pregnancy, one female lactated without pregnancy (Rood 1980). In this case, spontaneous lactation is due to pseudopregnancy, an endocrine state similar to pregnancy but without an implanted fetus (Nalbandov 1976). Nonbreeding lactators are usually closely related to the offspring that they suckle, and in the case of dwarf mongooses, their contribution is associated with enhanced survival of the young (Creel et al. 1991). This was less obvious in Ethiopian wolves, where the number of pups emerging from litters nursed by two females was actually fewer than were weaned on average by females that suckled alone (Sillero-Zubiri et al. 2004a). Intriguingly, considering that it might have seemed obvious that "helpers" would prove helpful, Kruchenkova et al. (in press) report findings that Arctic foxes on Mednyi Island rear fewer cubs with helpers than without them.

In some cases, having helpers does appear to be essential to survival of the young. Dwarf mongooses are an example of obligate cooperative breeders; females invest so heavily in reproduction that they simply cannot raise their young alone or in pairs (Rood 1978; Creel and Creel 1991). Females produce up to four litters in six months, each averaging 22% of the mother's body mass. Dominant females forage more than other pack members during the breeding season, probably because of the high-energy requirements involved, and they rely on subordinate pack members of both sexes to carry out most of the parental care other than suckling (e.g., grooming, guarding, feeding, and moving between dens; Rasa 1977, 1989; Rood 1978, 1986). Where helpers are more than a luxury, dependence on them can lead to the so-called Allee Effect (Allee 1931)—when group sizes fall below a certain threshold they perform progressively worse and thus spiral toward extinction. Courchamp et al. (2002) argue that this applies to African wild dogs whose pack size falls below five members. Indeed, these African wild dogs illustrate the tough choices pack members face—larger packs regularly leave a babysitter on guard with pups when they go hunting, whereas members of smaller packs generally do not leave a guard—presumably because, in the face of the labor shortage, they need the full pack's force for the hunt.

Membership in a society does not affect each individual equally. On the profit-and-loss account of membership, dominant breeding individuals may gain from the help of other group members in raising their offspring; the helpers, however, bear the debit of forging reproduction themselves. This demands an explanation. Reproductive suppression of subordinate females has been recorded in at least 44% of 25 canid species for which there is information (Moehlman and Hofer 1997). The degree of suppression, however, varies (Creel and Waser 1991, 1994) and subordinates often do reproduce, albeit at lower rates than dominants (Packard et al. 1983; Macdonald 1987; Fuller and Kat 1990). For example, subordinate female dwarf mongooses and meerkats may also breed, usually in synchrony with the dominant, so that both females can nurse the young from both litters (Rood 1980; Keane et al. 1994; Doolan and Macdonald 1997). When groups include more than one female, they are typically closely related (e.g., African lions, Packer et al. 1991; white-nosed coatis, Gompper et al. 1998). Kin selection and the supposed high relatedness between individuals in social groups was historically thought to be the primary mechanism underlying cooperative behavior within carnivore social groups (Rodman 1981; Macdonald and Moehlman 1982). However, the diversity of degrees of relatedness found by modern genetic studies of the Carnivora has cast doubt on the universality of this idea (Geffen et al. 1996). Cooperation between unrelated individuals may actually be common and kin selection only one mechanism at work in a process that may also include reciprocity and mutualism.

REPRODUCTIVE SUPPRESSION AND INFANTICIDE

Three linked facets of the private lives of carnivores are the suppression of male and female reproduction, infanticide, and multiple male mating. Among some carnivore families—notably canids and herpestids—it is common for the reproduction of subordinate females to be suppressed (Packard et al. 1983). Such females often act as helpers (see above). In parallel, aggressive competition for access to females in the same families restricts the mating opportunities for subordinate males. Sneaky subordinate males often achieve some reproductive success, however, and multiple paternity has been documented within groups and even within a single litter (e.g., in Eurasian badgers, Evans et al. 1989; dwarf mongooses, Keane et al. 1994; Domingo-Roura et al. 2003; American mink, Yamaguchi et al. 2004a). Recent noninvasive monitoring of hormone metabolites (reviewed by Wildt et al. 2001) has revealed that the likely means of suppressing reproduction in subordinate dwarf mongooses is excretion of excess estrogen by the alpha female (the alpha male had indistinguishable androgen profiles from subordinates). A similar mechanism is suggested for African wild dogs of the Selous Game Reserve, in which reproductive behaviors and dominance hierarchies are positively correlated with estrogen excretion (in females) but not androgen excretion (in males) (Monfort et al. 1997).

The extent of reproductive suppression varies among even closely related species—for example, most adult females may rear litters in groups of banded mongooses (Rood 1980), more than one may do so among bands of meerkats (Doolan and Macdonald 1997, 1999), and generally only one does so in groups of dwarf mongooses (Rasa 1986). The degree of suppression varies both among and within species (Creel and Waser 1991). For example, only 6% of African wild dog packs in Kruger National Park had more than one breeding female (Reich 1981), and the subordinates produced only about 9% of all pups (Girman et al. 1997). Yet in the Masai Mara Reserve several females bred in 38% of packs (Fuller et al. 1992; see also Creel et al. 2004).

The last chance for a dominant female to suppress a subordinate's reproduction is to kill the newborns. This phenomenon was famously filmed for African wild dogs by van Lawick (1974) and may be widespread in social canids (e.g., dingoes *[Canis familiaris dingo]*, Corbett 1988). It is also closely linked with the mismothering that appears commonplace among subordinate breeding females (e.g., red foxes, Macdonald 1981; bush dogs *[Speothos venaticus]*, Macdonald 1996b). Infanticide by males appears instead to be aimed at accelerating access to a mating opportunity by bringing the bereaved mothers into estrus sooner than would have otherwise been the case—for example, in lions (Bertram 1975). In a recent review of mammalian literature, Wolff and Macdonald (2004) conclude that the risk of such infanticide favors females mating promiscuously, as a mechanism for confusing paternity.

Multiple paternity within a litter is emerging as commonplace. An extremely interesting case is illustrated by the American mink. These mink are induced ovulators and exhibit delayed implantation (Thom et al. 2004), together with superfecundation (during one estrus eggs may be fertilized by more than one male) and superfoetation (ovulating after the initial fertilization, and kits can be sired by more than one male, which fertilize the ova of different ovulations during a single breeding season). The ranges of each male overlap those of several females, and during the mating season the resident males leave their territories, range widely, and then settle again in a new range. Yamaguchi et al. (2004a) tested two hypotheses to explain the males' behavior: perhaps they vacated their early mating season territories only after they had sired litters by the females therein and before setting off to seek additional females farther afield; alternatively, perhaps the traveling males were settling down in areas where they had successfully inseminated females whose kits they then wanted to protect. It seemed that neither hypothesis was correct. Not only were mixed-paternity litters born, but the kits born in a territory were often all unrelated to the males that had either occupied it early in the previous mating season or settled there after that mating season. Yamaguchi et al. (2004a) interpret these findings in terms of an adaptation to reduce the risk of male infanticide; the short period of delayed implantation, together with superfoetation and superfecundation, makes it hard for any one male to monopolize a female. The resulting uncertainty of paternity may diminish the risk of infanticide; alternative possibilities are that polyandry leads to indirect genetic benefits through offspring diversity, or heterozygosity, or the facilitation of sperm competition or sperm selection (Jennions and Petrie 2000). Indeed, in the case of the American mink, Thom et al. (2004b) provided females with

White-nosed coati *(Nasua narica)* taking a banana. Photo by Michael and Patricia Fogden/Minden Pictures.

the opportunity to mate promiscuously with several males and found that every female solicited copulations from more than one male. This turns on its head the earlier presumption that in sexually dimorphic carnivores like mustelids, females mated with multiple males only because they had no option.

Male white-nosed coatis interact with related females on a daily basis, but they wander more widely to find mates in their short synchronous breeding season (Gompper et al. 1998). There is evidence that in brown hyenas *(Hyaena brunnea)* reproduction is the prerogative of peripatetic wandering males rather than those resident in the group (Mills 1990).

"SOLITARY" SOCIAL SPECIES, HIDDEN ALLIANCES, AND OTHER CRYPTIC SOCIALITY

Although some long-term studies have unveiled intricacies of society that would have been unimaginable fifty years ago, the bottom line is the same: little is known of the social lives of most of the 271 species of carnivores. Even for those that have been studied in depth, our understanding of their societies remains frustratingly superficial. Part of the reason is that understanding nonhuman societies is both theoretically and philosophically complicated, but part is also, more prosaically, that it is simply very difficult to conduct such studies. Carnivores are often elusive, scarce, nocturnal, and, in many cases, they spend much of their time alone, although doubtless in touch with one another through social odors—a medium that humans are poorly adapted to explore. Notwithstanding the resonant title of Leyhausen's paper in 1968—"The social organization of solitary Carnivores"—the fact remains that social (i.e., gregarious) carnivores have been the focus of more sociobiological study than have their "solitary" cousins (Leyhausen 1965). However, the persistence and new technology of field biologists is starting to pay off. Recent glimpses into the social life of the most secretive species are revealing a number of surprises.

Many of these "solitary" species are found to live in not-so-obvious social groups. For example, the frugivorous, tree-living kinkajou was thought to be solitary until telemetry, nocturnal observation, and genetic analyses revealed a more social creature. While individuals spent most active time alone, they regularly met up at large feeding trees and slept together in their day dens. The pattern of these social interactions and home-range overlap resulted in polyandrous social groups and a promiscuous/monogamous mating system (Kays et al. 2000; Kays and Gittleman 2001; Kays 2003). Surprisingly, males were more social than females, perhaps reflecting the importance of food to the females (who bear the cost of reproduction alone) and their intolerance of feeding competition from other kinkajous.

Surprising levels of male-male sociality have also been described in a number of other species throughout the order including raccoons (Gehrt and Fritzell 1998), cheetahs (Caro 1994), mongooses (Rood 1989), and polar bears (Derocher and Stirling 1990). In many of these cases, male-male coalitions probably obtain benefits of better resource defense than individuals, and they are possible because of a lower cost of males sharing food (compared with reproductive females).

Communication (and Senses)

Nobody will be surprised to read that carnivores have considerable communicative powers—indeed, people and two domestic carnivores, dogs and cats, communicate across the species barrier with notable success. As Clutton-Brock (1996) observes, "a person may be able to tell a dog to sit, but the dog can just as easily dominate a cat with a look, and a lion can fully express its feelings to a hyaena," and there are several remarkable instances of interspecific communication involving at least one Carnivoran participant—a notable example being Rasa's (1986) discovery of the symbiotic responses of dwarf mongooses and hornbills to each other's calls, particularly in the context of their shared vigilance for predators.

DISPLAYS AND POSTURES

Although all carnivore families adopt expressive body and tail (except the almost tail-less bears) postures and movements, they differ markedly in the importance of facial expressions. For example, most of the Mustelidae exhibit limited facial expressions, although their body language can be flamboyant (e.g., the contorted dancing of displaying polecats). In contrast, people are especially attuned to the body language of domestic dogs (described by Lorenz 1954; Abrantes 1987), which is scarcely a dialect of that documented in ethograms for gray wolves (Zimen 1981) or golden jackals (Golani and Keller 1975) and is clearly part of the same "linguistic family" as that of foxes (Tembrock 1962; Fox 1971). Despite these canid similarities, communication is still a very species-specific language; for example, sinuous lashing of the vulpine tail is a clearly different action to the wagging of a lupine tail (Macdonald 1987).

VOCALIZATIONS

Although domestication has somewhat affected the domestic dog's repertoire (Coppinger and Coppinger 2002), the barks, growls, whines, and howls heard daily in our backyards are essentially the same as in wild wolves. Among wolf packs, howling serves to maintain or increase distance, helping to establish and maintain exclusive territories and reducing the probability of encountering strange wolves or packs in areas of border overlap (Harrington and Mech 1979). As in other mammals, pitch and quality of voice are apparently characteristics used to express and assess an individual's fighting or resource-holding potential. Harrington (1987) suggests that lower-pitched and harsher howls in wolves reflect greater hostility. Some canids produce less familiar sounds: the squeak of bush dogs (Kleiman 1972) and whistling of dholes (Fox 1984) are probably adaptations to keeping a hunting pack in coordinated contact in dense forest, and the oddly undoglike twittering of a social scrum of African wild dogs (van Lawick and van Lawick-Goodall 1970) appears to get the whole pack "tuned-in" to a forthcoming collective action.

Individual recognition of acoustic signals is a widespread phenomenon in animal communication. The role of long-distance vocalizations (e.g., loud calls of primates, howls of wolves or coyotes, whoops of spotted hyenas, roars of African lions) as a means of individual recognition is complicated by signal degradation associated with distance and the different acoustic characteristics of habitats (Wiley and Richards 1978). Nonetheless, individual recognition of long-distance signals has been reported in lions (McComb et al. 1994) and spotted hyenas (Mills 1990). Communication of collective strength is also revealed in McComb's (1992) and McComb et al.'s (1993, 1994) studies of roaring lions. Playbacks of taped voices illustrate that lionesses can estimate the number of voices they hear and decide whether or not to advance to an encounter depending on their strength of numbers relative to that of the perceived intruders. Similar playback experiments revealed that the Arctic foxes of Mednyi Island were able to distinguish the barks of their own family group from those of other individuals. Acoustic analysis revealed considerable variation in the spectral parameters of fox barks and, remarkably, evidence that the voices of group members were more like one another's than like those of other foxes (Frommolt et al. 2003). Both males and females bark in a variety of contexts: during territorial boundary patrols, during movements through their own territories, in response to barking by fellow group members, and in response to barking by their neighbors. The dens where pups are reared become a focus of barking activity, but barking is generally concentrated near the territorial boundary. These experiments showed that foxes responded differently to playbacks of group-member versus alien calls. The calls of group members never evoked territorial behavior but did elicit greetings (which never occurred in response to alien calls).

OLFACTORY COMMUNICATION

Few mammalian orders are as committedly olfactory as are the Carnivora. Their social odors stem from sebaceous and apocrine glandular secretions produced in diverse bodily nooks and crannies, together with urine and feces (reviewed by Kruuk and Macdonald 1985). A cocked leg is surely emblematic of canidness, providing a visual as well as olfactory signal (Bekoff 1978). Studies of gray wolves (Mech and Peters 1977), Ethiopian wolves (Sillero-Zubiri and Macdonald 1998), coyotes (Bekoff and Wells 1982), and red foxes (Macdonald 1979) reveal that all douse their territories with token urinations at very high rates and deposit their feces at such strategic sites as trail junctions and home-range borders (Macdonald 1980). In a surprising twist, dominant female bush dogs douse overhanging vegetation, which then drips their urine onto their trailing companions (Macdonald 1996b).

One of the most remarkable collections of olfactory adaptations is that displayed by the four members of the family Hyaenidae, which reaches its azimuth in the bicolored secretions produced by the intricate anal pouch of the brown hyena (Mills et al. 1980). The spotted hyena, the most conspicuously sociable member of the family, is also famously vocal and engages in extravagant postural communication.

The Eurasian badger marks its territory with conspicuous latrines, making these badgers an excellent case study in olfactory communication. In some parts of their geographic range, badgers live in large social groups that occupy territories ringed with a border of latrines to which each adult group member contributes feces (Kruuk 1978b; Stewart et al. 2001, 2002). The number of latrines deposited at any particular section of a group's territorial border is proportional to the number of its neighbors along that section (Stewart et al. 2001). Secretions of the large subcaudal pouch (Gorman et al. 1984) are used to mark the environment (object-marking) (Buesching and Macdonald 2001; Buesching and Macdonald, in press) and conspecifics (allomarking) (Buesching et al. 2004). Two types of allomarking have been distinguished: sequential allomarking, whereby a badger marks the flank or rump of another individual, and mutual allomarking, in which both badgers press their subcaudal glands together (probably thereby swapping the bacterial communities that determine the nuances of their olfactory signature). The volume and color, and thus presumably messages wafted therein, of the subcaudal secretion vary with season and with the age, sex, body condition, and reproductive status of the individual (Buesching et al. 2002c). The chemical properties of the subcaudal secretion encode information on age, sex, body condition, reproductive status, and group membership of the individual, as well as the age of the secretion itself (Buesching et al. 2002a; Buesching et al. 2002b).

Two features of olfactory communication are particularly noteworthy among the Carnivora. First, most species have various methods of deploying scents of diverse provenances. Second, in addition to the enormous interspecific variation in olfactory mechanisms, from the perineal glands of civets to the notorious anal sacs of skunks and the backwardly projected urine of cheetahs, there is also substantial intraspecific variation in the patterns of scent marking. Some of the latter variation may relate simply to the demands of production: latrines ringing a territory are associated with populations arranged as larger social groups in part, perhaps, because only a large social group can generate the necessary quantity of feces to fuel these signposts.

Conservation

WHY PROTECT CARNIVORES

On the asset side, carnivores are stunningly beautiful and conspicuously interesting, so it is increasingly widely accepted that their existence has high value (Gittleman et al. 2001b). They also are valued for commercial products, such as pelts of spotted cats and foxes, secretions from perineal glands of African civets and from bile ducts of bears, and bones of tigers *(Panthera tigris)* (Kenney et al. 1995). They also have value for recreation, which includes both photo-tourism and hunting for sport and trophies.

On the cost side, the biological nature of carnivores causes them to need relatively large amounts of space and to consume prey (which for some species occasionally includes *Homo sapiens*). These attributes lead to conflict with people, in terms of access to land and prey. The latter is a clear case of the ecological phenomenon of competition—carnivores and people often compete for wild prey or domestic livestock, and the extent to which this justifies killing carnivores has given rise to almost as much scientific debate as it has to human wrath and carnivore mortality. A further problem is that carnivores are sometimes involved in the spread of some infectious diseases that threaten humans or their livestock.

It is arguable that, on the continuum of biodiversity, carnivores, as predators, are particularly important in maintaining the biodiversity, stability, and integrity of various communities (Terborgh et al. 1999; Crooks and Soulé 1999). Although the functioning of ecosystems is itself far from fully understood, Johnson et al. (2001c) used this argument to conclude that "to protect carnivores . . . means to protect vegetation, other animals, water supplies, soil, and much more, and thus to sustain the quality of human life as well." The same conclusion can be reached simply on the grounds that carnivore conservation is generally difficult, so if it can be successfully addressed, so too can the conservation of many other organisms. Likewise, the emblematic importance of carnivores is generally unsurpassable. Various labels highlight species' roles: indicator species reflect critical environmental damage; keystone species are essential to ecosystem function; umbrella species require large areas and therefore also provide protection for other species. As Gittleman et al. (2001b) point out, carnivore species are often all these things and, what's more, they are also often vulnerable to extinction. In short, there are good arguments for conserving all species, but the arguments are particularly good when it comes to conserving carnivores.

The high profile of carnivores can work both ways. The fame and fortune of Nepal's tigers in Chitwan National Park have stimulated such intensive protection, antipoaching policies, and official scrutiny that not only have the tigers been protected, they have contributed to the triumphant restoration of the entire park ecosystem, leading to a financially promising community-based ecotourism (Dinerstein et al. 1999). Johnson et al. (2001c) argue, in contrast, that the razzmatazz associated with giant panda conservation has had only a superficial (and perhaps distracting) influence on thc root problems for their preservation. Dilemmas such as these, and the question of why carnivores and their conservation are special, are considered in detail by Macdonald (2001b) and other chapters in Gittleman et al. (2001a).

DECLINES IN GEOGRAPHIC RANGE AND POPULATION SIZE

The susceptibility of carnivores to decline is all too evident by their prominence in the list of local and global extinctions during historical times (Hummel and Pettigrew 1991). There is no satisfaction in mustering dramatic examples of this truth: the demise of the Falkland's wolf *(Dusicyon australis)* and the North American sea mink *(Mustela vison marcodon)* was uncontroversially human caused. Perhaps even more worrying is the long list of species whose distribution maps, formerly filled with broad brush strokes of color, have now faded to a rash of fragmentary refugia. Among the most obvious examples are the African wild dog, the lion, and the cheetah. Woodroffe et al's. (1997) action plan for the conservation of the wild dog reveals that perhaps only three populations are convincingly viable—those in the Kruger, Okavango, and Selous parks.

Nobody really knows how many lions used to live across their transcontinental, prehistoric range. However, it was certainly orders of magnitude more than our surviving population, estimated at circa 20,000 by the African Lion Working Group (Bauer and van der Merwe 2002). The trend began when lions were exterminated from Europe and the Middle East 2,000–3,000 years ago, a process that accelerated in Africa with the twentieth-century spread of firearms. In Asia only a small, genetically homogeneous population remains in the Gir Forest of India (Wildt et al. 1987).

The recent history of the cheetah is disquietingly similar: its world population is estimated to have halved in the last 25 years, with 20% of the remainder living in Namibian farmlands, where it is in conflict with stockmen (Marker et al. 2003).

A parallel saga describes the widespread eradication of the gray wolf from large parts of the Northern Hemisphere, fragmenting a range that was once the most expansive of that of any contemporary wild carnivore (although less extensive than the phenomenal range of Pleistocene lions, which spanned Africa, Eurasia, North and northern South America; Yamaguchi et al., in press). The bad news continues throughout the order, with carnivore species increasingly threatened and restricted in distribution.

On the other hand, more adaptable species, such as coyotes, jackals, red foxes, and three *Pseudalopex* fox species in the southern cone of South America, are all thriving despite tremendous hunting pressure due to the pelt trade or eradication campaigns. In fact, some carnivores are actually flourishing in new, manmade environments. There are urban populations of red foxes in the United Kingdom (Macdonald 1981; Lavin et al. 2003) and of coyotes (Grinder and Krausman 2001; Tigas et al. 2002), raccoons (Gehrt et al. 2002; Chamberlain et al. 2003), and even fishers (Kays and Bogan 2002) in North America, and palm civets *(Paradoxurus hermaphroditus)* now roam Hong Kong. Perhaps even more encouraging than this evolution in carnivore habitat preference is an evolution of public opinion, legal protection, and habitat recovery that has allowed the gray wolf to return to areas in Europe and North America where it was long ago hunted to extinction (Phillips et al. 2004).

EXTINCTION RISK

Just as the motives for conserving carnivores are numerous, so too are the types of threat that carnivores face (e.g., habitat loss, effects of introduced species, natural rarity, persecution, pathogens, exploitation). Indeed, these may be more numerous than the threats faced by any other mammalian order (Purvis et al. 2001). However, in the context of mammals as a group, no single carnivore taxon emerges as unusually threatened, nor does any carnivore family have an unusually high level of threatened species (Mace and Balmford 2000). Furthermore, at least in the case of canids, Macdonald and Sillero-Zubiri (in press) argue that, despite what at first seems an overwhelming array of different conservation issues, in reality the number of categories of problems and the trajectories open that might lead to a solution are rather few. Their argument can probably be extended to other carnivore families, distilling the problems to common strands that run through almost all of them, namely, competition for land, depredations on game and stock, overexploitation, and transmission of infectious disease. There are solutions to these problems (e.g., research, community involvement, education, and adaptive management, including restoration), which offer hope that lessons can be learned from previous experience and built upon in an approach to conservation that is truly evidence based.

The obvious fact that large, and thus generally wide-ranging, carnivores are the most likely to travel beyond the boundaries of protected areas and thus into dangerous encounters with people has been given quantitative weight by the analyses of Woodroffe and Ginsberg (1998; Purvis et al. 2001; Woodroffe 2001; Gittleman et al. 2001b). For example, many Iberian lynx *(Felis pardina)* radio-collared in Donana National Park, Spain, were killed on neighboring private land (Ferreras et al. 1992), while conflict with livestock farmers beyond the park borders caused nearly half the lion deaths recorded in Etosha and Nairobi National Parks (Rudnai 1979; Stander 1991). Although it seems obvious that large protected areas are a good thing, Leader-Williams and Albon (1988) point out that a big reserve may exceed the management and financial capacity of those running it. Despite nervousness about the hazards of small populations, Sunquist and Sunquist (2001) suggest that carnivores such as pumas, bears, tigers, and lynx can survive in small, isolated populations provided that habitat and prey are protected and some genetic exchange occurs (Beier 1996).

Susceptibility to extinction is not easily reflected by any one classification system. Wide-ranging carnivores are certainly at high risk because most reserves are small, and beyond them lies peril. However, given a chance, some of these species have a great capacity to bounce back, whereas others seem inherently rare. African lions in Kruger Park, for example, rapidly recovered from a massive cull in the 1970s (Smuts 1982), whereas it seems that African wild dogs are rare by nature (Creel and Creel 2003). Factors that contribute to resilience are high productivity under favorable conditions (e.g., large litter sizes, high proportions of females breeding), good dispersal ability (i.e., facilitating recolonization and population recovery) (Gese and Mech 1991; Clark and Fritzell 1992), dietary eclecticism, and a tolerance of human disturbance.

Differences in propensity for extinction may explain why there are still tigers in small reserves and in regions from which lions, cheetahs, and dholes have long since disappeared (Sunquist et al. 1999). In this vein, Purvis et al. (2001) argue

Tiger *(Panthera tigris)* leaping across water. Photo by Tim Fitzharris/Minden Pictures.

that several species of "least concern" actually embody a suit of characteristics that render them prone to extinction; their list includes the black-footed cat *(Felis nigripes),* brown hyena, sea otter, and culpeo *(Pseudalopex culpaeus).* As Woodroffe (2001) observes, these discrepancies between status classification and risk reflect the difference between a species' past record of extinction and the future it is likely to face. Tigers, for example, are a primarily tropical species that occupies threatened forest habitat in nations with high-density, rapidly growing human populations. By contrast, brown bears have a circumboreal distribution, with most remaining populations occurring at high northern latitudes, where human density is extremely low. Thus, by virtue of their distribution, brown bears may currently be less likely to become globally extinct than are tigers, despite their having been more extinction-prone over the last few hundred years.

SOURCES OF CONFLICT: ATTACKS ON HUMANS

Sillero-Zubiri and Laurenson (2001) conclude that, first and foremost, local communities view large carnivores as a direct threat to human life. In reality, large carnivores are fearful of humans and prefer to avoid them. Large cats, particularly tigers, African lions, leopards, and pumas, account for most human deaths by predators (Sunquist and Sunquist 2002). Large cats are not the only hunters of *Homo sapiens;* human mortality has also been documented from polar bears (Gjertz and Persen 1987), American black bears, and grizzly bears. Gray wolves caused no human deaths in North America during the twentieth century, but they are still widely feared (Kellert et al. 1996). For comparison, there are reported to be around 20 fatal dog attacks in the United States per year (see Lockwood 1995). However, many more people are killed by wolves in India and Eastern Europe, with elderly men or young children being the most vulnerable. In the last 20 years, 273 children have been reported killed by wolves in the Indian states of Andhra Pradesh, Bihar, and Uttar Pradesh, areas where wolf ecology brings them into close contact with shepherd children (Jhala and Sharma 1997).

Unsustainable levels of use lie behind one category of threats to carnivores. The obvious case is the fur trade in spotted cats. Jaguars *(Panthera onca)* once ranged from the southwestern United States throughout the Americas to central Argentina, but the fur trade, combined with habitat destruction, direct persecution, and declining prey have rendered them among the most imperiled big cats (Weber and Rabinowitz 1996), of which snow and clouded leopards *(Neofelis nebulosa)* may be even more precarious. A parallel can be drawn where hunters overharvested, for example, tigers during colonial times in India or, as is argued without much evidence, otters in the United Kingdom (although the decline of Britain's otters does dramatically illustrate an instance of incidental poisoning by agrochemicals) (Jefferies 1989). In parts of Africa trophy hunting for lions generates substantial foreign exchange; in Zimbabwe, the license to shoot a male lion may cost $5,000, and this trophy would often be the main motivator for spending about $30,000 on a safari. Considering that lions are part of complex societies, the disproportionate death rate of trophy males may trigger a cascade of sociobiological effects (the "perturbation effect," Tuyttens and Macdonald 2000; but see Whitman et al. 2004), and there is evidence that in some places this means that quotas have been set too high (Yamazaki 1996; Loveridge et al. 2002).

The trade in carnivore furs has fluctuated with supply and demand (Johnson et al. 2001c). As Sillero-Zubiri et al. (2004) note, much of the pioneering history of northern latitudes was built on trade in fur, much of it originally worn by foxes. A complication is that some of these species were killed also because of their perceived pest status. Thus, the Falkland Island wolf was extinguished in 1876, owing partly to fur traders and partly to poisoning by settlers to control sheep predation. This blend is commonplace; Novaro (1995) points out that farmhands kill culpeos and chillas *(Pseudalopex griseus)* in Argentina ostensibly to reduce predation on livestock or poultry, but the sales of their pelts may account for as much as 26% of the workers' annual income in Patagonia. It is therefore often not easy to attribute responsibility for a decline to a single motivation, and for any one carnivore the motivations of its persecutors may vary from place to place, as may the effect on their numbers.

SOURCES OF CONFLICT: COMPETITION

The obvious source of conflict between carnivores and people is the predatory habits of both. Red foxes kill chickens, lions kill cattle, brown bears kill sheep (and raid apiarists' hives), stoats *(Mustela erminea)* and weasels kill game birds, and wolverines *(Gulo gulo)* kill reindeer *(Rangifer tarandus)*. Some of the larger predators may even kill people. The response, to kill carnivores, has been relentless, ingenious, and often well planned, nationally organized, and state funded.

Hostility between people and carnivores may be deep rooted. Jorgensen and Redford (1993) point out that the diets of jaguar, puma, and tribal people in the neotropics overlap widely and deduce therefrom that we humans have always been in competition with big cats. Kruuk (2002) argues that our fearful, and often hostile, responses to carnivores are innate. On the other hand, the attention of people must surely have been a major selective pressure on some carnivores. Coyotes, wolves, and red foxes are all noted to be less diurnal and shyer in regions where they are persecuted. The history of human interactions with carnivores is

Female African lions *(Panthera leo)* killing a Grant's zebra *(Equus burchellii)* on the Serengeti Plains. Photo by Mitsuaki Iwago/ Minden Pictures.

enthralling. Long as this history is, the nub of the relationship has altered little. The emperor Charlemagne employed professional wolf hunters as early as AD 800 (Boitani 1995). Bounties or other control schemes were instigated by colonial governments to reduce predator numbers in Africa (Kenya Game Department 1958) and in North America (Leopold et al. 1964). Persecution was not even confined to areas of human habitation: wolves, puma, and African wild dogs were all persecuted inside designated national parks, with the aim of protecting "game" populations (Bere 1955; Phillips et al. 2004).

SOURCES OF CONFLICT: PREDATION ON LIVESTOCK

The history of conflict over domesticated stock traces back to the development and spread of herding societies (Reynolds and Tapper 1996). Although farmers and ranchers consistently express the most negative attitudes toward large carnivores (Reading and Kellert 1993), the damage these animals cause may be less than the cumulative effects of larger numbers of smaller carnivores, such as foxes, jackals, or coyotes. Macdonald and Sillero-Zubiri (2002) review reports on livestock raiding by big cats. For example, in the Gir Forest Sanctuary, India, approximately 2,000 domestic animals are killed annually by lions (Singh and Kampoj 1996), and in Nepal losses amounting to half the average annual per capita income were attributed by some villagers to wolves and snow leopards (Oli et al. 1994). Livestock predation by wild carnivores is by no means restricted to the developing world—for example, wolf and brown bear predation in parts of the Italian Abruzzo (Cozza et al. 1996) and brown bear and wolverine predation on free-ranging sheep in Norway (Landa and Tommeras 1997; Sagor et al. 1997; Landa et al. 1999).

Livestock depredation can produce costly cumulative tallies, and canids are often the main culprits. In the United States, 79% of sheep and 83% of cattle depredations are attributed to canids, with 15% and 18% of these losses (respectively) due to domestic dogs and the rest due mostly to coyotes (www.usda.gov/nass/). Estimates of sheep losses to wild canids in the United States were US$19–38 million in 1977; $75–150 million in 1980; $83 million in 1987; and $16 million in 1999 (Knowlton et al. 1999). Cattle and calf losses to predators in the United States represented $52 million in 2000. These economic losses promote high investments in control efforts. However, while these sums seem horrendous, it is well to remember that they are spread over immense areas and that losses are very patchy. Fifty percent of sheep producers in the United States, for example, reported annual losses to coyotes that were less than 5% of their stocks, but nearly a quarter of producers reported losses greater than 15% (Knowlton et al. 1999). Similarly, in the United Kingdom, lamb losses to red foxes are typically estimated or reported to be 1–2% of lambs born (Macdonald et al. 2000), though reported loss for individual flocks can reach 15% (Heydon and Reynolds 2000). Many factors affect this variation: Stahl et al. (2002) reported that lynx in the French Jura kill fewer than 0.5% of the available sheep, but there are, each year, half a dozen hotspots of killing; indeed, of flocks kept adjacent to large blocks of forest 39% suffered attacks. Similarly, the risk posed by Arctic foxes to lambs in Iceland increases with distance from the farmstead (Hersteinsson and Macdonald 1996). In Wisconsin, more than two-thirds of wolf packs did not attack nearby livestock (Wydeven et al. 2003); the crucial unknown is why they did not.

Of course, the relationship between carnivores and stockmen will reflect larger economic circumstances. In Botswana improved schooling following the discovery of diamonds has diminished the availability of small boys to tend livestock and thereby contributed to increased losses of cattle to African lions (G. Hemson, pers. comm.). In Argentina, the Patagonian steppe was devoted to sheep ranching during most of the twentieth century, and the densities of sheep predators were reduced by hunting throughout the region; the annual tally in culpeo skins averaged 15,000–20,000 during the 1970s and 1980s (Novaro 1995). Culpeos were reported to kill 7–15% of lambs annually and were second only to starvation in causes of lamb mortality. During the last two decades 90% of sheep ranches in southern Neuquén Province have switched to cattle production, resulting in a 51% decline of the total sheep stock. These changes, combined with lower fur prices, have lead to reduced hunting pressure on culpeos (as evidenced by a 70% decline in annual fox numbers killed) and increased fox densities (from 0.5–0.1/km^2 in 1989–92 to 1.0–0.2/km^2 in 2000–2002) and worsened depredation problems (24–40% of lamb stock killed per year) on those flocks that remain (Novaro et al. 2004).

Macdonald et al. (2003) explore the bioeconomics of red fox population control in the United Kingdom. For an arable farmer with no livestock or game interests, the balance is in favor of tolerating red foxes because they reduce rabbit grazing on crops, which would otherwise lead to reduced yields. This predation saves £1.6–9.3/ha in year 1, rising to £3.1–38.6 by year 3 (the range representing the outcome of different model choices). At the other extreme, a rural community dependent on an extensive low-husbandry sheep-farming system obtains a net break-even point by maintaining historical suppression of fox numbers on a regional basis, at a cost of £0.28/ewe (for comparison, this is one-tenth of routine veterinary costs), even though average lamb losses to foxes are 1–2%.

SOURCES OF CONFLICT: PREDATION ON GAME SPECIES

In Alaska, increasing gray wolf numbers have been blamed by hunters for declining moose populations and the resulting reduction in hunting quotas (Gasaway et al. 1992), whereas in the United Kingdom one of the main activities of professional gamekeepers is to reduce red fox, stoat, and weasel numbers to thus increase the harvestable surplus of game birds (Reynolds and Tapper 1996). Historically, gamekeepers were responsible for the widespread elimination in the early twentieth century of polecats from England, but with changing attitudes this species is now spreading eastward to recolonize England from its last refuge in Wales (Davison et al. 1999). Tapper et al. (1996) demonstrated by controlled experiment that local and seasonally targeted red fox culling, combined with similarly focused culling of other common predator species, allowed gray partridge *(Perdix perdix)* productivity to increase 3.5-fold over three years. Currently, an estimated annual tally of 70,000–80,000 red foxes is killed by a force of circa 3,500 professional gamekeepers, chiefly motivated to preserve game birds for sport shooting (Reynolds and Tapper 1996; Macdonald et al. 2000). Paradoxically, because there are far fewer gamekeepers now than a century ago (90% fewer), the overall current onslaught on red foxes is much reduced, which may partly explain why their numbers are higher and why the few remaining gamekeepers now encounter and kill many more foxes than they would have done a hundred years ago.

CONFLICT WITH THREATENED WILDLIFE

There is a temptation to caricature conservation dilemmas as polarized debates between good guys and bad guys, between the conservation cognoscenti and the environmentally, and ethically, illiterate. In reality, things are rarely that simple. Increasingly, instances arise in which the dilemma is between contrasting conservation outcomes. Spotted hyena prey on endangered black rhino *(Diceros bicornis)* calves in the Aberdare Mountains of Kenya (Sillero-Zubiri and Gottelli 1991); three separate reintroductions of captive-bred African wild dogs in Etosha National Park in Namibia failed when the dogs were killed by lions (Scheepers and Venzke 1995). Similarly, three male lions killed three subadult black rhinos within four months in 1995 in Etosha National Park (Brain et al. 1999). Dholes were held responsible for a decline in endangered banteng *(Bos javanicus)* in Java (Hedges and Tyson 1996), and gray wolves prey on endangered blackbuck *(Antilope cervicapra)* in India (Yadvendradew 1994). Where they have naturally invaded islands, Arctic fox have had a large impact on several Arctic seabird colonies (Birkenhead and Nettleship 1995). In the United Kingdom, there is a debate over whether control of indigenous red foxes is necessary to ensure the persistence of endangered indigenous bird populations such as capercaillie *(Tetrao urogallus)*, stone curlew *(Burhinus oedicnemus)*, and sandwich terns *(Sterna sandvicensis)*. Perhaps the most telling example concerns the endangered Island foxes of the California Channel Islands, where unpalatable trades-off seem unavoidable between the interests of the fox, a local shrike, two species of eagle, feral pigs *(Suidae)*, and various groups of local people (Roemer et al. 2002; Courchamp et al. 2003).

INTERSPECIFIC CONSIDERATIONS IN MANAGEMENT

To illustrate the complexity of managing predators it is necessary to look at the cascade of effects not only on their prey but also within their guild (Tuyttens and Macdonald 2000). Due to their broad range of sizes, carnivore species may be top, or meso-, predators, and thus can have dramatically different ecological roles. Killing them, therefore, may also have different impacts on their communities (Henke and Bryant 1999; Berger et al. 2001). Because larger carnivores may interfere with smaller ones, changes in the abundance of larger carnivores affect not only other predators and their prey but smaller carnivores as well (Palomares and Caro 1999; Linnell and Strand 2000; Tannerfeldt et al. 2002). Coyotes play a keystone role; their removal can lead to increased abundances of mesopredators, including gray foxes, and in turn to reductions in the abundances and diversity of prey, such as small rodents and jackrabbits *(Lepus californicus)* (Soulé et al. 1988; Henke and Bryant 1999). These effects have to be weighed against attempts to reduce coyotes with a view to protecting livestock and game. The removal of wolves from the Yellowstone ecosystem in the early twentieth century, for example, triggered a cascade effect by allowing increased density of moose, whose increased foraging pressure on riparian vegetation negatively affected the diversity of neotropical migrant birds (Berger et al. 2001). Examples of interspecific aggression are now numerous and often involve canids and felids as the aggressors and canids, felids, and mustelids as the victims (Palomares and Caro 1999). For example, intraguild killing is well documented among North American canids, with wolves killing coyotes and red foxes (Berg and Chesness 1978; Carbyn 1982; Paquet 1992), coyotes killing red foxes (Sargeant and Allen 1989; Voigt and Earle 1983), swift foxes (Scott-Brown et al. 1987), and kit foxes *(Vulpes macrotis)* (Ralls and White 1995), and red foxes killing kit foxes (Ralls and White 1995) and Arctic foxes (Bailey 1992).

Interspecific relationships have especial relevance to reintroductions. Gray wolf reintroductions (e.g., in Yellowstone) may result in reductions in both coyote and red fox populations (in addition to the suspected declines in the large prey species—ungulates, Phillips et al., in press). Thirty-one wolves were released in Yellowstone in 1995 and 1996 (Phillips et al., in press), and the population now numbers about 130. Of these, 40–50 wolves reside in the park's Northern Range, where coyotes were under study prior to the wolf reintroduction (Gese et al. 1996). The coyote population then comprised 80 individuals in 12 packs (Crabtree 1988; Crabtree and Sheldon 1999). Post-wolf, this population was reduced to 36 coyotes in 9 packs. The mean size of surviving packs dropped by 40%, from 6 to 3.6 adults and yearlings. Wolves were seen to kill 13 coyotes, with 19 additional deaths inferred from coyote carcasses with injuries and tracks attributable to wolves. For the post-wolf period, 25–33% of annual mortality has been attributed to wolves (Crabtree 1988; Crabtree and Sheldon 1999). Purvis et al. (2001) reveal that coyote mortality due to intraguild predation is largely additive, because population size has declined by 55% over a period of three years. One of the first accounts of this sort of hostility was suggested by Hersteinsson and Macdonald (1982) to explain the changing distribution of red and Arctic foxes—climate change enabling the larger red fox to oust the endangered Arctic foxes from high-quality, low-elevation breeding habitat (Tannerfeldt et al. 2002).

Interspecific hostility can be molded to conservation purposes. Sterile red foxes were introduced to the Pribiloff Islands to eradicate Arctic foxes (themselves earlier introductions) from imperilled sea-bird colonies (Bailey 1992). In this case, an alien species was used to fight another alien. Another example is how the reintroduction of a native species can be used to fight an alien species. Adopting an experimental approach, Bonesi and Macdonald (2004) monitored how the reintroduction of native Eurasian otters impacted a population of alien American mink in an area of the Upper Thames in England. A total of 17 otters were reintroduced to an area previously populated only by mink. Within the span of a year the mink population in this area was more than halved, showing a strong effect of interspecific competition, while a mink population of a nearby area, still unoccupied by otters, remained constant throughout the same period. Even though there is no evidence that otters are able to eradicate mink completely from an area, there is evidence that they can force mink to adopt a different lifestyle. As otter densities increase, mink have been observed to rely more on terrestrial rather than aquatic prey in order to avoid competition with the predominantly aquatic otter (Clode and Macdonald 1995; Bonesi et al. 2004).

Swift foxes were extirpated from Canada by the 1930s owing to native prairie habitat loss, poisoning, and trapping. Reintroductions using captive-bred and translocated animals have been ongoing since 1983, within a severely altered ecosystem (Moehrenschlager and Sovada, in press). Bison *(Bison bison)* have been replaced by cattle, while coyotes, which were previously rare, increased after the eradication of wolves. Since coyotes are smaller in body size, more generalist in diet, and exist at higher densities than wolves, swift foxes are now faced with a significant intraguild competitor and predator, which can account for up to 87% of mortalities (White and Garrott 1997). Coyotes will not necessarily drive swift fox populations to extinction, but red foxes, which may increase where coyotes are eradicated, could be an even greater threat. The ironic game unfolds where swift foxes can either be killed beyond sustainable levels if coyotes are abundant or be excluded by red foxes if coyotes are too rare to limit them. Swift fox persistence seems dependent on the availability of preferred habitats and on limited habitat conditions that disproportionately favor red foxes or coyotes (Moehrenschlager et al., in press). Large-scale habitat factors might determine the likelihood of canid encounters while small-scale variables, such as the availability of dens and badger burrows, which swift foxes use to elude predators, may be key in determining swift fox survival when they meet their canid cousins (Tannerfeldt et al. 2003). Clearly, swift fox movements in this landscape are risky, as the survival and reproduction of translocated foxes decreases with increased dispersal distances from release sites. Nevertheless, Canadian reintroduction efforts have gone well, and 88% of 125 live-trapped foxes caught at random sites were wild born (Moehrenschlager and Macdonald 2003).

These data highlight the, to some, unpalatable reality that many conservation decisions are consumer choices—do you prefer swift foxes or coyotes? There are also implications for the construction of networks of protected areas which, as Purvis et al. (2001) emphasize, reveals a flaw in the argument that where the top predator is safe so too will be everything else. Parks that support high densities of lions will probably support low densities of wild dogs, or none at all (Vucetich and Creel 1999). In Tanzania, national parks allow photo-tourism but not hunting and are generally in areas of high ungulate density, which support high densities of lions and hyenas. Because of the density of dominant competitors, wild dogs and cheetahs do not fare well in most of the national parks. In contrast, Tanzanian game reserves (which are used for trophy hunting instead of photo-tourism) are generally located in areas that support lower densities of ungulates, and thus lower densities of lions and hyenas. Wild dogs fare better in the game reserves. The same factors make Namibian ranch lands important to cheetah conservation—the cheetahs suffer intraguild persecution from larger carnivores in game parks, but these larger predators have been wiped out on ranch land, which is, paradoxically, a refuge for the cheetah. However, as Marker et al. (2003) point out, this makes the attitude of the ranchers crucial to the cheetah's future. In the light of these examples one might wonder how all these different predators co-existed in the past, and the answer is that they probably didn't, at least on a fine scale. In a prehistoric world the mosaic on habitats probably enabled incompatible species to shuffle around each other in a system of dynamic avoidance. Today, the island effect of protected areas surrounded by expanses of human activity offers the refugee species only a Hobson's choice.

INFECTIOUS DISEASE

An increasing cause of conflict between people and carnivores is the latter's role in transmitting infectious disease to livestock or humans. This is particularly evident for canids (Woodroffe et al. 2004), especially because of their involvement in rabies (Macdonald 1980). But rabies epizootiology also involves skunks and raccoons in North America (Charlton et al. 1988; Jenkins et al. 1998) and yellow mongooses *(Cynictis penicillata)*, bat-eared foxes, and jackals in southern Africa (Bingham et al. 1999). Hitherto, the main concern was that abundant wild carnivores acted as vectors of infectious disease, threatening people or livestock. A controversial example would be the European badger as a reservoir of bovine tuberculosis in cattle (Krebs et al. 1997). More recently, it has become clear that a second category of concern is the threat of disease from an abundant species spreading to a population of a rare species (Macdonald 1996a). Often, the abundant carnivore is an introduced domestic—for example, domestic dogs may have been the source of the rabies and distemper that devastated wild dogs and lions, respectively, in the Serengeti (Roelke-Parker et al. 1996) and Ethiopian wolves in the Bale Mountains (Sillero-Zubiri et al. 1996). Likewise, domestic cats may pose threats by transmitting diseases to wild felids (Yamaguchi et al. 1996).

INTRODUCED SPECIES

The translocation or introduction of animals to new ecosystems and the various impacts that result contend for second place among important causes of extinction and endangerment of native animals (e.g., Pimentel et al. 2000). Macdonald and Thom (2001) review this issue and identify the threats posed to native species worldwide by introduced carnivores as involving predation, hybridization, intraguild aggression, and the spread of disease. Interestingly, the mechanisms leading to loss of native biodiversity may differ when carnivores or herbivores are introduced (Macdonald et al. 2001). Australia's small mammal population has been decimated since European settlement, largely through predation by introduced carnivores such as red foxes and domestic cats and dogs (Smith and Quin 1996). Introduced predators are implicated in about half of all island bird extinctions (Diamond 1989; Milberg and Tyrberg 1993; Balmford 1996).

The dingo, introduced to the Australian continent perhaps 3,600 (or perhaps as many as 11,000) years ago, may have displaced by competition both the thylacine *(Thylacinus cynocephalus)* and the Tasmanian devil *(Sarcophilus harrisi)* (Lever 1994). Comparison of kangaroo *(Macropus)* and emu *(Dromaius novaehollandiae)* populations inside and outside the "dingo fence" in South Australia suggests that dingoes limit and probably regulate these prey species (Newsome et al. 1989; Pople et al. 2000). Similarly, red fox control by culling in the wheat belt of Western Australia allowed two rock wallaby *(Petrogale lateralis)* populations to increase by 138% and 223%, compared with 14% and 85% declines at nearby sites without fox control (Kinnear et al. 1998). In California, introduced red foxes are threatening rare clapper rails *(Rallus longirostris)* and salt marsh harvest mice *(Reithrodontomys raviventris)* (Reynolds and Tapper 1996), and they also kill endangered San Joaquin kit foxes *(Vulpes macrotis mutica)* (Ralls and White 1995).

In terms of carnivores being unwitting culprits, Ebenhard (1988) drew attention to the fact that they constitute 23 (19%) of the 118 recorded introduced mammalian species, more than three times their share of the total mammalian fauna (see also Boitani 2001). Five carnivore families are represented among in-

A dingo *(Canis familiaris dingo)* standing along a dirt road in Australia. Photo by Mitsuaki Iwago/Minden Pictures.

troduced faunas around the world, the exceptions being the Hyaenidae and the Ursidae. Of course, two domestic species, the dog and cat, rank high on these lists, respectively accounting for 59 and 19 of the 151 recorded "successful" Carnivoran introductions. Lever (1994) listed 153 native prey affected by introduced carnivores, of which 42% had been affected by cats, 25% by small Indian mongooses, and 11% by feral dogs, with the remaining 22% primarily affected by various foxes and mustelids. Jackson (1977) estimated that 61 bird taxa had been extirpated or become extinct due to introduced predators, of which cats and mongooses were responsible for 33 and 9 of these cases, respectively. Of nondomestics, the small Indian mongoose has impacted native species in 10 locations (Eben-hard 1988), and two mustelids, the American mink (Thom et al. 2004) and the stoat (King 1989), have also seriously affected local biodiversity.

Carnivores have also been victims of invasive species, and the mechanisms differ. Genetic introgression has been a threat to Ethiopian wolves interbreeding with domestic dogs (Gottelli et al. 1994) and to Scottish wild cats *(Felis silvestris)* interbreeding with domestic cats (Daniels et al. 2001); in the case of the Scottish wild cat, probably few individuals survive without at least some domestic cat genes (Beaumont et al. 2001). The European mink *(Mustela lutreola)*, once widespread throughout continental Europe, has almost disappeared, most plausibly because of interspecific aggression from the American mink—deliberately introduced into the countryside of the former Soviet Union with a view to bolstering the local fur-trapping economy. It seems that American mink attack their somewhat smaller European counterparts and drive them to suboptimal habitats (where they probably perish). Remarkably, in one area of Belarus, the body size of the few surviving European mink became significantly larger than that of their ancestors, which had first faced the invading American mink a decade earlier (Sidorovich et al. 2001).

Interestingly, introduced carnivores are not always problematic to the local predator community. So far, raccoon dogs *(Nyctereutes procyonoides)* from the Far East appear to have spread widely and in huge numbers in Finland, without obvious impact on numbers of local Eurasian badgers and red foxes; perhaps they have filled what was previously a genuinely empty niche (Kauhala and Saeki, in press). While domestic cats can devastate the local fauna when introduced on islands with no native carnivores (Bloomer and Bester 1992; Arnaud et al. 1993) or when kept at very high densities in "cat colonies" (Clarke and Pacin 2002), the typical house cat may not be such a threat. For example, Kays and DeWan (2004) found that inside/outside pet cats rarely hunted in a neighboring nature preserve and had no effect on populations of native mice in the preserve. They speculate that healthy coyote and fisher populations in this preserve probably limit the movement of cats into the forest, as was also suspected in Southern California (Crooks and Soulé 1999).

SOLVING CARNIVORE CONFLICT

In practice, what most affects peoples' actions toward carnivores is not the facts of their behavior (which are often unknown) but peoples' perceptions; questionnaire surveys concerning carnivores ranging from cheetahs in Namibia to red foxes in the United Kingdom reveal idiosyncrasies of perception (Macdonald and Johnson 2003; Marker et al. 2003). Perceptions of gray wolves vividly illustrate this (Kellert et al. 1996). Although they seldom attack people in North America and documented livestock losses are very low, wolves are often blamed for livestock attacks and are still widely feared (Gipson et al. 1998). Frank (1998) explores contradictions in the attitudes of Kenyan ranchers to large carnivores. For instance, the spotted hyena has the smallest per capita impact on livestock of any large predator, and damage by hyenas is largely preventable through livestock management. The big cats, by comparison, kill more stock and those depredations are more difficult to prevent. However, most ranchers admire the cats and avoid killing them when possible, while expressing antipathy for hyenas and sometimes shooting them on sight, even in the absence of serious depredation problems. These ranchers would like to see greater populations of lions, leopards, and cheetahs but fewer or no hyenas. Happily, more tolerant attitudes are emerging, but, as Sillero-Zubiri and Laurenson (2001) point out, these may be least evident among those people living closest to wild carnivores.

While some species seem frail in the proximity of humans, others can thrive among us: leopards and spotted hyenas can thrive where what Woodroffe (2001) terms the "critical human density" far exceeds that tolerable to African wild dogs and cheetahs. Furthermore, humans are not equal in this respect, and the distribution of gray wolves in North America in 1900 suggested a critical density of 13 people/km^2 above which wolves did not survive; yet, today, wolves persist in Cantabria, Spain (99 people/km^2), Abruzzo, Italy (118 people/km^2), and Rajasthan and Gujarat, India (129 and 211 people/km^2, respectively).

One relatively new tool in the development of carnivore management strategies is the GIS-based computer model (Macdonald and Rushton 2003). For example, models have clarified the importance of refugia of adequate size for American black bear populations (Powell et al. 1997) and bobcats (Knick 1990). Woodroffe and Ginsberg (1998) derived a measure of critical reserve size by using fitted regression models to predict the area at which populations persisted with a probability of 50%. This is analogous to an LD50, the dose of a drug which, when administered to experimental subjects, kills half of them. Critical reserve sizes range from 36 km^2 for the American black bear to 3,981 km^2 for its congener, the grizzly bear. Woodroffe and Ginsberg (1998) compared species' critical reserve sizes with average local population density and average female home-range size within the regions for which they investigated population extinction. After controlling for the potentially confounding effects of phylogeny, they concluded that average female home-range size was a good predictor of critical reserve size, while population density had no effect.

HUMAN DIMENSIONS

Twenty-first-century carnivore conservation heavily emphasizes the human dimension, including inputs from the social sciences, and it generally aspires to community-based programs. At the least, it is obvious that the goodwill of most stakeholders is essential if programs are to run smoothly. Contradictory elements of the human dimension are illustrated by the opposing views stimulated by proposals to reintroduce gray wolves in the United States: some saw them as a threat to livestock, others as a boost to ecotourism. In particular, Enck and Brown (2002) pointed to the hazards of unrealistic expectations; some of the communities in northern New York State that hoped to reap the benefits of ecotourism were socially and economically ill equipped to do so.

It is also essential to understand the organization of programs. Clark et al. (2001) review how the black-footed ferret *(Mustela nigripes)* reintroduction program was blighted by organizational mistakes. Antagonism among human factions threatens the conservation of grizzly bears, of which fewer than a thousand exist in the contiguous United States (Primm 1996). While the biological sciences are necessary for conservation, they are not sufficient; an interdisciplinary approach, embracing the social sciences and mindful of the realities of organizational and societal structures, is essential. Sillero-Zubiri and Laurenson (2001) emphasize community-based conservation and list various approaches to ameliorating conflict between people and carnivores, including fencing areas, changing human behavior to avoid contact with or attack by predators, reducing demand for carnivores and their products, changing livestock husbandry patterns to reduce predation, or modifying the predators' behavior by nonlethal methods, such as chemical deterrents. In the United States, Defenders of Wildlife has created the Proactive Carnivore Conservation Fund to share with ranchers the costs of actions to prevent livestock depredation. This has included buying livestock guardian dogs, erecting electric fencing, and hiring "wolf guardians" to monitor radio-collared wolves in sheep range, chasing them away when they get close to livestock.

CONTROL OF CARNIVORES

The logic and full interdisciplinary nature of carnivore control is complicated and rarely explored by a single study in a fully comprehensive manner. There are a number of key questions. How great is the damage caused by the carnivore, and what mismatch exists between this reality and its perception? What is the most effective means of diminishing that damage? Do the benefits of implementing control sufficiently outweigh the costs?

As early as 1848 the famous French lion hunter Jules Gérard expected that 20,000 francs to set up and maintain an African lion eradication team in Algeria would be easy to obtain because lions were causing damage worth an estimated 200,000 francs annually (Yamaguchi and Haddane 2002). However, aside from the considerable difficulty of measuring all the costs and benefits, the awkward reality is that each cost and benefit is likely to be measured in different and incommensurable units, and some of them will be open to different evaluation by different sections of society. Costs and benefits may be unevenly spread across society (Primm 1996). So, in Botswana, the costs of stock-raiding lions are born by a few hundred farmers adjoining national parks, tariffs returned to their community from ecotourism are spread among hundreds of thousands of people, and the real profits of ecotourism go to a largely different population (G. Hemson, pers. comm.).

Some governments have recognized that living with carnivores can be expensive. In Italy, for example, the local government compensates farmers for livestock killed by wolves, bears, and even feral dogs (Cozza et al. 1996). In France, recolonization by wolves has been closely monitored, with the provision of community guards and official damage assessors. In 2001, some 372 attacks by wolves on livestock killed 1,830 sheep, at a cost of more than €300,000 in compensation (Dahier 2002). Ignoring the economics, we can also ask the ethical question as to whether it was appropriate for about 3.5 million coyotes to have been killed in the United States during the past 50 years (Bekoff 2001). Questions of ethics are interwoven with science in many debates about carnivores (such as the hunting of red foxes with hounds for sport; Grandy et al. 2003).

NONLETHAL CONTROL

Traditionally, people have responded to perceived predator problems by killing the animals thought to be responsible. This might not be the most effective approach (Baker and Macdonald 1999). First, culling can be time-consuming, so it is often expensive and therefore not cost effective (Cowan et al. 2000). Second, intensive removal of territorial animals has been shown (at least in some cases) to produce no more than local, short-term reductions in predation as a result of both density-dependent compensations within the remaining population and swift replacement by conspecifics (Chesness et al. 1968; Macdonald et al. 2000; Reynolds et al. 1993). Third, lethal control can actually prove counterproductive; recently, Tuyttens and Macdonald (2000) demonstrated that removal of badgers, which are territorial, caused perturbation of their social systems, potentially impacting on the capacity of remaining population members to spread disease. Finally, there is a risk that, despite great efforts at population reduction or eradication, a single remaining predator can have a large impact on a prey species (Musgrave 1993). Quite apart from the problems of efficacy, there are ethical concerns over culling, and legislation surrounding lethal control is becoming tighter (Baker and Macdonald 1999). Yet another set of issues arise when the carnivore in question is endangered (see Sillero-Zubiri and Laurenson 2001).

Considering the various difficulties and doubts associated with controlling carnivores through culling, one might have expected human effort and ingenuity to have advanced further in the quest for effective alternatives. There have been some encouraging nonlethal successes. Fences have been used to increase the hatch rate of ground-nesting ducks (Greenwood et al. 1990). Wire-mesh fences successfully excluded Arctic foxes from the nests of Alaskan pectoral sandpipers *(Calidris melanotos)* (Estelle et al. 1996). Electric fences proved 99% effective against badgers (Poole and McKillop 1999), and mesh-grids safeguard Loggerhead turtle *(Caretta caretta)* nests from red foxes (Yerli et al. 1997). Andelt et al. (1999) used an electronic dog-training collar to deter captive coyotes from killing domestic lambs. A traditional means of stock protection, now pleasingly revived, is the use of guard animals (Coppinger and Coppinger 2002). Several breeds of dog, such as the Kuvasz of Slovakia, the Maremmano-Abruzzese of Italy, and the Anatolian Shepherds of Turkey, have proven effective in protecting livestock from wolves, coyotes, and cheetah (Rigg 2001; Marker et al. 2003).

Chemical approaches, such as repellents and Conditioned Taste Aversion (CTA), have also been explored (Baker and Macdonald 1999; Cowan et al. 2000). CTA is a natural behavioral response to the ingestion of poison; an animal consumes food containing a toxin, feels ill, and subsequently forms an aversion to the taste of that particular type of food (Nicolaus and Nellis 1987). The same process can be exploited for management purposes. Early trials with coyotes and gray wolves demonstrated the power of CTA to reduce predation in captivity (Gustavson et al. 1974, 1976). Unfortunately, early enthusiasm encouraged researchers to rush into ambitious and poorly designed large-scale field trials in an attempt to protect free-ranging domestic sheep from coyotes. Results were ambiguous, the work generated substantial controversy, and farmers lost confidence in the technique (Reynolds 1999). Subsequently, Lowell Nicolaus and coworkers have made real advances with CTA in wild predators, including raccoons (with eggs and tethered chickens) and mongooses (with eggs) (Nicolaus et al. 1982; Nicolaus and Nellis 1987; Semel and Nicolaus 1992). Macdonald and Baker (2003) demonstrated that a bitter-tasting substance, Bitrex, could be used to induce "generalized aversions" (falling somewhere between repellency and CTA) capable of protecting untreated foods from captive foxes. It remains to be demonstrated that learned food aversions, such as CTA and generalized aversion, can deter predation on free-ranging prey. Various methods of fertility control have been shown to suppress fertility in at least some species of carnivores. However, no single method has demonstrated the potential to cause substantial decline of a large natural population, and numerous questions remain to be answered regarding the effectiveness, humaneness, and ecological safety of fertility control methods (Tuyttens and MacDonald 1998).

Nonlethal control can potentially protect a target predator, as well as its prey (Gustavson 1977). A likely advantage of using certain types of nonlethal control (e.g., CTA or fertility control) with a territorial carnivore is that damage caused by the resident population may be reduced while leaving their other ecological relationships intact, including continued exclusion of untreated conspecifics through territorial defense (Cowan et al. 2000; Nicolaus and Nellis 1987; Nicolaus et al. 1992; Reynolds 1999). In other words, the "poacher" becomes the "gamekeeper." Further advantages of the resultant population stability include avoidance of social perturbation (Tuyttens and Macdonald 2000) and potential capacity for social facilitation of altered food preferences from adults to offspring (Semel and Nicolaus 1992).

REINTRODUCTION

Reintroductions are lengthy, costly, and complex processes (IUCN/SCC 1998) that often do not result in a clear success (Griffith et al. 1989; Beck et al. 1994). Almost invariably, reintroduced populations start small (Sarrazin and Barbault 1996), meaning that preserving genetic variation in released populations is difficult (Sarrazin and Barbault 1996) and protecting them from disease is imperative (Breitenmoser-Würsten et al. 2002). Costs of captive breeding for reintroductions increase with the size of the animal (Balmford 1996), and complex behavior is difficult to preserve in captivity (Shepherdson 1994; Mallinson 1995). Animal behavior cannot be ignored, because reintroductions have failed owing to behavioral deficiencies in released animals (Snyder et al. 1996).

Despite these complications, reintroduction is sometimes the only option, and the number of carnivore reintroductions has significantly increased over time. Breitenmoser et al. (2001) review recent carnivore reintroductions from an interdisciplinary viewpoint. Not surprisingly, large, continuous blocks of undeveloped land are helpful to large carnivore reintroduction, as is the capacity to release large numbers of individuals.

Public support is also essential. The first planned reintroduction of the red wolf *(Canis rufus)*, to the Kentucky-Tennessee border in the late 1970s, was eventually called off owing to public hostility, not for biological reasons. The second attempt, following extensive public relations work, went ahead in 1987 in North Carolina (Reading and Clark 1996). This extra effort paid off, as there were an estimated one hundred wolves ranging over 6,000 km^2 in the fall of 2003 (USFWS 2003), making this the first successful large-scale reintroduction of a large predator. Similar problems were encountered in the Florida panther recovery program following an experimental release of five panthers into Osceola National Forest in 1988 (Beldon et al. 1989) and in the recent reintroduction of Mexican wolves *(Canis lupus baileyi)* in Arizona, where some released animals were shot (Clark et al. 2001). The reintroduction of gray wolves into Yellowstone National Park was preceded by massive attention to local values and attitudes (Clark et al. 2001). Both the red wolf and gray wolf programs appear to be succeeding (Phillips et al., in press).

Acknowledgments

We warmly acknowledge the helpful comments made by Jorie Favreau, Dom Johnson, Chris Newman, Claudio Sillero-Zubiri, and Nobby Yamaguchi and the assistance of Lauren Harrington throughout, together with Sandra Baker and Tom Moorhouse.

Literature Cited

Abrantes, R. A. B. 1987. The expression of emotions in man and canid. Journal of Small Animal Practice 28:1030–36.

Alexander, R. D. 1974. The evolution of social behaviour. Annual Review of Ecology and Systematics 5:325–83.

Allee, W. C. 1931. Animal Aggregations: A Study in General Sociology. University of Chicago Press, Chicago.

Andelt, W. F., R. L. Phillips, K. S. Gruver, and J. W. Guthrie. 1999. Coyote predation on domestic sheep deterred with electronic dog-training collar. Wildlife Society Bulletin 27:12–18.

Arnaud, G., A. Rodriguez, R. A. Ortega, and C. S. Alvarez. 1993. Predation by cats on the unique endemic lizard of Socorro Island *(Urosaurus auriculatus)*, Revillagigedo, Mexico. Ohio Journal of Science 93:101–4.

Atkinson, R. P. D., C. J. Rhodes, D. W. Macdonald, and R. M. Anderson. 2002. Scale-free dynamics in the movement patterns of jackals. Oikos 98:134–40.

Bailey, E. P. 1992. Red foxes, *Vulpes vulpes*, as biological control agents for introduced Arctic foxes, *Alopex lagopus*, on Alaskan islands. Canadian Field-Naturalist 106:200–205.

Baker, P. J., S. M. Funk, S. Harris, and P. C. L. White. 2000. Flexible spatial organization of urban foxes, *Vulpes vulpes*, before and during an outbreak of sarcoptic mange. Animal Behaviour 59:127–46.

Baker, R. R. 1978. The Ecology of Animal Migration. Hodder and Stoughton, London.

Baker, S. E., and D. W. Macdonald. 1999. Non-lethal predator control: exploring the options. *In* P. D. Cowan and C. J. Feare (eds.), Advances in Vertebrate Pest Management, 251–66. Filander Verlag, Furth, Germany.

Ballard, W. B., J. S. Whitman, and C. L. Gardner. 1987. Ecology of an exploited wolf population in south-central Alaska. Wildlife Monographs 98:1–54.

Balmford, A. 1996. Extinction filters and current resilience: the significance of past selection pressures for conservation biology. Trends in Ecology and Evolution 11:193–96.

Bauer, H., and S. van der Merwe. 2002. The African Lion Database. IUCN/SSC African Lion Working Group. www.african-lion.org/ALD_2002.prf.

Beaumont, M., E. M. Barratt, D. Gottelli, A. C. Kitchener, M. J. Daniels, J. K. Pritchard, and M. W. Bruford. 2001. Genetic diversity and introgression in the Scottish wildcat. Molecular Ecology 10:319–36.

Beck, B. B., L. G. Rapaport, M. R. Stanley-Price, and A. C. Wilson. 1994. Reintroduction of captive-born animals. *In* P. J. S. Olney, G. M. Mace, and A. T. C. Feistner (eds.), Creative Conservation: Interactive Management of Wild and Captive Animals, 265–87. Chapman and Hall, London.

Beier, P. 1996. Metapopulations models, tenacious tracking, and cougar conservation. *In* D. R. McCullough (ed.), Metapopulations and Wildlife Conservation, 293–323. Island Press, Washington, D.C.

Bekoff, M. 1978. Scent-marking by free-ranging domestic dogs. Biology of Behaviour 4:123–39.

———. 2001. Human-carnivore interactions: adopting proactive strategies for complex problems. *In* J. L. Gittleman, S. M. Funk, D. Macdonald, and R. K. Wayne (eds.), Carnivore Conservation, 179–96. Cambridge University Press, Cambridge.

Bekoff, M., and M. C. Wells. 1980. The social ecology of coyotes. Scientific American 242:130–51.

———. 1982. The behavioral ecology of coyotes: social organization, rearing patterns, space use, and resource defense. Zeitschrift für Tierphysiologie 60:281–305.

Bekoff, M., T. J. Daniels, and J. L. Gittleman. 1984. Life history patterns and the comparative social ecology of carnivores. Annual Review of Ecology and Systematics 15:191–232.

Beldon, R. C., B. W. Hagdorn, and W. B. Frankenberger. 1989. Panther captive breeding/reintroduction feasibility. Final performance report. Florida Game and Freshwater Fish Commission, Tallahassee.

Bere, R. M. 1955. The African wild dog. Oryx 3:180–82.

Berger, J., P. B. Stacey, L. Bellis, and M. P. Johnson. 2001. A mammalian predator-prey imbalance: grizzly bear and wolf extinction affect avian neotropical migrants. Ecological Applications 11:947–60.

Bertram, B. C. R. 1975. Social factors influencing reproduction in wild lions. Journal of Zoology (London) 177:463–82.

Biknevicius, A. R., and B. Van Valkenburgh. 1996. Design for killing: craniodental adaptations of predators. *In* J. L. Gittleman (ed.), Carnivore Behavior, Ecology, and Evolution, 2:393–428. Cornell University Press, Ithaca, N.Y.

Bingham, J., C. L. Schumacher, F. W. G. Hill, and A. Aubert. 1999. Efficacy of SAG-2 oral rabies vaccine in two species of jackal (*Canis adustus* and *Canis mesomelas*). Vaccine 17:551–58.

Bininda-Emonds, O. R. P., J. L. Gittleman, and A. Purvis. 1999. Building large trees by combining phylogenetic information: a complete phylogeny of the extant Carnivora (Mammalia). Biological Reviews 74:143–75.

Bininda-Emonds, O. R. P., J. L. Gittleman, and C. K. Kelly. 2001. Flippers versus feet: comparative trends in aquatic and non-aquatic carnivores. Journal of Animal Ecology 70:386–400.

Bininda-Emonds, O. R. P., J. L. Gittleman, and M. A. Steel. 2002. The (super) tree of life: procedures, problems, and prospects. Annual Review of Ecology and Systematics 33:265–89.

Birkenhead, T. R., and D. N. Nettleship. 1995. Arctic fox influence on seabird community in Labrador: a natural experiment. Wilson Bulletin 107:397–412.

Bloomer, J. P., and M. N. Bester. 1992. Control of feral cats on sub-Antarctic Marion Island, Indian Ocean. Biological Conservation 60:211–19.

Boitani, L. 1995. Ecological and cultural diversities in the evolution of wolf-human relationships. *In* L. N. Carbyn, S. H. Fritts, and D. R. Seip (eds.), Ecology and Conservation of Wolves in a Changing World, 3–11. Canadian Circumpolar Institute, Edmonton.

———. 2001. Carnivore introductions and invasions: their success and management options, 123–44. *In* J. L. Gittleman, S. M. Funk, D. Macdonald, and R. K. Wayne (eds.), Carnivore Conservation, 123–44. Cambridge University Press, Cambridge.

Bonesi, L., and D. W. Macdonald. 2004. Impact of released Eurasian otters on a population of American mink: a test using an experimental approach. Oikos 106:9–18.

Bonesi, L., P. Chanin, and D. W. Macdonald. In press. Competition between Eurasian otter *Lutra lutra* and American mink *Mustela vison* probed by niche shift. Oikos.

Born, E., O. Wiig, and J. Thomassen. 1997. Seasonal and annual movements of radio-collared polar bears (*Ursus maritimus*) in Northeast Greenland. Journal of Marine Systems 10:67–77.

Bowen, W. D. 1981. Variation in coyote social organization: the influence of prey size. Canadian Journal of Zoology 59:639–52.

Brain, C., O. Forge, and P. Erb. 1999. Lion predation on black rhinoceros (*Diceros bicornis*) in Etosha National Park. African Journal of Ecology 37:107–9.

Breitenmoser, U., C. Breitenmoser-Würsten, L. Carbyn, and S. Funk. 2001. Assessment of carnivore reintroductions. *In* J. L. Gittleman, S. M. Funk, D. Macdonald, and R. K. Wayne (eds.), Carnivore Conservation, 241–81. Cambridge University Press, Cambridge.

Breitenmoser-Würsten, C., H. Posthouse, L. Bacciarini, and U. Breitenmoser. 2002. Causes of mortality in reintroduced Eurasian lynx in Switzerland. Journal of Wildlife Diseases 38:84–92.

Brocke, R. H., and K. A. Gustafson. 1992. Lynx (*Lynx canadensis*) in New York State. Reintroduction News 4.

Buesching, C. D., and D. W. Macdonald. 2001. Scent-marking behaviour of the European badger (*Meles meles*): resource defence or individual advertisement. *In* A. Marchlewska-Koj, J. L. Lepri, and D. Müller-Schwarze (eds.), Chemical Signals in Vertebrates 9:321–27. Kluwer Academic/Plenum Press, New York.

———. 2004. Variations in object-marking activity and over-marking behaviour of European badgers (*Meles meles*) in the vicinity of their setts. Acta Theriologica 49:235–46.

Buesching, C. D., C. Newman, and D. W. Macdonald. 2002c. Variations in colour and volume of the subcaudal gland secretion of badgers *(Meles meles)* in relation to sex, season and individual specific parameters. Mammalian Biology 67:147–56.

Buesching, C. D., P. Stopka, and D. W. Macdonald. 2004. The social function of allomarking behaviour in the European badger *(Meles meles)*. Behaviour 140:965–80.

Buesching, C. D., J. P. Waterhouse, and D. W. Macdonald. 2002a. Gas chromatographic analysis of the subcaudal gland secretion of the European badger *(Meles meles)*. Part I: Chemical differences related to individual-specific parameters. Journal of Chemical Ecology 28:41–56.

———. 2002b. Gas chromatographic analysis of the subcaudal gland secretion of the European badger *(Meles meles)*. Part II: Time-related variation in the individual-specific composition. Journal of Chemical Ecology 28:57–69.

Burt, W. H. 1943. Territoriality and home range concepts as applied to mammals. Journal of Mammalogy 24:346–52.

Buskirk, S. W., and R. A. Powell. 1994. Habitat ecology of fishers and American martens. *In* S. W. Buskirk, A. S. Harestad, M. G. Raphael, and R. A. Powell (eds.), Martens, Sables, and Fishers: Biology and Conservation, 283–96. Cornell University Press, Ithaca, N.Y.

Camenzind, F. J. 1978. Behavioral ecology of coyotes on the National Elk Refuge, Jackson, Wyoming. *In* M. Bekoff (ed.), Coyotes: Biology, Behavior, and Management, 267–94. Academic Press, New York.

Caraco, T., and L. L. Wolf. 1975. Ecological determinants of group sizes in foraging lions. American Naturalist 109:343–52.

Carbone, C., and J. L. Gittleman. 2002. A common rule for the scaling of carnivore density. Science 295:2273–76.

Carbone, C., G. M. Mace, S. C. Roberts, and D. W. Macdonald. 1999. Energetic constraints on the diet of terrestrial carnivores. Nature 402:286–88.

Caro, T. M. 1994. Cheetahs of the Serengeti Plains: Group Living in an Asocial Species. University of Chicago Press, Chicago.

Carr, G. M., and D. W. Macdonald. 1986. The sociality of solitary forager: a model based on resource dispersion. Animal Behaviour 34:1540–79.

Chamberlain, M. J., L. M. Conner, B. D. Leopold, and K. M. Hodges. 2003. Space use and multi-scale habitat selection of adult raccoons in central Mississippi. Journal of Wildlife Management 2003 67:334–40.

Charles-Dominique, C. 1978. Ecologie et vie sociale de *Nandinia binotata* (Carnivores, viverrides): comparaison avec les prosimiens sympatriques du Gabon. La Terre et la Vie 32:477–528.

Charlton, K. M., W. A. Webster, G. A . Casey, and C. E. Rupprecht. 1988. Skunk rabies. Reviews of Infectious Diseases 10:626–28.

Cheeseman, C. L., W. J. Cresswell, S. Harris, and P. J. Mallinson. 1988. Comparison of dispersal and other measurements in two badger *(Meles meles)* populations. Mammal Review 18:1–59.

Clark, T. W., D. J. Mattson, R. P. Reading, and B. J. Miller. 2001. Interdisciplinary problem solving in carnivore conservation: an introduction. *In* J. L. Gittleman, S. M. Funk, D. Macdonald, and R. K. Wayne (eds.), Carnivore Conservation, 223–40. Cambridge University Press, Cambridge.

Clark, W. R., and E. K. Fritzell. 1992. A review of population dynamics of furbearers. *In* D. R. McCullough and R. H. Barrett (eds.), Wildlife 2001: Populations, 899–910. Elsevier, London.

Clarke, A. L., and T. Pacin. 2002. Domestic cat "colonies" in natural areas: a growing exotic species threat. Natural Areas Journal 22:154–59.

Clode, D., and D. W. Macdonald. 1995. Evidence for food competition between mink and otter on Scottish islands. Journal of Zoology (London) 237:435–44.

Clutton-Brock, J. 1996. Competitors, companions, status symbols, or pests: a review of human associations with other carnivores. *In* J. L. Gittleman (ed.), Carnivore Behavior, Ecology, and Evolution, 2:375–392. Cornell University Press, Ithaca, N.Y.

Clutton-Brock, T. H. 1991. The Evolution of Parental Care. Princeton University Press, Princeton.

Clutton-Brock, T. H., A. Maccoll, P. Chadwick, D. Gaynor, R. Kansky, and J. D. Skinner. 1999. Reproduction and survival of suricates *(Suricata suricatta)* in the southern Kalahari. African Journal of Ecology 37:69–80.
Coppinger, R., and L. Coppinger. 2002. Dogs: A New Understanding of Canine Origin, Behavior and Evolution. University of Chicago Press, Chicago.
Corbett, L. 1988. Social dynamics of a captive dingo pack: population by dominant female infanticide. Ethology 78:177–78.
Cott, H. B. 1940. Adaptive Coloration in Animals. Methuen, London.
Courchamp, F., M. Langlais, and G. Sugihara. 1999. Cats protecting birds: modelling the mesopredator release effect. Journal of Animal Ecology 68:282–92.
Courchamp, F., G. S. A. Rasmussen, and D. W. Macdonald. 2002. Small pack size imposes a trade-off between hunting and pup-guarding in the painted hunting dog *Lycaon pictus.* Behavioral Ecology 13:20–27.
Courchamp, F., R. Woodroffe, and G. Roemer. 2003. Removing protected populations to save endangered species. Science 302:1532.
Cowan, D. P., J. C. Reynolds, and E. L. Gill. 2000. Reducing predation through conditioned aversion. *In* L. M. Gosling and W. J. Sutherland (eds.), Behaviour and Conservation, 281–99. Cambridge University Press, Cambridge.
Cox, R., P. D. Stewart, and D. W. Macdonald. 1999. The ectoparasites of the European badger, *Meles meles,* and the behavior of the host-specific flea, *Paraceras melis.* Journal of Insect Behavior 12:245–65.
Cozza, K., R. Fico, M. L. Battistini, and E. Rogers. 1996. The damage-conservation interface illustrated by predation on domestic livestock in central Italy. Biological Conservation 78:329–36.
Crabtree, R. 1988. Total impact. Tracker 5:12.
Crabtree, R. L., and J. W. Sheldon. 1999. Coyotes and canid coexistence in Yellowstone. *In* T. W. Clark, A. P. Curlee, S. C. Minta, and P. M. Kareiva (eds.), Carnivores in Ecosystems, 127–63. Yale University Press, New Haven.
Creel, S., and N. M. Creel. 1995. Communal hunting and pack size in African wild dogs *Lycaon pictus.* Animal Behaviour 50:1325–39.
———. 2002. Cooperative hunting and evolution of sociality. *In* J. R. Krebs and T. Clutton-Brock (eds.), The African Wild Dog: Behavior, Ecology, and Conservation, 67–102. Princeton University Press, Princeton.
Creel, S. R., and N. M. Creel. 1991. Energetics, reproductive suppression and obligate communal breeding in carnivores. Behavioural Ecology and Sociobiology 28:263–70.
Creel, S., and D. W. Macdonald. 1995. Sociality, group-size, and reproductive suppression among carnivores. Advances in the Study of Behavior 24:203–57.
———. 2003. The African Wild Dog: Behavior, Ecology, and Conservation. Princeton University Press, Princeton.
Creel, S. R., and P. M. Waser. 1991. Failures of reproductive suppression in dwarf mongooses: accident or adaptation? Behavioral Ecology 2:7–15.
———. 1994. Inclusive fitness and reproductive strategies in dwarf mongooses. Behavioral Ecology 5:339–48.
Creel, S., N. M. Creel, and S. L. Monfort. 1998. Birth order, estrogens and sex-ratio adaptation in African wild dogs *(Lycaon pictus).* Animal Reproduction Science 53:315–20.
Creel, S. R., M. G. L. Mills, and J. McNutt. 2004. Demography and population dynamics of African wild dogs in three critical populations. *In* D. W. Macdonald and C. Sillero-Zubiri (eds.), Canid Biology and Conservation, 337–52. Oxford University Press, Oxford.
Creel, S. R., S. L. Montfort, D. E. Wildt, and P. M. Waser. 1991. Spontaneous lactation is an adaptive result of pseudopregnancy. Nature 351:660–62.
Crooks, D. R., and M. E. Soulé. 1999. Mesopredator release and avifaunal extinctions in a fragmented system. Nature 400:563–66.
Crooks, K. 2002. Relative sensitivities of mammalian carnivores to habitat fragmentation. Conservation Biology 16:488–502.
Dahier, T. 2002. Année 2001: bilan des dommages sur les troupeax domestiques. L'Infoloups 10:11.

Daniels, M. J., M. A. Beaumont, P. J. Johnson, D. Balharry, D. W. MacDonald, and E. Barratt. 2001. Ecology and genetics of wild-living cats in the north-east of Scotland and the implications for the conservation of the wildcat. Journal of Applied Ecology 38:146–61.

Davison, A., J. D. S. Birks, R. C. Brookes, T. C. Braithwaite, and J. E. Messenger. 2002. On the origin of faeces: morphological versus molecular methods for surveying rare carnivores from their scats. Journal of Zoology (London) 257:141–43.

Davison, A., J. D. S. Birks, H. I. Griffiths, A. C. Kitchener, D. Biggins, and R. K. Butlin. 1999. Hybridization and the phylogenetic relationship between polecats and domestic ferrets in Britain. Biological Conservation 87:155–61.

Dayan, T., D. Simberloff, E. Tchernov, and Y. Yomtov. 1991. Calibrating the paleothermometer: climate, communities and the evolution of size. Paleobiology 17:189–99.

Debinski, D. M., and R. D. Holt. 2000. A survey and overview of habitat fragmentation experiments. Conservation Biology 14:342–55.

Derocher, A. E., and I. Stirling. 1990. Observation of aggregating behaviour in adult male polar bears. Canadian Journal of Zoology 68:1390–94.

Diamond, J. M. 1989. The present, past and future of human-caused extinctions. Philosophical Transactions of the Royal Society of London, Series B, 325:469–77.

Dijak, W. D., and F. R. I. Thompson. 2000. Landscape and edge effects on the distribution of mammalian predators in Missouri. Journal of Wildlife Management 64:209–16.

Dinerstein, E., A. Rijal, M. Bookbinder, B. Kattel, and A. Rajuria. 1999. Tigers as neighbours: efforts to promote local guardianship of endangered species in lowland Nepal. *In* J. Siedensticker, S. Christie, and P. Jackson (eds.), Riding the Tiger: Tiger Conservation in Human-Dominated Landscapes, 316–33. Cambridge University Press, Cambridge.

Domingo-Roura, X., D. W. Macdonald, M. S. Roy, J. Marmi, J. Terradas, R. Woodroffe, T. Burke, and R. K. Wayne. 2003. Confirmation of low genetic diversity and multiple breeding females in a social group of Eurasian badgers from microsatellite and field data. Molecular Ecology 12:533–39.

Doolan, S. P., and D. W. Macdonald. 1996. Dispersal and extra-territorial prospecting by slender-tailed meerkats *(Suricata suricatta)* in the south-western Kalahari. Journal of Zoology (London) 240:59–73.

———. 1997. Band structure and failures of reproductive suppression in a cooperatively breeding carnivore, the slender-tailed meerkat *(Suricata suricatta)*. Behaviour 134:827–48.

———. 1999. Co-operative rearing by slender-tailed meerkats *(Suricata suricatta)* in the southern Kalahari. Ethology 105:851–66.

Dragoo, J. W., and R. L. Honeycutt. 1997. Systematics of mustelid-like carnivores. Journal of Mammalogy 78:426–43.

East, M. L., and H. Hofer. 1991. Loud calling in a female-dominated mammalian society: I. Structure and composition of whooping bouts of spotted hyaenas, *Crocuta crocuta*. Animal Behaviour 42:637–49.

Ebenhard, T. 1988. Introduced birds and mammals and their ecological effects. Swedish Wildlife Research 13:1–107.

Eisenberg, J. F. 1981. The Mammalian Radiations. Chicago University Press, Chicago.

Eloff, F. C. 1964. On the predatory habits of lions and hyaenas. Koedoe 7:105–12.

Enck, J., and T. L. Brown. 2002. New Yorkers' attitudes toward restoring wolves to the Adirondack Park. Wildlife Society Bulletin 30:16–28.

Estelle, V. B., T. J. Mabee, and A. H. Farmer. 1996. Effectiveness of predator exclosures for pectoral sandpiper nests in Alaska. Journal of Field Ornithology 67:447–52.

Estes, J. A., N. S. Smith, and J. F. Palmisano. 1978. Sea otter predation and community organization in the western Aleutian Islands, Alaska. Ecology 59:822–33.

Estes, R. D., and J. Goddard. 1967. Prey selection and hunting behavior of the African wild dog. Journal of Wildlife Management 31:52–70.

Evans, P. G. H., D. W. Macdonald, and C. L. Cheeseman. 1989. Social structure of the Eurasian badger *(Meles meles)*: genetic evidence. Journal of Zoology (London) 218:587–95.

Ewer, R. F. 1973. The Carnivores. Comstock Publishing Associates, Ithaca, N.Y.
Fanshawe, J. H., and C. D. Fitzgibbon. 1993. Factors influencing the hunting success of an African wild dog pack. Animal Behaviour 45:479–90.
Fedriani, J. M., T. K. Fuller, and R. M. Sauvajot. 2001. Does availability of anthropogenic food enhance densities of omnivorous mammals? An example with coyotes in southern California. Ecography 24:325–31.
Fedriani, J. M., T. K. Fuller, R. M. Sauvajot, and E. C. York. 2000. Competition and intraguild predation among three sympatric carnivores. Oecologia 125:258–70.
Fentress, J., and J. Ryon. 1982. A long-term study of distributed pup feeding in captive wolves. *In* F. Harrington and P. C. Paquet (eds.), Wolves of the World, 238–61. Noyes Publications, Park Ridge, NJ.
Ferreras, P., J. J. Aldama, J. F. Beltran, and M. Delibes. 1992. Rates and causes of mortality in a fragmented population of Iberian lynx, *Felis pardina* Temminck, 1824. Biological Conservation 61:197–202.
Ford, L. S., and R. S. Hoffmann. 1988. *Potos flavus.* Mammalian Species 321:1–9.
Fox, M. W. 1971. Behaviour of Wolves, Dogs and Related Canids. Cape, London.
———. 1984. The Whistling Hunters: Field Studies of the Asiatic Wild Dog *(Cuon alpinus).* State University of New York Press, Albany.
Frame, L. H., J. R. Malcom, G. W. Frame, and H. van Lawick. 1979. Social organization of African wild dogs *(Lycaon pictus)* on the Serengeti plains, Tanzania 1967–1978. Zeitschrift für Tierphysiologie 50:225–49.
Frank, L. G. 1998. Living with lions: carnivore conservation and livestock in Laikipia District, Kenya. Development Alternatives, Unpublished Report. Bethesda, Md.
Frommolt, K. H., M. E. Goltsman, and D. W. Macdonald. 2003. Barking foxes, *Alopex lagopus:* field experiments in individual recognition in a territorial mammal. Animal Behaviour 65:509–18.
Fuller, T. K. 1989. Population dynamics of wolves in north-central Minnesota. Wildlife Monographs 105:1–41.
Fuller, T. K., and P. W. Kat. 1990. Movements, activity, and prey relationships of African wild dogs *(Lycaon pictus)* near Aitong, southwestern Kenya. African Journal of Ecology 28:330–50.
Fuller, T. K., and P. R. Sievert. 2001. Carnivore demography and the consequences of changes in prey availability. *In* J. L. Gittleman, S. M. Funk, D. Macdonald, and R. K. Wayne (eds.), Carnivore Conservation, 163–78. Cambridge University Press, Cambridge.
Fuller, T. K., M. G. Mills, M. Borner, K. Laurenson, and P. W. Kat. 1992. Long-distance dispersal by African wild dogs in East and South Africa. Journal of African Zoology 106:535–37.
Fuller, T. K., P. W. Kat, J. B. Bulger, A. H. Maddock, J. R. Ginsberg, R. Burrows, J. W. McNutt, and M. G. L. Mills. 1992. Population dynamics of African wild dogs. *In* D. R. McCullough and R. H. Barrett (eds.), Wildlife 2001: Population, 1125–39. Elsevier Applied Science, London.
Gasaway, W. C., R. D. Boertje, V. Grangaard, D. G. Kelleyhouse, R. O. Stephenson, and D. G. Larsen. 1992. The role of predation in limiting moose at low densities in Alaska and Yukon and implications for conservation. Wildlife Monographs 120:1–59.
Geffen, E., M. E. Gompper, J. L. Gittleman, H. K. Luh, D. W. MacDonald, and R. K. Wayne. 1996. Size, life-history traits, and social organization in the Canidae: a reevaluation. American Naturalist 147:140–60.
Gehrt, S. D., and E. K. Fritzell. 1998. Resource distribution, female home range dispersion and male spatial interactions: group structure in a solitary carnivore. Animal Behaviour 55:1211–27.
Gehrt, S. D., G. F. J. Hubert, and J. A. Ellis. 2002. Long-term population trends of raccoons in Illinois. Wildlife Society Bulletin 30:457–63.
Gese, E. M., and L. D. Mech. 1991. Dispersal of wolves *(Canis lupus)* in northeastern Minnesota, 1969–1989. Canadian Journal of Zoology 69:2946–55.
Gese, E. M., R. L. Ruff, and R. L. Crabtree. 1996. Social and nutritional factors influencing the dispersal of resident coyotes. Animal Behaviour 52:1025–43.

Gilbert, D., C. Packer, A. E. Pusey, J. C. Stephens, and S. J. O'Brien. 1992. Analytical DNA fingerprinting in lions: parentage, genetic diversity, and kinship. Journal of Heredity 82:378–86.

Gipson, P. S., W. B. Ballard, and R. M. Nowak. 1998. Famous North American wolves and the credibility of early wildlife literature. Wildlife Society Bulletin 26:808–16.

Girman, D. J., M. G. L. Mills, E. Geffen, and R. K. Wayne. 1997. A molecular genetic analysis of social structure, dispersal, and interpack relationships of the African wild dog *(Lycaon pictus)*. Behavioral Ecology and Sociobiology 40:187–98.

Gittleman, J. L., and P. H. Harvey. 1982. Carnivore home-range size, metabolic needs and ecology. Behavioral Ecology and Sociobiology 10:57–63.

Gittleman, J. L., and S. D. Thompson. 1988. Energy allocation in mammalian reproduction. American Zoologist 28:863–75.

Gittleman, J. L., S. M. Funk, D. Macdonald, and R. W. Wayne. 2001a. Carnivore Conservation. Cambridge University Press, Cambridge.

———. 2001b. Why carnivore conservation? *In* J. L. Gittleman, S. M. Funk, D. Macdonald, and R. W. Wayne (eds.), Carnivore Conservation, 1–8. Cambridge University Press, Cambridge.

Gjertz, I., and E. Persen. 1987. Confrontation between humans and polar bears in Svalbard (Norway). Polar Research 5:253–56.

Golani, I., and A. A. Keller. 1975. A longitudinal field study of the behavior of a pair of golden jackals. *In* M. W. Fox (ed.), The Wild Canids: Their Systematics, Behavioral Ecology and Evolution, 303–35. Van Nostrand, Reinhold, New York.

Gompper, M. E. 2002. Top carnivores in the suburbs? Ecological and conservation issues raised by colonization of northeastern North America by coyotes. Bioscience 52:185–90.

Gompper, M. E., and J. L. Gittleman. 1991. Home range scaling: intraspecific and comparative trends. Oecologia 87:343–48.

Gompper, M. E., J. L. Gittleman, and R. K. Wayne. 1998. Dispersal, philopatry, and genetic relatedness in a social carnivore: comparing males and females. Molecular Ecology 7:157–63.

Gorman, M. L., H. Kruuk, and A. Leitch. 1984. Social functions of the sub-caudal scent gland of the European badger *(Meles meles)* (Carnivora: Mustelidae). Journal of Zoology (London) 203:549–59.

Gosselink, T. E., T. R. Van Deleen, R. E. Warner, and M. G. Joselyn. 2003. Temporal habitat partitioning and spatial use of coyotes and red foxes in east-central Illinois. Journal of Wildlife Management 67:90–103.

Gossow, H. 1970. Vergleichende Verhaltensen an Marderartigen. I. Über Lautausserungen und zum Beuteverhalten. Zeitschrift für Tierphysiologie 27:405–80.

Goszczynski, J. 1999. Fox, raccoon dog and badger densities in Eastern Poland. Acta Theriologica 44:413–20.

Gottelli, D., C. Sillero-Zubiri, G. D. Applebaum, M. S. Roy, D. J. Girman, J. Garcia-Moreno, E. A. Osrander, and R. K. Wayne. 1994. Molecular genetics of the most endangered canid: the Ethiopian wolf, *Canis simensis*. Molecular Ecology 3:301–12.

Grandy, J. W., E. Stallman, and D. W. Macdonald. 2003. The science and sociology of hunting: shifting practices and perceptions in the United States and Great Britain. *In* D. J. Salem and A. N. Rowan (eds.), The State of the Animals II: 2003, 107–30. Humane Society Press, Washington, D.C.

Greenwood, P. J. 1980. Mating systems, philopatry, and dispersal in birds and mammals. Animal Behaviour 28:1140–62.

Greenwood, R. J., P. M. Arnold, and B. G. McGuire. 1990. Protecting duck nests from mammalian predators with fences, traps and a toxicant. Wildlife Society Bulletin 18:75–82.

Griffith, B., J. M. Scott, J. W. Carpenter, and C. Reed. 1989. Translocation as a species conservation tool: status and strategy. Science 245:477–80.

Grinder, M. I., and P. R. Krausman. 2001. Home range, habitat use, and nocturnal activity of coyotes in an urban environment. Journal of Wildlife Management 65:887–98.

Gros, P. M., M. Kelly, and T. M. Caro. 1996. Estimating carnivore densities for conservation purposes: indirect methods compared to baseline demographic data. Oikos 77:197–206.

Gustavson, C. R., J. Garcia, W. G. Hankins, and K. W. Rusiniak. 1974. Coyote predation control by aversive conditioning. Science 184:581–83.

Gustavson, C. R., D. J. Kelly, M. Sweeney, and J. Garcia. 1976. Prey-lithium aversions I: coyotes and wolves. Behavioral Biology 17:61–72.

Hamilton, W. J. I. 1973. Life's Color Code. McGraw-Hill, New York.

Hansson, B., and L. Westerberg. 2002. On the correlation between heterozygosity and fitness in natural populations. Molecular Ecology 11(12):2467–74.

Harrington, F. H. 1987. Aggressive howling in wolves. Animal Behaviour 35:7–12.

Harrington, F. H., and L. D. Mech. 1979. Wolf howling and its role in territory maintenance. Behaviour 68:207–49.

Harrington, F. H., L. D. Mech, and S. H. Fritts. 1983. Pack size and wolf survival: their relationship under varying ecological conditions. Behavioural Ecology and Sociobiology 13:19–26.

Harris, S., and W. J. Cresswell. 1987. Dynamics of a suburban badger *(Meles meles)* population. Symposia of the Zoological Society of London 58:295–311.

Harris, S., and P. C. L. White. 1992. Is reduced affiliative rather than increased agonistic behaviour associated with dispersal in red foxes? Animal Behaviour 44:1085–89.

Hedges, S., and M. Tyson. 1996. Is predation by ajag (Asiatic wild dog, *Cuon alpinus*) a threat to the banteng *(Bos javanicus)* population in Alas Purow? Review of the evidence and discussion of management solutions. Report to the Directorate General of Forest Protection and Nature Conservation, Indonesia.

Heidt, G. A. 1972. Anatomical and behavioural aspects of killing and feeding by the least weasel, *Mustela nivalis* L. Proceedings of the Arkansas Academy of Sciences 26:53–54.

Henke, S. E., and F. C. Bryant. 1999. Effects of coyote removal on the faunal community in western Texas. Journal of Wildlife Management 63:1066–81.

Henschel, J. R., and J. D. Skinner. 1987. Social relationships and dispersal patterns in a clan of spotted hyaenas *Crocuta crocuta* in the Kruger National Park. South African Journal of Zoology 22:18–24.

Hernandez, L., R. R. Parmenter, J. W. Dewitt, D. C. Lightfoot, and J. W. Laundre. 2002. Coyote diets in the Chihuahuan Desert: more evidence for optimal foraging. Journal of Arid Environments 51:613–24.

Hersteinsson, P., and D. W. Macdonald. 1982. Some comparisons between red and Arctic foxes, *Vulpes vulpes* and *Alopex lagopus,* as revealed by radiotracking. Symposium of Zoological Society of London 49:259–89.

Hersteinsson, P., and D. Macdonald. 1996. Diet of Arctic foxes *(Alopex lagopus)* in Iceland. Journal of Zoology (London) 240:457–74.

Heydon, M. J., and J. C. Reynolds. 2000. Fox *(Vulpes vulpes)* management in three contrasting regions of Britain, in relation to agricultural and sporting interests. Journal of Zoology (London) 251:237–52.

Hill, R. A., and R. I. M. Dunbar. 1998. An evaluation of the roles of predation rate and predation risk on selective pressures on primate grouping behavior. Behaviour 135:411–30.

Hofer, H., and M. L. East. 1993. The commuting system of Serengeti spotted hyaenas: how a predator copes with migratory prey. I. Social organization. Animal Behaviour 46:547–57.

Holekamp, K. E., J. Ogutu, L. G. Frank, H. T. Dublin, and L. Smale. 1993. Fission of a spotted hyena clan: consequences of female absenteeism and causes of female emigration. Ethology 93:285–99.

Hummel, M., and S. Pettigrew. 1991. Wild Hunters: Predators in Peril. Key Porter Books, Toronto.

Hunt, R. M., Jr. 1996. Biogeography of the order Carnivora. *In* J. L. Gittleman (ed.), Carnivore Behavior, Ecology, and Evolution, 2:485–541. Cornell University Press, Ithaca, N.Y.

Isbell, L. A. 1994. Predation on primates: ecological patterns and evolutionary consequences. Evolutionary Anthropology 3:61–71.

IUCN. Species Survival Commission. 1998. Reintroduction Specialist Group 1998. IUCN Guidelines for Reintroduction. IUCN, Gland, Switzerland.

Jackson, J. A. 1977. Alleviating problems of competition, predation, parasitism, and disease in endangered birds. *In* S. A. Temple (ed.), Endangered Birds, 75–84. University of Wisconsin Press, Madison.

Gaffers, D. J. 1989. The changing otter population of Britain 1700–1989. Biological Journal of the Linnean Society 38, 61–69.

Jenkins, S. R., B. D. Perry, and W. G. Winkler. 1998. Ecology and epidemiology of raccoon rabies. Reviews of Infectious Diseases 10:620–25.

Jennions, M. D., and M. Petrie. 2000. Why do females mate multiply? A review of the genetic benefits. Biology Review 75:21–64.

Jhala, Y. V., and D. K. Sharma. 1997. Childlifting by wolves in eastern Uttar Pradesh, India. Journal of Wildlife Research 2:94–101.

Johnson, D., R. Kays, P. Blackwell, and D. Macdonald. 2002a. Does the resource dispersion hypothesis explain group living? Trends in Ecology and Evolution 17:563–70.

Johnson, D. D. P., S. Baker, M. D. Morecroft, and D. W. Macdonald. 2001a. Long-term resource variation and group size: a large-sample field test of the Resource Dispersion Hypothesis. BMC Ecology 1.

Johnson, D. D. P., W. Jetz, and D. W. Macdonald. 2002b. Environmental correlates of badger social spacing across Europe. Journal of Biogeography 29:411–25.

Johnson, D. D. P., D. W. Macdonald, and A. J. Dickman. 2000. An analysis and review of models of the sociobiology of the Mustelidae. Mammal Review 30:171–96.

Johnson, D. D. P., D. W. Macdonald, C. Newman, and M. D. Morecroft. 2001b. Group size versus territory size in group-living badgers: a large-sample field test of the Resource Dispersion Hypothesis. Oikos 95:265–74.

Johnson, D. D. P., P. Stopka, and D. W. Macdonald. 2004. Ideal flea constraints on group living: unwanted public goods and the emergence of cooperation. Behavioural Ecology 15:181–86.

Johnson, W. E., E. Eizirik, and G. M. Lento. 2001c. The control, exploitation and conservation of carnivores. *In* J. L. Gittleman, S. M. Funk, D. Macdonald, and R. K. Wayne (eds.), Carnivore Conservation, 196–221. Cambridge University Press, Cambridge.

Johnson, W. E., T. K. Fuller, and W. L. Franklin. 1996. Sympatry in canids: a review and assessment. *In* J. L. Gittleman (ed.), Carnivore Behavior, Ecology, and Evolution, 2:189–218. Cornell University Press, Ithaca, N.Y.

Jorgensen, J. P., and K. H. Redford. 1993. Humans and big cats as predators in the Neotropics. Symposium Zoological Society of London 65:367–90.

Kauhala, K., and M. Saeki. 2004. Finnish and Japanese raccoon dogs—on the road to speciation. *In* D. W. Macdonald and C. Sillero-Zubiri (eds.), Canid Biology and Conservation, 217–26. Oxford University Press, Oxford.

Kays, R. 2003. Social polyandry and promiscuous mating in a primate-like carnivore: the kinkajou *(Potos flavus)*. *In* U. H. Reichard and C. Boesch (eds.), Monogamy: Mating Strategies and Partnerships in Birds, Humans and Other Mammals, 125–37. Cambridge University Press, Cambridge.

Kays, R. W. 1999. Food preferences of kinkajous *(Potos flavus):* a frugivorous carnivore. Journal of Mammalogy 80:589–99.

Kays, R. W., and D. Bogan. 2002. Coyotes, fishers and housecats in a North East suburban forest preserve. Proceedings of Carnivores 2002, 2.

Kays, R. W., and A. A. DeWan. 2004. The ecological impact of inside/outside house cats around a suburban nature preserve. Animal Conservation 7:1–11.

Kays, R. W., and J. G. Gittleman. 2001. The social organization of the kinkajou *Potos flavus* (Procyonidae). Journal of Zoology (London) 253:491–504.

Kays, R. W., and D. E. Wilson. 2002. Mammals of North America. Princeton University Press, Princeton.

Kays, R. W., J. G. Gittleman, and R. K. Wayne. 2000. Microsatellite analysis of kinkajou social organization. Molecular Ecology 9:743–51.

Keane, B., P. M. Waser, S. R. Creel, N. M. Creel, L. F. Elliott, and D. J. Minchella. 1994. Subordinate reproduction in dwarf mongooses. Animal Behaviour 47:65–75.

Kellert, S. R., M. Black, C. R. Rush, and A. J. Bath. 1996. Human culture and large carnivore conservation in North America. Conservation Biology 10:997–1090.

Kenney, J., J. Smith, A. Starfield, and C. McDougal. 1995. The long-term effects of tiger poaching on population viability. Conservation Biology 9:1127–33.

Kenya Game Department. 1958. Colony and Protectorate of Kenya—game department annual report. Game Department, Nairobi.

King, C. M. 1989. The Natural History of Weasels and Stoats. Cornell University Press, Ithaca, N.Y.

Kinnear, J. E., M. L. Onus, and R. Sumner. 1998. Fox control and rock-wallaby population dynamics: II. An update. Wildlife Research 25:81–88.

Kitchener, A. 1991. The Natural History of the Wild Cats. Comstock Publishing Associates, Ithaca, N.Y.

Kleiman, D. G. 1972. Social behaviour of the maned wolf *(Chrysocyon brachyurus)* and bush dog *(Speothos venaticus):* a study in contrast. Journal of Mammalogy 53:791–806.

Kleiman, D. G., and J. F. Eisenberg. 1973. Comparisons of canid and felid social systems from an evolutionary perspective. Animal Behavior 21:637–59.

Knick, S. T. 1990. Ecology of bobcats relative to exploitation and a prey decline in southeastern Idaho. Wildlife Monographs 108:1–42.

Knowlton, F. F., E. M. Gese, and M. M. Jaeger. 1999. Coyote depredation control: an interface between biology and management. Journal of Range Management 52:398–412.

Kowalczyk, R., A. N. Bunevich, and B. Jedrzejewska. 2000. Badger density and distribution of setts in Bialowieza Primeval Forest (Poland and Belarus) compared to other Eurasian populations. Acta Theriologica 45:395–408.

Krebs, J. R., R. Anderson, T. Clutton-Brock, I. Morrison, D. Young, C. Donnelly, S. Frost, and R. Woodroffe. 1997. Bovine tuberculosis in cattle and badgers. The Ministry of Agriculture, Fisheries and Food Publications, London.

Kruchenkova, E. P., M. Goltsman, S. Sergeev, and D. W. Macdonald. 1973. Ineffective helpers and a disinclination to disperse in two island subspecies of Arctic fox, *Alopex lagopus.* Klaeiman and Eisenberg.

Kruuk, H. 1972. The Spotted Hyaena. University of Chicago Press, Chicago.

———. 1978a. Foraging and spatial organisation of the European badger, *Meles meles* L. Behavioural Ecology and Sociobiology 4:75–89.

———. 1978b. Spatial organisation and territorial behaviour of the European badger *Meles meles.* Journal of Zoology (London) 184:1–19.

———. 2002. Hunter and Hunted: Relationships between Carnivores and People. Cambridge University Press, Cambridge.

Kruuk, H., and D. W. Macdonald. 1985. Group territories of carnivores: empires and enclaves. *In* R. M. Sibly and R. H. Smith (eds.), Behavioural Ecology: Ecological Consequences of Adaptive Behaviour, 521–36. Blackwell Scientific Publishing, Oxford.

Kruuk, H., and T. Parrish. 1982. Factors affecting the population density, group size and territory size of the European badger, *Meles meles,* in relation to earthworm population, Scotland. Journal of Zoology (London) 196:31–39.

Kruuk, H., and M. Turner. 1967. Comparative notes on predation by lion, leopard, cheetah and wild dog in the Serengeti area, East Africa. Mammalia 31:1–27.

Lamprecht, J. 1978. On diet, foraging behaviour and interspecific food competition of jackals in the Serengeti National Park, East Africa. Zeitschrift für Säugetierkunde 43:210–23.

Landa, A., and B. A. Tommeras. 1997. A test of aversive agents on wolverines. Journal of Wildlife Management 61:510–16.

Landa, A., K. Gudvangen, J. E. Swenson, and E. Roskaft. 1999. Factors associated with wolverine *(Gulo gulo)* predation on domestic sheep. Journal of Applied Ecology 36:963–73.

Lavin, S. R., T. R. Van Deelen, P. W. Brown, R. E. Warner, and S. H. Ambrose. 2003. Prey use by red foxes *(Vulpes vulpes)* in urban and rural areas of Illinois. Canadian Journal of Zoology 81:1070–82.

Leader-Williams, N., and S. D. Albon. 1988. Allocation of resources for conservation. Nature 336:533–35.

Lehman, N. A., K. Eisenhawer, L. Hansen, D. Mech, R. O. Peterson, P. J. P. Gofan, and R. K. Wayne. 1991. Introgression of coyote mitochondrial DNA into sympatric North American gray wolf populations. Evolution 45:104–19.

Leopold, A. S., S. A. Cain, C. M. Cottam, I. N. Gabrielson, and T. L. Kimball. 1964. Predator and rodent control in the United States. Transactions of the North American Wildlife and Natural Resources Conference 29:27–49.

Lever, C. 1994. Naturalized Animals: The Ecology of Successfully Introduced Species. T. and A. D. Poyser, London.

Leyhausen, P. 1965. The communal social organisation of solitary mammals. Symposium of the Zoological Society of London 14:249–63.

———. 1979. Cat Behaviour: The Predatory and Social Behavior of Domestic and Wild Cats. Garland STPM Press, New York.

Linnell, J. D. C., and O. Strand. 2000. Interference interactions, co-existence and conservation of mammalian carnivores. Diversity and Distributions 6:169–76.

Lockwood, R. 1995. The ethology and epidemiology of canine aggression. *In* J. Serpell (ed.), The Domestic Dog: Its Evolution, Behaviour and Interactions with People, 131–38. Cambridge University Press, Cambridge.

Lorenz, K. Z. 1954. Man Meets Dog. Methuen and Co., London.

Loveridge, A. J., A. Lynam, and D. W. Macdonald. 2002. Lion Conservation Research Workshop 2: Modelling Conflict. WildCRU and Darwin Initiative, Oxford.

Lucas, P. W., and D. A. Luke. 1984. Chewing it over: basic principles of food breakdown. *In* D. J. Chivers, B. A. Wood, and A. Bilsborough (eds.), Food Acquisition and Processing in Primates, 283–301. Plenum Press, New York.

Maas, B., and D. W. Macdonald. 2004. Bat-eared foxes, insectivory and luck: lessons from an extreme canid. *In* D. W. Macdonald and C. Sillero-Zubiri (eds.), Canid Biology and Conservation, 227–42. Oxford University Press, Oxford.

Macdonald, D. W. 1976. Food caching by red foxes and some other carnivores. Zeitschrift für Tierphysiologie 42:170–85.

———. 1977. On food preference in the red fox. Mammal Review 7:7–23.

———. 1979. The flexible social system of the golden jackal, *Canis aureus.* Behavioural Ecology and Sociobiology 5:17–38.

———. 1980. Rabies and wildlife: a biologist's perspective. Oxford University Press, Oxford.

———. 1983. The ecology of carnivore social behaviour. Nature 301:379–84.

———. 1987. Running with the Fox. Unwin Hyman, London.

———. 1992. The Velvet Claw: A Natural History of the Carnivores. BBC Books, London.

———. 1996a. Dangerous liaisons and disease. Nature 379:400–401.

———. 1996b. Social behaviour of captive bush dogs *(Speothos venaticus).* Journal of Zoology (London) 239:525–43.

———. 2001a. The New Encyclopaedia of Mammals. Oxford University Press, Oxford.

———. 2001b. Postscript: science, compromise and tough choices. *In* J. L. Gittleman, S. M. Funk, D. Macdonald, and R. K. Wayne (eds.), Carnivore Conservation, 524–38. Cambridge University Press, Cambridge.

Macdonald, D. W., and C. J. Amlaner. 1981. Resource dispersion and the social organisation of the red fox *(Vulpes vulpes). In* J. A. Chapman and D. Pursley (eds.), The First International Worldwide Furbearer Conference, 918–49. Worldwide Furbearer Conference, Frostburg, Md.

Macdonald, D. W., and S. E. Baker. 2004. Non-lethal control of fox predation: the potential of generalised aversion. Animal Welfare 13:77–85.

Macdonald, D. W., and G. M. Carr. 1989. Food security and the rewards of tolerance. *In* V. and R. A. Foley (eds.), Comparative Socioecology. Standen, 75–99. Blackwell Scientific Publications, Oxford.

Macdonald, D. W., and O. Courtenay. 1996. Enduring social relationships in a population of crab-eating zorros, *Cerdocyon thous,* in Amazonian Brazil (Carnivora, Canidae). Journal of Zoology (London) 239:329–55.

Macdonald, D. W., and D. P. Johnson. 2001. Dispersal in theory and practice: consequences for conservation biology. *In* J. Clobert, E. Danchin, A. A. Dhondt, and J. D. Nichols (eds.), Dispersal, 358–72. Oxford University Press, Oxford.

Macdonald, D. W., and P. J. Johnson. 2003. Farmers as conservation custodians: links between perception and practice. *In* F. H. Tattersall, W. Manley, and J. Yorkshire (eds.), Conservation and Conflict: Mammals and Farming in Britain, 2–16. Westbury Publishing.

Macdonald, D. W., and P. D. Moehlman. 1982. Cooperation, altruism and restraint in the reproduction of carnivores. Perspectives in Ecology 5:433–67.

Macdonald, D. W., and C. Newman. 2002. Population dynamics of badgers *(Meles meles)* in Oxfordshire, UK: numbers, density and cohort life histories, and a possible role of climate change in population growth. Journal of Zoology (London) 256:121–38.

Macdonald, D. W., and S. Rushton. 2003. Modelling space use and dispersal of mammals in real landscapes: a tool for conservation. Journal of Biogeography 30:607–20.

Macdonald, D. W., and C. Sillero-Zubiri. 2002. Large carnivores and conflict: lion conservation in context. *In* A. J. Loveridge, T. Lynam, and D. W. Macdonald (eds.), Lion Conservation Research. Workshop 2: Modelling Conflict, 1–8. Wildlife Conservation Research Unit, Oxford.

———. 2004a. Canid Biology and Conservation. Oxford University Press, Oxford.

———. 2004b. Wild canids: an introduction and *dramatis personae. In* D. W. Macdonald and C. Sillero-Zubiri (eds.), Canid Biology and Conservation, 3–38. Oxford University Press, Oxford.

Macdonald, D. W., and M. D. Thom. 2001. Alien carnivores: unwelcome experiments in ecological theory. *In* J. L. Gittleman, S. M. Funk, D. Macdonald, and R. K. Wayne (eds.), Carnivore Conservation, 93–122. Cambridge University Press, Cambridge.

Macdonald, D. W., J. M. Bryce, and M. D. Thom. 2001. Introduced mammals: do carnivores and herbivores usurp native species by different mechanisms? *In* H. Pelz, J. Cowan, D. P. Feare, and C. J. Fürth (eds.), Advances in Vertebrate Pest Management 2:11–44. Filander Verlag.

Macdonald, D. W., C. Newman, J. Dean, C. D. Buesching, and P. J. Johnson. In press. The distribution of Eurasian badger *Meles meles* setts in a high-density area: field observations contradict the Sett Dispersion Hypothesis. Oikos.

Macdonald, D. W., J. C. Reynolds, C. Carbone, F. Mathews, and P. J. Johnson. 2003. The bioeconomics of fox control. *In* F. H. Tattersall and W. J. Manley (eds.), Conservation and Conflict: Mammals and Farming in Britain, 220–36. Westbury Publishing, Yorkshire, UK.

Macdonald, D. W., F. H. Tattersall, P. J. Johnson, C. Carbone, J. Reynolds, J. Langbein, S. P. Rushton, and M. Shirley. 2000. Managing British Mammals: Case Studies from the Hunting Debate. WildCRU, Oxford.

Macdonald, D. W., P. D. Stewart, P. J. Johnson, J. Porkert, and C. Buesching. 2002. No evidence of social hierarchy amongst feeding badgers, *Meles meles.* Ethology 108:613–28.

Mace, G. M., and A. Balmford. 2000. Patterns and processes in contemporary mammalian extinction. *In* A. Entwhistle and N. Dunstone (eds.), Priorities for the Conservation of Mammalian Biodiversity, 27–52. Cambridge University Press, Cambridge.

Maehr, D. S., and G. B. Caddick. 1995. Demographics and genetic introgression in the Florida panther. Conservation Biology 9:1295–98.

Malcom, J. R., and K. Marten. 1982. Natural selection and the communal rearing of pups in African wild dogs *Lycaon pictus.* Behavioural Ecology and Sociobiology 10:1–13.

Mallinson, J. J. C. 1995. Conservation breeding programs: an important ingredient for species survival. Biodiversity and Conservation 4:615–35.

Marean, C. W., L. M. Spencer, R. J. Blumenschine, and S. D. Capaldo. 1992. Captive hyaena bone choice and destruction: the schlepp effect and Olduvai archaeofaunas. Journal of Archaeological Science 18:101–21.

Marker, L. L., M. G. L. Mills, and D. W. Macdonald. 2003. Factors influencing perceptions and tolerance toward cheetahs on Namibian farmlands. Conservation Biology 17:1290–98.

McComb, K. 1992. Playback as a tool for studying contests between social groups. *In* P. K. McGreggor (ed.), Playback and Studies of Animal Communication, 111–19. Plenum Publishing, New York.

McComb, K., C. Packer, and A. E. Pusey. 1994. Roaring and numerical assessment in contests between groups of female lions, *Panthera leo.* Animal Behaviour 47:379–87.

McComb, K., A. Pusey, C. Packer, and J. Grinnell. 1993. Female lions can identify potentially infanticidal males from their roars. Proceedings of the Royal Society of London, B, 252:59–64.

McDonald, R. A. 2002. Resource partitioning among British and Irish mustelids. Animal Ecology 71:185–200.

McLaren, B. E., and R. O. Peterson. 1994. Wolves, moose, and tree rings on Isle Royal. Science 266:1555–58.

McLellan, B. 1994. Density-dependent population regulation of brown bears. International Conference on Bear Research and Management Monograph Series 3:16–24.

McNab, B. 1989. Basal rate of metabolism, body size, and food habits in the order Carnivora. *In* J. L. Gittleman (ed.), Carnivore Behaviour, Ecology, and Evolution, 2:335–54. Cornell University Press, Ithaca, N.Y.

McNutt, J. W. 1996. Sex-biased dispersal in African wild dogs, *Lycaon pictus.* Animal Behaviour 52:1067–77.

Mech, D. L., and G. Peters. 1977. The study of chemical communication in free-ranging mammals. *In* D. Müller-Schwarze and M. M. Mozell (eds.), Chemical Signals in Vertebrates, 321–31. Plenum Press, New York

Mech, L. D. 1970. The Wolf: The Ecology and Behavior of an Endangered Species. University of Minnesota Press, Minneapolis.

———. 1977. Productivity, mortality and population trends of wolves on northeastern Minnesota. Journal of Mammalogy 58:559–74.

———. 1987. Age, season, distance, direction and social aspects of wolf dispersal from a Minnesota pack. *In* B. D. Chepko-Sade and Z. T. Halpin (eds.), Mammalian Dispersal Patterns: The Effects of Social Structure on Population Genetics, 55–74. University of Chicago Press, Chicago.

Méndez, E., F. Delgado, and D. Miranda. 1981. The coyote *(Canis latrans)* in Panama. International Journal for the Study of Animal Problems 2:252–55.

Messier, F. 1985. Solitary living and extraterritorial movements of wolves in relation to social status and prey abundance. Canadian Journal of Zoology 63:239–45.

———. 1995. On the functional and numerical responses of wolves to changing prey density. *In* L. N. Carbyn, S. H. Fritts, and D. R. Seip (eds.), Ecology and Conservation of Wolves in a Changing World, 187–97. Canadian Circumpolar Institute, Edmonton, Alberta.

Messier, F., and C. Barrette. 1982. The social system of the coyote *(Canis latrans)* in a forested habitat. Canadian Journal of Zoology 60:1743–53.

Milberg, P., and T. Tyrberg. 1993. Naive birds and noble savages: a review of man-caused prehistoric extinctions of island birds. Ecography 16:229–41.

Mills, M. G. L. 1990. Kalahari Hyaenas: The Comparative Behavioural Ecology of Two Species. Unwin Hyman, London.

Mills, M. G. L., M. L. Gorman, and M. E. J. Mills. 1980. The scent marking behaviour of the brown hyena *(Hyaena brunnea)* in the Southern Kalahari. South African Journal of Zoology 15:240–48.

Mittermeier, R. A., N. Myers, P. R. Gill, and C. G. Mittermeier. 1999. Hotspots. CEMEX, Mexico.

Mladenoff, D. J., T. Sickley, and A. P. Wydeven. 1999. Predicting gray wolf landscape recolonization: logistic regression models vs. new field data. Ecological Applications 9:37–44.

Moehlman, P. 1979. Jackal helpers and pup survival. Nature 277:382–83.

Moehlman, P., and H. Hofer. 1997. Cooperative breeding, reproductive suppression and body mass in canids. *In* N. G. Solomon and J. A. French (eds.), Cooperative Breeding in Mammals, 76–128. Cambridge University Press, Cambridge.

Moehrenschlager, A., B. Cypher, K. Ralls, R. List, and M. A. Sovada. In press. Comparative ecology and conservation priorities of swift and kit foxes. *In* D. W. Macdonald and C. Sillero-Zubiri (eds.), Canid Biology and Conservation. Oxford University Press, Oxford.

Moehrenschlager, A., and D. W. Macdonald. 2003. Movement and survival parameters of translocated and resident swift foxes. Animal Conservation 6:199–206.

Moehrenschlager, A., and M. A. Sovada. In press. *Vulpes velox*. *In* D. W. Macdonald and C. Sillero-Zubiri (eds.), Canid Species Status and Conservation Action Plan. IUCN, Gland, Switzerland.

Monfort, S. L., S. K. Wasser, K. L. Mashburn, M. Burke, B. Brewer, and S. R. Creel. 1997. Steroid metabolism and validation of noninvasive endocrine monitoring in the African wild dog *(Lycaon pictus)*. Zoo Biology 16:533–48.

Moruzzi, T. L., T. K. Fuller, R. M. DeGraaf, R. T. Brooks, and W. Li. 2002. Assessing remotely triggered cameras for surveying carnivore distribution. Wildlife Society Bulletin 30:380–86.

Musgrave, M. 1993. Outfoxing the foxes. Enact 1:6–9.

Nalbandov, A. V. 1976. The estrous cycle. *In* A. V. Nalbanov (ed.), Reproductive Physiology of Mammals and Birds: The Comparative Physiology of Domestic and Laboratory Animals and Man, 98–124. W. H. Freeman and Co., San Francisco.

Nel, J. A. J., and A. J. Mackie. 1990. Food and foraging behaviour of bat-eared foxes in the south-eastern Orange Free State. South African Journal of Wildlife Research 20:162–66.

Newsome, A. E., I. Parer, and P. C. Cattling. 1989. Prolonged prey suppression by carnivores: predator removal experiments. Oecologia 78:458–67.

Nicolaus, L. K., and D. W. Nellis. 1987. The first evaluation of the use of conditioned taste aversion to control predation by mongooses on eggs. Applied Animal Behaviour Science 17:329–46.

Nicolaus, L. K., T. E. Hoffman, and C. R. Gustavson. 1982. Taste aversion conditioning in free-ranging raccoons, *Procyon lotor*. Northwest Science 56:165–69.

Nielsen, S. E., M. S. Boyce, G. B. Stenhouse, and R. H. M. Munro. 2002. Modeling grizzly bear habitats in the Yellowhead ecosystem of Alberta: taking autocorrelation seriously. Ursus 13:45–56.

Novaro, A. J. 1995. Sustainability of harvest of culpeo foxes in Patagonia. Oryx 29:18–22.

Novaro, A. J., M. C. Funes, and J. E. Jiménez. 2004. Selection for introduced prey and conservation of culpeo and chilla zorros in Patagonia. *In* D. W. Macdonald and C. Sillero-Zubiri (eds.), Canid Biology and Conservation, 243–54. Oxford University Press, Oxford.

Nowak, R. M. 2003. Walker's Marine Mammals of the World. Johns Hopkins University Press, Baltimore.

Oli, M. K. 1994. Snow leopards and blue sheep in Nepal: densities and predator:prey ratio. Journal of Mammalogy 75:998–1004.

Oli, M. K., I. R. Taylor, and M. E. Rodgers. 1994. Snow leopard *Panthera uncia* predation of livestock: an assessment of local perceptions in the Annapurna Conservation Area, Nepal. Biological Conservation 68:63–68.

Ortolani, A., and T. M. Caro. 1996. The adaptive significance of color patterns in carnivores: phylogenetic tests of classic hypotheses. *In* J. L. Gittleman (ed.), Carnivore, Behaviour, Ecology and Evolution, 2:132–88. Cornell University Press, Ithaca, N.Y.

Packard, J., L. Mech, and U. Seal. 1983. Social influences on reproduction in wolves. *In* L. N. Carbyn (ed.), Wolves in Canada and Alaska: Their Status, Biology and Management, 78–85. Canadian Wildlife Service, Edmonton.

Packer, C. 1986. The ecology of sociality in felids. *In* D. I. Rubenstein and R. W. Wrangham (eds.), Ecological Aspects of Social Evolution, 429–51. Princeton University Press, Princeton.

Packer, C., and A. E. Pusey. 1982. Cooperation and competition within coalitions of male lions: kin selection or game theory? Nature 296:740–42.

———. 1987. Intrasexual cooperation and the sex ratio in African lions. American Naturalist 130:636–42.

Packer, C., D. A. Gilbert, A. E. Pusey, and S. J. O'Brien. 1991. A molecular genetic analysis of kinship and cooperation in African lions. Nature 351:562–65.

Palomares, F., and T. M. Caro. 1999. Interspecific killing among mammalian carnivores. American Naturalist 153:492–508.

Palomares, F., J. A. Godoy, A. Piriz, S. J. O'Brien, and W. E. Johnson. 2002. Faecal genetic analysis to determine the presence and distribution of elusive carnivores: design and feasibility for the Iberian lynx. Molecular Ecology 11:2171–82.

Parker, G. 1995. Eastern Coyote: The Story of Its Success. Nimbus Publishing, Halifax, N.S.

Persson, I. L., S. Wikan, J. E. Swenson, and I. Mysterud. 2001. The diet of the brown bear *Ursus arctos* in the Pasvik Valley, northeastern Norway. Wildlife Biology 7:27–37.

Phillips, M. K., E. Bangs, L. D. Mech, B. T. Kelly, and B. B. Fazio. 2004. Extermination and recovery of red wolf and grey wolf in the contiguous United States. *In* D. W. Macdonald and C. Sillero-Zubiri (eds.), Canid Biology and Conservation, 297–310. Oxford University Press, Oxford.

Pigozzi, G. 1992. Frugivory and seed dispersal by the European badger in a Mediterranean habitat. Journal of Mammalogy 73:630–39.

Pimentel, D., L. Lach, R. Zuniga, and D. Morrison. 2000. Environmental and economic costs of non-indigenous species in the United States. Bioscience 50:53–65.

Poole, D. W., and I. G. McKillop. 1999. Comparison of the effectiveness of two types of electric fences to exclude badgers. Crop Protection 18:61–66.

Pople, A. R., G. C. Grigg, S. C. Cairns, L. A. Beard, and P. Alexander. 2000. Trends in the numbers of red kangaroos and emus on either side of the South Australian dingo fence: evidence for predator regulation? Wildlife Research 27:269–76.

Powell, R. A., J. W. Zimmerman, and D. E. Seaman. 1997. Ecology and Behaviour of North American Black Bears. Chapman and Hall, London.

Primm, S. A. 1996. A pragmatic approach to grizzly bear conservation. Conservation Biology 10:1036–45.

Purvis, A., G. M. Mace, and J. L. Gittleman. 2001. Past and future carnivore extinctions: a phylogenetic perspective. *In* J. L. Gittleman, S. M. Funk, D. Macdonald, and R. K. Wayne (eds.), Carnivore Conservation, 11–34. Cambridge University Press, Cambridge.

Radinsky, L. B. 1981. Evolution of skull shape in carnivores. 1. Representative modern carnivores. Biological Journal of the Linnean Society 15:368–88.

Ralls, K., and P. J. White. 1995. Predation on San Joaquin kit foxes by larger canids. Journal of Mammalogy 76:723–29.

Rasa, O. A. E. 1977. The ethology and sociology of the dwarf mongoose *(Helogale undulata rufula)*. Zeitschrift für Tierpsychologie 43:337–406.

———. 1986. Coordinated vigilance in dwarf mongoose family groups: the "watchman's song" hypothesis and the costs of guarding. Zeitschrift für Tierphysiologie 71:340–44.

———. 1989. Behavioral parameters of vigilance in the dwarf mongoose: social acquisition of a sex-biased role. Behaviour 110:125–45.

Reading, R. P., and T. W. Clark. 1996. Carnivore reintroductions: an interdisciplinary examination. *In* J. L. Gittleman (ed.), Carnivore Behavior, Ecology, and Evolution, 2:296–336. Cornell University Press, Ithaca, N.Y.

Reading, R. P., and S. R. Kellert. 1993. Attitudes towards a proposed reintroduction of black-footed ferret *(Mustela nigripes)*. Conservation Biology 7:569–80.

Reich, A. 1981. The behaviour and ecology of the African wild dog, *Lycaon pictus*, in the Kruger National Park. PhD dissertation, Yale University, New Haven.

Reynolds, J. C., M. N. Goddard, and M. H. Brocklers. 1993. The impact of local fox *(Vulpes*

vulpes) removal on fox populations at two sites in Southern England. Gibier faune Sauvage. Game Wildlife 10:319–34.

Reynolds, J. C., and S. C. Tapper. 1996. Control of mammalian predators in game management and conservation. Mammal Review 26:127–56.

Reynolds, J. D. 1999. Animal breeding systems. Trends in Ecology and Evolution 11:68–72.

Richardson, P. R. K. 1987. The most highly specialized myrmecophagus mammal? South African Journal of Science 83:643–46.

Rigg, R. 2001. Livestock guarding dogs: their current use world wide. IUCN/SSC Canid Specialist Group Occasional Paper 1:1–133.

Riley, S. J., and R. A. Malecki. 2001. A landscape analysis of cougar distribution and abundance in Montana, USA. Environmental Management 28:317–23.

Rode, K. D., and C. T. Robbins. 2000. Why bears consume mixed diets during fruit abundance. Canadian Journal of Zoology 78:1640–45.

Rodman, P. S. 1981. Inclusive fitness and group size with a reconsideration of group sizes in lions and wolves. American Naturalist 118:275–83.

Rodríguez-Bolaños, A., A. Cadena, and P. Sánches. 2000. Trophic characteristics in social groups of the mountain coati, *Nasuella olivaceai* (Carnivora: Procyonidae). Small Carnivore Conservation Newsletter 23:1–6.

Roelke-Parker, M. E., L. Munson, C. Packer, R. Kock, S. Cleaveland, M. Carpenter, B. S. J. O, A. Pospischil, L. R. Hofmann, H. Lutz, G. L. M. Mwamengele, M. N. Mgasa, G. A. Machange, B. A. Summers, and M. J. G. Appel. 1996. A canine distemper virus epidemic in Serengeti lions *(Panthera leo).* Nature 379:441–45.

Roemer, G. W., C. J. Donlan, and F. Courchamp. 2002. Golden eagles, feral pigs, and insular carnivores: how exotic species turn native predators into prey. Proceedings of the National Academy of Science (U.S.A.) 99:791–96.

Rogers, C. M., and M. J. Caro. 1998. Song sparrows, top carnivores and nest predation: a test of the mesopredator release hypothesis. Oecologia 116:227–33.

Rogers, L. M., R. Delahay, C. L. Cheeseman, S. Langton, G. C. Smith, and R. S. Clifton-Hadley. 1998. Movement of badgers *(Meles meles)* in a high-density population: individual, population and disease effects. Proceedings of the Royal Society of London, B, 265:1269–76.

Rood, J. P. 1978. Dwarf mongoose helpers at the den. Zeitschrift für Tierphysiologie 48:277–87.

———. 1980. Mating relationships and breeding suppression in dwarf mongoose. Animal Behaviour 28:143–50.

———. 1986. Ecology and social evolution in the mongooses. *In* D. I. Rubenstein and R. W. Wrangham (eds.), Ecological Aspects of Social Evolution, 131–52. Princeton University Press, Princeton.

———. 1989. Male associations in a solitary mongoose. Animal Behaviour 38:725–28.

Rowe-Rowe, D. T. 1976. Food of the black-backed jackal in nature conservation and farming areas in Natal. East African Wildlife Journal 14:345–48.

Rudnai, J. 1979. Ecology of lions in Nairobi National Park and the adjoining Kitengela Conservation Unit in Kenya. African Journal of Ecology 17:85–95.

Sagor, J. T., J. E. Swenson, and E. Roskaft. 1997. Compatibility of brown bear, *Ursus arctos,* and free-ranging sheep in Norway. Biological Conservation 81:91–95.

Sandell, M. 1989. The mating tactics and spacing patterns of solitary carnivores. *In* J. L. Gittleman (ed.), Carnivore Behavior, Ecology, and Evolution, 1:164–82. Cornell University Press, Ithaca, N.Y.

Sargeant, A. B., and S. H. Allen. 1989. Observed interactions between coyotes and red foxes. Journal of Mammalogy 70:631–33.

Sarrazin, F., and R. Barbault. 1996. Reintroductions: challenges and lessons for basic ecology. Trends in Ecology and Evolution 11:474–78.

Schaller, G. B. 1972. The Serengeti Lion: A Study of Predator-Prey Relations. University of Chicago Press, Chicago.

Scheepers, J. L., and K. A. E. Venzke. 1995. Attempts to reintroduce Africa wild dogs *Ly-*

caon pictus into Etosha National Park, Namibia. South African Journal of Wildlife Research 25:138–40.
Schmidt, P. A., and L. D. Mech. 1997. Wolf pack size and food acquisition. American Naturalist 150:513–17.
Scott, J. M., P. J. Heglund, M. L. Morrison, J. B. Haufler, M. G. Raphael, W. A. Wall, and F. B. Samson. 2002. Predicting Species Occurrences: Issues of Accuracy and Scale. Island Press, Washington, D.C.
Sechrest, W., T. M. Brooks, G. A. B. da Fonseca, W. R. Konstant, R. A. Mittermeier, A. Purvis, A. B. Rylands, and J. L. Gittleman. 2002. Hotspots and the conservation of evolutionary history. Proceedings of the National Academy of Sciences (U.S.A.) 99: 2067–71.
Semel, B., and L. K. Nicolaus. 1992. Estrogen-based aversion to eggs among free-ranging raccoons. Ecological Applications 2:439–49.
Shankaranarayanan, P., M. Banerjee, R. K. Kacker, R. K. Aggarwal, and L. Singh. 1997. Genetic variation in Asiatic lions and Indian tigers. Electrophoresis 18:1693–1700.
Shepherdson, D. 1994. The role of environmental enrichment in the captive breeding and reintroduction of endangered species. *In* P. J. S. Olney, G. M. Mace, and A. T. C. Feistner (eds.), Creative Conservation: Interactive Management of Wild and Captive Animals, 167–77. Chapman and Hall, London
Shivaji, S., D. Jayaprakash, and S. B. Patil. 1998. Assessment of inbreeding depression in big cats: testosterone levels and semen analysis. Current Science Bangalore, 75, 923–930.
Sidorovich, V. E., D. W. MacDonald, M. M. Pikulik, and H. Kruuk. 2001. Individual feeding specialization in the European mink, *Mustela lutreola* and the American mink, *M. vison* in North-Eastern Belarus. Folia Zoologica 50:27–42.
Sillero-Zubiri, C., and D. Gottelli. 1991. Aberdare rhinos: predation versus poaching. Pachyderm 14:37–38.
———. 1995. Diet and feeding behavior of Ethiopian wolves (*Canis simensis*). Journal of Mammalogy 76:531–41.
Sillero-Zubiri, C., and M. K. Laurenson. 2001. Interactions between carnivores and local communities: conflict or co-existence? *In* J. L. Gittleman, S. M. Funk, D. Macdonald, and R. K. Wayne (eds.), Carnivore Conservation, 282–312. Cambridge University Press, Cambridge.
Sillero-Zubiri, C., and D. W. Macdonald. 1998. Scent-marking and territorial behaviour of Ethiopian wolves *Canis simensis*. Journal of Zoology (London) 245:351–61.
Sillero-Zubiri, C., A. A. King, and D. W. Macdonald. 1996. Rabies and mortality in Ethiopian wolves *(Canis simensis)*. Journal of Wildlife Diseases 32:80–86.
Sillero-Zubiri, C., J. Marino, D. Gottelli, and D. W. Macdonald. 2004a. Afroalpine ecology, solitary foraging and intense sociality amongst Ethiopian wolves. *In* Macdonald, D. W. and C. Sillero-Zubiri (eds.), Canid Biology and Conservation, 311–22. Oxford University Press, Oxford.
Sillero-Zubiri, C., J. Reynolds, and A. Novaro. 2004b. Management and control of wild canids alongside people. *In* D. W. Macdonald and C. Sillero-Zubiri (eds.), Canid Biology and Conservation, 107–22. Oxford University Press, Oxford.
Sillero-Zubiri, C., D. Gottelli, and D. W. Macdonald. 1996. Male philopatry, extra-pack copulations and inbreeding avoidance in Ethiopian wolves (*Canis simensis*). Behavioral Ecology and Sociobiology 28:331–40.
Singh, H. S., and R. D. Kampoj. 1996. Predation patterns of the Asiatic lion on domestic livestock. Indian Forester 122:869–76.
Slattery, P. J., and S. J. O'Brien. 1995. Molecular phylogeny of the red panda *(Ailurus fulgens)*. Journal of Heredity 86:413–22.
Smale, L., L. G. Frank, and K. E. Holekamp. 1993. Ontogeny of dominance in free-living spotted hyaenas: juvenile rank relations with adults. Animal Behaviour 46:467–77.
Smith, A. P., and D. G. Quin. 1996. Patterns and causes of extinction and decline in Australian conilurine rodents. Biological Conservation 77:243–67.
Smuts, G. L. 1982. Lion. Macmillan South Africa, Johannesburg.

Snyder, N. F. R., S. R. Derrickson, S. R. Beissinger, J. W. S. Wiley, W. D. Toome, and B. Miller. 1996. Limitations of captive breeding in endangered species recovery. Conservation Biology 10:338–48.

Soulé, M. E., D. T. Bolger, A. C. Alberts, J. Wright, M. Sorice, and S. Hill. 1988. Reconstructed dynamics of rapid extinctions of chaparral-requiring birds in urban habitat islands. Conservation Biology 2:75–92.

Stahl, P., J. M. Vandel, S. Ruette, L. Coat, Y. Coat, and L. Balestra. 2002. Factors affecting lynx predation on sheep in the French Jura. Journal of Applied Ecology 39:204–16.

Stander, P. E. 1991. Demography of lions in the Etosha National Park. Madoqua 18:1–9.

Stander, P. E., P. J. Haden, Kaqece, and Ghau. 1997. The ecology of asociality in Namibian leopards. Journal of Zoology (London) 242:343–64.

Stewart, P. D., D. W. Macdonald, C. Newman, and C. L. Cheeseman. 2001. Boundary faeces and matched advertisement in the European badger *(Meles meles):* a potential role in range exclusion. Journal of Zoology (London) 255:191–98.

Stewart, P. D., D. W. Macdonald, C. Newman, and F. H. Tattersall. 2002. Behavioural mechanisms of information transmission and reception by badgers, *Meles meles,* at latrines. Animal Behaviour 63:99–107.

Storm, R. D., R. L. Andrews, R. A. Phillips, D. B. Bishop, D. B. Siniff, and J. R. Tester. 1976. Morphology, reproduction, dispersal, and mortality of midwestern red fox populations. Wildlife Monographs 49:1–82.

Stratman, M. R., and M. R. Pelton. 1999. Feeding ecology of black bears in northwest Florida. Florida Field Naturalist 27:95–102.

Sunquist, M., and F. Sunquist. 2002. Wild Cats of the World. University of Chicago Press, Chicago.

Sunquist, M. E., and F. Sunquist. 2001. Changing landscapes: consequences for carnivores. *In* J. L. Gittleman, S. M. Funk, D. Macdonald, and R. K. Wayne (eds.), Carnivore Conservation, 399–418. Cambridge University Press, Cambridge.

Sunquist, M., K. Karanth, and F. Sunquist. 1999. Ecology, behavior and resilience of the tiger and its conservation needs. *In* J. P. Seidensticker, P. Jackson, and S. Christie (eds.), Riding the Tiger: Tiger Conservation in Human-Dominated Landscapes, 5–18. Cambridge University Press, Cambridge.

Tannerfeldt, M., B. Elmhagen, and A. Angerbjorn. 2002. Exclusion by interference competition? The relationship between red and Arctic foxes. Oecologia 132:213–20.

Tannerfeldt, M., A. Moehrenschlager, and A. Angerbjörn. 2003. Den ecology of swift, kit and Arctic foxes: a review. *In* M. Sovada and L. Carbyn (eds.), Ecology and Conservation of Swift Foxes in a Changing World, 167–81. Canadian Plains Research Center, University of Regina, Saskatchewan.

Tapper, S. C., G. R. Potts, and M. H. Brockless. 1996. The effect of an experimental reduction in predation pressure on the breeding success and population density of grey partridges *(Perdix perdix).* Journal of Applied Ecology 33:965–78.

Tembrock, G. 1962. Zur strukturanalyse des kampfverhaltens bei *Vulpes.* Behaviour 19, 261–82.

Terborgh, J., J. A. Estes, P. Paquet, K. Ralls, D. Boyde-Heger, B. J. Miller, and R. F. Noss. 1999. The role of top carnivores in regulating terrestrial ecosystems. *In* M. E. Soulé and J. Terborgh (eds.), Continental Conservation: Scientific Foundations of Regional Reserve Networks, 39–64. Island Press, Washington, D.C.

Thom, M. D., L. A. Harrington, and D. W. Macdonald. 2004a. Why are American mink sexually dimorphic? A role for niche separation. Oikos 105:525–35.

Thom, M. D., D. D. P. Johnson, and D. W. Macdonald. 2004. The evolution and maintenance of delayed implantation in the Mustelidae (Carnivora: Mammalia). Evolution 58:175–83.

Thom, M. D., D. W. Macdonald, G. J. Mason, V. Pedersen, and P. Johnson. 2004b. Simultaneously polyandrous female mink *(Mustela vison)* mate randomly in a free-choice environment. Animal Behaviour 67:975–84.

Tigas, L. A., D. H. Van Vuren, and R. M. Sauvajot. 2002. Behavioral responses of bobcats and coyotes to habitat fragmentation and corridors in an urban environment. Biological Conservation 108:299–306.

Tuyttens, F. A. M., and D. W. Macdonald. 1998. Fertility control: an option for non-lethal control of wild carnivores? Animal Welfare 7:339–64.

———. 2000. Consequences of social perturbation for wildlife management and conservation. *In* L. M. Gosling and W. J. Sutherland (eds.), Behaviour and Conservation, 315–329. Cambridge University Press, Cambridge.

Unites States Fish and Wildlife Service. 2003. Red Wolf News 4:1.

Valenzuela, D., and D. W. Macdonald. 2002. Home-range use by white-nosed coatis *(Nasua narica):* limited water and a test of the resource dispersion hypothesis. Journal of Zoology (London) 258:247–56.

van Lawick, H. 1974. Solo, the Story of an African Wild Dog. Houghton Mifflin, Boston.

van Lawick, H., and van Lawick-Goodall, K. 1970. Innocent Killers. Houghton Mifflin, Boston.

Van Orsdol, K. G., J. P. Hanby, and J. D. Bygott. 1985. Ecological correlates of lion social organisation *(Panthera leo)*. Proceedings of a symposium on Lions and Leopards as Game Ranch Animals. Onderstepoort, 177–83.

van Schaik, C. P., and J. A. R. A. M. van Hooff. 1983. On the ultimate cause of primate social systems. Behaviour 85:91–117.

Van Valkenburgh, B., and C. Ruff. 1987. Canine tooth strength and killing behaviour in large carnivores. Journal of Zoology (London) 212:379–97.

Voight, D. R., and B. D. Earle. 1983. Avoidance of coyotes by red fox families. Journal of Wildlife Management 47:852–57.

Vucetich, J. A., and S. Creel. 1999. Ecological interactions, social organization, and extinction risk in African wild dogs. Conservation Biology 13:1172–82.

Vucetich, J. A., R. O. Peterson, and T. A. Waite. 2003. Raven scavenging favours group foraging in wolves. Animal Behaviour 67:1117–26.

Waser, P. M. 1981. Sociality or territorial defence? The influence of resource renewal. Behavioural Ecology and Sociobiology 8:231–37.

———. 1996. Patterns and consequences of dispersal in gregarious carnivores. *In* J. L. Gittleman (ed.), Carnivore Behaviour, Ecology and Evolution, 2:267–95. Cornell University Press, Ithaca, N.Y.

Waser, P. M., B. Keane, S. R. Creel, L. F. Elliott, and D. J. Minchella. 1994. Possible male coalitions in a solitary mongoose. Animal Behaviour 47:289–94.

Wayne, R. K., N. Lehman, M. W. Allard, and R. L. Honeycutt. 1992. Mitochondrial DNA variability of the gray wolf: genetic consequences of population decline and habitat fragmentation. Conservation Biology 6:559–80.

Weber, W., and A. Rabinowitz. 1996. A global perspective on large carnivore conservation. Conservation Biology 10:1046–54.

Wei, F., Z. Feng, Z. Wang, and M. Li. 1999. Feeding strategy and resource partitioning between giant and red pandas. Mammalia 63:417–30.

Werdelin, L., and N. Solounias. 1991. The Hyaenidae: taxonomy, systematics and evolution. Fossils and Strata 30:1–104.

White, P. J., and R. A. Garrott. 1997. Factors regulating kit fox populations. Canadian Journal of Zoology 75:1982–88.

Wildt, D., M. Bush, K. L. Goodrowe, C. Packer, A. E. Pusey, J. L. Brown, P. Joshin, and S. J. O'Brien. 1987. Reproductive and genetic consequences of founding isolated lion populations. Nature 329:328–31.

Wildt, D. M., J. G. Howard, and J. Brown. 2001. Role of reproductive science in carnivore conservation. *In* J. L. Gittleman, S. M. Funk, D. Macdonald, and R. K. Wayne (eds.), Carnivore Conservation, 359–71. Cambridge University Press, Cambridge.

Wiley, R. H., and D. G. Richards. 1978. Physical constraints on acoustic communication in the atmosphere: implications for the evolution of animal vocalizations. Behavioral Ecology and Sociobiology 3:69–94.

Wilson, D. E., and D. Reeder. 1993. Mammal Species of the World. Smithsonian Institution Press, Washington, D.C.

Wilson, E. O. 1987. The little things that run the world: the importance of conservation of invertebrates. Conservation Biology 1:344–46.

Wisely, S. M., J. J. Ososky, and S. W. Buskirk. 2002. Morphological changes to black-footed ferrets *(Mustela nigripes)* resulting from captivity. Canadian Journal of Zoology 80: 1562–68.

Wolff, J. O., and D. W. Macdonald. 2004. Promiscuous females protect their offspring. Trends in Ecology and Evolution 19:127–34.

Wolff, J. O., and J. A. Peterson. 1998. An offspring-defense hypothesis for territoriality in female mammals. Ethology, Ecology, and Evolution 10:227–39.

Woodroffe, R. 2001. Strategies for carnivore conservation: lessons from contemporary extinctions. *In* J. L. Gittleman, S. M. Funk, D. Macdonald, and R. K. Wayne (eds.), Carnivore Conservation, 61–92. Cambridge University Press, Cambridge.

Woodroffe, R., and J. R. Ginsberg. 2000. Ranging behaviour and vulnerability to extinction in carnivores. *In* L. M. Gosling and W. J. Sutherland (eds.), Behaviour and Conservation, 125–40. Cambridge University Press, Cambridge.

———. 1998. Edge effects and the extinction of populations inside protected areas. Science 280:2126–28.

Woodroffe, R., and A. Vincent. 1994. Mother's little helpers: patterns of male care in mammals. Trends in Ecology and Evolution 9:294–97.

Woodroffe, R. B., J. R. Ginsberg, and D. W. Macdonald. 1997. The African Wild Dog: status survey and conservation action plan. IUCN/SSC, Canid Specialist Group, Gland, Switzerland.

Woodroffe, R., D. W. Macdonald, and J. D. Silva. 1995. Dispersal and philopatry in the European badger, *Meles meles.* Journal of Zoology (London) 237:227–39.

Woodroffe, R., S. Cleaveland, O. Courtenay, K. Laurenson, and M. Artois. 2004. Infectious disease in the management and conservation of wild canids. *In* D. W. Macdonald and C. Sillero-Zubiri (eds.), Canid Biology and Conservation, 123–42. Oxford University Press, Oxford.

Wozencraft, W. C. 1993. Order Carnivora. *In* D. E. Wilson and D. M. Reeder (eds.), Mammal Species of the World, 279–348. Smithsonian Institution Press, Washington, D.C.

Wrangham, R. W. 1980. An ecological model of female-bonded primate groups. Behaviour 75:262–97.

Wrangham, R. W., J. L. Gittleman, and C. A. Chapman. 1993. Constraints on group size in primates and carnivores: population density and day-range as assays of exploitation competition. Behavioral Ecology and Sociobiology 32:199–209.

Wydeven, A. P., A. Treves, B. Brost, and J. E. Wiedenhoeft. In press. Characteristics of wolf packs depredating on domestic animals in Wisconsin, USA. *In* N. Fasciore, A. Delach, and M. Smith (eds.), Predators and People: From Conflict to Conservation. Island Press, Washington, D.C.

Wyman, J. 1967. The jackals of the Serengeti. Animals 1967:79–83.

Yadvendradew, V. J. 1994. Predation on blackbuck by wolves in Velvadar National Park, Gujarat, India. Conservation Biology 7:874–81.

Yamaguchi, N., A. Cooper, L. Werdelin, and D. W. Macdonald. In press. The evolution of the lion: from fossil record to social biology. Journal of Zoology (London).

Yamaguchi, N., and B. Haddane. 2002. The North African Barbary lion and the Atlas Lion Project. International Zoo News 49:465–81.

Yamaguchi, N., D. W. Macdonald, W. C. Passanisi, D. A. Harbour, and C. D. Hopper. 1996. Parasite prevalence in free-ranging farm cats, *Felis silvestris catus.* Epidemiology and Infection 116:217–23.

Yamaguchi, N., S. Rushton, and D. W. Macdonald. 2003. Habitat preferences of feral American mink in the Upper Thames. Journal of Mammalogy 84:1356–73.

Yamaguchi, N., R. J. Sarno, W. E. Johnson, S. J. O'Brien, and D. W. Macdonald. 2004. Multiple paternity and reproductive tactics of free-ranging American minks, *Mustela vison.* Journal of Mammalogy 85:432–39.

Yamazaki, K. 1996. Social variation of lions in a male-depopulated area in Zambia. Journal of Wildlife Management 60:490–97.

Yerli, S., A. F. Canbolat, L. J. Brown, and D. W. Macdonald. 1997. Mesh grids protect log-

gerhead turtle *Caretta caretta* nests from red fox *Vulpes vulpes* predation. Biological Conservation 82:109–11.
Yoder, A. D., M. M. Burns, S. Zehr, T. Delefosse, G. Veron, S. M. Goodman, and J. J. Flynn. 2003. Single origin of Malagasy Carnivora from an African ancestor. Nature 421:734–37.
Zielinski, W. J., and T. E. Kucera. 1995. American marten, fisher, lynx, and wolverine: survey methods for their detection. Pacific Southwest Research Station, Forest Service, USA, Albany, Calif.
Zimen, E. 1976. On the regulation of pack size in wolves. Zeitschrift für Tierpsychologie 40:300–41.
———. 1981. The Wolf: His Place in the Natural World. Souvenir Press, London.

Order Carnivora

Dogs, Bears, Raccoons, Weasels, Civets, Mongooses, Hyenas, and Cats

This order of 8 Recent families, 97 genera, and 246 species occurs naturally throughout the world except in Australia, New Guinea, New Zealand, Antarctica, and many oceanic islands. One species, *Canis familiaris,* apparently was introduced into Australia by human agency in prehistoric time and subsequently established wild populations on that continent. Although recently there has been much reassessment of carnivore systematics, most authorities (e.g., E. R. Hall 1981; Hunt and Tedford 1993; Stains 1984; Wayne et al. 1989; Werdelin and Solounias 1991; Wozencraft 1989*b;* Wyss and Flynn 1993) continue to recognize two basic phylogenetic divisions: (1) the suborder Caniformia or superfamily Arctoidea (or Canoidea), with the families Canidae, Ursidae, Procyonidae, and Mustelidae; and (2) the suborder Feliformia or superfamily Aeluroidea (or Feloidea), with the families Viverridae, Herpestidae, Hyaenidae, and Felidae. A major point of contention involves the interrelationships of the Aeluroidea; some evidence indicates that the Hyaenidae and Felidae are most closely related and that the other two families represent earlier evolutionary divergences; other data suggest a dichotomy, with the Herpestidae and Hyaenidae forming one related group and the Viverridae and Felidae forming another.

Many, perhaps most, authorities also now include the Pinnipedia within the Carnivora. Simpson (1945) treated the Pinnipedia as a suborder of the Carnivora and united all the terrestrial families in another suborder, the Fissipedia. Other authorities (e.g., Rice 1977; Tedford 1976; Wozencraft 1989*b* and *in* Wilson and Reeder 1993; Wyss and Flynn 1993) place the pinnipeds in the suborder Caniformia and/or the superfamily Arctoidea. For further information on questions regarding the classification of the Pinnipedia, see the companion volume *Walker's Marine Mammals of the World.*

The smallest living carnivore is the least weasel *(Mustela nivalis),* which has a head and body length of 135–85 mm, a tail length of 30–40 mm, and a weight of 35–70 grams. The largest is the grizzly or brown bear, some individuals of which, particularly along the coast of southern Alaska, attain a head and body length of 2,800 mm and a weight of 780 kg.

Carnivores have four or five clawed digits on each limb. The first digit (pollex and hallux) is not opposable and sometimes is reduced or absent. Some carnivores, including canids and felids, are digitigrade, walking only on their toes. Others, such as ursids, are plantigrade, walking on their soles with the heels touching the ground. The brain has well-developed cerebral hemispheres, and the skull is heavy, with strong facial musculature. The articulation of the lower jaw permits only open-and-shut (not side-to-side) movements. The stomach is simple. Males have a baculum. The number of mammae in females is variable; they are located on the abdomen, except that in the Ursidae some are pectoral.

The teeth are rooted. The small, pointed incisors number 3/3 in all species except *Ursus ursinus,* which has 2/3, and *Enhydra lutris,* which has 3/2. The first incisor is the smallest, and the third is the largest, the difference in size being most marked in the upper jaw. The canine teeth are strong, recurved, pointed, elongate, and round to oval in section. The premolars are usually adapted for cutting, and the molars usually have four or more sharp, pointed cusps. The last upper premolar and the first lower molar, the carnassials, often work together as a specialized shearing mechanism. The carnassials are most highly developed in the Felidae, which have a diet consisting almost entirely of meat, and are least developed in the omnivorous Ursidae and Procyonidae.

Most carnivores are terrestrial or climbing animals. Two genera, *Potos* and *Arctictis,* have prehensile tails. Apparently, all carnivores can swim if necessary, but the polar bear *(Ursus maritimus)* and the river otters *(Lutra, Lontra, Lutrogale, Aonyx, Pteronura)* are semiaquatic, and the sea otter *(Enhydra)* spends practically its entire life in the water. Land-dwelling carnivores shelter in caves, crevices, burrows, and trees. They may be either diurnal or nocturnal.

Most species of the Canidae, Mustelidae, Viverridae, Herpestidae, and Felidae live solely or mainly on freshly killed prey. Their whole body organization and manner of living are adapted for predation. The diet may vary by season and locality. Hunting is done by scent and sight, and the prey is captured by a surprise pounce from concealment *(Panthera pardus),* a stalk followed by a swift rush *(Mustela frenata),* or a lengthy chase *(Canis lupus).* Some species regularly eat carrion. *Arctictis* is largely frugivorous, and the diet of certain other viverrids consists partly of fruit. The Hyaenidae include one genus *(Crocuta)* that is primarily a hunter of large animals, two *(Hyaena, Parahyaena)* that feed mainly on carrion, and one *(Proteles)* that is largely insectivorous. The Procyonidae and Ursidae, except for the carnivorous polar bear, are omnivorous, eating a wide variety of plant and animal life.

Carnivores are solitary or associate in pairs or small

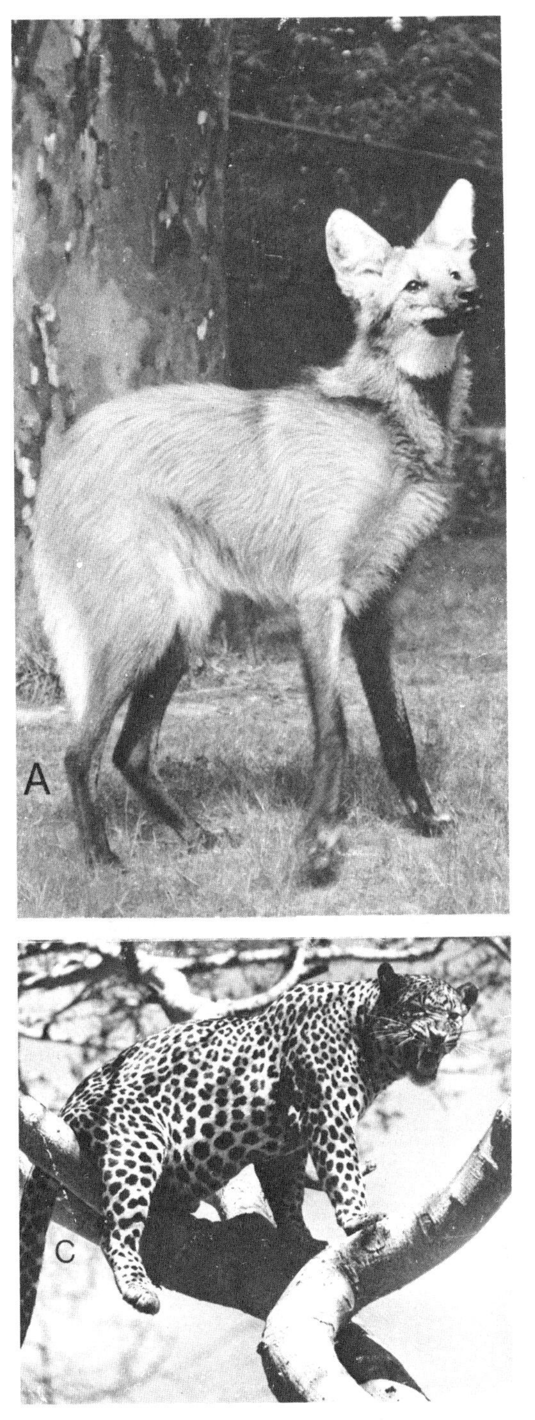

A. Maned wolf *(Chrysocyon brachyurus).* B. Linsang *(Prionodon linsang).* C. Leopard *(Panthera pardus).* D. Giant panda *(Ailuropoda melanoleuca).* Photos by Bernhard Grzimek.

groups. Females commonly produce a single litter each year, but those of a few species may give birth two or three times annually, and those of some large species usually mate at intervals of several years. Most species have gestation periods of about 49–113 days. Delayed implantation of the fertilized egg occurs in ursids and some mustelids, so the period from mating to birth is considerably longer than average. Litter size commonly ranges from 1 to 13. The offspring usually are born blind and helpless but with a covering of hair. They are cared for solicitously by the mother and, in some species, by the father. There is often a lengthy period of parental care and instruction.

The Carnivora were once considered to include the Creodonta, an extinct group dating back to the late Cretaceous, as a suborder. It now appears, however, that the Carnivora evolved independently, perhaps from ancestral insectivores or from the same basal stock that gave rise to the Primates and Chiroptera (Novacek 1992; Wozencraft 1989*a;* Wyss and Flynn 1993). The oldest known groups usually referred to the Carnivora are the Viverravidae, an arboreal family that lived from the early Paleocene to the middle Eocene, and the Miacidae, another small, viverridlike family of the Eocene (L. D. Martin 1989). Heinrich and Rose (1995) reported the oldest known miacid skeleton, from the early Eocene of Wyoming, and indicated that it represented a highly arboreal animal that weighed about 1.3 kg. Wyss and Flynn (1993) suggested that the Viverravidae and Miacidae be included not within the Carnivora but in a larger, supraordinal grouping called the Carnivoramorpha.

CARNIVORA; **Family CANIDAE**

Dogs, Wolves, Coyotes, Jackals, and Foxes

This family of 16 Recent genera and 36 species has a natural distribution that includes all land areas of the world except the West Indies, Madagascar, Taiwan, the Philippines, Borneo and islands to the east, New Guinea, Australia, New Zealand, Antarctica, and most oceanic islands. There are wild populations of the species *Canis familiaris* in Australia and New Guinea, but these apparently originated through introduction by human agency. The living Canidae traditionally have been divided, mainly on the basis of dentition, into three subfamilies: the Caninae, with the genera *Canis, Alopex, Vulpes, Fennecus, Urocyon, Nyctereutes, Dusicyon, Cerdocyon, Atelocynus,* and *Chrysocyon;* the Simocyoninae, with *Speothos, Cuon,* and *Lycaon;* and the Otocyoninae, with *Otocyon.*

Recent studies have indicated that subfamilial distinction for the Simocyoninae and the Otocyoninae is not warranted and have revealed considerable controversy regarding the systematics of the Caninae. Langguth (1975) referred most species of the South American *Dusicyon* to a subgenus *(Pseudalopex)* of *Canis,* referred one species *(D. vetulus)* to the genus *Lycalopex,* and retained only one species *(D. australis)* in *Dusicyon.* Clutton-Brock, Corbet, and Hills (1976) considered the genus *Vulpes* to include *Fennecus* and *Urocyon* and the genus *Dusicyon* to include *D. australis,* the species referred by Langguth to *Pseudalopex* and *Lycalopex,* and also *Cerdocyon* and *Atelocynus.* Van Gelder (1978) expanded the genus *Canis* to contain the following as subgenera: *Dusicyon,* with the single species *D. australis; Pseudalopex,* with most species traditionally assigned to *Dusicyon; Lycalopex,* with the species formerly called *Dusicyon vetulus; Cerdocyon; Atelocynus; Vulpes,* including *Fennecus* and *Urocyon;* and *Alopex.* Berta (1987) considered *Otocyon* and *Urocyon* to be closely related to *Vulpes, Lycalopex* to be part of *Pseudalopex, D. australis* to be the only modern species of *Dusicyon, Nyctereutes* to be the closest living relative of *Cerdocyon,* and *Speothos* to be the closest living relative of *Atelocynus.* Based on analyses of mitochondrial DNA, Geffen, Mercure, et al. (1992) considered *Alopex* to be very closely related to *Vulpes velox* and *Fennecus* to form a mono-

A. Maned wolf pup *(Chrysocyon brachyurus),* photo from Los Angeles Zoological Society. B. Red fox pup *(Vulpes vulpes),* photo by Leonard Lee Rue III.

Bush dogs *(Speothos venaticus)*, photo by Bernhard Grzimek.

phyletic unit with *V. cana.* They recommended that all be included in the single genus *Vulpes* but also concluded that *Urocyon* is not closely related to any of the other genera.

Because of the controversy, a conservative position has been taken here, and all traditionally recognized genera have been maintained. In addition, *Pseudalopex* and *Lycalopex* have been given generic rank. Wozencraft (*in* Wilson and Reeder 1993) took much the same position but included *Lycalopex* within *Pseudalopex.* The sequence of genera presented here is based partly on information given by Langguth (1975) and Nowak (1978, 1979) indicating that *Vulpes* and the other foxes are more primitive than *Canis.* The sequence also gives some consideration to the molecular analyses of Wayne (1993) and Wayne and O'Brien (1987), which suggest the existence of a wolflike (including all *Canis*), a foxlike, and a South American group of genera. That scheme, however, indicates that *Lycaon* and possibly *Speothos* are associated with the wolflike group and that the genera *Urocyon, Nyctereutes,* and *Otocyon* show no close affinity with any of the groups. Also by that approach *Chrysocyon* is associated distantly with the South American group, but a chromosomal analysis (Vitullo and Zuleta 1992) suggested northern affinity for *Chrysocyon.* Further systematic comments are given in the generic accounts. Two additional genera, *Cubacyon* and *Paracyon,* described from subfossil material found on Cuba, almost certainly represent domestic *Canis familiaris* introduced by Amerindian peoples (Morgan and Woods 1986) and are not discussed further here.

In wild species head and body length is 357–1,600 mm, tail length is 125–560 mm, and weight is 1–80 kg. *Fennecus zerda* is the smallest species and *Canis lupus* is the largest. In a given population males generally are larger than females. Most species are uniformly colored or speckled, but one species of jackals *(Canis adustus)* has stripes on the sides of its body, and *Lycaon* is covered with blotches.

Canids have a lithe, muscular, deep-chested body; usually long, slender limbs; a bushy tail; a long, slender muzzle; and large, erect ears. There are four digits on the hind foot and five on the forefoot, except in *Lycaon,* which has four on both the front and the back foot. The claws are blunt. Males have a well-developed baculum, and females generally have three to seven pairs of mammae.

The skull is elongate. The bullae are prominent but usually are not highly inflated. The dental formula in all but three species is: (i 3/3, c 1/1, pm 4/4, m 2/3) × 2 = 42. The molars are 1/2 in *Speothos,* 2/2 in *Cuon,* and 3/4 or 4/4 in *Otocyon.*

Canids occur from hot deserts *(Fennecus)* to arctic ice fields *(Alopex).* For dens they may use burrows, caves, crevices, or hollow trees. These alert, cunning animals may be diurnal, nocturnal, or crepuscular. They are generally active throughout the year. They walk, trot tirelessly, amble, or canter, either entirely on their digits or partly on more of the foot. At full speed they gallop. The gray foxes *(Urocyon)* often climb trees, an unusual habit for canids. The senses of smell, hearing, and sight are acute. Prey is captured by an open chase or by stalking and pouncing. The diet may vary by season, and vegetable matter is important to some species at certain times.

Some canids, especially the larger species, occur in packs of up to 30 members and seek prey animals that are larger than themselves. Most smaller canids hunt alone or in pairs, preying on rodents and birds. There is usually a regular home range, part or all of which may be an exclusive territory. Females generally give birth once a year. Litters usually contain 2–13 young. Gestation averages around 63 days. The offspring are blind and helpless at birth but are covered with hair. They are cared for solicitously by the mother and often by the father and other group members as well. Sexual maturity comes after 1 or 2 years. Potential longevity is probably at least 10 years in all species.

The geological range of this family is late Eocene to Recent in North America and Europe, early Oligocene to Recent in Asia, late Pliocene to Recent in South America, Pliocene to Recent in Africa, and late Pleistocene to Recent in Australia (Berta 1987; Langguth 1975; Macintosh 1975).

CARNIVORA; CANIDAE; **Genus VULPES**
Frisch, 1775

Foxes

There are 10 species (Clutton-Brock, Corbet, and Hills 1976; Coetzee *in* Meester and Setzer 1977; Corbet 1978; Corbet and Hill 1992; Ellerman and Morrison-Scott 1966; Geffen et al. 1993; E. R. Hall 1981; Mendelssohn et al. 1987; Roberts 1977):

V. vulpes (red fox), Eurasia except the southeastern tropical zone, northern Africa, most of Canada and the United States;
V. corsac (corsac fox), dry steppe and subdesert zone from the lower Volga River to Manchuria and Tibet;
V. ferrilata (Tibetan sand fox), high plateau country of Tibet, Nepal, and north-central China;
V. cana (Blanford's fox), dry mountainous regions of southern Turkmenistan, Iran, Pakistan, Afghanistan, southern and western Arabian Peninsula, Israel, and Sinai;
V. velox (swift fox), southern Alberta and North Dakota to northwestern Texas;
V. macrotis (kit fox), southern Oregon to Baja California and north-central Mexico;
V. bengalensis (Bengal fox), Pakistan, India, Nepal, Bangladesh;
V. rueppellii (sand fox), desert zone from Morocco and Niger to Afghanistan and Somalia;
V. pallida (pale fox), savannah zone from Senegal to northern Sudan and Somalia;
V. chama (Cape fox), dry areas of southern Angola, Namibia, Botswana, western Zimbabwe, and South Africa.

The reason for using Frisch, 1775, rather than Bowdich, 1821, as the authority for this generic name was explained by Corbet and Hill (1992). Treatment of *Vulpes* as a distinct genus that does not include *Fennecus*, *Urocyon*, or *Alopex* is in keeping with the arrangements of such authorities as Coetzee (*in* Meester and Setzer 1977), E. R. Hall (1981), Jones et al. (1992), and Rosevear (1974). Other views have been to consider *Vulpes* a full genus that includes *Alopex* (Youngman 1975), a full genus that includes *Fennecus* and *Urocyon* but not *Alopex* (Clutton-Brock, Corbet, and Hills 1976), a full genus that includes *Fennecus* but not *Urocyon* and *Alopex* (Wozencraft *in* Wilson and Reeder 1993), a full genus that includes *Fennecus* and *Alopex* but not *Urocyon* (Geffen, Mercure, et al. 1992), and a subgenus of *Canis* that includes *Fennecus* and *Urocyon* but not *Alopex* (Van Gelder 1978). The North American red fox has sometimes been designated a separate species, *V. fulva*, but most authorities now consider it to be conspecific with the Palaearctic *V. vulpes*.

E. R. Hall (1981) treated *V. macrotis* as being conspecific with *V. velox* because Rohwer and Kilgore (1973) had reported interbreeding between these two kinds of fox where their ranges meet in eastern New Mexico and western Texas. Combining the two species was supported by Dragoo et al. (1990) based on genic data, though morphometric data were inconclusive, and Wozencraft (*in* Wilson and Reeder 1993) took the same position. Additional studies using morphological characters (Stromberg and Boyce 1986; Thornton and Creel 1975) and analyses of mitochondrial DNA (Mercure et al. 1993) have argued for continued recognition of the two as separate species.

Vulpes is characterized by a rather long, low body; relatively short legs; a long, narrow muzzle; large, pointed ears; and a bushy, rounded tail that is at least half as long, and often fully as long, as the head and body. The pupils of the eyes generally appear elliptical in strong light. Some species have a pungent "foxy" odor, arising mainly from a gland located on the dorsal surface of the tail, not far from its base. Females usually have six or eight mammae. Additional information is provided separately for each species.

Vulpes vulpes (red fox)

Head and body length is 455–900 mm, tail length is 300–555 mm, and weight is 3–14 kg. Average weights in North America are 4.1–4.5 kg for females and 4.5–5.4 kg for males (Ables 1975). The usual weight in central Europe is 8–10 kg (Haltenorth and Roth 1968). The typical coloration ranges from pale yellowish red to deep reddish brown on the upper parts and is white, ashy, or slaty on the underparts. The lower part of the leg is usually black, and the tail is generally tipped with white or black. Color variants, known as the "cross fox" and the "silver fox," represent, respectively, about 25 percent and 10 percent of the species. The cross fox is reddish brown in color and gets its name from the cross formed by one black line down the middle of the back and another across the shoulders. The color of the silver fox, whose fur is the most prized among foxes, ranges from strong silver to nearly black. The general color effect depends on the proportion of white or white-tipped hairs to black hairs. An individual with only a few white hairs is sometimes called a "black fox." In Europe such black individuals occur only in the north and represent at most 1 percent of the population; in North America they are more common (Krott 1992). Aberrant individuals known as "samson foxes" sometimes appear in a population and may occur in substantial numbers for various periods (Voipio 1990). They lack the guard hairs of the normal pelage and have other unusual morphological and behavioral characters.

The red fox rivals the gray wolf *(Canis lupus)* for having the greatest natural distribution of any living terrestrial mammal besides *Homo sapiens*. Habitats range from deep forest to arctic tundra, open prairie, and farmland, but the red fox prefers areas of highly diverse vegetation and avoids large homogeneous tracts (Ables 1975). Elevational range is sea level to 4,500 meters (Haltenorth and Roth 1968). Daily rest may be taken in a thicket or any other protected spot, but each individual or family group usually has a main earthen den and one or more emergency burrows within the home range. An especially large den may be constructed during the late winter and subsequently used to give birth and rear the young. Some dens are used for many years by one generation of foxes after another. The preferred site is a sheltered, well-drained slope with loose soil. Often a marmot burrow is taken over and modified. Tunnels are up to 10 meters long and lead to a chamber 1–3 me-

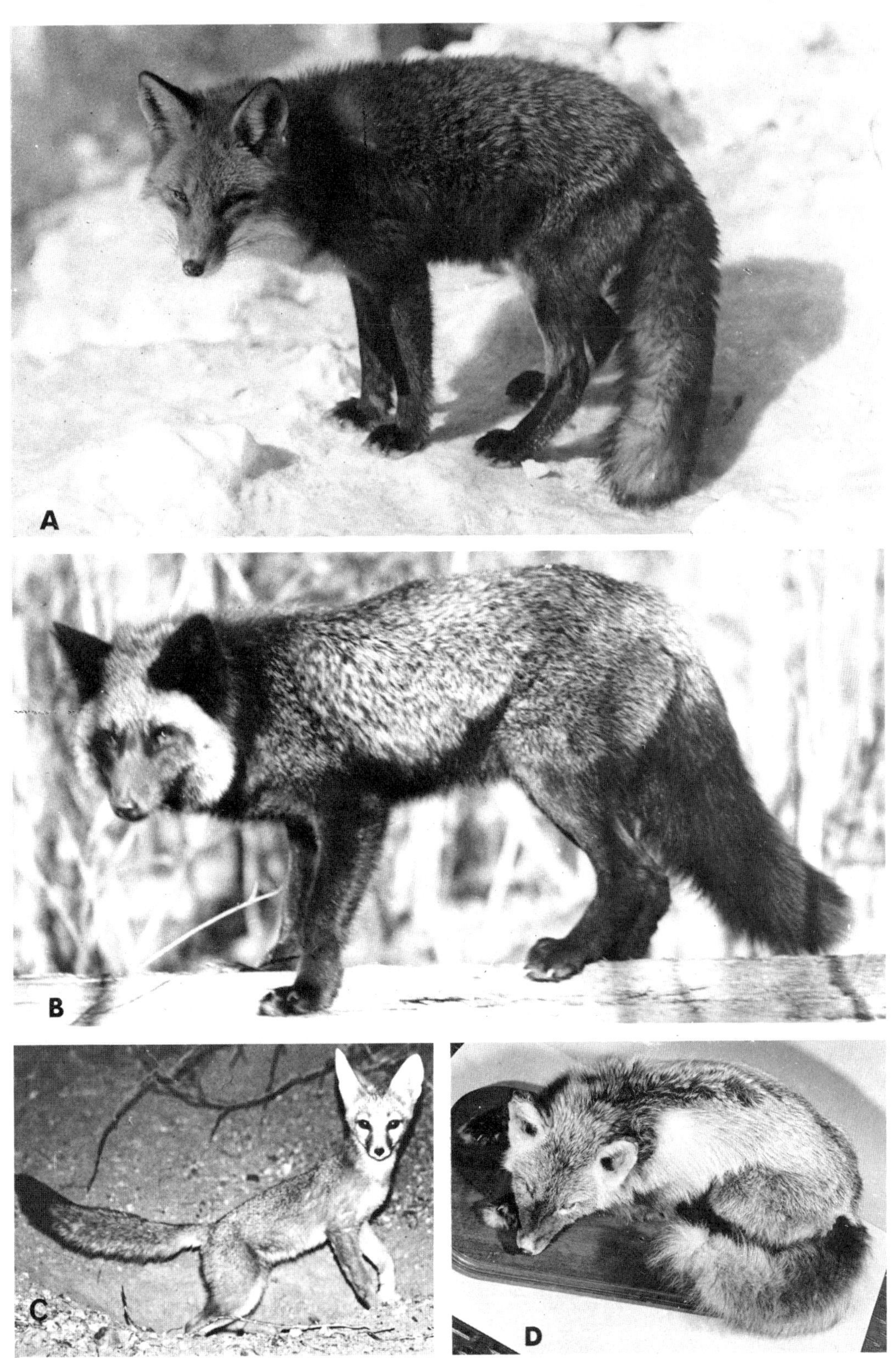

A. Red fox *(Vulpes vulpes)*, photo from New York Zoological Society. B. Silver fox *(V. vulpes)*, photo from Fromm Brothers, Inc. C. Kit fox *(V. macrotis)*, photo by O. J. Reichman. D. Cross fox *(V. vulpes)*, photo by Howard E. Uible of mounted specimen in U.S. National Museum of Natural History.

ters below the surface. There is sometimes only a single entrance, but there may be as many as 19. A system of pathways connects the dens, other resting sites, favored hunting areas, and food storage holes (Ables 1975; Banfield 1974; Haltenorth and Roth 1968; Stanley 1963).

The red fox is terrestrial, normally moving by a walk or trot. It has great endurance and can gallop for many kilometers if pursued. It can run at speeds of up to 48 km/hr, can leap fences 2 meters high, and can swim well (Haltenorth and Roth 1968). It has keen senses of sight, smell, and hearing. Its ability to survive in the close proximity of people and often to elude human hunters and their dogs has given it a reputation for cunning and intelligence. Most activity is nocturnal and crepuscular. Individuals cover up to 8 km per night as they move on circuitous routes through the home range (Banfield 1974). During the autumn the young born the previous spring disperse from the parental home range. The usual distance traveled at this time is about 40 km for males and 10 km for females; the maximum known distance traveled is 394 km (Ables 1975; Storm et al. 1976). Once the young animals establish themselves in a new area, they generally remain there for life.

The diet is omnivorous, consisting mostly of rodents, lagomorphs, insects, and fruit. To hunt mice, the red fox stands motionless, listens and watches intently, and then leaps suddenly, bringing its forelegs straight down to pin the prey. Rabbits are stalked and then captured with a rapid dash (Ables 1975). Daily consumption is around 0.5–1.0 kg. Sometimes a hole is dug and excess prey placed therein and covered over, to be eaten at a later time (Haltenorth and Roth 1968).

The most favorable areas usually support an average of one or two adults per sq km (Ables 1975; Haltenorth and Roth 1968; Insley 1977). Home range size varies with habitat conditions and food availability; it becomes larger in winter and smallest around the time of the arrival of newborn (Ables 1975). According to Zimen (*in* Grzimek 1990), the home range under natural conditions is usually 1–10 sq km but may be as small as 10 ha. in suburban areas. Jones and Theberge (1982) reported home range to average about 16 sq km in tundra habitat of northwestern British Columbia, larger than in temperate environments. *V. vulpes* is apparently territorial. There is little overlap of home ranges, and individuals on different ranges avoid one another (Storm and Montgomery 1975). Captive males were found continually to harass and chase foxes newly introduced to the enclosure, but females seldom became involved in such interaction (Preston 1975). In the breeding season, however, females do exhibit territorial behavior (Haltenorth and Roth 1968).

A home range is typically occupied by an adult male, one or two adult females, and their young (Storm and Montgomery 1975). Occasionally two females have litters in the same den (Pils and Martin 1978). Males may fight one another during the breeding season. A vixen sometimes mates with several males, but she later establishes a partnership with just one of them (Haltenorth and Roth 1968). For a period extending from shortly before birth to several weeks thereafter the female remains in or very near the den. The male then brings her food but does not actually enter the maternal den.

The mating season varies with latitude. In Europe it is December–January in the south, January–February in central regions, and February–April in the north (Haltenorth and Roth 1968). In North America mating occurs over about the same period (Ables 1975; Storm et al. 1976). Females are monestrous; estrus is 1–6 days and gestation is 49–56, usually 51–53, days. Litter size is 1–13 young but averages about 5 throughout the fox's range (Ables 1975; H. G. Lloyd 1975). The young weigh 50–150 grams at birth, open their eyes after 9–14 days, emerge from the den at 4–5 weeks, and are weaned at 8–10 weeks. They may be moved to a new den at least once. The family remains together until the autumn. Sexual maturity is reached at about 10 months. Potential natural longevity is around 12 years, though few individuals live more than 3–4 years, at least where the species is heavily hunted and trapped (Ables 1975). The oldest individual reliably recorded was a male aged 10 years and 8 months taken in Labrador (Chubbs and Phillips 1996).

The red fox is killed by people for sport, to protect domestic animals and game, to prevent the spread of rabies, and to obtain the valuable pelt. Sport hunting may involve an elaborate daytime chase by large numbers of riders and dogs or a nocturnal effort by one person to lure the fox with a call imitating that of a wounded rabbit. In Great Britain *V. vulpes* is traditionally valued as a game animal, but H. G. Lloyd (1975) noted that it was also the only mammal in the country subject to a government-approved bounty. It has become common in parts of London and other cities, and control efforts there have not substantially reduced its numbers (Harris and Smith 1987). The red fox is often considered to be a threat to poultry, but depredations are generally localized and many of the birds eaten are taken in the form of carrion. Studies have indicated that the red fox has little effect on wild pheasant populations (Ables 1975). Rabid foxes are said to be a serious menace in some areas, especially Europe, and intensive persecution there may be threatening the species in certain parts of its range; 200,000 individuals are taken every year in Germany alone. Such direct killing has had little effect in preventing the disease, but Switzerland, which has an extensive program of spreading oral vaccine baits, has become virtually free of rabies (Zimen *in* Grzimek 1990). A rabies epizootic, the main vector of which is the fox, spread from Poland across much of Europe from the 1940s to 1970s; however, only 5–10 percent of reported cases of rabies in domestic animals in the involved region have resulted from this epizootic (Steck and Wandeler 1980). The epizootic reached a peak in 1989 but subsequently declined in association with greater use of oral vaccine baits (Barrat and Aubert 1993). Most cases of rabies in Canada from 1958 to 1986 were reported from Ontario, and most of those occurrences (17,982) were in the red fox (Rosatte 1988). Problems caused by the fox are perhaps more than balanced by its control of rodent populations, which might otherwise multiply and damage human interests.

From 1900 to 1920 in North America, and to some extent in other parts of the world, catching wild foxes and raising them in captivity developed into an important industry. In the early stages of the breeding effort choice animals often sold for more than $1,000 each. Through selective breeding, strains were developed that nearly always produce silver-colored offspring. The number of foxes being raised for their fur now exceeds that of all other normally wild animals except possibly the mink *(Mustela vison)*. One fox farm permanently employed about 400 people and sold pelts worth more than $18 million annually. The fur is used in coats, stoles, scarves, and trimming. The value of fox pelts has varied widely depending on fashions, availability, and economic conditions. According to Banfield (1974), the average price of a silver fox skin was $246.46 in 1919–20 but only $17.94 in 1971–72. The average price of a wild-caught U.S. red fox skin rose from $12.00 in the 1970–71 season to about $48.00 in 1976–77. The reported number of red foxes trapped for their fur during the latter season in the United States and Canada was 421,705 (Deems and Pursley 1978). The number of pelts taken annually rose to more than 500,000 during the early 1980s, with average prices peaking at more than $60.00, but had declined to less than $20.00 by 1984 (Voigt 1987). In the 1991–92 season 217,257 skins were taken in the United States and sold for an average price of only $10.75 (Linscombe 1994). Considering inflation, this price was far less than what it had been one to two decades earlier. The drop may have been associated with growing social disdain for the use of wild-caught furs in fashion and with a pending ban by the European Community on the importation of pelts derived from the use of leghold traps.

Despite human persecution, *V. vulpes* has maintained or even increased its numbers in many parts of its range. There are now probably more in Great Britain than there were in medieval times because of improved habitat conditions resulting from the establishment of hedgerows and crop rotation (H. G. Lloyd 1975). This species is able to carry on its mode of life in intensively farmed areas and sometimes even in large cities (Ables 1975; Grzimek 1975). It also has been successfully introduced in some areas, especially by persons of English background who desired to continue traditional fox hunting. The species was brought to Australia in 1868 and subsequently spread over much of that continent, to the lasting detriment of the native fauna (Clutton-Brock, Corbet, and Hills 1976; Ride 1970).

Introductions from England also were made in eastern North America in colonial times. The species was naturally present in this region but apparently was not abundant. It subsequently increased in numbers and became established in areas not previously occupied, mainly because of the breaking up of the homogeneous forests by people and continuous introduction by hunting clubs. In the twentieth century the red fox has greatly extended its range in the southeastern United States and has occupied Baffin Island and moved as far north as the southern coast of Ellesmere Island. It has spread westward across the Great Plains, possibly in response to a human-caused reduction of coyote *(Canis latrans)* numbers (Banfield 1974; E. R. Hall 1981; Hatcher 1982; Lowery 1974). The only major North American population that may be in trouble from a conservation viewpoint is that of the Sierra Nevada of California, where surveys indicate that the native subspecies *(V. v. necator)* is very rare and evidently declining (Schempf and White 1977).

Vulpes corsac (corsac fox)

Head and body length is 500–600 mm and tail length is 250–350 mm. The fur is thick and soft. The general coloration of the upper parts is pale reddish gray, or reddish brown with silvery overtones. The underparts are white or yellow. *V. corsac* is externally similar to *V. vulpes* but has relatively longer legs. Its ears are large, pointed, and very broad at the base (Novikov 1962).

The corsac fox is a typical inhabitant of steppes and semidesert. It avoids forests, thickets, plowed fields, and settled areas. It lives in a burrow, often taken over from another mammal, such as a marmot or badger. Self-excavated burrows, sometimes found in groups, are simple and usually very shallow (Novikov 1962). Although usually reported to be nocturnal in the wild, *V. corsac* is active by day in captivity; it is said to be an excellent climber (Grzimek 1975). It runs with only moderate speed and can be caught by a slow dog, but it has excellent senses of vision, hearing, and smell (Stroganov 1969). Most reports indicate that it is nomadic and does not keep to a fixed home range (Ognev 1962). It may migrate southward when deep snow and ice make hunting difficult (Stroganov 1969). The diet consists mostly of small rodents but also includes pikas, birds, insects, and plant material.

This species is more social than other foxes, with sever-

Corsac foxes *(Vulpes corsac)*, photo from Amsterdam Zoo.

al individuals sometimes living together in the same burrow (Ognev 1962). Small hunting packs are said to form in the winter (Stroganov 1969), though perhaps these represent mated pairs and their grown young of the previous spring. Males fight one another during the breeding season but then remain with the family (Grzimek 1975; Novikov 1962). Mating occurs from January to March, gestation lasts 50–60 days, and litters usually contain 2–11 young (Stroganov 1969). Females in the Berlin Zoo did not reach sexual maturity until their third year of life (Grzimek 1975).

The corsac fox lacks the penetrating odor of most *Vulpes* and was frequently kept as a pet in eighteenth-century Russia (Grzimek 1975). Its warm and beautiful fur led to large-scale commercial trapping; up to 10,000 pelts were sold annually in the western Siberian city of Irbit in the late nineteenth century. For this reason, and also because of the settlement and plowing of the steppes, the corsac fox has disappeared in much of its range (Ognev 1962; Stroganov 1969).

Vulpes ferrilata (Tibetan sand fox)

Head and body length is 575–700 mm, tail length is 400–475 mm, and males weigh up to 7 kg (Mitchell 1977). The fur is soft and thick and the tail is bushy. The upper parts are pale gray agouti or sandy, with a tawny band along the dorsal region. The underparts are pale, the front of the leg is tawny, and the tip of the tail is white. The skull is peculiarly elongated and has a very narrow maxillary region (Clutton-Brock, Corbet, and Hills 1976).

Mitchell (1977) found this fox on barren slopes and in streambeds at 3,000–4,000 meters in the Mustang district of Nepal. In this area dens are made in boulder piles or in burrows under large rocks. The diet consists of rodents, lagomorphs, and ground birds. Mitchell observed pairs hunting along streambeds, on boulder heaps, and in wheat fields. Mating occurs in late February, and two to five young are born in April or May.

Vulpes cana (Blanford's fox)

Specimens from Afghanistan and Iran had a head and body length of 400–500 mm and a tail length of 330–410 mm (Geffen 1994). A series from Israel averaged about 420 mm in head and body length, 320 mm in tail length, and 1 kg in weight (Geffen, Hefner, et al. 1992*b*). Clutton-Brock, Corbet, and Hills (1976:155) described *V. cana* as "a small fox with extremely soft fur and a long very bushy tail. The colouring is blotchy black, grey and white with a dark tip to the tail and a dark patch over the tail gland. There is an almost black mid-dorsal line and the hind legs may be dark. . . . The underparts are white, the ears are grey, and there is a small dark patch between the eyes and nose."

According to Roberts (1977), the habitat of *V. cana* is mountain steppe. It is reportedly more frugivorous than the other foxes of Pakistan, being fond of ripe melons and seedless grapes, and sometimes damages crops. Geffen et al. (1993) found it relatively common in the hot, rocky habitats of the Negev and Judean deserts. Mendelssohn et al. (1987) added that it has an astonishing jumping ability and can move upward among cliffs by pushing itself from one vertical wall to another. It is strictly nocturnal and shows little change in temporal activity throughout the year (Geffen and Macdonald 1993). Its diet in Israel consists mainly of arthropods, fruit, and other plant material (Geffen, Hefner, et al. 1992*a*), and it apparently has no need to drink free water (Geffen, Degen, et al. 1992).

Reported population densities in Israel are 0.5/sq km and 2.0/sq km (Geffen 1994). A radio-tracking study there showed that individuals moved 7–11 km per night and utilized home ranges of about 1.6 sq km. These ranges were occupied by strictly monogamous pairs and overlapped only minimally with the ranges of other individuals (Geffen and Macdonald 1992). Females captured in that area gave birth to litters of 1 and 3 young in February and April (Mendelssohn et al. 1987). The gestation period is 50–60 days, pups weigh about 29 grams at birth, lactation lasts 30–45 days, and age at sexual maturity is 8–12 months (Geffen 1994).

The skin of *V. cana* is valued in commerce and is heavily hunted. Novikov (1962) called this species one of the rarest predators of the old Soviet Union. It is on appendix 2 of the CITES.

Vulpes velox (swift fox)

Head and body length is 375–525 mm, tail length is 225–350 mm, and weight is 1.8–3.0 kg. Males average larger than females. The winter coat is long and dense; the upper parts are dark buffy gray; the sides, legs, and lower surface of the tail are orange tan; and the underparts are buff to pure white. In summer the coat is shorter, harsher, and more reddish. *V. velox* differs from the closely related *V. macrotis* in having smaller ears, a broader snout, and a shorter tail (Egoscue 1979).

The swift fox inhabits prairies, especially those with grasses of short and medium height. For shelter it depends on burrows, which are either self-excavated or taken over from another mammal. The burrows are usually simple and located on high, well-drained ground. The tunnels may be 350 cm long and lead to a chamber as much as 150 cm below the surface. There are one to seven or more entrances. The swift fox is primarily nocturnal but sometimes suns itself near the den. Its diet consists mostly of lagomorphs and also includes rodents, birds, lizards, and insects (Egoscue 1979; Kilgore 1969).

A radio-tracking study in Nebraska indicated an average nightly movement of about 13 km and found the average home range of 11 individuals to be about 32 sq km, larger than that of any other species of *Vulpes* (Hines and Case 1991). The usual social unit is a mated pair and their young, but occasionally a male will live with two adult females. The mating season in Oklahoma is late December to early January, and most young are born in March or early April. Females are monestrous. Litters consist of three to six young. Their eyes open after 10–15 days, weaning occurs after 6–7 weeks, and they probably remain with the parents until August or early September. A captive lived for 12 years and 9 months (Egoscue 1979; Kilgore 1969).

The swift fox is not as cautious as *V. vulpes* and seems to take poison baits readily. In the mid- and late nineteenth century intensive poisoning was carried out on the Great Plains, mainly to eliminate wolves, coyotes, and other predators, and many swift foxes were accidentally killed. Subsequently much habitat was lost as the prairies were converted to agriculture. The swift fox was also taken for its fur. From 1853 to 1877 in Canada the Hudson's Bay Company reportedly sold more than 100,000 pelts. By the 1920s the northern subspecies, *V. velox hebes*, apparently had disappeared, though occasional reports continued in Canada, and the southern subspecies, *V. velox velox*, survived only in Colorado, New Mexico, western Texas, and possibly western Kansas. For reasons not fully understood the species reappeared in Oklahoma, much of Kansas, Nebraska, and Wyoming in the 1950s and in South Dakota, North Dakota, and Montana in the 1960s and 1970s (Carbyn, Armbruster, and Mamo 1994; Egoscue 1979; Floyd and Stromberg 1981; Kilgore 1969; Moore and Martin 1980; Zumbaugh and Choate 1985). In the last few years it has been recognized that *V. velox* is again declining, mainly because of destruction and fragmentation of native habitat by agricultural activity and perhaps excessive fur trapping; the entire species is estimated to occupy only 10 percent of its original range, and numbers are very low in the northern part of the range (Allardyce 1995; Smeeton 1993). *V. velox* is classified as conservation dependent by the IUCN.

V. velox hebes is on appendix 1 of the CITES. The USDI lists the subspecies as endangered, but this designation officially applies only in Canada. The swift fox populations now on the northern plains of the United States may be descended from animals that moved north from the range of *V. velox velox*. There has been considerable controversy during the last decade regarding both the systematic status of the swift fox subspecies and a reintroduction program in Canada. The latter project, which began in 1983, involves the capture of foxes in Colorado, Wyoming, and South Dakota and the release of them or their offspring in Alberta and Saskatchewan. Stromberg and Boyce (1986) argued that *hebes* probably is not a valid subspecies but that there is significant geographic variation in *V. velox* and that gene flow from the transplanted animals might adversely affect the viability of natural populations on the U.S. side of the border. Herrero, Schroeder, and Scott-Brown (1986) replied that the transplanted stock was taken from the northernmost readily available populations and that more northerly animals in the United States may be recently descended from the same stock. Despite the USDI listing, the original swift fox population of Canada evidently had disappeared by the 1930s. Carbyn, Armbruster, and Mamo (1994) reported that the program had released 569 animals by the end of 1992, that perhaps 13 percent were known to have survived at least 1–2 years, and that reproduction was occurring. In late 1991 an estimated breeding population of 150–225 foxes occupied about 1,200 sq km of southern Alberta and Saskatchewan.

Vulpes macrotis (kit fox)

The account of this species is based in large part on McGrew (1979). Head and body length is 375–500 mm and tail length is 225–323 mm. Average weight is 2.2 kg for males and 1.9 kg for females. The back is generally light grizzled or yellowish gray, the shoulders and sides are buffy to orange, and the underparts are white. The ears, which are proportionally the largest among the North American canids, are set close together.

The kit fox is closely associated with steppe and desert habitat, generally with a covering of shrubs or grasses. Dens usually have multiple entrances, the number varying from 2 to 24. There are groups of dens in favorable areas, and a fox family may move from one to another during the year, leaving most vacant at any given time. *V. macrotis* is nocturnal and may travel several kilometers per night during hunts. Its diet consists largely of rodents, such as kangaroo rats, and lagomorphs.

Optimal habitat in Utah was found to support two adults per 259 ha. Other areas had densities of one fox per 471–1,036 ha. In the San Joaquin Valley, Morrell (1972) found that each fox apparently spent its entire life in an area of 260–520 ha. Home ranges overlap extensively, and there apparently is no definite territory. Usually an adult male and female live together, though not necessarily permanently, and a second female is sometimes present. When a female is nursing young it rarely leaves the den, and the male supplies it with food. Several vocalizations are known, including a bark by mothers to recall the young. Females are monestrous. Mating occurs from December to February, and the young are born in February and March. There are usually four or five offspring weighing about 40 grams each. The young emerge from the den at 1 month and begin to accompany their parents at 3–4 months. The family splits up in the autumn, with the young dispersing beyond the parental home range. According to Jones (1982), one specimen was still living after 20 years in captivity.

The kit fox is not a particularly cautious animal, and its numbers have been greatly reduced in some areas by poisoning, trapping, and shooting. Habitat disruption also has led to declines, especially in California. The subspecies *V. m. macrotis*, of the southwestern corner of the state, had disappeared by 1910. The San Joaquin Valley subspecies, *V. m. mutica*, is classified as endangered by the USDI and as rare by the California Department of Fish and Game (1978). It has declined because of conversion of areas of natural veg-

Kit fox *(Vulpes macrotis)*, photo from San Diego Zoological Society.

etation to irrigated agriculture. Most states classify *V. macrotis* as a fur bearer and allow trapping. O'Farrell (1987) reported that the annual legal kill was 5,400–7,800 and that pelts sold for $12 to $23. In the 1991–92 season 1,519 skins were taken in the United States and sold for an average price of only $4.24 (Linscombe 1994).

Vulpes bengalensis (Bengal fox)

The account of this species is based largely on Roberts (1977). Head and body length is 450–600 mm and tail length is 250–350 mm. Males weigh 2.7–3.2 kg and females weigh less than 1.8 kg. The upper parts are yellowish gray or silvery gray, the underparts are paler, and the backs of the ears and the tip of the tail are dark.

The Bengal fox is generally found in open country with a scattering of trees. It avoids deserts and mountains. It digs its own burrow and hunts mainly at night. A captive could climb low trees and a vertical wire net. The diet is omnivorous and includes small vertebrates, insects, and fruit. *V. bengalensis* is adept at catching frogs and digging lizards out of their burrows. The gestation period is 50–51 days (Hayssen, Van Tienhoven, and Van Tienhoven 1993). The young, usually four to a litter, are born from February to April.

Although once extremely common in India, the Bengal fox is rapidly declining and has disappeared from much of its range. It is not an agricultural pest and its pelt has no commercial value; it is being killed simply for sport (Ginsberg and Macdonald 1990).

Vulpes rueppellii (sand fox)

Head and body length is 400–520 mm and tail length is 250–350 mm (Roberts 1977). Weight is about 1.5–3.0 kg. The coat is very soft and dense. The upper parts are silvery gray, the sides are grayish buff, and the underparts are whitish. *V. rueppellii* is much more lightly built than *V. vulpes;* it has rather short legs and broad ears (Dorst and Dandelot 1969).

The usual habitat is stony or sandy desert. Activity is mainly nocturnal. In a study in Oman, Lindsay and Macdonald (1986) found individuals to spend the day in underground dens and to change their den site on an average of once every 4.7 days. By night these animals hunted over a large home range calculated to average 30.4 sq km by one method and 69.1 sq km by another. The diet in that area consisted mostly of small mammals and also included lizards, insects, and grass. There were three monogamous pairs; each pair shared a den site and a portion of the hunting range, but there also was some overlap in the ranges of different pairs. Two of the pairs had cubs during January–April. Other reported information (Dorst and Dandelot 1969; Roberts 1977) is that a captive female had a litter of three young, and a litter of two young evidently was born in March.

Vulpes pallida (pale fox)

Head and body length is 406–55 mm, tail length is 270–86 mm, and weight is 1.5–3.6 kg. The upper parts are pale sandy fawn variably suffused with blackish, the flanks are paler, and the underparts are buffy white. The tail is long and bushy (Dorst and Dandelot 1969; Rosevear 1974).

The habitat is savannah. Burrows are large, with tunnels extending 10–15 meters and opening into chambers lined with dry vegetation. Activity is mainly nocturnal. The diet includes rodents, small reptiles, birds, eggs, and vegetable matter. The pale fox is gregarious (Dorst and Dandelot 1969; Coetzee *in* Meester and Setzer 1977). Three captive adults, a female and two males, seemed to get along amicably. The female gave birth to a litter of four young in June 1965 (Bueler 1973).

Vulpes chama (Cape fox)

Head and body length is about 560 mm, tail length is about 330 mm, and weight is about 4 kg. The upper parts are silvery gray and the underparts are pale buff. The very bushy tail has a black tip. The ears are pointed. The muzzle is short but pointed (Dorst and Dandelot 1969).

The Cape fox inhabits dry country, mainly open plains

Sand fox *(Vulpes rueppelli)*, photo from Antwerp Zoo.

Cape fox *(Vulpes chama)*, photo from San Diego Zoological Society.

and karroo. It is nocturnal, hiding by day under rocks or in burrows in sandy soil. The diet consists mainly of small vertebrates and insects. This species lives alone or in pairs. Average population density is 0.3/sq km in the Orange Free State. Home range is 1.0–4.6 sq km and there may be overlap where prey density is high. The call is a yell followed by several yaps. The breeding season is September–October, gestation lasts 51–52 days, and litters contain three to five young. Human persecution has been responsible for numerical and distributional declines, though studies have shown that *V. chama* is not a harmful predator (Bekoff 1975; Bothma 1966; Dorst and Dandelot 1969; Ginsberg and Macdonald 1990).

CARNIVORA; CANIDAE; **Genus FENNECUS**
Desmarest, 1804

Fennec Fox

The single species, *F. zerda,* occurs in the desert zone from southern Morocco and Niger to Egypt and Sudan (Coetzee *in* Meester and Setzer 1977). There also are records from Sinai, southern Iraq, Kuwait, and the southeastern Arabian Peninsula (Gasperetti, Harrison, and Büttiker 1985). *Fennecus* was included in the genus *Vulpes* by Clutton-Brock, Corbet, and Hills (1976), Corbet (1978), Geffen, Mercure, et al. (1992), and Wozencraft (*in* Wilson and Reeder 1993) and in the subgenus *Vulpes* of the genus *Canis* by Van Gelder (1978).

With the possible exception of *Vulpes cana* this is the smallest canid, but it has the proportionally largest ears in the family. Head and body length is 357–407 mm, tail length is 178–305 mm, and weight is about 1.0–1.5 kg. The ears are 100–150 mm long. The coloration of the upper parts, the palest of any fox's, is reddish cream, light fawn, or almost white. The underparts are white, and the tip of the tail is black. The coat is thick, soft, and long. The tail is heavily furred. The bullae are exceedingly large, and the dentition is weak (Clutton-Brock, Corbet, and Hills 1976). The feet have hairy soles, enabling the animal to run in loose sand (Bekoff 1975).

Fennecus occurs in arid regions and usually lives in burrows several meters long in the sand. It digs so rapidly that it has gained the reputation of being able to sink into the ground. It is nocturnal and quite agile; a captive could spring 60–70 cm upward from a standing position and could jump about 120 cm horizontally. Some food apparently is obtained by digging, as evidenced by the pronounced scratching or raking habit of captives. The diet includes plant material, small rodents, birds and their eggs, lizards, and insects, such as the noxious migratory locusts. Although an abundance of fennec tracks around some water holes indicates that this fox drinks freely when the opportunity arises, travelers also report tracks in the desert far from oases. Laboratory studies suggest that *Fennecus* can survive without free water for an indefinite period (Banholzer 1976).

The fennec lives in groups of up to 10 individuals. Males mark their territory with urine and become aggressive during the breeding season. One captive male became dominant over a group at the age of 4 years and then killed its 8-year-old father. Females are aggressive and defend the nest site when they have newborn offspring. Males remain with their mate after the young are born and defend them but do not enter the maternal den. Mating occurs in January and February in captivity, and the young are born in late winter and early spring. Females normally give birth once a year, but if the first litter is lost, another may be produced 2.5–3 months later. The gestation period is about 50–52 days, and there are 2–5 young per litter. The young are weaned at 61–70 days and become sexually mature at 11 months (Bekoff 1975; Koenig 1970). One captive lived 14 years and 7 months (Jones 1982).

Although it does no harm to human interests, the fennec is intensively hunted by the native people of the Sahara. It has disappeared or become rare in many parts of northern Africa (Zimen *in* Grzimek 1990). It is on appendix 2 of the CITES.

Fennec foxes *(Fennecus zerda)*, photo from New York Zoological Society.

CARNIVORA; CANIDAE; **Genus ALOPEX**
Kaup, 1829

Arctic Fox

The single species, *A. lagopus*, occurs on the tundra and adjacent lands and ice-covered waters of northern Eurasia, North America, Greenland, and Iceland (Banfield 1974; Chesemore 1975; Corbet 1978; E. R. Hall 1981). *Alopex* was included within the genus *Vulpes* by Geffen, Mercure, et al. (1992) and Youngman (1975) and was considered a subgenus of *Canis* by Van Gelder (1978) but was treated as a full genus by Clutton-Brock, Corbet, and Hills (1976) and all other authorities cited in this account. Although *Alopex* once was thought to have some distant affinity to *Lycalopex*, the work of Geffen, Mercure, et al. (1992), Wayne (1993), Wayne and O'Brien (1987), and other authorities suggests that *Alopex* is closely related to *Vulpes*, far more so than is *Urocyon*. In any event, *Lycalopex* is really not like *Alopex*.

Head and body length is 458–675 mm, tail length is 255–425 mm, shoulder height is about 280 mm, and weight is 1.4–9.0 kg. The dense, woolly coat gives *Alopex* a heavy appearance. There are two color phases. Individuals of the "white" phase are generally white in winter and brown in summer but may remain fairly dark throughout the year in areas of less severe climate. Individuals of the "blue" phase are pale bluish gray in winter and dark bluish gray in summer. The blue phase constitutes less than 1 percent of the arctic fox population of most of mainland Canada and less than 5 percent of that of Baffin Island but makes up 50 percent of the population of Greenland. In Iceland most foxes are blue, perhaps because of increased camouflage value in coastal areas (Hersteinsson 1989). The winter pelage develops in October and is shed in April. In addition to its coloration, *Alopex* differs from *Vulpes* in having short, rounded ears and long hairs on the soles of its feet.

Alopex is found primarily in arctic and alpine tundra, usually in coastal areas. It generally makes its den in a low mound, 1–4 meters high, on the open tundra. Dens usually have 4–12 entrances and a network of tunnels covering about 30 sq meters. Some dens may be used for centuries, by many generations of foxes, and eventually become very large, with up to 100 entrances. The organic matter that accumulates in and around a den stimulates a much more extensive growth of vegetation than is found on most of the tundra. *Alopex* sometimes dens in a pile of rocks at the base of a cliff. It may use its den throughout the year but often seems not to have a fixed home site except when rearing young. It is active at any time of the day and throughout the year. It moves easily over snow and ice and swims readily (Banfield 1974; Chesemore 1975; Macpherson 1969; Stroganov 1969). During blizzards it may shelter in burrows dug in the snow. Several individuals were noted in winter on the Greenland icecap, more than 450 km from the nearest ice-free land and with the temperature below −50° C. Captives have survived experimental temperatures as low as −80° C.

According to Prestrud (1991), the adaptations of *Alopex* to low temperatures and winter food scarcity include: the best insulative fur of any mammal, with a seasonal increase in fur depth of nearly 200 percent; short muzzle, ears, and legs and a short, rounded body; increased blood flow to a capillary rete in the skin of the pads, which keeps the feet from freezing, and probably a countercurrent vascular exchange mechanism to avoid heat loss when warm blood flows into the legs; storage of substantial body fat reserves, with individuals commonly having a fat content of 20–40 percent during winter; and capability to reduce the basal metabolic rate by a remarkable 40–50 percent during periods of limited food availability. At such times the fox possibly seeks shelter for an extended period but remains relatively alert and able to immediately resume foraging when conditions improve. Hersteinsson and Macdonald (1992) observed that the fur of the arctic fox provides 50 percent better insulation than that of the red fox, but they did not think the differing distributions of the two species resulted directly from relative adaptation to extreme cold. Rather, the northern limit of the red fox seems to be determined by resource availability, and the southern limit of the arctic fox is a factor of interspecific competition with the red fox.

Arctic foxes *(Alopex lagopus):* A. White and blue phases; B. Blue phase; C. Summer coat; photos by A. Pedersen. D. Winter coat, photo by Ernest P. Walker.

The arctic fox makes the most extensive movements of any terrestrial mammal other than *Homo sapiens.* Over much of its range, including Alaska, there is a basic seasonal pattern, probably associated with food availability (Chesemore 1975). In the autumn and early winter the animals shift toward the shore and out onto the pack ice. They are capable of remarkably long travels across the sea ice, having been sighted 640 km north of the coast of Alaska. One individual reportedly reached a latitude of 88° N, at a point 800 km from the nearest land in Russia. A fox tagged on 8 August 1974 on Banks Island was trapped on 15 April 1975 on the northeastern mainland of the Northwest Territories, having covered a straight-line distance of 1,530 km (Wrigley and Hatch 1976). And a fox tagged in northeastern Alaska reached a point on Banks Island, 945 km away (Eberhardt and Hanson 1978). Some foxes have been carried by ice floes as far as Cape Breton and Anticosti islands in the Gulf of St. Lawrence (Banfield 1974).

In some areas, though apparently not in Alaska, there are inland migrations to the forest zone during the winter, in addition to or instead of a shift onto the sea ice. These movements sometimes involve large numbers of foxes and seem especially extensive following a crash in populations of lemmings, a major food source (Banfield 1974; Chesemore 1975; Pulliainen 1965). *Alopex* has occasionally traveled overland to south-central Ontario and the St. Lawrence River. The deepest known penetration in North America was made by an individual taken in December

1974 in southern Manitoba, nearly 1,000 km south of the tundra. In Russia, however, inland movements have extended as far as 2,000 km (Wrigley and Hatch 1976).

Alopex takes any available animal food, alive or dead, and frequently stores it for later use. It may follow polar bears on the pack ice and wolves on land in order to obtain carrion (Banfield 1974). Its winter diet includes the remains of marine mammals, invertebrates, sea birds, and fish (Chesemore 1975). It also is known to be an important predator of the ringed seal *(Phoca hispida)*, digging through the snow to reach the pups in their subnivean lairs (Smith 1976).

In winter for those populations that move inland and in summer throughout most of the range of *Alopex* the diet consists mainly of lemmings. Other small mammals, ground-nesting birds, and stranded marine mammals are also utilized, but arctic fox populations seem to be associated with those of lemmings in a three- to five-year cycle (Chesemore 1975; Wrigley and Hatch 1976). Following a lemming crash, population density has been estimated at only 0.086 foxes per sq km (Banfield 1974).

Although a number of dens may be found in a favorable area, at any one time most are not utilized. Occupied dens are at least 1.6 km apart and are usually distributed such that there is only one family per every 32–70 sq km (Macpherson 1969). In a nonmigratory population on the coast of Iceland home ranges varied from 8.6 to 18.5 sq km and evidently represented defended territories. The foxes there lived in social groups comprising one adult male, two vixens, and young of the year. In such situations one of the vixens apparently is a nonbreeding animal born the previous year that stays on to help care for the next litter before dispersing (Hersteinsson and Macdonald 1982). *Alopex* is monogamous and may mate for life (Chesemore 1975). A number of adults sometimes gather temporarily around a food source, such as a stranded whale, but they then may fight one another. Vocalizations include barks, screams, and hisses (Banfield 1974).

Mating occurs from February to May, and births take place from April to July. Females are monestrous and have an estrus of 12–14 days and an average gestation period of 52 (49–57) days. The number of young per litter varies depending on environmental conditions but ranges from 2 to as many as 25. The usual number seems to be about 6–12. The young weigh an average of 57 grams at birth. They emerge from the den and are weaned at around 2–4 weeks. Subsequently they are brought food by both parents for a brief period, but by autumn they have dispersed. They are capable of breeding at 10 months. Most young do not survive their first 6 months of life, and few animals live more than several years in the wild (Banfield 1974; Chesemore 1975; Macpherson 1969; Stroganov 1969). One individual, however, lived in captivity for 16 years and 2 months (Marvin L. Jones, Zoological Society of San Diego, pers. comm., 1995).

Raising blue-phase foxes has been an important industry; the undressed skins have sometimes sold for up to $300 each. Some operations have pens, where selective breeding is practiced, but most were on small islands, where the animals ran at liberty. The history of the industry in Alaska has been covered in detail by E. Bailey (1993). Foxes originally did not occur on most islands of the state, and despite some assertions to the contrary, *Alopex* evidently is not indigenous to any part of the Aleutian chain. The Russians began releasing arctic foxes in the Aleutians in 1750, and the process continued after the American purchase in 1867. Introductions also were made on islands off the southern coast of Alaska and in the Alexander Archipelago, but they were generally unsuccessful in the latter region. There are known records of introduction on 455 Alaskan islands, with formal leasing for the purpose of fox farming starting in 1882. The industry grew rapidly as fur prices rose in the early twentieth century, and by 1925 there were 391 farms with over 36,000 blue foxes valued at about $6 million. The Great Depression effectively killed the industry, with average prices falling from $185 in 1919 to $108 in 1929 and only $26 in 1945. Although a few people still raise foxes in pens, no farms remain on the islands. Feral foxes persist, however, on 46 islands, mostly in the Aleutians and off the Alaska Peninsula. Even in the early nineteenth century these animals were known to be adversely affecting the native avifauna of the islands, and as the foxes increased in numbers many populations of waterfowl and seabirds were nearly or entirely exterminated. Some of these populations recovered after the foxes died out or were eliminated by trappers.

White-phase foxes also have desirable furs and have been subject to intensive trapping. The skins may be left the natural white or dyed one of many colors, especially "platinum" or "blue" imitation. The arctic fox has long been important to the economy of the native people living within its range. Fluctuations in fox numbers and fur prices have frequently combined to cause hardship for native trappers. Trade in white fox skins developed in northern Alaska during the nineteenth century, in conjunction with the whaling industry. In the 1920s fox trapping was the most important source of income in the area, the price per pelt averaging $50. The Depression destroyed the market, and by 1931 skins sold for $5 or less (Chesemore 1972). Recently prices began to rise again. During the 1976–77 season 4,261 skins were marketed in Alaska at an average price per pelt of $36; the figures for Canada were 36,482 and $54.20 (Deems and Pursley 1978). The kill fell to 720 in Alaska and 16,405 in Canada during the 1983–84 season, when the average price in Canada was only $6 (Novak, Obbard, et al. 1987).

The arctic fox still is generally common in the wild, there being a total annual harvest of 100,000–150,000 pelts throughout the North American and Eurasian parts of its range (Prestrud 1991). However, there has been a general northward contraction in distribution during the twentieth century, perhaps in association with long-term climatic fluctuation (Ginsberg and Macdonald 1990; Hersteinsson and Macdonald 1992). In northern Alaska there also is concern about the effects of petroleum exploitation and other human development on the limited number of long-term favorable den sites (Garrott, Eberhardt, and Hanson 1983). In Iceland the arctic fox is persecuted because of its reputed depredations on sheep and lambs. Legislation there has promoted its destruction since at least as early as A.D. 1295. State-subsidized hunting has been so efficient that large parts of the island are now devoid of the species (Hersteinsson and Macdonald 1982). Overall numbers in Iceland fell to about 1,000 in the early 1970s but subsequently rebounded. A drastic decline occurred during the early twentieth century in Scandinavia, probably because of overhunting, and only a few hundred individuals now are present throughout Norway, Sweden, and Finland. Although the arctic fox has been legally protected in those three countries for decades, it has failed to recover, perhaps because of competition and predation from red fox populations that were able to occupy its range (Hersteinsson and Macdonald 1992; Hersteinsson et al. 1989; Frafjord, Becker, and Angerbjörn 1989).

Gray fox *(Urocyon cinereoargenteus)*, photo from San Diego Zoological Society.

CARNIVORA; CANIDAE; **Genus UROCYON**
Baird, 1858

Gray Foxes

There are two species (Cabrera 1957; E. R. Hall 1981):

U. cinereoargenteus, Oregon and southeastern Canada to western Venezuela;
U. littoralis, San Miguel, Santa Rosa, Santa Cruz, Santa Catalina, San Nicolas, and San Clemente islands off southwestern California.

Urocyon was included in the genus *Vulpes* by Clutton-Brock, Corbet, and Hills (1976) and in the subgenus *Vulpes* of the genus *Canis* by Van Gelder (1978). All other authorities cited in this account treated *Urocyon* as a full genus. *U. littoralis* often is considered to be conspecific with *U. cinereoargenteus*, but a series of recent investigations involving morphology, genetics, biochemical analysis, paleontology, and biogeography indicate that it is a distinct species that originated in the later Pleistocene (Collins 1991*a*, 1991*b*, 1992; Moore and Collins 1995; Wayne, George, et al. 1991). It also is possible that the population now assigned to *U. cinereoargenteus* on Tiburon Island in the Gulf of California is a separate species (Collins 1992).

In *U. cinereoargenteus* head and body length is 483–685 mm and tail length is 275–445 mm. In *U. littoralis* head and body length is 480–500 mm and tail length is 110–290 mm. Usual weight in the genus is 1.8–7.0 kg. The face, upper part of the head, back, sides, and most of the tail are gray. The throat, insides of the legs, and underparts are white. The sides of the neck, lower flanks, and ventral part of the tail are rusty. The hairs along the middle of the back and top of the tail are heavily tipped with black, which gives the effect of a black mane. Black lines also occur on the legs and face of most individuals. A concealed mane of stiff hairs occurs on top of the tail. The pelage is coarse. The skull of *Urocyon* is distinguished from that of *Vulpes* in having a deeper depression above the postorbital process, much more pronounced and more widely separated ridges extending from the postorbital processes to the posterior edges of the parietals, and a conspicuous notch toward the rear of the lower edge of the mandible (Lowery 1974).

Gray foxes frequent wooded and brushy country, often in rocky or broken terrain, and are possibly most common in the arid regions of the southwestern United States and Mexico. The preferred habitat in Louisiana is mixed pine-oak woodland bordering pastures and fields with patches of weeds (Lowery 1974). Several sheltered resting sites may be used on different days. The main den is in a pile of brush or rocks, a crevice, a hollow tree, or a burrow that is either self-excavated or taken over from another animal. Dens in hollow trees have been found up to 9.1 meters above the ground. Dens used for giving birth may be lined with vegetation (Trapp and Hallberg 1975).

Urocyon is sometimes called a "tree fox" because it frequently climbs trees, a rather unusual habit for a canid. It lacks the endurance of the red fox, and when pursued it often will seek refuge in a tree. It also climbs without provocation, shinnying up the trunk and then leaping from branch to branch. Most activity is nocturnal and crepuscular. Nightly movements cover about 200–700 meters (Trapp and Hallberg 1975). *U. littoralis* tends to be diurnal (Moore and Collins 1995). Juvenile gray foxes have dispersed as far as 84 km (Fritzell and Haroldson 1982). The diet includes many kinds of small vertebrates as well as insects and vegetable matter. *Urocyon* seems to take plant food more than do other foxes, and its diet may consist mostly of fruits and grains at certain seasons and places.

Reported population densities are about 0.4–10.0/sq km (Trapp and Hallberg 1975). Reported home range size varies from 0.13 to 7.7 sq km. Four females in central California followed by radio tracking for various periods from January to July used ranges of 0.3–1.85 sq km (Fuller 1978). Range size increases in autumn and winter. Apparently, each family group uses a separate area and the normal social unit is an adult pair and their young, though there is some conflicting evidence on these points (Fritzell and Haroldson 1982). There is one litter per year. Mating occurs from late December to March in the southeastern United States and from mid-January to late May in New York. The gestation period, never precisely determined, has been variously estimated at 50–63 days. Litters usually contain about 4 young but may have from 1 to 10. The young weigh around 100 grams at birth, are blackish in color, and open their eyes after 9–12 days. They can climb vertical tree trunks after 1 month and begin to take solid food in 6 weeks. They forage independently by late summer or early autumn but apparently remain in the parental home range until January or February. Most females breed in their first year of life (Trapp and Hallberg 1975). A captive lived for 13 years and 8 months (Jones 1982).

If captured when small, gray foxes tame readily, are as affectionate and playful as domestic dogs, and make more satisfactory pets than do red foxes. *U. littoralis,* in particular, is said to be docile and to readily approach humans (Moore and Collins 1995). However, deliberate removal from the wild is often illegal and may be dangerous to both the animals and the persons involved. The attractive skins of *Urocyon* are used commercially but are not classed as fine furs. For much of the twentieth century the price per pelt averaged around $0.50, but it began to rise in the 1960s (Lowery 1974). For the 1970–71 season the reported harvest in the United States was 26,109 skins with an average value of $3.50. By the 1976–77 season the reported take had grown to 225,277 pelts, and the average price to $34 (Deems and Pursley 1978). In the 1991–92 season 90,604 skins were taken in the United States and sold at an average price of $9.09 (Linscombe 1994).

The island fox, *U. littoralis,* is classified as conservation dependent by the IUCN and as threatened by the California Department of Fish and Game (Crooks 1994) and is fully protected by California state law. Each of the six Channel Islands has its own recognized subspecies, whose validity recent studies have upheld. Effective population size ranges from 150 individuals on the smallest island, San Miguel, to approximately 1,000 on the largest island, Santa Cruz. It is thought that *Urocyon* initially reached one of the three northern islands (Santa Cruz, Santa Rosa, and San Miguel) prior to 24,000 years ago, perhaps by accidental rafting, and subsequently spread to all three islands when they were joined at certain times during the late Pleistocene. A fossil fox collected on Santa Rosa shows that *U. littoralis* had evolved its diminutive size by at least 16,000 years ago. Native Americans arrived on the Channel Islands about 10,000 years ago, and apparently it was they who transported foxes from the northern to the southern islands (Santa Catalina, San Clemente, and San Nicolas). The first occurrence in the south dates from about 3,400–4,300 years ago on San Clemente. Numerous remains of foxes have been found at archaeological sites and indicate that the animals were kept as pets and had an important role in religious and ceremonial practices. They probably were brought to the various islands in the course of trade and subsequently became feral. Their very small population sizes and lack of genetic variability now make them especially vulnerable to environmental disruption and introduction of disease (Collins 1991*a*, 1991*b*, 1992; Garcelon, Wayne, and Gonzales 1992; George and Wayne 1991; Wayne, George, et al. 1991).

Hoary fox *(Lycalopex vetulus)*, photo by Luiz Claudio Marigo.

CARNIVORA; CANIDAE; **Genus LYCALOPEX**
Burmeister, 1854

Hoary Fox

Langguth (1975) considered *Lycalopex* to be a full genus with a single species, *L. vetulus,* occurring in the states of Mato Grosso, Goias, Minas Gerais, and Sao Paulo in south-central Brazil. *Lycalopex* was treated as a subgenus of *Dusicyon* by Cabrera (1957) and as a subgenus of *Canis* by Van Gelder (1978). *L. vetulus* was included within *Dusicyon* by Clutton-Brock, Corbet, and Hills (1976), who noted, however, that it was the most foxlike member of that genus. Berta (1987) included *Lycalopex* within *Pseudalopex* and suggested a relationship to *P. sechurae.* Wozencraft (*in* Wilson and Reeder 1993) also included *Lycalopex* in *Pseudalopex.*

Head and body length is about 585–640 mm, tail length is 280–320 mm, and weight is about 2.7–4.0 kg. The coat is short. The upper parts are a mixture of yellow and black, giving an overall gray tone. The ears and the outsides of the legs are reddish or tawny. The tail has a black tip and a marked dark stripe along the dorsal line. The underparts are cream to fawn. Compared with *Dusicyon, Lycalopex* has a short muzzle, a small skull and teeth, reduced carnassials, and broad molars (Bueler 1973; Clutton-Brock, Corbet, and Hills 1976; Röhrs *in* Grzimek 1990).

The habitat is grassy savannah on smooth uplands, or savannahs with scattered trees. A deserted armadillo burrow may be used for shelter and for the natal nest. The diet consists of small rodents, birds, and insects, especially grasshoppers. The dentition suggests substantial dependence on insects, but *Lycalopex* is persecuted by people because of presumed predation on domestic fowl (Langguth 1975). It is usually timid but courageously defends itself and its young. Births occur in the austral spring (September), and there are usually two to four young per litter (Bueler 1973; Röhrs *in* Grzimek 1990).

CARNIVORA; CANIDAE; **Genus PSEUDALOPEX**
Burmeister, 1856

South American Foxes

There are four species (Berta 1987; Clutton-Brock, Corbet, and Hills 1976; Duckworth 1992; Ginsberg and Macdonald

South American fox *(Pseudalopex culpaeus)*, photo by Ernest P. Walker.

1990; Langguth 1975; Pacheco et al. 1995; Redford and Eisenberg 1992; Van Gelder 1978):

P. gymnocercus, humid grasslands of southern Brazil, Paraguay, northern Argentina, and Uruguay;
P. culpaeus, Andes and adjacent highlands from Ecuador and possibly Colombia to Patagonia and Tierra del Fuego;
P. griseus, plains and low mountains of Chile and Argentina, including Patagonia, and possibly up the Pacific coast to Peru and southwestern Ecuador;
P. sechurae, arid coastal zone of southwestern Ecuador and northwestern Peru.

Pseudalopex was considered a subgenus of *Canis* by Langguth (1975) and Van Gelder (1978). The four species listed above were included in the genus *Dusicyon* along with the type species of that genus, *D. australis,* by Clutton-Brock, Corbet, and Hills (1976). Berta (1987) recognized *Pseudalopex* as a full genus comprising the above four species and considered *D. australis* to be the only living species of *Dusicyon.* Information provided by these various authorities suggests that *D. australis* is at least as different from the species of *Pseudalopex* as it is from *Canis* and that *Pseudalopex* is at least as different from *Ca-nis* as it is from *Vulpes.* Therefore, since *Dusicyon* (for the species *D. australis*) and *Vulpes* are here being maintained as separate genera, it is also advisable to give generic rank to *Pseudalopex.* Medel et al. (1990) and Miller et al. (1983) treated *P. fulvipes,* known from Chiloe Island and a mainland area of Chile about 600 km to the north, as a species distinct from *P. griseus.*

Head and body length is 530–1,200 mm, tail length is 250–500 mm, and weight is 4–13 kg. *P. culpaeus* is the largest species and *P. sechurae* is the smallest. The coat is usually heavy, with a dense underfur and long guard hairs. The upper parts are generally gray agouti with some ochraceous or tawny coloring (Clutton-Brock, Corbet, and Hills 1976). The head, ears, and neck are often reddish. The underparts are usually pale. The tail is long, bushy, and black-tipped. There is some resemblance to a small coyote *(Canis latrans).* The dentition, however, is more foxlike than doglike; the molars are well developed, and the carnassials are relatively short (Clutton-Brock, Corbet, and Hills 1976).

Habitats include sandy deserts for *P. sechurae;* low, open grasslands and forest edge for *P. griseus;* pampas, hills, deserts, and open forests for *P. gymnocercus;* and dry rough country and mountainous areas up to 4,500 meters in elevation for *P. culpaeus* (Crespo 1975; Langguth 1975; Röhrs *in* Grzimek 1990). Dens are usually among rocks, under bases of trees and low shrubs, or in burrows made by other animals, such as viscachas and armadillos. Most activity is nocturnal, but some individuals are occasionally active during the day. *P. gymnocercus* sometimes collects and stores objects, such as strips of leather and cloth. This species may freeze and remain motionless upon the appearance of a human being; in one case it reportedly did not move even when approached and struck with a whip handle. The voice of *Pseudalopex* has been described as a howl or a series of barks and yaps. It is heard mainly at night, especially during the breeding season.

The omnivorous diet of *Pseudalopex* includes rodents, lagomorphs, birds, lizards, frogs, insects, fruit, and sugar cane. Studies of stomach contents indicate that *P. gymnocercus* takes an equal amount of plant and animal food. *P. culpaeus* seems to be more carnivorous than other species and reportedly sometimes preys heavily on introduced sheep and European hares. In western Argentina during the spring, part of the population of *P. culpaeus* shifts 15–20 km into the higher mountains in response to the seasonal movements of the sheep and hares. Its normal home range in that area is 4 km in diameter (Crespo 1975). In areas where *P. culpaeus* is sympatric with the smaller *P. griseus* there is no overlap of local territories, and the latter species is able to survive in poorer habitat by supplementing its diet with beetles and plant material, especially from spring to autumn (Johnson 1992; Johnson and Franklin 1994). *P. griseus* has been found to occur at an average density of 1/43 ha. in southern Chile (Durran, Cattan, and Yáñez 1985). Observations of *P. gymnocercus* in the Paraguayan Chaco usually are of a single individual and indicate a density of about 1/100 ha. (Brooks 1992). In a study of *P. sechurae* in northwestern Peru, Asa and Wallace (1990) found mostly nocturnal activity, a diet consisting largely of plant matter and invertebrates, and two home ranges, one occupied by a single male, the other apparently by an adult female and two juveniles.

Observations of *P. griseus* in Patagonia (Johnson 1992) indicate that a monogamous pair maintains an exclusive year-round territory, though occasionally a second female is present in the area and assists in rearing the young; births, usually of four to five pups, occur by mid-October. According to Crespo (1975), *P. culpaeus* and *P. gymnocercus* mate from August to October and give birth from October to December (the austral spring). Females are monestrous. The gestation period is 55–60 days. Embryo counts range from one to eight, with averages of about four in *P. gymnocercus* and five in *P. culpaeus.* The male helps to provide food to the family. At 2–3 months the young begin to hunt with the parents. Sexual maturity apparently is attained by 1 year in *P. culpaeus* and *P. griseus* (Ginsberg and Macdonald 1990). Few individuals live more than several years in the wild, but a captive *P. gymnocercus* lived 13 years and 8 months (Jones 1982).

These canids are killed by people because they are alleged to prey on domestic fowl and sheep and because their fur is desirable. Populations of *Pseudalopex* have thus declined in some areas, such as in Buenos Aires Province, Argentina. *P. griseus* was introduced to Tierra del Fuego in 1951 to control the previously introduced European rabbit *(Oryctolagus)* and also has been released on several small islands in the Falklands (Ginsberg and Macdonald 1990). In southern Chile *P. griseus* is legally protected but is threatened by persecution for alleged depredations and environmental disturbance (Durran, Cattan, and Yáñez 1985; Miller et al. 1983). The subspecies (or possibly species) *P. griseus fulvipes* of Chiloe Island is probably the rarest canid in South America (Medel et al. 1990). Crespo (1975) noted, however, that *P. culpaeus* occurred in relatively low numbers in Neuquen Province, western Argentina, until the early twentieth century and then greatly increased in response to the introduction of sheep and European hares *(Lepus europaeus). P. griseus, P. gymnocercus,* and *P. culpaeus* are on appendix 2 of the CITES.

Records of the CITES indicate an annual export during the 1980s of about 100,000 skins of *P. griseus* and several thousand of *P. culpaeus,* virtually all from Argentina (Broad, Luxmoore, and Jenkins 1988). Such uncontrolled hunting is thought to have resulted in a decline of about 80 percent in populations of *P. griseus* since 1970 (Roig 1991). Numbers of *P. culpaeus* seem to have been less seriously affected overall but are declining on Tierra del Fuego (Novaro 1993). *P. gymnocercus* also is heavily hunted and trapped for its fur; up to 30,000 skins may be exported annually from Paraguay even though the species is officially protected there (Ginsberg and Macdonald 1990).

CARNIVORA; CANIDAE; **Genus DUSICYON**
Hamilton-Smith, 1839

Falkland Island Wolf

Berta (1987) considered this genus to include a single modern species, *D. australis,* which formerly occurred on West and East Falkland Islands, off the southeastern coast of Argentina. She added that another species, *D. avus,* occurred in the late Pleistocene of southern Chile and survived into the Recent of southern Argentina. Additional species usually have been assigned to *Dusicyon.* Cabrera (1957) recognized two subgenera: *Dusicyon,* with the species here placed in the genus *Pseudalopex;* and *Lycalopex,* which is here treated as a full genus. Clutton-Brock, Corbet, and Hills (1976) included within *Dusicyon* all of those species that are here placed in the genera *Lycalopex, Pseudalopex, Dusicyon, Cerdocyon,* and *Atelocynus.* Those authorities indicated, however, that *D. australis* is at least as close, systematically, to some species of *Canis* as it is to the species here assigned to *Pseudalopex.* They even discussed the possibility that *D. australis* is a form of *Canis familiaris.* Both Langguth (1975) and Van Gelder (1978) considered *Dusicyon* to be a subgenus of *Canis* and to include only *D. australis.*

Only 11 specimens of *D. australis* are known, and not all include skins. In one specimen head and body length was 970 mm and tail length was 285 mm. The upper parts are brown with some rufous and a speckling of white, and the underparts are pale brown. The coat is soft and thick. The tail is short, bushy, and tipped with white. The face and ears are short and the muzzle is broad. The skull is large and has inflated frontal sinuses, more like the situation of *Canis* than that of *Pseudalopex* (Allen 1942; Clutton-Brock, Corbet, and Hills 1976).

Dusicyon was the only terrestrial mammal found on the Falkland Islands by the early explorers. Its natural diet consisted mainly of birds, especially geese and penguins, and also included pinnipeds. Its presence on the islands, about 400 km from the mainland, is something of a mystery.

Clutton-Brock, Corbet, and Hills (1976) thought it most likely that *Dusicyon* had been taken to the Falklands as a domestic animal by prehistoric Indians. Possible descent from either *Pseudalopex* or *Canis* was suggested. Berta (1987), however, pointed out that a lowered sea level during the Pleistocene would have facilitated natural movement and that the distinguishing characters of *Dusicyon* probably result from subsequent isolation rather than from domestication.

Dusicyon demonstrated remarkable tameness toward people. Individuals waded out to meet landing parties. Later they came into the camps in groups, carried away articles, pulled meat from under the heads of sleeping men, and stood about while their fellow animals were being killed. The dogs were often killed by a man holding a piece of meat as bait in one hand and stabbing the dog with a knife in the other hand.

Although *Dusicyon* was discovered in 1690, it was still common, and still behaving in a very tame manner, when Darwin visited the Falklands in 1833. In 1839, however, large numbers were killed by fur traders from the United States. In the 1860s Scottish settlers began raising sheep on the islands. *Dusicyon* preyed on the sheep and was therefore intensively poisoned. The genus was very rare by 1870, and the last individual is said to have been killed in 1876 (Allen 1942). The IUCN classifies *D. australis* as extinct.

Crab-eating fox *(Cerdocyon thous)*, photo from San Diego Zoological Garden.

CARNIVORA; CANIDAE; **Genus CERDOCYON**
Hamilton-Smith, 1839

Crab-eating Fox

The single species, *C. thous,* has been recorded from Colombia, Venezuela, Guyana, Surinam, eastern Peru, Bolivia, Paraguay, Uruguay, northern Argentina, and most of Brazil outside of the lowlands of the Amazon Basin (Grimwood 1969; Husson 1978; Langguth 1975). *Cerdocyon* was recognized as a distinct genus by Berta (1982, 1987), Cabrera (1957), and Langguth (1975), as a part of *Dusicyon* by Clutton-Brock, Corbet, and Hills (1976), and as a subgenus of *Canis* by Van Gelder (1978).

Head and body length is 600–700 mm, tail length is about 300 mm, and weight is 5–8 kg. The coloration is variable, but the upper parts are usually grizzled brown to gray, often with a yellowish tint, and the underparts are brownish white. The short ears are ochraceous or rufous. The tail is fairly long, bushy, and either totally dark or black-tipped (Clutton-Brock, Corbet, and Hills 1976). The relatively short and robust legs may be tawny (Brady 1979).

Except as noted, the information for the remainder of this account was taken from Brady (1978, 1979). *Cerdocyon* inhabits woodlands and savannahs and is mainly nocturnal. On the llanos of Venezuela during the wet season it uses high ground and shelters by day under brush. During the dry season it occupies lowlands and spends the day in clumps of matted grass. These grass shelters have several entrances, are used repeatedly, and may serve as natal dens. *Cerdocyon* forages from about 1800 hours to 2400 hours. It stalks and pounces on small vertebrates and apparently listens for crabs in tussocks of grass. In the dry season the percentage composition of its diet is: vertebrates, 48; crabs, 31; insects, 16; carrion, 3; and fruit, 2. In the wet season the percentage breakdown is: insects, 54; vertebrates, 20; fruit, 18; and carrion, 7.

Population densities of 4/sq km and group territory sizes of 5–10 sq km have been reported (Ginsberg and Macdonald 1990). However, three pairs studied by Brady occupied home ranges of approximately 54, 60, and 96 ha. The ranges overlapped to some extent but were regularly marked with urine. Tolerance of neighbors was shown in the wet season, but aggression increased in the dry season. Mated pairs form a lasting bond and commonly travel together but do not usually hunt cooperatively. There is a variety of vocalizations, including a siren howl for long-distance communication between separated family members (Brady 1981). Breeding may take place throughout the year, but births peak in January and February on the llanos of Venezuela. Captive females produced two litters annually at intervals of about 8 months. The gestation period is 52–59 days and litter size is three to six young. The young weigh 120–60 grams at birth, open their eyes at 14 days, begin to take some solid food at 30 days, and are completely weaned by about 90 days. Both parents guard and bring food to the young. Independence comes at 5–6 months, and sexual maturity at about 9 months. According to Smielowski (1985), one specimen lived in captivity for 11 years and 6 months.

Cerdocyon is now on appendix 2 of the CITES. Ginsberg

and Macdonald (1990), however, noted that it is widespread and common. It is hunted, but its pelt has little value.

CARNIVORA; CANIDAE; **Genus NYCTEREUTES**
Temminck, 1839

Raccoon Dog

The single species, *N. procyonoides,* originally occurred in the woodland zone from southeastern Siberia to northern Viet Nam as well as on all the main islands of Japan (Corbet 1978). Although their present distributions may suggest otherwise, *Nyctereutes* and *Cerdocyon* evidently are closely related. This view is supported by the fossil record, which suggests the presence of a common ancestor that ranged over Eurasia and North America 4–10 million years ago (Berta 1987). Molecular analyses show no affinity of *Nyctereutes* with any other genus (Wayne 1993; Wayne and O'Brien 1987).

Head and body length is 500–680 mm and tail length is 130–250 mm. Weight is 4–6 kg in the summer but 6–10 kg prior to winter hibernation (Novikov 1962). The pelage is long, especially in winter. The general color is yellowish brown. The hairs of the shoulders, back, and tail are tipped with black. The limbs are blackish brown. The facial markings resemble those of raccoons *(Procyon).* There is a large dark spot on each side of the face, beneath and behind the eye. *Nyctereutes* is somewhat like a fox in external appearance but has proportionately shorter legs and tail.

According to Novikov (1962), the raccoon dog occurs mainly in forests and in thick vegetation bordering lakes and streams. It usually dens in a hole initially made by a fox or badger or in a rocky crevice but sometimes digs its own burrow. *Nyctereutes* is the only canid that hibernates, though the process is neither profound for individuals nor universal for the species. In northern parts of the range well-nourished animals hibernate from as early as November to as late as March but may awaken occasionally to forage on warm days. In the southern parts of the range there is no winter sleep. Poorly nourished individuals do not hibernate, even in the north. Successful hibernation may be preceded by a period of intensive eating that increases weight by nearly 50 percent. The summer diet may consist largely of frogs and also includes rodents, reptiles, fish, insects, mollusks, and fruit. In the autumn, vegetable matter such as berries, seeds, and rhizomes becomes important. Northern individuals that do not hibernate have a difficult time in the winter but may subsist on small mammals, carrion, and human refuse. *Nyctereutes* usually is reported to be primarily nocturnal, but Ward and Wurster-Hill (1989) found it active both by day and by night at two study sites in Japan. The greatest distance traveled in one night was 8 km.

In the Ussuri region of southeastern Siberia, Kucherenko and Yudin (1973) found population density to average 1–3/1,000 ha. and to reach 20/1,000 ha. in the best habitat. According to Kauhala, Helle, and Taskinen (1993), reported home range size of the introduced European populations has varied from 0.4 sq km to 20.0 sq km. Studies in Finland found an average range size of 9.5 sq km, including a core area of 3.4 sq km. Such ranges were generally shared by an evidently monogamous adult male-female pair. The ranges, especially those of the males, were larger in autumn than in summer. The core areas of most adjacent pairs did not overlap in the pup-rearing season but did overlap to some extent in the autumn. Juveniles of both sexes left their natal ranges in their first autumn, many traveling more than 10 km.

Much smaller ranges have been recorded in Japan. Ward and Wurster-Hill (1989) found averages of 49 ha. on Kyushu and 59 ha. on Honshu during the autumn. The ranges were not exclusive, though there was no evidence of

Raccoon dog *(Nyctereutes procyonoides),* photo by Ernest P. Walker.

social grouping. During the spring on Kyushu, Ikeda (1986) found home ranges of only 8–48 ha., which overlapped extensively, and observed *Nyctereutes* to undergo much amicable social activity. At major food sources, such as fruiting trees and garbage dumps, up to 20 individuals were seen to intermingle without belligerence. There also was communal use of latrines, which may serve as sites to pass on information. Antagonism increased during the breeding season. Male-female pairs formed in late winter, and both parents cared for the young, though each sex foraged separately. The family dissolved at the end of summer, but pair bonds may have persisted from year to year. Vocalizations include growls and whines but not barks.

Ward and Wurster-Hill (1990) indicated some question concerning whether *Nyctereutes* is monogamous or polygamous, noting that an estrous female may be courted by several males. Mating occurs from January to April, estrus lasts about 4 days, and the gestation period is usually 59–64 days. There are commonly 4–10 young, but as many as 19 have been reported. They weigh 60–90 grams at birth, open their eyes after 9–10 days, and nurse for up to 2 months. Both parents, however, begin to bring them solid food after only 25–30 days. They are capable of an independent existence by 4–5 months and attain sexual maturity at 9–11 months (Helle and Kauhala 1995; Novikov 1962; Valtonen, Rajakoski, and Mäkelä 1977). Maximum longevity reported in a study of wild individuals in Finland was 7–8 years (Helle and Kauhala 1993). A captive specimen lived 10 years and 8 months (Jones 1982).

In Japan the flesh of the raccoon dog has been used for human consumption, and the bones for medicinal preparations. The skin, known commercially as "Ussuri raccoon," is used widely in the manufacture of such items as parkas, bellows, and decorations on drums. Although about 70,000 are taken every year in Japan for the fur trade, *Nyctereutes* is still common there and sometimes is found within large cities (Ikeda 1986). Populations have declined in southeastern Siberia through overhunting and habitat disturbances (Kucherenko and Yudin 1973).

From 1927 to 1957 more than 9,000 raccoon dogs were released by people in regions to the west of the natural range, especially in European Russia. The hope was to create new and valuable fur-producing populations. *Nyctereutes* did become established, eventually spreading almost throughout European Russia, and its importance in the fur trade increased. It also, however, extended its range westward, reaching Finland in 1935, Sweden in 1945, Romania in 1951, Poland in 1955, Slovakia in 1959, Germany and Hungary in 1962, and France in 1979 (Artois and Duchêne 1982; Kubiak 1965; Mikkola 1974; Novikov 1962). It now is very common in the Baltic states and many parts of eastern Europe (Kauhala 1994) and occurs throughout Germany (Roben 1975). The rate of expansion has slowed in recent years, though the prognosis is that the Balkans, Scandinavia, and France will be colonized (Nowak 1984). An individual was recently caught in England (*Oryx* 13 [1977]: 434), and the USDI designates the genus as "injurious," thereby prohibiting importation of live specimens. *Nyctereutes* is generally considered a nuisance to the west of Russia. It destroys small game animals and fish, and its fur, which does not become as long as in the native habitat, is nearly worthless in most areas. However, in Finland about 80,000 pelts are derived annually from farmed animals and another 40,000–60,000 from wild-trapped individuals (Ginsberg and Macdonald 1990; Helle and Kauhala 1991).

CARNIVORA; CANIDAE; **Genus ATELOCYNUS**
Cabrera, 1940

Small-eared Dog

The single species, *A. microtis,* inhabits the Amazon, upper Orinoco, and upper Parana basins in Brazil, Peru, Ecuador, Colombia, and probably Venezuela. *Atelocynus* was recognized as a distinct genus by Berta (1986, 1987), Ca-

Small-eared dog *(Atelocynus microtis)*, photo from Field Museum of Natural History.

brera (1957), and Langguth (1975), as a part of *Dusicyon* by Clutton-Brock, Corbet, and Hills (1976), and as a subgenus of *Canis* by Van Gelder (1978).

Head and body length is 720–1,000 mm, tail length is 250–350 mm, shoulder height is about 356 mm, and weight is around 9–10 kg. The ears, rounded and relatively shorter than those of any other canid, are only 34–52 mm in length. The upper parts are dark gray to black, and the underparts are rufous mixed with gray and black. The thickly haired tail is black except for the paler basal part on the underside. It has been reported to sweep the ground when hanging perpendicularly; however, when captives at the Brookfield Zoo, in Chicago, were standing, they curved the tail forward and upward against the outer side of a hind leg, so that the terminal hairs did not drag on the ground. The temporal ridges of the skull are strongly developed, and the frontal sinuses and cheek teeth are relatively large (Clutton-Brock, Corbet, and Hills 1976).

Atelocynus inhabits tropical forests from sea level to about 1,000 meters. It moves with a catlike grace and lightness not observed in any other canid. Observations in the wild indicate that *Atelocynus* hunts individually and captures primarily small and sometimes medium-sized rodents, whereas the sympatric *Speothos* hunts in packs and commonly kills larger animals (Peres 1991). A male in captivity in Bogota ate raw meat, shoots of grass, and foods commonly eaten by people. This male and a female were brought to the Brookfield Zoo. They proved to be completely different in temperament: the male was exceedingly friendly and docile, but the female exhibited constant hostility. Although the male was shy in captivity before being sent to Brookfield, and growled and snarled when angry or frightened, it did not show any unfriendly actions at the Chicago zoo and, in fact, became very tame. It permitted itself to be hand fed and petted by persons it recognized, responding to petting by rolling over on its back and squealing. This male came to react to attention from familiar people by a weak but noticeable wagging of the back part of its tail. The female, on the other hand, emitted a continuous growling sound, without opening its mouth or baring its teeth, when under direct observation.

The odor from the anal glands was strong and musky in the male but scarcely detectable in the female. In both sexes the eyes glowed remarkably in dim light. The male, though smaller, was dominant in most activities. Although some snapping was observed between the two animals, no biting or fighting was noted, and they occupied a common sleeping box when not active. According to Jones (1982), one captive lived 11 years.

According to Thornback and Jenkins (1982), *Atelocynus* seems to be rare in some countries, though its status is difficult to assess because of its nocturnal, solitary habits. It may be at risk from indirect influence of human intrusion throughout its range. It is protected by law in Brazil and Peru.

CARNIVORA; CANIDAE; **Genus SPEOTHOS**
Lund, 1839

Bush Dog

The single living species, *S. venaticus*, occurs in Panama, Colombia, Ecuador, Venezuela, the Guianas, eastern Peru, Brazil, eastern Bolivia, Paraguay, and extreme northeastern Argentina (Cabrera 1957; Ginsberg and Macdonald 1990; E. R. Hall 1981; Thornback and Jenkins 1982). *Speothos* was first described from fossils collected in caves in Brazil. Although *Speothos* sometimes has been united with *Cuon*

Bush dog *(Speothos venaticus)*, photo from San Diego Zoological Garden.

and *Lycaon* in the subfamily Simocyoninae, Berta (1984) showed that its true affinities lie with other South American canids, especially *Atelocynus.* She reported also that a second species, *S. pacivorus,* is known from late Pleistocene and early Recent deposits in Minas Gerais, Brazil.

Head and body length is 575–750 mm, tail length is 125–50 mm, and shoulder height is about 300 mm. Two males weighed 5 and 7 kg. The head and neck are ochraceous fawn or tawny, and this color merges into dark brown or black along the back and tail. The underparts are as dark as the back, though there may be a light patch behind the chin on the throat (Clutton-Brock, Corbet, and Hills 1976). The body is stocky, the muzzle is short and broad, and the legs are short. The tail is short and well haired but not bushy.

Speothos inhabits forests and wet savannahs, often near water. It seems to be mainly diurnal and to retire to a den at night, in either a burrow or a hollow tree trunk. It is reportedly semiaquatic: one captive could dive and swim underwater with great facility. Wild individuals have been observed to swim across large rivers and to chase prey into water. The bush dog captures mainly relatively large rodents, such as *Agouti* and *Dasyprocta* (Husson 1978; Kleiman 1972; Langguth 1975; Strahl, Silva, and Goldstein 1992).

Speothos is a highly social canid, living and hunting cooperatively in packs of up to 10 individuals. Several captives of the same or opposite sex can stay confined together without fighting, though a dominance hierarchy may be established. There are a number of vocalizations, the most common of which is a high-pitched squeak that appears to help maintain contact as the group moves about in dense forest (Clutton-Brock, Corbet, and Hills 1976; Kleiman 1972).

Husson (1978) wrote that litters of 2–3 young are produced during the rainy season in the wild. Extended observations in captivity (Porton, Kleiman, and Rodden 1987), however, indicate that the reproductive pattern of *Speothos* is aseasonal and uninterrupted and is partly influenced by social factors. Young females typically do not experience estrus when living with their mother or sisters but quickly do so when paired with males. Estrus and births have been recorded throughout the year. Estrus periods average 4.1 days, and some females show polyestrous activity, which has not been recorded in any other canid. The interbirth interval averages 238 days, and gestation, 67 days. Litter size averages 3.8 young and ranges from 1 to 6. The earliest age of conception in a female was 10 months. Hayssen, Van Tienhoven, and Van Tienhoven (1993) listed newborn weights of 130–90 grams and a nursing period of 4–5 months. According to Marvin L. Jones (Zoological Society of San Diego, pers. comm., 1995), a captive lived 13 years and 4 months.

Speothos is classified as vulnerable by the IUCN and is on appendix 1 of the CITES. It is still widespread but is scarce throughout its range, and it seems to disappear as settlement progresses and forests are cleared (Thornback and Jenkins 1982).

CARNIVORA; CANIDAE; **Genus CANIS**
Linnaeus, 1758

Dogs, Wolves, Coyotes, and Jackals

There are eight species (Clutton-Brock, Corbet, and Hills 1976; Coetzee *in* Meester and Setzer 1977; Corbet 1978; Ferguson 1981; Ginsberg and Macdonald 1990; E. R. Hall 1981; Kingdon 1977; Nowak 1979; Sillero-Zubiri and Gottelli 1994; Thomas and Dibblee 1986; Vaughan 1983):

C. adustus (side-striped jackal), open country from Senegal to Somalia, and south to northern Namibia and eastern South Africa;

Simien jackal or Ethiopian wolf *(Canis simensis)*, photo by Claudio Sillero-Zubiri.

Dingos *(Canis familiaris)*, photos by Lothar Schlawe.

- *C. mesomelas* (black-backed jackal), open country from Sudan and Ethiopia to Tanzania and from southwestern Angola to Mozambique and South Africa;
- *C. aureus* (golden jackal), Balkan Peninsula to Thailand and Sri Lanka, Morocco to Egypt and northern Tanzania;
- *C. simensis* (Simien jackal or Ethiopian wolf), mountains of central Ethiopia;
- *C. latrans* (coyote), Alaska to Nova Scotia and Panama;
- *C. rufus* (red wolf), central Texas to southern Pennsylvania and Florida;
- *C. lupus* (gray wolf), Eurasia except tropical forests of southeastern corner, Egypt and Libya, Alaska, Canada, Greenland, conterminous United States except southeastern quarter and most of California, highlands of Mexico;
- *C. familiaris* (domestic dog), worldwide in association with people, extensive feral populations in Australia and New Guinea.

Van Gelder (1977, 1978) considered *Canis* to comprise *Vulpes* (including *Fennecus* and *Urocyon*), *Alopex, Lycalopex, Pseudalopex, Dusicyon, Cerdocyon,* and *Atelocynus* as subgenera. *C. simensis* sometimes has been placed in a separate genus or subgenus, *Simenia* Gray, 1868, but recent mitochondrial DNA analysis indicates that it is phylogenetically closer to the coyote and wolves than to the other jackals (Sillero-Zubiri and Gottelli 1994). There also is molecular evidence that *C. aureus* has closer affinity to the coyote-wolf group than to the jackal group even though it has demonstrated some morphological convergence with *C. adustus* and *C. mesomelas* in East Africa (Van Valkenburgh and Wayne 1994). Lawrence and Bossert (1967, 1975) suggested that *C. rufus* is not more than subspecifically distinct from *C. lupus*. In contrast, recent analyses of mitochondrial and nuclear DNA have led to arguments that *C. rufus* is a hybrid population that originated subsequent to European settlement through interbreeding between *C. latrans* and *C. lupus* (Jenks and Wayne 1992; Roy et al. 1994; Roy, Girman, and Wayne 1994; Wayne 1992; Wayne and Gittleman 1995; Wayne and Jenks 1991). That view has been questioned based on other interpretations of genetic data (Cronin 1993; Dowling, DeMarais, et al. 1992; Dowling, Minckley, et al. 1992), behavioral and ecological data (Phillips and Henry 1992), and morphological and paleontological evidence indicating that *C. rufus* is a primitive kind of wolf that has been present in the Southeast since the mid-Pleistocene (Nowak 1992). The cited presence of *C. lupus* in Egypt and Libya is taken from Ferguson's (1981) remarkable determination that the population formerly known as *C. aureus lupaster* actually consists of small wolves rather than large jackals. However, Spassov (1989) argued that *lupaster* shows closer affinity to *C. aureus* than to *C. lupus* and may even be a separate species. *C. familiaris* was included in *C. lupus* by Wozencraft (*in* Wilson and Reeder 1993), and that view has been widely accepted. The feral populations of *C. familiaris* in Australia and apparently related populations in parts of southern Asia and the East Indies sometimes are considered to represent a distinct species, *C. dingo*.

Canis is characterized by a relatively high body, long legs, and a bushy, cylindrical tail. The pupils of the eyes generally appear round in strong light. Although most species have a scent gland near the base of the tail, it does not produce as strong an odor as that of *Vulpes*. The skull has large frontal sinuses and temporal ridges that are close together, often uniting to form a sagittal crest. The facial region of the skull, except in *C. simensis*, is relatively shorter than in *Vulpes* and *Pseudalopex* (Clutton-Brock, Corbet, and Hills 1976). Females have 8 or 10 mammae. Additional information is provided separately for each species.

Canis adustus (side-striped jackal)

Head and body length is 650–810 mm, tail length is 300–410 mm, shoulder height is 410–500 mm, and weight is 6.5–14.0 kg (Kingdon 1977). Males are larger than females. The coat is long and soft. The upper parts are generally mottled gray; on each side of the body is a line of white hairs below which is a line of dark hairs; and the underparts and tip of the tail are white (Clutton-Brock, Corbet, and Hills 1976).

According to Kingdon (1977), the side-striped jackal is

Side-striped jackal *(Canis adustus)*, photo by Cyrille Barrette.

widespread in the moister parts of savannahs, thickets, forest edge, cultivated areas, and rough country up to 2,700 meters in elevation. Favored denning sites include old termite mounds, abandoned aardvark holes, and burrows dug into hillsides. The animal is strictly nocturnal in areas well settled by people. It is a more omnivorous scavenger than other kinds of jackals, taking a variety of invertebrates, small vertebrates, carrion, and plant material. Social groups are well spaced and usually consist of a mated pair and their young. In East Africa births occur mainly in June–July and September–October, gestation lasts 57–70 days, and litters consist of three to six young. Ginsberg and Macdonald (1990) stated that lactation lasts 8–10 weeks, sexual maturity is attained at 6–8 months, and longevity is 10–12 years. They noted that the species is rare throughout its range. In a recent radio-tracking study of the three species of jackals in East Africa, Fuller et al. (1989) were able to obtain data on only a single *C. adustus;* in 24 days of monitoring it was found to use a home range of 1.1 sq km.

Canis mesomelas (black-backed jackal)

Head and body length is 450–900 mm, tail length is 260–400 mm, shoulder height is 300–480 mm, and weight is 6.0–13.5 kg. Males average about 1 kg heavier than females. A dark saddle extends the length of the back to the black tip of the tail. The sides, head, limbs, and ears are rufous, and the underparts are pale ginger. The build is slender, and the ears are very large (Bekoff 1975; Clutton-Brock, Corbet, and Hills 1976; Kingdon 1977).

The black-backed jackal is found mainly in dry grassland, brushland, and open woodland. It generally dens in an old termite mound or aardvark hole. It is partly diurnal and crepuscular in undisturbed areas but becomes nocturnal in places intensively settled by people. The diet includes a substantial proportion of plant material, insects, and carrion, but *C. mesomelas* frequently hunts rodents and is capable of killing antelopes as large as *Gazella thomsoni.* Some food may be cached (Bothma 1971*b;* Kingdon 1977; Smithers 1971).

In a study in southern Africa average daily movement was found to be about 12 km, but one radio-collared individual covered 87 km in four nights. Average home range size in this area was about 11 sq km for adults, but some subadults wandered over much greater areas (Ferguson, Nel, and de Wet 1983). In the same region, Rowe-Rowe (1982) estimated a population density of 1/2.5–2.9 sq km and an average home range size of 18.2 sq km. According to these and other investigations (Bekoff 1975; Kingdon 1977; Moehlman 1978, 1983, 1987), all or a major part of the adult home range is a territory that is marked with urine and defended by both sexes. Each sex seems to repel mainly members of the same sex. Although 20–30 individuals may gather around a lion kill, the basic social unit is a mated pair and their young. Some young remain with the parents, do not breed, and assist in feeding, grooming, and guarding the next litter. This tendency seems to be stronger in *C. mesomelas* than in *C. aureus,* perhaps because the former maintain larger territories and, because of a dependency on smaller game, must make a greater effort to provide for the newborn. Final dispersal of the young may not occur until they are about 2 years old. Mated pairs are known to have maintained a monogamous relationship for at least 6 years.

Births occur mainly from July to November. The gestation period is about 60 days. Average litter size is four young, ranging from one to eight. Weight at birth is about 159 grams (Hayssen, Van Tienhoven, and Van Tienhoven 1993). The young emerge from the den after 3 weeks and may then be moved several times to new sites. They are weaned at 8–9 weeks, and the adults then regurgitate food for them. After about 3 months they no longer use the den, and by the age of 6 months they are hunting on their own. Sexual maturity may come at 10–11 months, though young that remain with the parents do not breed. Captives have lived up to 14 years (Bekoff 1975; Bothma 1971*a;* Kingdon 1977; Moehlman 1978, 1987).

Unlike *C. adustus,* the black-backed jackal is considered a serious predator of sheep and is therefore intensively hunted and poisoned (Clutton-Brock, Corbet, and Hills 1976). According to Bothma (1971*b*), *C. mesomelas* causes more problems than any other animal in the sheep-farming areas of the Transvaal.

Black-backed jackal *(Canis mesomelas)*, photo by Ernest P. Walker.

Canis aureus (golden jackal)

Head and body length is 600–1,060 mm, tail length is 200–300 mm, shoulder height is 380–500 mm, and weight is 7–15 kg (Kingdon 1977; Lekagul and McNeely 1977). The fur is generally rather coarse and not very long. The dorsal area is mottled black and gray; the head, ears, sides, and limbs are tawny or rufous; the underparts are pale ginger or nearly white; and the tip of the tail is black (Clutton-Brock, Corbet, and Hills 1976).

The golden jackal is found mainly in dry, open country. It is strictly nocturnal in areas inhabited by people but may be partly diurnal elsewhere. Its opportunistic diet includes young gazelles, rodents, hares, ground birds and their eggs, reptiles, frogs, fish, insects, and fruit. It takes carrion on occasion but is a capable hunter (Kingdon 1977; Moehlman 1987; Van Lawick and Van Lawick–Goodall 1971).

The basic social unit is a mated pair and their young. Monogamy is the rule, though pair bonds do not seem to

Golden jackal *(Canis aureus)*, photo by Lothar Schlawe.

be as strong as in *C. mesomelas.* Sometimes the young of the previous year remain in the vicinity of the parents and even mate and bear their own litters. However, a more usual procedure is for one or two of the young to stay with the parents, refrain from breeding, and help care for the next litter. The resulting associations are probably responsible for the reports of large packs hunting together. The usual hunting range of a family is about 2–3 sq km. A portion of this area is a territory marked with urine and defended against intruders. The young are born in a den within the territory. Estrus lasts 3–4 days and newborn weight is about 201–14 grams (Hayssen, Van Tienhoven, and Van Tienhoven 1993). Births occur mainly in January–February in East Africa and in April–May in Central Asia but take place throughout the year in tropical Asia. The gestation period is 63 days. Litters contain one to nine, usually two to four, young. Their eyes open after about 10 days, and they begin to take some solid food at about 3 months. Both parents provide food and protection. Sexual maturity comes at 11 months. Captives have lived up to 16 years (Kingdon 1977; Lekagul and McNeely 1977; Moehlman 1983, 1987; Novikov 1962; Rosevear 1974; Van Lawick and Van Lawick–Goodall 1971).

As with other wild canids, the relationship of the golden jackal with people is controversial. In Bangladesh, for example, *C. aureus* plays an important scavenging role by eating garbage and animal carrion around towns and villages and may benefit agriculture by preventing increases in the numbers of rodents and lagomorphs. However, this jackal also raids such crops as corn, sugar cane, and watermelons. It may be involved in the spread of rabies, and in 1979 two young children were attacked and killed by jackals (Poché et al. 1987). Since about 1980 *C. aureus* has extended its European range into eastern Austria and northeastern Italy; it also has occupied parts of the Balkan Peninsula where it had not previously been seen, perhaps in association with a decline of the native wolf population (Ginsberg and Macdonald 1990; Hoi-Leitner and Kraus 1989; Krystufek and Tvrtkovic 1990). On the other hand, the intensive spreading of poison baits for wolves after World War II evidently also eliminated the jackal in some areas, such as Macedonia, where no more were collected until 1989 (Krystufek and Petkovski 1990).

Canis simensis (Simien jackal or Ethiopian wolf)

Head and body length is 841–1,012 mm, tail length is 270–396 mm, shoulder height is 530–620 mm, and weight is 11–19 kg; males average about 20 percent larger than females (Sillero-Zubiri and Gottelli 1994). The general coloration of the upper parts is tawny rufous with pale ginger underfur. The chin, insides of the ears, chest, and underparts are white. There is a distinctive white band around the ventral part of the neck. The tail is rather short, the facial region of the skull is elongated, and the teeth, especially the upper carnassials, are relatively small (Clutton-Brock, Corbet, and Hills 1976).

The present habitat is montane grassland and moorland at elevations of approximately 3,000–4,000 meters. *C. simensis* may have been found at lower elevations before becoming subject to severe human persecution (Yalden, Largen, and Kock 1980). In the Bale Mountains of south-central Ethiopia, Morris and Malcolm (1977) found the Simien jackal to be relatively numerous in grasslands supporting large rodent populations. It was uncommon in scrub and was not seen in forests at low altitudes. It was primarily diurnal, but there was some evidence of activity at night, especially in moonlight. Farther north, in the Simien Mountains, the few remaining jackals were reported to be almost entirely nocturnal, probably because of human disturbance. The animals made little effort to find shelter, and individuals slept in the open or in places with slightly longer grass than usual even when temperatures were as low as −7° C. Sillero-Zubiri and Gottelli (1994) noted that dens are dug in open ground or located in a rocky crevice

Simien jackal, or Ethiopian wolf *(Canis simensis)*, photo by Claudio Sillero-Zubiri.

and are used only by females to give birth. Rodent prey also commonly is dug out of the ground, and kills often are cached for future use. Morris and Malcolm (1977) found the diet of *C. simensis* in the Bale Mountains to consist mainly of diurnal rodents, especially *Tachyoryctes* and *Otomys*. Hunting is done mainly by walking slowly through areas of high rodent density, investigating holes and listening carefully, and then stealthily creeping toward the prey and capturing it with a final dash. Individuals generally were seen to hunt alone and to not come together during the day, though there was apparently considerable overlap in hunting ranges.

According to Gottelli and Sillero-Zubiri (1994*a*) and Sillero-Zubiri and Gottelli (1994), *C. simensis* also is a facultative, cooperative hunter. Small packs occasionally have been seen chasing and killing young antelopes, lambs, and hares. The species lives in cohesive packs of 3–13 adults that share and defend an exclusive territory. These animals include all males born into the pack during consecutive years and 1 or 2 females. Annual home ranges of eight packs monitored for four years had an average size of 6 sq km, and ranges in an area of less prey averaged 13.4 sq km. Population density has been found to vary from about 0.1 to 1.0 adults per sq km, also depending on prey availability. Pack members congregate for friendly social interaction and territorial patrols at dawn and noon and in the evening and rest together at night but commonly forage individually in the morning and afternoon. Territories are marked with urine and feces and sometimes are contested through aggressive behavior and chases. There are various vocalizations, including yelps and barks during interaction and howls that can be heard 5 km away. The males and females of a pack form separate dominance hierarchies, and only the highest-ranking pair mates. Births in the Bale Mountains occur from October to January. The gestation period evidently is 60–62 days and litters contain 2–6 young. They are regularly shifted between dens up to 1,300 meters apart and finally emerge on their own after 3 weeks. They begin to take some solid food at about 5 weeks, are fully weaned at 10 weeks, begin to hunt with the pack at 6 months, and attain sexual maturity in their second year.

If *C. simensis* actually is a wolf, as suggested by genetic analysis, its distribution may not seem so incongruous if consideration is given to the presence of another small wolf, *Canis lupus arabs*, just across the Red Sea to the east. *C. simensis* was reported from most Ethiopian provinces in the nineteenth century. It subsequently declined because of agricultural development in its range and environmental disruption that reduced prey populations. Also, it was incorrectly believed to be a predator of sheep and so was frequently shot. A newly recognized and potentially devastating threat is hybridization with domestic dogs *(C. familiaris)*. The latter species also is a source of harassment and diseases, such as rabies. At present only about 70–150 individuals of the subspecies *C. s. simensis* remain in the Simien Mountains and nearby parts of north-central Ethiopia. It is estimated that 270–370 individuals of the subspecies *C. s. citerni* remain in the Bale Mountains and that there are a few others in neighboring highlands. There are national parks in the Simien and Bale mountains, but these areas are still being used for grazing (Ginsberg and Macdonald 1990; Gottelli and Sillero-Zubiri 1992, 1994*a*, 1994*b*; Gottelli et al. 1994; Sillero-Zubiri and Gottelli 1994). *C. simensis* is protected by law in Ethiopia and is classified as critically endangered by the IUCN and as endangered by the USDI.

Canis latrans (coyote)

Head and body length is 750–1,000 mm and tail length is 300–400 mm. In a given population males are generally

Coyote *(Canis latrans)*, photo from U.S. Fish and Wildlife Service.

larger than females. Bekoff (1977) listed weights of 8–20 kg for males and 7–18 kg for females. Northern animals are usually larger than those to the south. Gier (1975) gave average weight as 11.5 kg in the deserts of Mexico and 18 kg in Alaska. The largest coyotes of all are found in the northeastern United States; there is considerable debate whether this development has resulted from hybridization with *C. lupus* or other genetic factors or is simply a phenotypic response to enhanced nutrition (Larivière and Crête 1993; Nowak 1979; Peterson and Thurber 1993). Pelage characters are variable, but usually the coat is long, the upper parts are buffy gray, and the underparts are paler. The legs and sides may be fulvous, and the tip of the tail is usually black (Clutton-Brock, Corbet, and Hills 1976). From *C. lupus* the coyote is distinguished by its smaller size, narrower build, proportionally longer ears, and much narrower snout.

The coyote is found in a variety of habitats, mainly open grasslands, brush country, and broken forests. The natal den is located in such places as brush-covered slopes, thickets, hollow logs, rocky ledges, and burrows made by either the parents themselves or other animals. Tunnels are about 1.5–7.5 meters long and lead to a chamber about 0.3 meter wide and 1 meter below the surface. Activity may take place at any time of day but is mainly nocturnal and crepuscular. The average distance covered in a night's hunting is 4 km. The coyote is one of the fastest terrestrial mammals in North America, sometimes running at speeds of up to 64 km/hr. There may be a migration to the high country in the summer and a return to the valleys in the autumn. Some of the young disperse from the parental range in the autumn and winter, generally covering a distance of 80–160 km (Banfield 1974; Bekoff 1977; Gier 1975). In a tagging study in Iowa, Andrews and Boggess (1978) found individuals to travel an average of about 31 km, but up to 323 km, from the point of capture. Carbyn and Paquet (1986) reported that a tagged individual in south-central Canada moved a record 544 km.

About 90 percent of the diet is mammalian flesh, mostly jack rabbits, other lagomorphs, and rodents. Such small animals are usually taken by stalking and pouncing (Bekoff 1977; Gier 1975). Coyotes have been seen to enter shallow streams and snatch fish from the water (Springer 1980). Larger mammals, especially deer, are also eaten, often in the form of carrion but sometimes after a chase in which several coyotes work together. In northwestern Wisconsin deer represent 21.3 percent of the diet (Niebauer and Rongstad 1977), and in northern Minnesota deer is the coyote's main food, mostly as carrion (Berg and Chesness 1978). In Alberta half of the food is carrion (Nellis and Keith 1976). Analyses of coyote predation on large mammals generally indicate that most of the individuals taken are immature, aged, or sick (Bekoff 1977). Livestock is killed by some coyotes, but the diet of the species seems to be largely neutral or beneficial with respect to human interests. *C. latrans* sometimes forms a "hunting partnership" with the badger *(Taxidea)*. The two move together, the coyote apparently using its keen sense of smell to locate burrowing rodents and the badger digging them up with its powerful claws. Both predators then share in the proceeds. Recent observations by Minta, Minta, and Lott (1992) have confirmed that such associations do commonly exist, that they are mutually beneficial, and that the two species interact in a friendly manner.

Population density is generally 0.2–0.4/sq km but may be as high as 2.0/sq km under extremely favorable conditions (Bekoff 1977; Knowlton 1972). Reported home range size is 8–80 sq km. The ranges of males tend to be large and to overlap one another considerably; the ranges of females are smaller and do not overlap (Bekoff 1977). In northern Minnesota Berg and Chesness (1978) found home range size to average 68 sq km for males and 16 sq km for females. In Alberta, however, Bowen (1982) found home ranges to be shared by an adult male and female and to average about 14 sq km for both sexes. Laundré and Keller (1984) reviewed various reports of home range size of about 10–100 sq km and concluded that most were based on inadequate data.

Social structure and territoriality vary depending on habitat conditions and food supplies. *C. latrans* is found alone, in pairs, or in larger groups but generally is less social than *C. lupus* (Bekoff 1977). Several males may court a female during the mating season, but apparently she eventually selects one for a lasting relationship. The pair then live and hunt together, sometimes for years. Gier (1975) suggested that pair territories tend to be bordered by natural features, cover about 1 sq km or less, and are defended only during the denning season. More recent work suggests that territories are larger and are exclusive throughout the year, though the ranges of young or transient animals may be superimposed (Andelt 1985).

Some of the offspring of a pair disperse, but others may remain with the parents and thus form the basis of a pack. These helpers, whose main function apparently is to provide increased protection to the next litters, may not leave until they are 2–3 years old. Studies in northwestern Wyoming indicate that packs are larger and more stable if a major, clumped food resource, such as ungulate carrion, is readily available (Bekoff and Wells 1980, 1982). In this area, on the National Elk Refuge, Camenzind (1978) found that 61 percent of the coyotes composed resident packs of 3–7 members, 24 percent were resident mated pairs, and 15 percent were nomadic individuals. The latter ranged through the territories of the resident animals and were continually chased away. The packs had well-defined social hierarchies and marked and defended territories. The sizes of two pack territories were 5 sq km and 7.2 sq km. Sometimes more than one pair of a pack mated and bore young. Occasionally up to 22 individuals would come together around carrion, but such aggregations usually lasted less than an hour.

The coyote has a great variety of visual, auditory, olfactory, and tactile forms of communication (Lehner 1978). At least 11 different vocalizations have been identified; these are associated with alarm, threats, submission, greeting, and contact maintenance. The best-known sounds are the howls given by one or more individuals apparently to announce location. Group howls may have a territorial and spacing function.

Mating usually occurs from January to March, and births take place in the spring. Females are monestrous. Estrus averages 10 (4–15) days, gestation averages 63 (58–65) days, and litter size averages about 6 (2–12) young. Larger numbers of offspring are sometimes found in one den, but these apparently represent the litters of more than one female. The young weigh about 250 grams at birth, open their eyes after 14 days, emerge from the den at 2–3 weeks, and are fully weaned at 5–6 weeks. The mother, and perhaps the father and older siblings, begins to regurgitate some solid food for the young when they are about 3 weeks old. Adult weight is attained at about 9 months. Some individuals mate in the breeding season following their birth, but others wait another year. Maximum known longevity in the wild is 14.5 years, but few animals survive that long. In one unexploited population 70 percent of the animals were less than 3 years old (Bekoff 1977; Gier 1975; Kennelly 1978). A captive lived 21 years and 10 months (Jones 1982).

Coyote *(Canis latrans)*, photo by David M. Shackleton.

The coyote long has been considered a serious predator of domestic animals, especially sheep. It is also sometimes alleged to spread rabies and to damage populations of game birds and mammals. It generally is considered harmless to people, though during the 1980s several children reportedly were seriously injured, and one killed, in attacks in California and western Canada (Carbyn 1989). Bounties against the coyote were first offered in the United States in 1825; they are still paid in some areas but have never achieved lasting prevention of depredations. In 1915 the U.S. government initiated a large-scale program of predator control, mainly in the West. The resulting known kill of *C. latrans* often exceeded 100,000 individuals per year. From 1971 to 1976 the annual take in 13 western states averaged 71,574 (Evans and Pearson 1980). Some state governments and private organizations also have control programs. Killing methods include trapping, poisoning, shooting from aircraft, and destroying the young in dens. In 1972 the federal government banned the general use of poison on its lands and in its programs.

Few wildlife issues have been as controversial as coyote control (Bekoff 1977; Bekoff and Wells 1980; Hall 1946; Sterner and Shumake 1978; Wade 1978). No authorities deny that predation is a problem to some farmers and ranchers, but almost since the beginning of the federal program there have been serious doubts concerning the extent of claimed livestock losses and the appropriate response. There was particular resentment against the widespread application of poison on western ranges, which was thought to kill not only far more coyotes than necessary but also many other carnivorous mammals. It has been argued that sheep raising is declining in the United States regardless of the coyote and its control and that damage by predators represents a relatively small part of overall losses to the industry. Surveys by trained biologists generally have indicated that losses are substantial in some areas but lower than often claimed. The stated intention of most current control operations is elimination of the specific individual coyotes that may be causing problems, not the destruction of entire populations.

Information compiled by the U.S. Department of Agriculture (Gee et al. 1977) indicates that in 1974 losses attributed to the coyote in the western states amounted to 728,000 lambs (more than 8 percent of those born and one-third of the total lamb deaths) and 229,000 adult sheep (more than 2 percent of inventory and one-fourth of all adult deaths), valued at $27 million. In that same year the amount expended for coyote control in the West was $7 million (Gum, Arthur, and Magleby 1978). In 1974, 71,522 coyotes were killed by the federal control program. Data compiled by Pearson (1978) suggest that the total kill in the 17 western states in 1978 (including the take by fur trappers and sport hunters) was around 300,000 individuals, approximately the same as in 1946. Data compiled by Terrill (1986) indicate that the rate of loss of sheep and lambs to coyote predation through 1985 generally remained about the same as given above, though there was a temporary decline in 1980, apparently caused by the spread of canine parovirus in the coyote population. Subsequent losses have been subject to conflicting reports, though one estimate sets the value of sheep and lambs killed by coyotes in the 17 western states during 1989 at $18 million (Connolly 1992). About 89,000 coyotes were killed during fiscal year 1995 by the federal control program (unpublished data from U.S. Department of Agriculture, 1996).

Coyote fur has varied in price. In 1974 skins taken in the United States sold for an average of $17.00 each. For the 1976–77 season 30 states reported a total harvest of 320,323 pelts with an average value of $45.00; in 6 Canadian provinces the take was 65,819 pelts with an average value of $59.76 (Deems and Pursley 1978). Coyote numbers evidently rebounded in the 1980s, but fur prices dropped. The total known kill in the United States during the 1983–84 season was 439,196. In Canada 68,975 pelts with an average value of $15.90 were taken (Novak, Obbard, et al. 1987). In the 1991—92 season 158,001 skins were taken in the United States and sold at an average price of $13.53 (Linscombe 1994).

It is sometimes said that *C. latrans*, as a species, can easily sustain human-inflicted losses because of its adaptability, cunning, and high reproductive potential. That may not necessarily be the case. In the past, intensive control programs eliminated the species from central Texas, much of North Dakota, and certain large sheep-raising sections of Colorado, Wyoming, Montana, Utah, and Nevada (Gier 1975; Nowak 1979).

It does appear that *C. latrans* has substantially extended its range since the arrival of European colonists in North America (Brady and Campbell 1983; Gier 1975; Hill, Sumner, and Wooding 1987; Hilton 1978; Holzman, Conroy, and Pickering 1992; Monge-Nájera and Brenes 1987; Moore and Millar 1984; Nowak 1978, 1979; Sabean 1989; Thomas and Dibblee 1986; Vaughan 1983; Weeks, Tori, and Shieldcastle 1990; Wooding and Hardisky 1990). The newly occupied regions may include Alaska, the Yukon, and the southern part of Central America, though some evidence suggests that the species was already there, at least intermittently, in prehistoric times. It is known to have moved through Costa Rica and into western Panama since 1960. The main expansion was eastward, the two main factors apparently being (1) the creation of favorable habitat through the breaking up of the forests by settlement and (2) human elimination of the wolves *(C. rufus* and *C. lupus)* that might have competed with the coyote. In the late nineteenth and early twentieth centuries the coyote moved from its prairie homeland, through the Great Lakes region, and into southeastern Canada. From the 1930s to the 1960s the species established itself in New England and New York. It now has occupied Ohio and Kentucky, moved as far east as Nova Scotia and Prince Edward Island, and spread down the Appalachians to southern Virginia. Other coyote populations pushed into southern Missouri and Arkansas in the 1920s, overran Louisiana in the 1950s, crossed into Mississippi by the 1960s, and subsequently became established in Tennessee and as far as Florida and the Carolinas. The expanding coyote populations were modified by hybridization with the remnant pockets of wolves that they encountered, *C. lupus* in southeastern Canada and *C. rufus* in the south-central United States, with the result that the coyotes of eastern North America are larger and generally better adapted for forest life than their western relatives.

Canis rufus (red wolf)

Head and body length is 1,000–1,300 mm, tail length is 300–420 mm, shoulder height is 660–790 mm, and weight is 20–40 kg. While the red element of the fur is sometimes pronounced, the upper parts are usually a mixture of cinnamon buff, cinnamon, or tawny with gray or black; the dorsal area is generally heavily overlaid with black. The muzzle, ears, and outer surfaces of the limbs are usually tawny. The underparts are whitish to pinkish buff, and the tip of the tail is black. There was reported to be a locally common dark or fully black color variant in the forests of the Southeast. From *C. lupus* the red wolf is distinguished by its narrower proportions of body and skull, shorter fur, and relatively longer legs and ears.

Habitats include upland and bottomland forests, swamps, and coastal prairies. Natal dens are located in the trunks of hollow trees, stream banks, and sand knolls. Dens, which the animal either excavates itself or takes over from

Red wolf *(Canis rufus)*, photo from the *Washington Post.*

some other animal, average about 2.4 meters in length and usually extend no further than 1 meter below the surface. The red wolf is primarily nocturnal but may increase its daytime activity during the winter. It hunts over a relatively small part of its home range for about 7–10 days and then shifts to another area. Reported foods include nutria, muskrats, other rodents, rabbits, deer, hogs, and carrion (Carley 1979; Nowak 1972; Riley and McBride 1975).

Home range in southeastern Texas has been reported: (1) to average 44 sq km for 7 individuals (Shaw and Jordan 1977); (2) to cover 65–130 sq km over 1–2 years (Riley and McBride 1975); and (3) to average 116.5 sq km for males and 77.7 sq km for females (Carley 1979). Extensive observations of the reintroduced population in North Carolina indicate that pack home range is 50–100 sq km (Phillips 1994). The basic social unit is apparently a mated, territorial pair. Groups of 2–3 individuals are most common, though larger packs have frequently been reported. The normal age of dispersal appears to be 16–22 months (Phillips 1994). Vocalizations are intermediate to those of *C. latrans* and *C. lupus* (Paradiso and Nowak 1972). Mating occurs from January to March and offspring are produced in the spring. The gestation period is 60–63 days, and litters contain up to 12 young, usually about 4–7. Several of the individuals reintroduced in North Carolina are known to have lived more than 4 years in the wild (Phillips 1994), and potential longevity in captivity is at least 14 years (Carley 1979).

Like most large carnivores, the red wolf was considered to be a threat to domestic livestock, if not to people themselves. It was therefore intensively hunted, trapped, and poisoned after the arrival of European settlers in North America. In addition, disruption of its habitat and reduction of its numbers allowed *C. latrans* to invade its range from the west and north and evidently stimulated interbreeding between the two species. This process led to the genetic swamping of the small pockets of red wolves that survived human persecution. By the 1960s the only pure populations of *C. rufus* were in the coastal prairies and swamps of southeastern Texas and southern Louisiana. Conservation efforts, especially by the U.S. Fish and Wildlife Service, were not successful. The hybridization process continued to spread, and by 1975 the prevailing view was that the species could be saved only by securing and breeding some of the remaining animals that appeared to represent unmodified *C. rufus* (Carley 1979; McCarley and Carley 1979; Nowak 1972, 1974, 1979).

Many animals were captured and examined, and eventually 14 were selected for the breeding program. Since 1977 offspring have been produced from this stock, and by 1995 there were 289 living descendants (the last wild-caught wolf died in 1989). Most of these animals are still in captivity, the majority at the Point Defiance Zoo in Tacoma, Washington, but experimental reintroductions have been made on Bulls Island off South Carolina, St. Vincent Island off Florida, and Horn Island off Mississippi. Starting in 1987, pairs have been released as part of a large-scale effort to reestablish a permanent red wolf population at the Alligator River National Wildlife Refuge and adjacent lands in eastern North Carolina. By 1995 this population contained about 50 animals, which were behaving normally, reproducing freely, and readily taking deer and other natural prey. Another small group had been released to the west in Great Smoky Mountains National Park (Henry 1993, 1995; Phillips 1994; Rees 1989; Waddell 1995*a*, 1995*b*).

From a management standpoint the reintroductions have been highly successful; indeed, they serve as a model for future efforts involving other large predatory mammals. However, the projects are jeopardized by intense political controversy evidently generated by the concerns of a small minority of landowners in the affected areas (Bourne 1995*a*). Widely publicized views that the red wolf may have originated as a hybrid (see above), though not generally accepted by the scientific community, have provided further grounds for challenging the reintroductions and related conservation work. In August 1995 a measure that would have cut off funding for this work was brought before the U.S. Senate; it was narrowly defeated, largely through the efforts of one environmentally oriented Republican senator (Chafee 1995). Notwithstanding the controversy, the red wolf continues to be classified as endangered by the USDI and now is classified as critically endangered by the IUCN.

Canis lupus (gray wolf)

This species includes the largest wild individuals in the family Canidae. Head and body length is generally about 1,000–1,600 mm and tail length is 350–560 mm. However, there are several small subspecies along the southern edge of the range of the gray wolf, especially *C. lupus arabs* of the Arabian Peninsula; an adult male of that subspecies had a head and body length of only 820 mm and a tail length of 320 mm (Harrison 1968). Ferguson (1981) listed the following data: *C. lupus arabs* of Israel and Arabia, head and body length 1,140–1,440 mm, tail length 230–360 mm, and weight 14–19 kg; *C. lupus lupaster* of Egypt and Libya, head and body length 1,058–1,300 mm, tail length 283–355 mm, and weight 10–16 kg. In North America the largest animals are found in Alaska and western Canada, the smallest in Mexico. In a given population males are larger on average than females. Overall mean (and extreme) weights are about 40 (20–80) kg for males and 37 (18–55) kg for females (Mech 1970, 1974*a*). The pelage is long; the upper parts are usually light brown or gray, sprinkled with black, and the underparts and legs are yellow white. Entirely white individuals occur frequently in tundra regions and occasionally elsewhere. Black individuals are also common in some populations.

The gray wolf has the greatest natural range of any living terrestrial mammal other than *Homo sapiens.* It is found in all habitats of the Northern Hemisphere except tropical forests and arid deserts. Dens, used only for the rearing of young, may be located in rock crevices, hollow logs, or overturned stumps but are usually in a burrow, either dug by the parents themselves or initially made by another animal and enlarged by the wolves. Sometimes several such burrows, perhaps as far apart as 16 km, are excavated in the same season. A den may be used year after year. The animals prefer an elevated site near water. Tunnels are about 2–4 meters long and lead to an enlarged underground chamber with no bedding material. There may be several entrances, each marked by a large mound of excavated soil. When the young are about 8 weeks old they are moved to a rendezvous site, an area of about 0.5 ha., usually near water and marked by trails, holes, and matted vegetation. Here the young romp and play, and the other members of the pack gather for daily rest during the summer. Such sites are frequently changed, but some may be used for one or two months (Mech 1970, 1974*a;* Peterson 1977).

Movements are extensive and usually take place at night, but diurnal activity may increase in cold weather. During the summer the pack usually sets out in the early evening and returns to the den or rendezvous site by morning. In winter the animals wander farther and do not necessarily return to a particular location. They tend to move in single file along regular pathways, roads, streams, and

Gray wolf *(Canis lupus)*, photo from Zoological Society of London.

ice-covered lakes. The daily distance covered ranges from a few to 200 km (Mech 1970). In Finland the mean daily movement was determined to be 23 km (Pulliainen 1975). On Isle Royale, in Lake Superior, Peterson (1977) found that packs averaged 11 km per day or 33 km per kill. When individuals permanently disperse from a pack, they move much farther than normal; one traveled 206 km in 2 months (Mech 1974*a*), and another covered a straight-line distance of 670 km in 81 days (Van Camp and Gluckie 1979). Recent record movements are of a single animal that dispersed 886 km, from Minnesota to Saskatchewan (Fritts 1983), and a group of at least 2 and probably 4 wolves that moved at least 732 km across Alaska (Ballard, Farnell, and Stephenson 1983). Five dispersing wolves tracked by Fritts and Mech (1981) emigrated about 20–390 km, though others remained in the vicinity of the parental groups. Packs that depend on barren ground caribou make seasonal migrations with their prey and move as far as 360 km (Kuyt 1972; Mech 1970, 1974*b*).

The gray wolf usually moves at about 8 km/hr but has a running gait of 55–70 km/hr. It can cover up to 5 meters in a single bound and can maintain a rapid pursuit for at least 20 minutes. Prey is located by chance encounter, direct scenting, or following a fresh scent trail. Odors up to 2.4 km away can be detected. A careful stalk may be used to get as close as possible to the prey. If the objective is large and healthy and stands its ground, the pack usually does not risk an attack. Otherwise a chase begins, usually covering 100–5,000 meters. If the wolves cannot quickly close with the intended victim, they generally give up. There seems to be a continuous process of testing the individual members of the prey population to find those that are easily captured. Most hunts are unsuccessful. On Isle Royale, for example, only about 8 percent of the moose tested are actually killed (Mech 1970, 1974*b*). In one exceptional case the chase of a deer went for 20.8 km (Mech and Korb 1978). Once the wolves overtake an ungulate, they strike mainly at its rump, flanks, and shoulders.

The gray wolf is primarily a predator of mammals larger than itself, such as deer, wapiti, moose, caribou, bison, muskox, and mountain sheep. The smallest consistent prey is beaver. Following a drastic decline of deer in central Ontario, beaver remains were found in most wolf droppings (Voight, Kolenosky, and Pimlott 1976). Kill rates vary from about one deer per individual wolf every 18 days to one moose per wolf every 45 days (Mech 1974*a*). Usually a pack of several wolves will make a kill, consume a large amount of food, and then make another kill some days later. An adult can eat about 9 kg of meat in one feeding. Studies in Minnesota indicate that an average daily consumption is about 2.5 kg of deer per wolf and that pack members generally remain in the vicinity of a kill for several days, eventually utilizing nearly the entire carcass, including much of the hair and bones (Mech et al. 1971). Most analyses of predation on wild species show that immature, aged, and otherwise inferior individuals constitute most of the prey taken by wolves (Mech 1974*a*).

There long has been controversy regarding the wolf's effect on overall prey populations. It was once generally thought, even by some experienced zoologists, that the wolf could and did eliminate its prey. For example, Bailey (1930) wrote that "wolves and game animals can not be successfully maintained on the same range." Subsequently a popular view developed that the wolf and the animals

B

Gray wolf *(Canis lupus)*, photo from New York Zoological Society.

on which it depends are in precise balance, with the wolf taking just enough to prevent substantial population increases. Mech (1970) tentatively concluded that wolf predation is the major controlling factor where prey-predator ratios are about 11,000 kg of prey per wolf or less, as on Isle Royale. More recently it became apparent that the wolf is not regulating the moose herd on Isle Royale but merely cropping part of the annual surplus production (Mech 1974*b;* Peterson 1977). On the other hand, studies by Mech and Karns (1977) indicate that wolf predation was a major contributing factor in a serious decline of deer in the Superior National Forest of Minnesota from 1968 to 1974. Primary causes in the decline there, as well as in other parts of the Great Lakes region, were maturation of the forests that had been cut around the turn of the century and a series of severe winters.

Wolf population density varies considerably, being as low as 1/520 sq km in parts of Canada. In Alberta, Fuller and Keith (1980) found densities of 1/73 sq km to 1/273 sq km. Several studies in the Great Lakes region found apparently stable densities of around 1/26 sq km and suggested that this level was the maximum allowed by social tolerance (Mech 1970; Pimlott, Shannon, and Kolenosky 1969). Subsequent investigations determined that the wolf could attain densities nearly twice as great in areas where prey concentrate for the winter (Kuyt 1972; Parker 1973; Van Ballenberghe, Erickson, and Byman 1975). On Isle Royale, an area of 544 sq km, the number of wolves remained relatively stable at a mid-winter average of 22 from 1959 to 1973. There was an increase to 44 in 1976 evidently in response to rising moose numbers, thus indicating that food supply is the main determinant of density (Peterson 1977). The Isle Royale wolf population peaked at 50 in 1980 and then underwent a crash in association with falling moose numbers (Peterson and Page 1988). By 1990 only 14 wolves survived there even though the moose population again was on the rise, which led to concern about genetic viability (Wayne, Gilbert, et al. 1991).

Home range size depends on food availability, season, and number of wolves. The largest pack range, found in Alaska during winter, was 13,000 sq km, and the smallest, in southeastern Ontario during summer, was 18 sq km (Mech 1970; Pimlott, Shannon, and Kolenosky 1969). In Minnesota, ranges of 52–555 sq km have been reported (Fritts and Mech 1981; Harrington and Mech 1979; Van Ballenberghe, Erickson, and Byman 1975). One radio-collared pack in Minnesota used summer ranges of 117–32 sq km and winter ranges of 123–83 sq km (Mech 1977*a*). Alberta packs used ranges of 195–629 sq km in summer and 357–1,779 sq km in winter (Fuller and Keith 1980). Pack range in Russia and Kazakhstan varies from 30 sq km in areas of abundant food and good cover to more than 1,000 sq km in deserts and tundra (Bibikov, Filimonov, and Kudaktin 1983).

The home range of a wolf pack usually corresponds to a defended territory. There is generally little or no overlap between the ranges of neighboring packs. Lone wolves that split off from a pack may disperse a considerable distance or they will be continuously pursued by the resident packs and forced to shift about (Mech 1974*a;* Peterson 1977; Rothman and Mech 1979). Territories are relatively stable; some are known to have been used for at least 10 years (Mech 1979). Buffer zones, areas of little use, tend to develop between territories; lone wolves may sometimes center their activities there. In Minnesota deer have been found to have higher densities in such border areas and to survive longer there than elsewhere during times of general population declines (Mech 1977*e,* 1979; Rogers et al. 1980). Under normal conditions packs avoid areas where they might encounter other packs. When food shortages induce stress, however, wolves move into the buffer zones to hunt and eventually trespass into the territories of other packs. Meetings between packs are agonistic and may result in chases, savage fighting, and mortality (Harrington and Mech 1979; Mech 1970, 1977*b;* Van Ballenberghe and Erickson 1973).

Although packs are hostile to one another, the gray wolf is among the most social of carnivores. Groups usually contain 5–8 individuals but have been reported to have as many as 36 (Mech 1974*a*). Such associations are probably essential for consistent success in the pursuit and overpowering of large prey. The number of wolves in a pack tends to increase with the size of the usual prey, being 7 or less where deer is the only important food, 6–14 where both deer and wapiti are eaten, and 15–20 on Isle Royale, where moose is the primary prey (Peterson 1977). In certain southern parts of its range, such as Mexico, Italy, and Arabia, *C. lupus* seems to be less gregarious than in the north (Harrison 1968; McBride 1980; Zimen 1981). This situation may be partly unnatural, resulting from intensive human persecution and forced dependence on easily captured domestic animals and garbage. However, it has been demonstrated that a single female, and sometimes even an adult male, can successfully rear a litter of young (Boyd and Jimenez 1994).

A pack is essentially a family group, comprising an adult pair, which may mate for life, and their offspring of one or more years (Mech 1970). A few exceptions have been revealed by both field and genetic study (Lehman et al. 1992; Mech and Nelson 1990). The leader of a normal pack is usually a male, often referred to as the alpha male. He initiates activity, guides movements, and takes control at critical times, such as during a hunt. The males and females of a pack may have separate dominance hierarchies reinforced by aggressive behavior and elaborate displays of greeting and submission by subordinate members. Generally only the most dominant pair mate, and they inhibit sexual activity in the others. Social status is rather consistent, and a leader may retain its position for years, but roles can be reversed. Intragroup strife, perhaps resulting from increasing membership or declining food supplies, can result in a division into two packs or the splitting off of individuals. The latter may maintain a loose association with the parental group, sometimes following at a distance and feeding on scraps left behind, or disperse to a new area to seek a mate and begin a new pack (Mech 1970, 1974*a;* Peterson 1977; Wolfe and Allen 1973; Zimen 1975).

Studies by Fritts and Mech (1981) have contributed to our understanding of new pack formation in an area of an expanding wolf population. A young individual evidently does not remain with its parental pack past breeding age (about 22 months) unless it becomes a breeder itself upon the death of an alpha animal of the same sex. As it approaches maturity it may actively explore the fringes of the parental territory. After dispersal it either joins another lone wolf to search for a new territory or establishes itself in an area and awaits the arrival of an animal of the opposite sex. A new pack territory is relatively small but may incorporate a portion of the parental territory. Therefore, as population size increases, average territory size decreases. Data summarized by Gese and Mech (1991) indicate that most wolves disperse from their natal pack during February–April, when they are 11–12 months old, or during October–November, when they are 17–19 months old.

The gray wolf has a variety of visual, olfactory, and auditory means of communication. Vocalizations include growls, barks, and howls—continuous sounds usually lasting 3–11 seconds (Mech 1974*a*). Individuals have distinctive howls (Peterson 1977). Humans can hear howls 16 km away on the open tundra, and wolves probably respond to howls at distances of 9.6–11.2 km. Howling functions to bring packs together and as an immediate, long-distance form of territorial expression (Harrington and Mech 1979).

Territories are also maintained by scent marking via scratching, defecation, and especially urination. Scent marking differs from ordinary elimination in that there is a regular pattern of deposition at certain repeatedly used points. Peters and Mech (1975) determined that as a pack moves through its territory—visiting most parts at least every three weeks—signs are left at average intervals of 240 meters. Rothman and Mech (1979) found that scent marking also is important in bringing new pairs together for breeding and in helping established pairs to achieve reproductive synchrony.

Threats and attacks by the dominant members of a pack probably prevent sexual synchronization in subordinates; thus only the highest-ranking female normally bears a litter during the reproductive season (Zimen 1975). Some exceptions were reported by Van Ballenberghe (1983), and Mech and Nelson (1989) documented that during a single breeding season a male mated with two females, both of which raised litters. Mating may occur anytime from January, in low latitudes, to April, in high latitudes. Births take place in the spring. Courtship may extend for days or months, estrus lasts 5–15 days, and the gestation period is usually 62–63 days. Mean litter size is 6 young, and the range is 1–11. The young weigh about 450 grams at birth and are blind and deaf. Their eyes open after 11–15 days, they emerge from the den at 3 weeks, and they are weaned at around 5 weeks. The mother usually stays near the den for a period, during which time the father and other pack members hunt and bring food for both her and the pups. The young are commonly fed by regurgitation. At 8–10 weeks the young are shifted to the first in a series of rendezvous sites, each up to 8 km from the other. If they are in good condition, the young begin to travel with the pack in early autumn. Sexual maturity generally comes at around 22 months, but social restrictions often prevent mating at that time. A captive pair successfully bred at only 10 months. Mortality is highest among the young. In times of sharply declining food supplies all pups may be severely underweight or die from malnutrition, and reproduction may even cease. For adults in such a situation the primary mortality factor has been found to be intraspecific strife. Annual survival of adults in a population not under nutritional stress or human exploitation has been calculated at 80 percent. Wild females are known to have given birth when 10 years old and to have lived to an age of 13 years and 8 months. Potential longevity is at least 16 years (Mech 1970, 1974*a*, 1977*a*, 1977*b*, 1989; Medjo and Mech 1976; Van Ballenberghe, Erickson, and Byman 1975; Van Ballenberghe and Mech 1975).

The wolf is often believed to be a direct threat to people. In Eurasia attacks are unusual but evidently have occurred, sometimes resulting in death (Pulliainen 1980; Ricciuti 1978). In North America there appear to have been only four well-documented attacks by wild wolves (none resulting in death): (1) an injury inflicted by a female that may have been attracted to a camp by male dogs (Jenness 1985); (2) a prolonged assault involving a probably rabid individual (Mech 1970); (3) simply an aggressive leap that made contact with the person but caused no injury (Munthe and Hutchison 1978); and (4) several lunges by members of a pack at a group of three people (Scott, Bentley, and Warren 1985). Mech (1990) discussed a number of less detailed reports of attacks as well as several incidents involving apparently accidental wolf-human contact and rabid wolves (again without death or serious injury to people).

A far more substantive basis for the age-old warfare between people and the gray wolf is depredation by the latter on domestic animals, notably cattle, sheep, and reindeer. The wolf also has been persecuted, especially in the twentieth century, because of its alleged threat to populations of the wild ungulates that are desired by some persons for sport and subsistence hunting. The wolf was long taken by various kinds of traps and snares as well as by pursuit with packs of specially trained dogs. In the nineteenth century poison came into widespread use, and in the mid–twentieth century hunting from aircraft became popular, especially on the open tundra.

The last wolves in the British Isles were exterminated in the eighteenth century. By the early twentieth century the species, except for occasional wandering individuals, had disappeared in most of western Europe and in Japan. Modest comebacks occurred in Europe during World Wars I and II, but currently the only substantial populations on the Continent west of Russia are in the Balkans. Recent conservation efforts have prevented extirpation in Poland, Spain, and Italy, even allowing increases in distribution and numbers (Blanco, Reig, and de la Cuesta 1992; Boitani 1992; Okarna 1993). There are also remnant groups in Portugal, Slovakia, Hungary, and Scandinavia. A few individuals recently have entered Germany and France (*Oryx* 28 [1994]: 6). *C. lupus* survives over much of its former range in southwestern and south-central Asia but is generally rare. Only 500–800 survive in India.

There were estimated to be 150,000–200,000 wolves in the old Soviet Union after World War II, but these became subject to an intensive government control program. The annual kill was 40,000–50,000 individuals from 1947 to 1962 and subsequently dropped to about 15,000. In the mid-1970s the estimated number of wolves in the old Soviet Union was 50,000, about two-thirds of them in the Central Asian republics (Grzimek 1975; Pimlott 1975; Pulliainen 1980; Roberts 1977; Shahi 1982; Smit and Van Wijngaarden 1981; Zimen 1981). Wolf numbers subsequently again increased in the Soviet Union, leading to implementation of a bounty of up to 100 rubles (Bibikov 1980). There also is aerial hunting and poisoning by the government. In 1980 the number of wolf pelts taken there was 35,573 (Bibikov, Ovsyannikov, and Filimonov 1983).

The decline of the gray wolf was even more sweeping in the New World than in the Old. The species was largely eliminated along the east coast and in the Ohio Valley by the mid–nineteenth century. By 1914 the last gray wolves had been killed in Canada south of the St. Lawrence River and in the eastern half of the United States, except for northern parts of Minnesota and Wisconsin and the upper peninsula of Michigan. In the following year the federal government began a large-scale program to destroy predators. Partly as a result of this campaign resident populations of the gray wolf apparently had disappeared from the western half of the conterminous United States by the 1940s. The rate of decline subsequently slowed, mainly because the wolf had been eliminated from nearly all parts of the continent that were conducive to the raising of livestock. Conflict with agricultural interests does continue all along the lower edge of the current major range of the gray wolf, which extends across southern Canada, from coast to coast, and dips down into northern Minnesota (Nowak 1975*a*). Regular government control operations are still carried out in parts of Canada and even have intensified recently in some areas in response to reported predation on livestock and big game (Gunson 1983; Stardom 1983; Tompa 1983).

The wolf still occurs over much of the northern part of Minnesota, primarily in boreal forest and bog country that is not suitable for agriculture but also in adjoining lands that are well settled. The species is protected by federal law, except that individuals preying on domestic animals may be taken by government agents. Many persons in Minnesota have argued that current protective regulations are overly restrictive and do not allow adequate control of depredations and that the wolf should be subject to sport hunting and commercial trapping. Recent studies in northwestern Minnesota, however, indicate that wolves infrequently attack domestic livestock even when living nearby (Fritts 1982; Fritts and Mech 1981; Fritts et al. 1992; Mech 1977*d*). In 1983 the U.S. Fish and Wildlife Service actually attempted to open the Minnesota wolf to sport hunting, but this effort was defeated through litigation. The species thus remained under protection, and perhaps as a result, it continued to occupy more of Minnesota and even to extend its range into northern Wisconsin, upper Michigan, and the Dakotas (see below).

In Alaska the wolf is legally open to sport hunting and fur trapping, but there is also controversy (Harbo and Dean 1983). Shooting from aircraft was a major hunting method for many years. In 1972 the practice was banned by federal law, except for government predator-control programs. Starting in the mid-1970s the Alaska Department of Fish and Game authorized aerial hunting of wolves in an effort to increase the numbers of moose and caribou in certain parts of the state. There followed a series of legal battles involving opposing conservationist organizations, the state government, and the federal agencies that owned much of the land where the activity was to take place. Some of the hunts were halted, but others were allowed to proceed. The taking of the wolf in Alaska has been stimulated in part by rising fur prices. For the 1976–77 season the reported state harvest was 1,076 pelts with an average value of $200 (Deems and Pursley 1978). In the 1991–92 season 1,162 pelts were taken in Alaska and sold for an average price of $275 (Linscombe 1994).

The wolf population of the Alexander Archipelago and adjacent mainland of southeastern Alaska represents a subspecies distinct from that of the rest of Alaska and western Canada. This population preys largely on the black-tailed deer herds of the area, which in turn are dependent on the understory of old-growth forests (Ingle 1994). Planned logging of these forests could result in a drastic decline of the deer and ultimately extinction of the wolf. Nonetheless, the U.S. Fish and Wildlife Service recently rejected a petition to classify the wolf population as threatened (Beattie 1995), a measure that would have allowed protection of the forests, nearly all of which are on federal lands.

The total number of gray wolves in North America can only be guessed at. Overall populations do not appear to have declined since the 1950s, but there is concern that excessive hunting, as well as oil and mineral exploitation, could adversely affect prey species, especially the caribou of tundra regions. The species was exterminated in Greenland by the 1930s, but individuals subsequently wandered back in from arctic Canada. There now is a resident population of perhaps 50 animals, and breeding has occurred well down the east coast of the island (Dawes, Elander, and Ericson 1986; Higgins 1990). However, the wolves of the arctic islands of Canada may themselves be in jeopardy because of a crash in native caribou populations and increasing human harassment. There also is some suggestion that the wolves of that region have been affected by hybridization with domestic dogs (Clutton-Brock, Kitchener, and Lynch 1994). There are about 4,000–7,000 wolves in Alaska, and there are estimates of 30,000–60,000 in Canada (Carbyn 1983, 1987; Theberge 1991). The fur take in Canada was 5,000–7,000 annually in the late 1970s but fell to about 4,000 in 1984 (Carbyn 1987). Wolf numbers have continued to fall in Mexico, mainly because of persecution by cattle ranchers. There may still have been hundreds in the 1950s, but there were no more than 50 in the 1970s and only 10 by the 1990s (Ginsberg and Macdonald 1990; McBride 1980). Occasional individuals from Mexico crossed into the United States until the 1970s, but they usually were quickly killed. There is a captive breeding pool of about 140 Mexican wolves, and consideration is being given to reintroduction projects in Arizona and New Mexico (Parsons 1996; Siminski *in* Bowdoin et al. 1994).

In Minnesota the number of wolves has grown steadily under protection, from about 600 in the 1950s, to 1,200 in the 1970s, and to more than 1,800 in the early 1990s (Fuller, Berg, et al. 1992; Mech 1977*d*; Mech, Pletscher, and Martinka 1995). Numerous individuals from this expanding population evidently have dispersed into the Dakotas (Licht and Fritts 1994). Resident populations may have been eliminated in Wisconsin and Michigan in the 1950s, and an attempted reintroduction in Michigan failed in 1974 when all four of the involved wolves were quickly killed by human agency (Weise et al. 1975). Nonetheless, legal protection subsequently allowed dispersing wolves from Minnesota to move back into both northern Wisconsin and the adjacent upper peninsula of Michigan, and by the early 1990s a number of packs were established, with at least 50 animals in each state (Jensen, Fuller, and Robinson 1986; Mech 1995; Mech and Nowak 1981; Mech, Pletscher, and Martinka 1995; Robinson and Smith 1977; Thiel 1985).

There is increasing concern about the effects of hybridization on wolves in the Great Lakes region and possibly some other areas. There long has been morphological evidence indicating that the expanding population of *C. latrans* hybridized with the small wolf subspecies *C. lupus lycaon* in southeastern Ontario and southern Quebec (Nowak 1979). Recent studies have shown that all wolves in those areas and on Isle Royale have mitochondrial DNA derived from coyotes and that most wolves in Minnesota and southwestern Ontario have been similarly affected (Lehman et al. 1991). Additional investigation indicates a general decline in the genetic viability of the small, isolated population of wolves on Isle Royale (Wayne, Gilbert, et al. 1991). Fortunately, there is as yet no evidence suggesting any morphological, behavioral, or ecological modification of *C. lupus* in Minnesota, western Ontario, Isle Royale, or Algonquin National Park (Nowak 1992). Hybridization between *C. lupus* and *C. familiaris* has not yet been shown to be detrimental to wild populations of the former, despite the far greater abundance of the latter. Of possible concern is the growing popularity of captive wolf-dog hybrids, now estimated to number about 300,000 in the United States. If such animals become feral, they could facilitate a breakdown of genetic integrity, especially in sparse or recolonizing wolf populations. A few of these hybrids have escaped and attacked children, thereby contributing to a negative image of *C. lupus* in general (Willems 1995).

Since the 1940s there have been regular reports of wolves from the northwestern conterminous United States, especially along the Rocky Mountains between Glacier and Yellowstone national parks (Ream 1980; Weaver 1978). Such records probably represented individuals that wandered from Canada. Legal protection, research, and conservation measures provided pursuant to the U.S. Endangered Species Act of 1973 allowed some of these wolves to establish resident packs. A breeding population of about 100 animals now utilizes land on both sides of the border in and around Glacier National Park (Chadwick 1995). Some individuals from this population may have spread as far as Wyoming and Idaho. However, development of a federal program to reintroduce wolves directly into those states was initiated in 1988. In early 1995, following an expensive and controversial process, several packs of wolves that had been captured in western Canada were released in Yellowstone National Park, Wyoming, and wilderness areas of central Idaho. As of mid-1996 the reintroduced populations contained more than 100 wolves, breeding was occurring regularly, and predation on domestic livestock had not been a significant problem (Bangs and Fritts 1996; Cook 1993; Gerhardt 1995). And recently there has been a natural movement of wolves into the state of Washington (Mech 1995).

The IUCN classifies the population of *C. lupus* in Mexico as extinct in the wild, the population in Italy as vulnerable, and the population in Spain and Portugal as conservation dependent. The USDI lists all populations of *C. lupus* in the conterminous United States and Mexico as endangered, except for that in Minnesota, which is designated as threatened. There have been repeated claims that the Minnesota population should be completely removed from the U.S. List of Endangered and Threatened Wildlife. However, it now is known that in addition to direct human persecution and long-term habitat disturbance, this population is jeopardized by parasitic heartworms and the deadly disease canine parovirus, both evidently spread by domestic dogs (Mech and Fritts 1987). *C. lupus* is on appendix 2 of the CITES, except for the populations of Pakistan, India, Nepal, and Bhutan, which are on appendix 1.

Canis familiaris (domestic dog)

There are approximately 400 breeds of domestic dog, the chihuahua being the smallest and the Irish wolfhound the largest (Grzimek 1975). According to the National Geographic Society (1981), head and body length is 360–1,450 mm, tail length is 130–510 mm, shoulder height is 150–840 mm, and the normal weight range is 1–79 kg. The heaviest individual on record, a St. Bernard, weighs about 150 kg (*Washington Post,* 29 November 1987). In the wild subspecies of Australia, *C. f. dingo,* head and body length is 1,170–1,240 mm, tail length is 300–330 mm, shoulder height is about 500 mm, and weight is 10–20 kg. The dingo is usually tawny yellow in color, but some individuals are white, black, brown, rust, or other shades. The feet and tail tip are often white (Clutton-Brock, Corbet, and Hills 1976). From other forms of *C. familiaris* of comparable size and shape the dingo can be distinguished by its longer muzzle, larger bullae, more massive molariform teeth, and longer, more slender canine teeth (Newsome, Corbett, and Carpenter 1980).

There have been various ideas regarding the origin of the domestic dog. The current consensus is that it was derived from one of the small south Eurasian subspecies of *C. lupus* and subsequently spread throughout the world in association with people (Nowak 1979). Some authorities, including Clutton-Brock, Corbet, and Hills (1976), think that the direct ancestor is probably the Indian wolf, *C. lupus pallipes,* but Olsen and Olsen (1977) argued in favor of the Chinese wolf, *C. l. chanco.* The wolf and dog still hybridize readily, but only when brought together in captivity or under very unusual natural conditions. The oldest well-documented remains of *C. familiaris,* dating from about 11,000 and 12,000 years ago, were found in Idaho and Iraq, respectively. Beebe (1978) reported a specimen from the northern Yukon with a minimum age of 20,000 years, but Olsen (1985) considered that record doubtful.

At present, from the Balkans and North Africa to Japan and the East Indies dogs known as pariahs lead a semidomestic or even feral existence around villages (Bueler 1973; Fox 1978; Trumler *in* Grzimek 1990). They may take food from people, scavenge, or actively hunt deer and other an-

Dingo *(Canis familiaris)*, photo from New York Zoological Society.

imals. They are socially flexible, being found either alone or in packs, but there seems to be a dominance hierarchy in any given area. These dogs, generally primitive in physical appearance, are probably closely related to the earliest dogs as well as to the Australian dingo. In New Guinea and Timor also there are wild dog populations, which are related to the primitive pariah-dingo group (Troughton 1971). The oldest definitely known fossils of the dingo in Australia date from about 3,500 years ago. People arrived in Australia at least 30,000 years ago. The dingo evidently was brought in long afterward but before true domestication had been achieved, and it was able to establish wild populations (Macintosh 1975). It is possible that these dogs were spread by maritime peoples of south-central Asia rather than by migrating aboriginal peoples (Gollan 1984). There are many other populations of feral dogs, notably on islands and in Italy, but these animals are descendants of fully domesticated individuals (Lever 1985).

In a recent reassessment of the early history of dogs Corbett (1995) identified the entire complex of primitive, dingolike animals across southern Asia, the East Indies, New Guinea, and Australia as the single taxon *dingo.* This revisionist view, which was followed by Ginsberg and Macdonald (1990), holds that *dingo* evolved on mainland southern Asia in association with people and then about 3,500–4,000 years ago was spread by seafarers to other regions. Some of these animals may even have been brought across the Bering Strait to North America and persist today in the form of the Carolina hunting dog. Others are thought to have reached some of the Pacific islands, Madagascar, and southern Africa. It is difficult to determine how to approach these populations from the standpoint of research and conservation as they are not fully natural entities but may have been established components of their ecosystems for thousands of years. Outside of Australia little is known about *dingo;* it probably is declining in association with environmental disruption and is threatened by hybridization with true domestic dogs, but there apparently still are some pure, viable, and wild-living populations in Southeast Asia, especially Thailand.

The dingo population of New Guinea long has been recognized as a distinctive wild population and even has been given the specific name *Canis hallstromi.* It once was thought to be ancestral to the Australian dingo, but Flannery (1990) noted that remains from New Guinea are not as old as those found in Australia and that introduction to the latter region thus probably occurred first. The New Guinea dingo sometimes is called the "singing dog" because of its unique vocalizations, including a form of howling marked by an extraordinary degree of frequency modulation and a number of signals, such as a high-pitched rapid trill, not reported for other canids. Social behavior seems to be limited, and females show a single annual estrus. The wild population is now restricted to a few isolated sites and is jeopardized by hybridization with true domestic dogs (Brisbin et al. 1994).

The dingo was once found throughout Australia, in forests, mountainous areas, and plains (Bueler 1973). Studies by Corbett and Newsome (1975) were carried out in an arid region of deserts and grasslands in central Australia, where the dingo depends in part on water holes made for cattle. Natal dens were found in caves, hollow logs, and modified rabbit warrens, usually within 2–3 km of water. Most activity in this region is nocturnal. In good seasons the diet consists mainly of small mammals, especially the introduced European rabbit *(Oryctolagus).* In times of drought the dingo takes kangaroos and cattle, mostly calves. Studies by Robertshaw and Harden (1985) in a forested area of northeastern New South Wales found the diet to consist mostly of larger native mammals, particularly wallabies. In an open area of Western Australia, Thomson (1992*b*) found the major prey to be large kangaroos and wallaroos *(Macropus rufus* and *Macropus robustus).*

Harden (1985) found daily movements to be about 10–20 km. Periods of movement usually were short and separated by even shorter periods of rest. On the average, individuals were active for 15.25 hours and at rest for 8.75 hours each day. Average home range for adults was 27 sq km. Thomson (1992*c*) reported that in Western Australia packs occupy permanent territories of 45–113 sq km. These areas have very little overlap and there is minimal interaction between neighboring packs, the animals evidently being kept apart by howling and scent marking. According to

Domestic dogs *(Canis familiaris)*, photo from J. Nowak and E. Nowak.

Corbett (1995), territory size averages 39 sq km in the tropical north and is 10–21 sq km in the forested east. Most individuals remain in their natal area, but some disperse, especially young males. The longest recorded movement for a tagged individual is 250 km over a 10-month period.

The dingo may be basically solitary in some areas (Corbett and Newsome 1975), but where it is not subject to human persecution there are discrete and stable packs of 3–12 individuals (Corbett 1995). Rank order is determined and maintained largely by aggressive interaction, especially among males. The dingo is not a particularly vocal canid but has a variety of sounds. Howls, which probably function to locate friends and repel strangers, are heard frequently in the single annual breeding season. Corbett (1988) reported that a captive group had separate male and female dominance hierarchies; all the females became pregnant, but the alpha female killed the newborn of the others. Wild females have a single annual estrus and bear their pups mostly during the winter, following a gestation period of 63 days. Litter size is usually 4–5 young but ranges from 1 to 10 (Bueler 1973; Corbett 1995; Thomson 1992*a*). Yearlings may assist an older pair to raise their pups. Independence is generally achieved by 3–4 months, but the young animals often then associate with a mature male (Corbett and Newsome 1975). Maximum known longevity is 14 years and 9 months (Grzimek 1975).

Although the dingo is said to be regularly captured and tamed by the natives and other people of Australia, Macintosh (1975) argued that it has never been successfully domesticated. He noted that it does have a close relationship with some groups of aborigines but evidently is not used in hunting nor intentionally fed and that its main function may be to sleep in a huddle with persons and thereby provide protection from the cold. Corbett (1995), however, cited several records of the dingo's being used for the pursuit of game and otherwise living in close association with people.

Agriculturalists in Australia generally have a low opinion of the dingo mainly because of its predation on sheep. Intensive persecution began in the nineteenth century and has continued to the present. During the 1960s about 30,000 dingos were killed for bounty annually in Queensland alone. Nearly 10,000 km of fencing has been constructed in eastern Australia in an effort to keep the wild dogs off the sheep ranges. Nonetheless, Macintosh (1975) suggested that depredations have been greatly exaggerated, and Corbett (1995) pointed out that the dingo has actually aided agriculture by destroying introduced rabbits. Examination of stomach contents in Western Australia indicated that domestic stock was not a significant part of the dingo's diet even though sheep and cattle were common in the study area (Whitehouse 1977). Human persecution has caused a decline in dingo populations, but another serious problem is hybridization with the domestic dog, a phenomenon that evidently is spreading with human settlement (Newsome and Corbett 1982, 1985). Dingo populations in the tropical north and in the northwest are still pure, but those of southeastern Australia now seem to have hybridized to some extent (Corbett 1995; Jones 1990; Thomson 1992*a*).

Aside from the dingo, *C. familiaris* long was one of the least-known canids with respect to its behavior and ecology under noncaptive conditions, but there has been increasing study in recent years. Beck (1973, 1975) estimated that up to half of the 80,000–100,000 dogs in Baltimore, Maryland, are free-ranging, at least at times, with an average density of about 230/sq km. They shelter in vacant buildings and garages and under parked cars and stairways. They are active mainly from 0500 to 0800 and from 1900 to 2200 hours in the summer, remaining out of sight during the midday heat. Their diet, consisting mostly of garbage but also including rats and ground-nesting birds, seems adequate to maintain weight and good health. Studies of unrestrained pets (Berman and Dunbar 1983; Rubin and Beck 1982) indicate that movement is concentrated in the early morning and late evening but usually covers a very small area, probably because of the lack of a need to secure food. The distance a dog tends to move away from its home is positively correlated with the amount of time that it is allowed freedom.

Fully feral dogs sometimes occur in the countryside. Scott and Causey (1973) found Alabama packs to use moist floodplains in warm weather and dry uplands in cool weather and to cover distances of 0.5–8.2 km per day. Nesbitt (1975) determined that the females on a wildlife refuge in Illinois did not dig a den but gave birth in heavy cover. Activity in that area occurred anytime but was mainly nocturnal and crepuscular. The dogs traveled single file along roads, trails, and crop rows; if frightened, they took cover among trees and bushes. They fed on crippled waterfowl and deer, road kills and other carrion, small animals, some vegetation, and garbage.

In a study of free-ranging urban dogs in Newark, New Jersey, Daniels (1983*a*, 1983*b*) calculated a population density of about 150/sq km. Home range varied from 0.2 to 11.1 ha. in summer and from 0.1 to 5.7 ha. in winter. The relatively few groups that formed rarely contained more than two animals and often lasted only a few minutes; aggression was unusual, and there was no evidence of territoriality. However, presence of an estrous female in an area led to a congregation of males, usually about five or six, and to an increase in aggression and formation of a dominance hierarchy. Familiarity was observed to be an important basis for all social activity and for acceptance of a male by a female. At several sites in Mexico and the southwestern United States, Daniels and Bekoff (1989) found feral dogs to be more social than those in urban and rural areas. It was suggested that the formation of permanent groups offers little advantage to dogs dependent on people but that pack living facilitates feeding and protection for feral animals.

Home range was found to be 1.74 ha. for temporarily unrestrained suburban pets in Berkeley, California (Berman and Dunbar 1983); about 2.6 ha. each for 2 full-time free-ranging individuals in Baltimore (Beck 1975); 61 ha. for a group of 3 unowned dogs that lived in an abandoned building and used the streets and parks of St. Louis (Fox, Beck, and Blackman 1975); 444–1,050 ha. each for three feral packs of 2–5 dogs (Scott and Causey 1973); and 28,500 ha. for a feral pack of 5–6 (Nesbitt 1975). The last group was observed to have a dominance hierarchy and to be led by a female. Pets tend to be solitary and form social groups only randomly, whereas the three unowned urban animals maintained a long-term relationship, the single female member initiating movements. In the Baltimore study, half of the animals seen were solitary, 26 percent were in pairs, and the rest were in groups of up to 17 members. Spring and autumn breeding peaks there were suggested by fluctuations in reports of unwanted dogs.

Female dogs enter estrus twice a year, usually in late winter or early spring and in the autumn. Heat lasts about 12 days. At such times males tend to leave their owners' homes, mark territories, and fight rivals. The gestation period averages 63 days, litter size is usually 3–10 young, and nursing lasts about 6 weeks. Males often remain with the females and young. Sexual maturity comes after 10–24 months, and old age generally after 12 years, but a few individuals live for 20 years (Asdell 1964; Bueler 1973; Grzimek 1975).

There are an estimated 50 million owned dogs in the United States, and many more lack owners. Although these animals have abundant uses and values, they may cause problems for some people. In Baltimore, for example, dogs have been implicated in the spread of several diseases (in addition to rabies), they may benefit rats by overturning garbage cans, and there are about 7,000 reported attacks on people each year (Beck 1973, 1975). There are 1–3 million reported attacks annually for the whole United States (Rovner 1992). In 1986, 13 people, most of them children, were killed in the United States; in 7 of these cases the breed known as the pit bull was responsible (*Washington Post*, 1 September 1987, B-1, B-5). In 1990, 18 people were killed, 8 by pit bulls (ibid., 22 May 1991, A-29). With respect to rabies, dogs are traditionally viewed as the primary threat to people, but in recent decades other animals have been of far greater significance, at least in the United States and Europe (particularly *Lasionycteris, Eptesicus, Vulpes vulpes, Procyon*, and *Felis catus*). *C. familiaris* also often is considered to be a serious predator of livestock and game animals, especially deer; however, field studies have indicated that feral dogs do not significantly affect deer populations and may even have a sanitary function in eliminating carrion and crippled animals (Gipson and Sealander 1977; Nesbitt 1975; Scott and Causey 1973).

CARNIVORA; CANIDAE; **Genus CHRYSOCYON**
Hamilton-Smith, 1839

Maned Wolf

The single species, *C. brachyurus*, originally occurred in open country of central and eastern Brazil, eastern Bolivia, Paraguay, Argentina, and Uruguay (Cabrera 1957; Langguth 1975; Mones and Olazarri 1990; Roig 1991). There may be disjunct populations on the llanos of Colombia (Dietz 1985) and in southeastern Peru (Ginsberg and Macdonald 1990).

Head and body length is 950–1,320 mm, tail length is 280–490 mm, shoulder height is 740–900 mm, and weight is 20–26 kg. There is a general impression of a red fox *(Vulpes vulpes)* on stilts (Clutton-Brock, Corbet, and Hills 1976). The general coloration is yellow red. The hair along the nape of the neck and middle of the back is especially long and may be dark in color. The muzzle and lower parts of the legs are also dark, almost black. The throat and tail tuft may be white. The coat is fairly long, somewhat softer than that of *Canis*, and has an erectile mane on the back of the neck and top of the shoulders. The ears are large and erect, the skull is elongate, and the pupils of the eyes are round.

The maned wolf inhabits grasslands, savannahs, and swampy areas (Langguth 1975). The natal nest is located in thick, secluded vegetation (Bueler 1973). Activity is mainly nocturnal and crepuscular (Dietz 1984). It has been suggested that the remarkably long legs are an adaptation for fast running or for movement through swamps, but the actual function is probably to allow seeing above tall grass. *Chrysocyon* is not an especially swift canid, does not pursue prey for long distances, and generally stalks and pounces like a fox (Kleiman 1972). Its omnivorous diet includes rodents, other small mammals, birds, reptiles, insects, fruit, and other vegetable matter.

Chrysocyon is monogamous, with mated pairs sharing defended territories averaging 27 sq km. However, most activity is solitary; the male and female associate closely only during the breeding season (Dietz 1984). Captives some-

Maned wolf *(Chrysocyon brachyurus)*, photo by Bernhard Grzimek.

times can be kept together without apparent strife, though there is usually an initial period of fighting and then establishment of a dominance hierarchy (Brady and Ditton 1979). The father is known to regurgitate food for the young in captivity (Rasmussen and Tilson 1984) and may also have a significant parental role in the wild (Dietz 1984). The three main vocalizations are: a deep-throated single bark, heard mainly after dusk; a high-pitched whine; and a growl during agonistic behavior (Kleiman 1972).

Births in captivity have occurred in July and August in South America and in January and February in the Northern Hemisphere. Females are monestrous, heat lasts about 5 days, the gestation period is 62–66 days, and litter size is two to four young. The young weigh about 350 grams at birth, open their eyes after 8 or 9 days, begin to take some regurgitated food at 4 weeks, and are weaned by 15 weeks (Brady and Ditton 1979; Da Silveira 1968; Faust and Scherpner 1967). One captive lived 15 years and 8 months (Marvin L. Jones, Zoological Society of San Diego, pers. comm., 1995).

The maned wolf is not extensively hunted for its fur, but it is sometimes persecuted because of an unjustified belief that it kills domestic livestock. It often is killed because of alleged depredations on chickens. Resident populations disappeared from most of Argentina and from Uruguay in the nineteenth century, though one individual recently was taken in the latter country. It now is threatened in other regions by the annual burning of its grassland habitat, hunting, live capture, and disease (Ginsberg and Macdonald 1990; Roig 1991; Thornback and Jenkins 1982). It also, however, recently extended its range into a deforested zone of central Brazil (Dietz 1985). It is classified as near threatened by the IUCN and as endangered by the USDI and is on appendix 2 of the CITES.

CARNIVORA; CANIDAE; **Genus OTOCYON**
Müller, 1836

Bat-eared Fox

The single species, *O. megalotis,* is found from Ethiopia and southern Sudan to Tanzania and from southern Angola and Zimbabwe to South Africa (Coetzee *in* Meester and Setzer 1977). According to Ansell (1978), it does not occur in Zambia. *Otocyon* sometimes has been placed in a separate subfamily, the Otocyoninae, on the basis of its unusual dentition. Clutton-Brock, Corbet, and Hills (1976), however, considered *Otocyon* to be simply an aberrant fox with systematic affinities to *Urocyon* (which they included in *Vulpes*) and some behavioral similarities to *Nyctereutes.*

Head and body length is 460–660 mm, tail length is 230–340 mm, shoulder height is 300–400 mm, and weight is 3.0–5.3 kg. The upper parts are generally yellow brown with gray agouti guard hairs. The throat, underparts, and insides of the ears are pale. The outsides of the ears, mask, lower legs, feet, and tail tip are black. In addition to coloration, distinguishing characters include the relatively short legs and the enormous ears (114–35 mm long).

Otocyon has more teeth than any other placental mammal that has a heterodont condition (the teeth being differentiated into several kinds). Whereas in all other canids there are no more than two upper and three lower molars, *Otocyon* has at least three upper and four lower molars. This condition is sometimes held to be primitive, but it more likely represents the results of a mutation that caused the appearance of the extra molar teeth in what had been a fox population with normal canid dentition (Clutton-Brock, Corbet, and Hills 1976).

Bat-eared fox *(Otocyon megalotis)*, photo by Bernhard Grzimek.

The bat-eared fox is found in arid grasslands, savannahs, and brush country. It seems to prefer places with much bare ground or where the grass has been kept short by burning or grazing. When the grass again grows high, *Otocyon* may depart and wander about in search of a new place of residence. A capable digger, it either excavates its own den or enlarges the burrow of another animal. A family may have more than one den in its home range, each with multiple entrances and chambers and several meters of tunnels. In the Serengeti 85 percent of activity occurs at night, but in South Africa *Otocyon* is mainly diurnal in winter and nocturnal in summer. One female was observed to forage over about 12 km per night. The diet consists predominantly of insects, most notably termites, and also includes other arthropods, small rodents, the eggs and young of ground-nesting birds, and vegetable matter (Kingdon 1977; Lamprecht 1979; Nel 1978; Smithers 1971). In one part of the central Karoo about half of the food was plant material, especially wild fruits (Kuntzsch and Nel 1992).

In the Masai Mara Reserve of Kenya, Malcolm (1986) found an overall population density of 0.8–0.9/sq km. Group home ranges there averaged 3.53 sq km and overlapped; in one case there was no area of exclusive use. Groups consisted of 2–5 members and generally had amicable relations with other groups. In South Africa, Nel (1978) also found home ranges to overlap extensively and observed no territorial defense or marking. Up to 15 individuals, representing four groups, were seen foraging within less than 0.5 sq km. In the Serengeti, however, Lamprecht (1979) found resident families to occupy largely exclusive home ranges of 0.25–1.5 sq km and to mark them with urine. Groups usually consisted of a mated adult pair accompanied by their young of the year for a lengthy period. A few observations suggested that there may sometimes be 2 adult females with a male. Strangers of the same sex generally were hostile. Contact between members of a group was maintained by soft whistles. Nel and Bester (1983) reported that most communication involves visual signaling, mainly with the large ears and tail.

In both the Serengeti and Botswana, births occur mainly from September to November (Lamprecht 1979; Smithers 1971), but pups have been recorded in Uganda in March (Kingdon 1977), and reproduction may be year-round in some parts of East Africa (Malcolm 1986). Gestation is usually reported as 60–70 days, but Rosenberg (1971) calculated the period at 75 days for a birth in captivity. Litters contain 2–6 young. According to Skinner and Smithers (1990), the average litter size is 5, there sometimes are two litters per year, newborn weight is about 100–140 grams, and weaning occurs at about 10 weeks. The young then begin to forage with the parents; regurgitation is evidently rare. The young are probably full grown by 5 or 6 months and separate from the parents prior to the breeding season. According to Jones (1982), a captive lived 13 years and 9 months.

Otocyon has declined in settled parts of South Africa (Coetzee *in* Meester and Setzer 1977). Nonetheless, it is apparently extending its range eastward into Mozambique and into previously unoccupied parts of Zimbabwe and Botswana (Pienaar 1970). Rabies has recently been identified as a major problem to the genus in East Africa and was the main known cause of mortality in Serengeti National Park in the late 1980s (Maas 1994).

CARNIVORA; CANIDAE; **Genus CUON**
Hodgson, 1838

Dhole

The single species, *C. alpinus*, is found from southern Siberia and eastern Kazakhstan to India and the Malay Peninsula and on the islands of Sumatra and Java but not

Dhole *(Cuon alpinus)*, photo from New York Zoological Society.

Sri Lanka. In the Pleistocene, *Cuon* did occur in Sri Lanka and probably Borneo and also had a vast range extending from western Europe to Mexico (Corbet and Hill 1992; Kurten 1968; Kurten and Anderson 1980). Recently, Serez and Eroglu (1994) reported the presence of a living population of *C. alpinus* in northeastern Turkey, far from the remainder of the known modern range of the species. Except as otherwise noted, the information for this account was taken from the review papers by Cohen (1977, 1978) and Davidar (1975).

Head and body length is 880–1,130 mm, tail length is 400–500 mm, and shoulder height is 420–550 mm. Males weigh 15–21 kg and females, 10–17 kg. The coloration is variable, but generally the upper parts are rusty red, the underparts are pale, and the tail is tipped with black. In the northern parts of the range the winter pelage is long, soft, dense, and bright red; the summer coat is shorter, coarser, sparser, and less vivid in color. *Cuon* resembles *Canis* externally, but the skull has a relatively shorter and broader rostrum. Females have 12–16 mammae.

The dhole occupies many types of habitat but avoids deserts. In the Soviet Union it occurs mainly in alpine areas, and in India it is found in dense forest and thick scrub jungle. The preferred habitat in Thailand is dense montane forest at elevations of up to 3,000 meters (Lekagul and McNeely 1977). *Cuon* may excavate its own den, enlarge a burrow made by another animal, or use a rocky crevice. M. W. Fox (1984) found one earth den with six entrances leading to a labyrinth of at least 30 meters of interconnected tunnels and four large chambers; many generations of dholes probably had developed this complex. *Cuon* may be active at any time but mainly in early morning and early evening. Cohen et al. (1978) reported a major peak of activity at 0700–0800 hours and a lesser peak at 1700–1800 hours.

The dhole hunts in packs and is primarily a predator of mammals larger than itself. Prey is tracked by scent and then pursued, sometimes for a considerable distance. When the objective is overtaken, it is surrounded and attacked from different sides. Prey animals include deer, wild pigs, mountain sheep, gaur, and antelopes. The chital *(Axis axis)* is probably the major prey in India, though Cohen et al. (1978) found remains of this deer to occur less frequently than those of *Lepus* in the droppings of *Cuon*. The diet also includes rodents, insects, and carrion. Reports of predation on tigers, leopards, and bears generally are not well documented, but those carnivores are sometimes driven from their kills by packs of dholes. There are numerous records of leopards being treed.

In a study in southern India, Johnsingh (1982) found population densities of 0.35–0.90/sq km. A pack in this area contained an average of 8.3 adults and used a home range of 40 sq km. Other work suggests that there are usually 5–12 dholes in a pack, but up to 40 have been reported. This discrepancy was perhaps explained by M. W. Fox (1984), who pointed out that the larger groups are actually clans comprising several related packs. In parts of India the clans keep together during that part of the year when only juvenile and adult chital are available as prey but divide into smaller hunting packs when the chital have fawns. A pack apparently consists of a mated pair and their offspring. Although the social structure has not been closely studied, there seems to be a leader, a dominance hierarchy, and submissive behavior by lower-ranking animals. Intragroup fighting is rarely observed. More than one female sometimes den and rear litters together. Vocalizations include nearly all of those made by the domestic dog except loud and repeated barking; the most distinctive sound is a peculiar whistle that probably serves to keep the pack together during pursuit of prey.

In India mating occurs from September to November, and births from November to March. At the Arignar Anna Zoo in India captives first showed signs of sexual maturity at 11 months of age, females had estrus periods of 14–39 days, and normal gestation lasted 60–67 days (Paulraj et al. 1992). In the Moscow Zoo mating occurred in February and gestation lasted 60–62 days (Sosnovskii 1967). Litters usually contain four to six young, but up to nine embryos have been recorded. The young weigh 200–350 grams at birth and are weaned after about 2 months (Hayssen, Van Tienhoven, and Van Tienhoven 1993). In the wild both mother and young are provided with regurgitated food by other pack members. The pups leave the den at 70–80 days and participate in kills at 7 months (Johnsingh 1982). Longevity in the Moscow Zoo is 15–16 years.

Although the dhole only rarely takes domestic livestock and was only once reported to attack a person, it has been intensively poisoned and hunted throughout its range. This situation seems based mainly on dislike by hunters, who see *Cuon* as a competitor for game and who are repulsed by its method of predation. Some of the village people of India actually welcome the dhole, following it in order to expropriate its kills. Because of direct persecution, elimination of natural prey, and destruction of its forest habitat, the dhole has declined seriously in range and numbers. Recent surveys indicate that remnant populations in southern Asia are small, widely scattered, and difficult to locate even in protected parks (Stewart 1993, 1994). *Cuon* has nearly disappeared in Siberia and Kazakhstan, possibly because it is inadvertently killed in poisoning campaigns against *Canis lupus* (Ginsberg and Macdonald 1990). It is classified as vulnerable by the IUCN, which estimates total numbers at fewer than 10,000, and as endangered by the USDI and Russia and is on appendix 2 of the CITES.

CARNIVORA; CANIDAE; **Genus LYCAON**
Brookes, 1827

African Hunting Dog

The single species, *L. pictus,* originally occurred in most of Africa south of the Sahara Desert as well as in suitable parts of the Sahara and in Egypt (Coetzee *in* Meester and Setzer 1977; Kingdon 1977).

Head and body length is 760–1,120 mm, tail length is 300–410 mm, shoulder height is 610–780 mm, and weight is 17–36 kg (Kingdon 1977). Males and females are about the same size (Frame et al. 1979). There is great variation in pelage, the mottled black, yellow, and white occurring in almost every conceivable arrangement and proportion. In most individuals, however, the head is dark and the tail has a white tip or brush. The fur is short and scant, sometimes so sparse that the blackish skin is plainly visible. The ears are long, rounded, and covered with short hairs. The legs are long and slender, and there are only four toes on each foot. The jaws are broad and powerful. *Lycaon* has a strong, musky odor. Females have 12–14 mammae (Van Lawick and Van Lawick–Goodall 1971).

Lycaon inhabits grassland, savannah, and open woodland (Coetzee *in* Meester and Setzer 1977). Its den, usually an abandoned aardvark hole, is occupied only to bear young (Kingdon 1977). The pack does not wander very far from the den when pups are present (Frame et al. 1979). Otherwise movements are generally correlated with hunting success: if prey is scarce, the entire home range may be traversed in two or three days. Hunts take place in the morning and early evening. Schaller (1972) recorded peaks of activity from 0700 to 0800 and from 1800 to 1900 hours. Prey is apparently located by sight, approached silently, and then pursued at speeds of up to 66 km/hr for 10–60 minutes (Kingdon 1977). Van Lawick and Van Lawick–Goodall (1971) observed a pack to maintain a speed of about 50 km/hr for 5.6 km. Their investigation indicated this distance to be about the maximum that *Lycaon* would usually follow before giving up. They observed 91 chases, 39 of which were successful. In all but 1 of the latter cases the quarry was killed within five minutes of being caught. In his study, Schaller (1972) observed 70 percent of chases to be successful. Groups of *Lycaon* generally cooperate in hunting large mammals, but individuals sometimes pursue hares, rodents, or other small animals. The main prey seems to vary by area, being bushduiker and reedbuck in the Kafue Valley of Zambia, impala in Kruger National Park of South Africa, and Thomson's gazelle and wildebeest in the Serengeti of Tanzania (Kingdon 1977). Certain packs in the Serengeti, however, specialize in the capture of zebra (Malcolm and Van Lawick 1975). Some food may be cached in holes (Malcolm 1980*b*).

Reported population densities vary from 2 to 35 per 1,000 sq km (Fuller, Kat, et al. 1992). From 1970 to 1977 density in the Serengeti declined from 1 adult *Lycaon* per 35 sq km to 1/200 sq km. Pack home range in this area is generally 1,500–2,000 sq km (Frame et al. 1979). A pack in South Africa reportedly used a home range of about 3,900 sq km (Van Lawick and Van Lawick–Goodall 1971). Range contracts to only about 50–200 sq km when there are small pups at a den (Fuller, Kat, et al. 1992); at such time one Serengeti pack used an area of 160 sq km for 2.5 months (Schaller 1972). The home range of a pack overlaps by about 10–50 percent with those of several neighboring packs (Frame and Frame 1976). Territoriality does not seem to be well developed, and hundreds of individuals may once have gathered temporarily in response to migrations of the formerly vast herds of springbok in southern Africa (Kingdon 1977).

Studies on the Serengeti Plains of Tanzania have revealed that *Lycaon* has an intricate and unusual social structure (Frame and Frame 1976; Frame et al. 1979; Malcolm 1980*a*; Malcolm and Marten 1982; Schaller 1972; Van Lawick and Van Lawick–Goodall 1971). Groups were found to contain averages of 9.8 (1–26) individuals, 4.1 (0–10) adult males, and 2.1 (0–7) adult females. This sexual proportion is unlike the usual condition in social mammals. Some packs have as many as 8 adult males with only a single adult female. Moreover, in a reversal of the usual mammalian process, females emigrate from their natal group far more than do males. Commonly, several sibling females 18–24 months old leave their pack and join another that

African hunting dog *(Lycaon pictus)*, photo by Bernhard Grzimek.

lacks sexually mature females. Following the transfer, one of the females achieves dominance, whereupon her sisters may depart. Whereas no female seems to stay in its natal pack past the age of 2.5 years, about half of the young males do remain; the other males emigrate, usually in sibling groups. The typical pack thus consists of several related males, often representing more than one generation, and 1 or more females that are genetically related to each other but not to the males. Some male lineages within a pack are known to have lasted at least 10 years.

There are separate dominance hierarchies for each sex. Normally, only the highest-ranking male and female breed, and they inhibit reproduction by subordinates. There is intense rivalry among the females for the breeding position. If a subordinate female does bear pups, the dominant one may steal them. Females sometime fight savagely, and the loser may leave the group and perish. Aside from this aspect of the social life of *Lycaon*, packs are remarkably amicable, with little overt strife. Food is shared, even by individuals that do not participate in the kill. An animal with a broken leg was allowed to feed, when it hobbled up after the others, throughout the time required for its leg to mend. Pups old enough to take solid food are given first priority at kills, eating even before the dominant pair. Subordinate animals, especially males, help feed and protect the pups. There are several vocalizations, the most striking of which is a series of wailing hoots that probably serves to keep the pack together during pursuit of prey.

Studies at Masai Mara Reserve in Kenya and Kruger National Park in South Africa suggest some differences in behavior (Fuller, Kat, et al. 1992). Both males and females regularly disperse from their natal packs at 1–2 years, usually in groups composed only of one sex. They move up to 250 km, eventually joining with groups of the opposite sex that have dispersed from elsewhere to form a new breeding pack. Sometimes they join an established pack. Occasionally a female does not leave her natal pack and may even replace her mother. It is not unusual for two females of a pack to mate and successfully rear young.

Births may occur at any time of year but peak from March to June, during the second half of the rainy season. The interval between births is normally 11–14 months but may be as short as 6 months if all the young perish. The gestation period is 79–80 days. There is some confusion about litter size. Although up to 23 young have been seen at den sites, and although single females have given birth to as many as 21 pups, litter sizes at various captive facilities have averaged only 4–8. It is likely that the very large litters observed in the wild represent the synchronous offspring of two or even three females in the pack. The newborn weigh about 300 grams and open their eyes after 13 days. At about 3 weeks the young emerge from the den and begin to take some solid food. Weaning is normally completed by 11 weeks. All adult pack members regurgitate food to the young. Once, when a mother died, the males of the pack were able successfully to raise her 5-week-old pups. When the pack is hunting, 1 or 2 adults remain at the den to guard the pups. After 3 months the young begin to follow the pack, and at 9–11 months they can kill easy prey, but they are not proficient until about 12–14 months. Social restrictions blur the actual time of sexual maturity. Five males were observed to first mate at 21, 33, 36, 36, and 60 months. The youngest female to give birth was 22 months old at the time. Maximum observed longevity in the wild is 11 years (Frame et al. 1979; Fuller, Kat, et al. 1992; Kingdon 1977; Malcolm 1980*a;* Schaller 1972; Van Heerden and Kuhn 1985; Van Lawick and Van Lawick–Goodall 1971). A captive lived for about 17 years (Marvin L. Jones, Zoological Society of San Diego, pers. comm., 1995).

Lycaon has been reported only rarely to attack people but is intensively hunted and poisoned because of a largely undeserved reputation as an indiscriminate killer of livestock and valued game animals. Remnant and fragmented populations also are jeopardized by habitat loss, diseases spread by domestic animals, loss of genetic viability, and possibly even the stress caused by well-meaning research workers. The genus has been almost entirely extirpated in the Saharan region, only small and probably inviable populations survive in West and Central Africa, and it has been wiped out in South Africa except in the vicinity of Kruger National Park. It is thought to have vanished entirely even in Zaire and Congo. There still may be viable populations present from southern Ethiopia, through Kenya, Tanzania, Zambia, Zimbabwe, and Botswana, to Namibia. Even in those countries, however, there has been a great decline in distribution. The largest remaining population, with perhaps more than 1,000 animals, is located in the Selous Game Reserve and adjacent lands of southern Tanzania. Numbers in Zimbabwe have been estimated at 300–600 individuals, and persecution continues there except in national parks. Numbers throughout all of Africa were thought to be fewer than 7,000 in 1980. Estimates issued in the early 1990s varied from only 2,000 to about 5,000 (Buk 1994; Burrows, Hofer, and East 1994; Childes 1988; Fanshawe, Frame, and Ginsberg 1991; Fuller, Kat, et al. 1992; Ginsberg 1993; Ginsberg and Macdonald 1990; Kingdon 1977; Lensing and Joubert 1977; Malcolm 1980*a;* Skinner, Fairall, and Bothma 1977). There are another 300 in captivity (Brewer *in* Bowdoin et al. 1994).

Lycaon now is classified as endangered by the IUCN and USDI. With the establishment in 1992 of the Licaone Fund by concerned Italian biologists, as well as other international efforts, its plight is finally receiving substantive attention. Considering its immense former distribution and its scientific, cultural, and behavioral interest, the prospective disappearance of this genus from the wild at a time of supposed increasing emphasis on conservation values must rank as one of the great wildlife tragedies of the late twentieth century.

CARNIVORA; **Family URSIDAE**

Bears

This family of three Recent genera and eight species occurred historically almost throughout Eurasia and North America, in the Atlas Mountains of North Africa, and in the Andes of South America. E. R. Hall (1981) recognized three living subfamilies: Tremarctinae, with the genus *Tremarctos;* Ursinae, with *Ursus;* and Ailuropodinae, with *Ailuropoda.* O'Brien (1993) and Wayne et al. (1989), using molecular and karyological data, supported the same arrangement. Wozencraft (*in* Wilson and Reeder 1993) included *Tremarctos* in the Ursinae and placed *Ailuropoda,* together with *Ailurus,* in another ursid subfamily, the Ailurinae. The sequence of genera presented here basically follows that of Simpson (1945), though he, like many other authorities, recognized additional genera.

Head and body length is 1,000 to approximately 2,800 mm, tail length is 65–210 mm, and weight is 27–780 kg. Males average about 20 percent larger than females. The coat is long and shaggy, and the fur is generally unicolored, usually brown, black, or white. Some species have white or buffy crescents or semicircles on the chest. *Tremarctos,* the spectacled bear of South America, typically has a patch of white hairs encircling each eye. *Ailuropoda,* the giant panda, has a striking black and white color pattern.

Bears have a big head; a large, heavily built body; short, powerful limbs; a short tail; and small eyes. The ears are small, rounded, and erect. The soles are hairy in species that are mainly terrestrial but naked in species that climb considerably, such as *Ursus malayanus.* All limbs have five digits. The strong, recurved claws are used for tearing and digging. The lips are free from the gums.

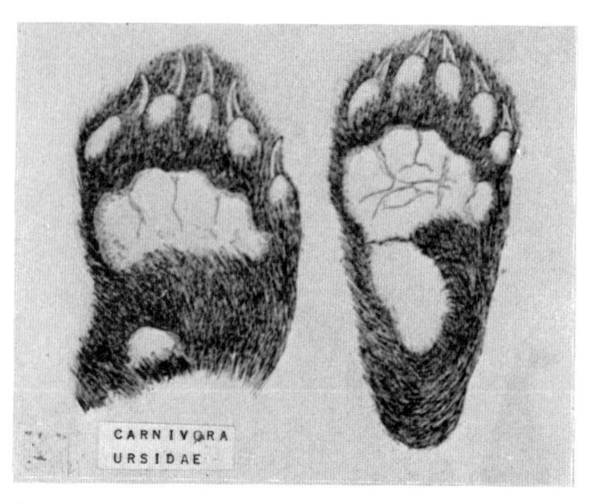

American black bear *(Ursus americanus)*, right forepaw and right hind foot, photo from *Proc. Zool. Soc. London.*

The skull is massive and the tympanic bullae are not inflated. In most genera the dental formula is: (i 3/3, c 1/1, pm 4/4, m 2/3) × 2 = 42. The species *Ursus ursinus,* however, has only 2 upper incisors and a total of 40 teeth. Ursid incisors are not specialized, the canines are elongate, the first 3 premolars are reduced or lost, and the molars have broad, flat, and tubercular crowns. The carnassials are not developed as such.

Habitats range from arctic ice floes to tropical forests. Those populations that occur in open areas often dig dens in hillsides. Others shelter in caves, hollow logs, or dense vegetation. Bears have a characteristic shuffling gait. They walk plantigrade, with the heel of the foot touching the ground. They are capable of walking on their hind legs for short distances. When need be, they are surprisingly agile and careful in their movements. Their eyesight and hearing are not particularly good, but their sense of smell is excellent. Bears are omnivorous, except that the polar bear *(Ursus maritimus)* feeds mainly on fish and seals.

During the autumn bears in most parts of the range of the family become fat. With the approach of cold weather they cease eating and go into a den that they have prepared in a protected location. Here they sleep through the winter, living mainly off stored fat reserves. With certain exceptions, especially pregnant females, the polar bear does not undergo winter sleep. Some authorities prefer not to call this process hibernation since body temperature is not substantially reduced, body functions continue, and the animals can usually be easily aroused. Sometimes they awaken on their own during periods of mild weather. Folk, Larson, and Folk (1976) found, however, that the heart rate of a hibernating bear drops to less than half of normal and that other physiological changes occur. They concluded that bears do experience true mammalian hibernation.

Except for courting pairs and females with young, bears live alone. Litters are produced at intervals of 1–4 years. In most regions births occur from November to February, while the mother is hibernating. The period of pregnancy is commonly extended 6–9 months by delayed implantation of the fertilized egg. Litter size is one to four young. The young are relatively tiny at birth, ranging from 225–680 grams each. They remain with the mother at least through their first autumn. They become sexually mature at 2.5–6 years and normally live 15–30 years in the wild.

Bears are usually peaceful animals that try to avoid con-

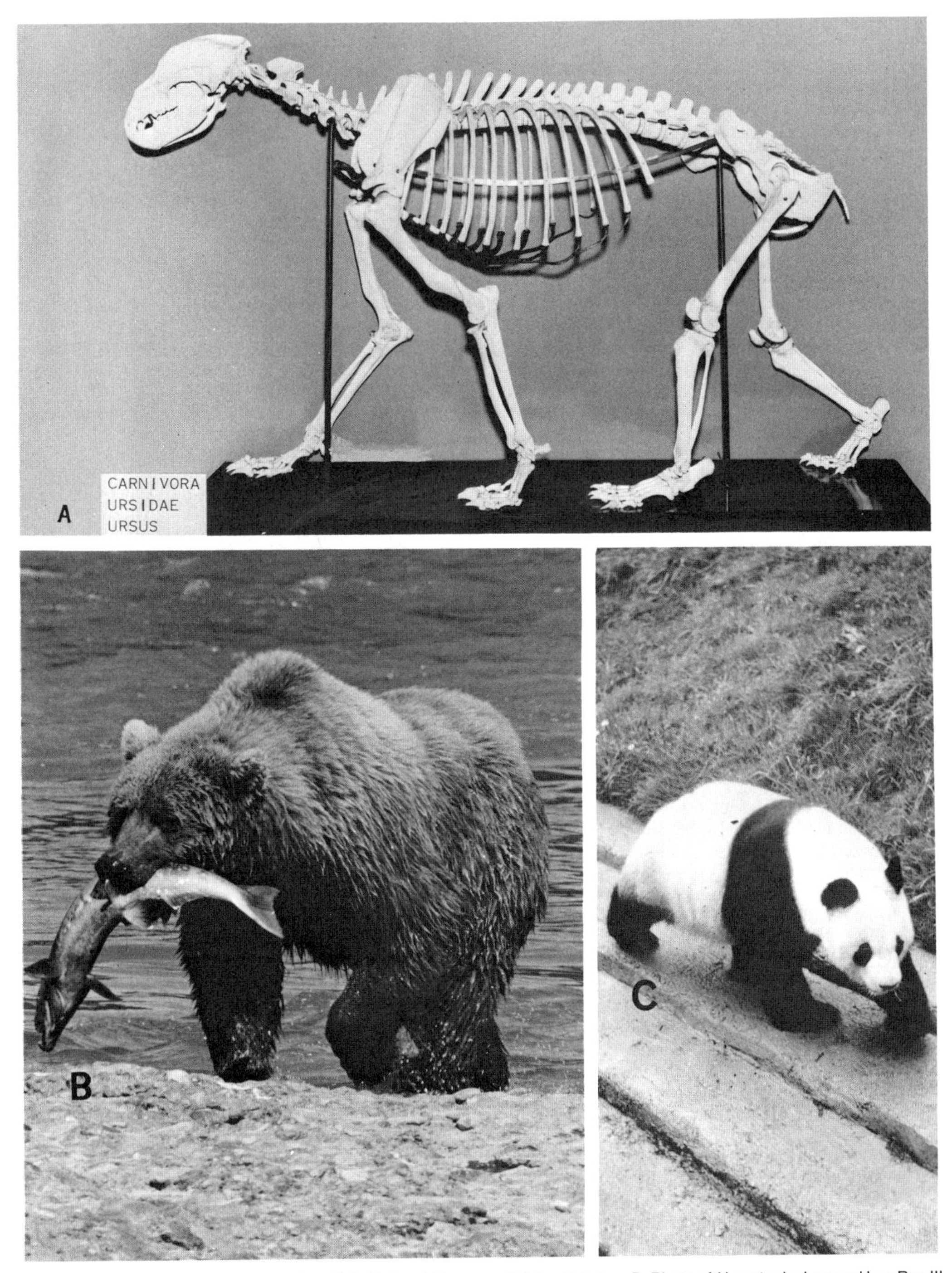

A. Brown bear *(Ursus arctos)*, skeleton from U.S. National Museum of Natural History. B. Photo of *U. arctos* by Leonard Lee Rue III. C. Giant panda *(Ailuropoda melanoleuca)*, photo by Elaine Anderson.

flict. However, if they consider themselves, their young, or their food supply threatened they can become formidable adversaries. Only a small proportion of the stories of unprovoked attacks by bears on people are true. When such cases are carefully investigated, it is usually found that there was provocation. Nonetheless, bears have been persecuted almost throughout their range because of alleged danger to humans and because they are sometimes considered to be serious predators of domestic livestock. These problems tend to increase dramatically when the natural habitat of bears is invaded and fragmented by people for purposes of agriculture, recreation, settlement, and logging, which inevitably leads to conflicts and also facilitates access by human hunters. Another longstanding problem that only recently has achieved widespread recognition is the killing of bears for their body parts, especially the gallbladder, for use as medicines and food. Most species of bears in Asia are now directly threatened by such exploitation, and bear populations around the world have been increasingly subject to consequent illegal hunting and commercial trade (Rose and Gaski 1995).

The geological range of the Ursidae is late Miocene to Recent in North America, late Pliocene to Recent in South America, late Eocene to Recent in Europe, early Miocene to Recent in Asia, Pleistocene to Recent in North Africa, and Pliocene in South Africa (Hendey 1977; Stains 1984). The cave bear *(Ursus spelaeus)*, which was common in the late Pleistocene of Europe, was equal in size to the modern brown bear of Alaska. Although generally thought to have vanished about 10,000 years ago, certain unfossilized remains suggest that the species persisted until more recent times (Geist 1993).

CARNIVORA; URSIDAE; **Genus TREMARCTOS**
Gervais, 1855

Spectacled Bear

The single species, *T. ornatus*, is known to inhabit the mountainous regions of western Venezuela, Colombia, Ecuador, Peru, and western Bolivia (Cabrera 1957). It also has been reported from eastern Panama and northern Argentina (Peyton 1986).

Head and body length is usually 1,200–1,800 mm, tail length is about 70 mm, and shoulder height is 700–800 mm. One male 1,740 mm in length weighed 140 kg. Peyton (1980) reported that a male 2,060 mm in total length weighed 175 kg. Grzimek (1975) gave the weight of females as 60–62 kg. The entire body is uniformly black or dark brown except for large circles or semicircles of white around the eyes and a white semicircle on the lower side of the neck, from which lines of white extend onto the chest. The common name is derived from the white around the eyes. The head and chest markings are variable, however, and may be completely lacking in some individuals.

In the Andes of Peru, Peyton (1980) found *Tremarctos* to occupy a wide variety of habitats from 457 to 3,658 meters in elevation. The preferred habitats are humid forests between 1,900 and 2,350 meters and coastal thorn forests when water is available. High-altitude grasslands are also utilized. *Tremarctos* apparently is mainly nocturnal and crepuscular. During the day it beds down between or under large tree roots, on a tree trunk, or in a cave. It frequently climbs large trees to obtain fruit. In a tree it may assemble a large platform of broken branches, on which it positions itself to eat and to reach additional fruit. Peyton found one such platform at a height of 15 meters. Goldstein (1991) observed similar platforms apparently used as resting places.

Tremarctos feeds extensively on fruit, moving about in response to seasonal ripening. It also depends on plants of the family Bromeliaceae, especially when ripe fruit is not available. It tears off the leaves of large bromeliads to feed on the white bases and obtains the edible hearts of small bromeliads by ripping the entire plant off the substrate. In addition, this bear climbs large cacti to get the fruits at the top, tears into the green stalks of young palms to eat the unopened inner leaves, and strips bark off trees to feed on the cortex. The diet also includes bamboo hearts, corn, rodents, and insects. Only about 4 percent of the food was found to be animal matter (Peyton 1980).

In Peru, Peyton (1980) received reports that a male often enters a cornfield with one or more females, and sometimes with yearling animals, during the months of March–July. According to Grzimek (1975), *Tremarctos* has a "striking shrill voice." In the Buenos Aires Zoo young were produced in July, while in European zoos births have occurred from late December to March. Pregnancy lasts 6.5–8.5 months and apparently involves delayed implantation. Litters contain one to three young weighing about 320 grams each (Bloxam 1977; Gensch 1965; Grzimek 1975). One captive lived 38 years and 8 months (Marvin L. Jones, Zoological Society of San Diego, pers. comm., 1995).

Spectacled bear *(Tremarctos ornatus)*, photo from San Diego Zoological Garden.

Mittermeier et al. (1977) reported that the meat of *Tremarctos* is highly esteemed in northern Peru and that this bear also is killed by people for its skin and fat. Servheen (1989) wrote that the body parts of bears are commonly used for alleged medicinal purposes in South America and that legal protection is ineffective. Grimwood (1969) warned that *Tremarctos* had become rare and endangered in Peru through intensive hunting by sportsmen and landowners, who consider it to be a predator of domestic livestock. Mondolfi (1989) regarded it as endangered and still declining in Venezuela, with remnant populations subject to hunting and habitat disruption even in national parks. Peyton (1980) did not believe *Tremarctos* to be in immediate danger of extinction because it is adapted to a diversity of habitats, some of which are largely inaccessible to people. He did note, however, that some bears become habituated to raiding cornfields and that these animals are frequently shot. Later, Peyton (1986) noted that vast parts of the original range have been replaced by agriculture and that surviving bear populations are fragmentary. Thornback and Jenkins (1982) stated that the spectacled bear is declining through much of its range because of habitat loss due to settlement, to human persecution that subsequently results from raids on crops and livestock, and to hunting for its meat and skin. *Tremarctos* is classified as vulnerable by the IUCN and is on appendix 1 of the CITES.

CARNIVORA; URSIDAE; **Genus URSUS**
Linnaeus, 1758

Black, Brown, Polar, Sun, and Sloth Bears

There are six species (Corbet 1978; Corbet and Hill 1992; Ellerman and Morrison-Scott 1966; E. R. Hall 1981; Kurten 1973; Laurie and Seidensticker 1977; Lay 1967; Lekagul and McNeely 1977; Ma 1983; Simpson 1945):

U. thibetanus (Asiatic black bear), Afghanistan, southeastern Iran, Pakistan, Himalayan region, Burma, Thailand, Indochina, China, Manchuria, Korea, extreme southeastern Siberia, Japan (except Hokkaido), Taiwan, Hainan;
U. americanus (American black bear), Alaska, Canada, conterminous United States, northern Mexico;
U. arctos (brown or grizzly bear), western Europe and Palestine to eastern Siberia and Himalayan region, Atlas Mountains of northwestern Africa, Hokkaido, Alaska to Hudson Bay and northern Mexico;
U. maritimus (polar bear), primarily on arctic coasts, islands, and adjacent sea ice of Eurasia and North America;
U. malayanus (Malayan sun bear), Assam southeast of Brahmaputra River, Sichuan and Yunnan in south-central China, Burma, Thailand, Indochina, Malay Peninsula, Sumatra, Borneo;
U. ursinus (sloth bear), India, Nepal, Bangladesh, Sri Lanka.

Each of these species has often been placed in its own genus or subgenus: *Selenarctos* Heude, 1901, for *U. thibetanus; Euarctos* Gray, 1864, for *U. americanus; Ursus* Linnaeus, 1758, for *U. arctos; Thalarctos* Gray, 1825, for *U. maritimus; Helarctos* Horsfield, 1825, for *U. malayanus;* and *Melursus* Meyer, 1793, for *U. ursinus.* E. R. Hall (1981), however, placed all of these names in the synonymy of *Ursus.* The latter arrangement is supported by molecular and karyological data (O'Brien 1993; Wayne et al. 1989) and in part by the captive production of viable offspring through hybridization between several of the above species (Van Gelder 1977). Wozencraft (*in* Wilson and Reeder 1993) treated *Helarctos* and *Melursus* as valid genera.

The systematics of the brown or grizzly bear have caused considerable confusion. Old World populations long have been recognized as composing a single species, with the scientific name *U. arctos* and the general common name "brown bear." In North America the name "grizzly" is applied over most of the range, while the term "big brown bear" is often used on the coast of southern Alaska and nearby islands, where the animals average much larger than those inland. E. R. Hall (1981) listed 77 Latin names that have been used in the specific sense for different populations of the brown or grizzly bear in North America. No one now thinks that there actually are so many species, but some authorities, such as Burt and Grossenheider (1976), have recognized the North American grizzly *(U. horribilis)* and the Alaskan big brown bear *(U. middendorffi)* as species distinct from *U. arctos* of the Old World. Other authorities (Erdbrink 1953; Kurten 1973; Rausch 1953, 1963), based on limited systematic work, have referred the North American brown and grizzly to *U. arctos.* This procedure is being used by most persons now studying or writing about bears and is followed here. Kurten (1973) distinguished three North American subspecies: *U. a. middendorffi,* on Kodiak and Afognak islands; *U. a. dalli,* on the south coast of Alaska and the west coast of British Columbia; and *U. a. horribilis,* in all other parts of the range of the species. Hall (1984), however, recognized nine North American subspecies of *U. arctos.*

From *Tremarctos, Ursus* is distinguished by its masseteric fossa on the lower jaw not being divided by a bony septum into two fossae. From *Ailuropoda* it is distinguished in having an alisphenoid canal (E. R. Hall 1981). Additional information is provided separately for each species.

Ursus thibetanus (Asiatic black bear)

Head and body length is 1,200–1,800 mm and tail length is 65–106 mm. Stroganov (1969) listed weight as 110–50 kg for males and 65–90 kg for females. Roberts (1977) stated that an exceptionally large male weighed 173 kg but that an adult female weighed only 47 kg. The coloration is usually black but is sometimes reddish brown or rich brown. There is some white on the chin and a white crescent or V on the chest.

The Asiatic black bear frequents moist deciduous forests and brushy areas, especially in the hills and mountains. It ascends to elevations as high as 3,600 meters in the summer and descends in the winter. It swims well. According to Lekagul and McNeely (1977), this bear is generally nocturnal, sleeping during the day in hollow trees, caves, or rock crevices. It is also seen abroad by day when favored fruits are ripening. It climbs expertly to reach fruit and beehives. It usually walks on all fours but often stands on its hind legs so that its forepaws can be used in fighting. The diet includes fruit, buds, invertebrates, small vertebrates, and carrion. Domestic livestock is sometimes taken, and animals as large as adult buffalo are killed by breaking their necks. Individuals become fat in late summer and early autumn before hibernation, but some populations either do not undergo winter sleep or do so only during brief periods of severe weather. Roberts (1977) stated that in the Himalayas *U. thibetanus* hibernates, sometimes in a burrow of its own making, but that in southern Pakistan there is no evidence of hibernation. According to Stroganov (1969), hibernation in Siberia begins in November and lasts four or five months. Dens in that area are usually in tree holes. The bears are easily aroused during the first month but sleep more deeply from December to February.

A radio-tracking study in Sichuan, central China (Reid et al. 1991), found the ecology of *U. thibetanus* to compare closely to that of *U. americanus* in Tennessee and to differ in a few ways from the pattern indicated above. Bears were generally active in daylight and far less active at night. Daily movements decreased from about 650 meters per day during the summer to 300 meters per day in the autumn, though there also was an increase in nocturnal activity in the autumn. The diet was almost exclusively vegetarian. In the autumn there was a shift to lower elevations to obtain acorns and other mast foods. Winter denning took place from late November to early April. Population density in

Asiatic black bear *(Ursus thibetanus)*, photo from New York Zoological Society.

the study area was 0.1–1.3/sq km, and minimum home range estimates for two individuals during a 30-month period were 16.4 and 36.5 sq km. Seasonal ranges were smaller, about 5–6 sq km.

In Siberia individual home range is 500–600 ha., only one-third or one-fourth the size of that of *U. arctos.* Mating in Siberia occurs in June or July, and births take place from late December to late March, mostly in February (Stroganov 1969). In Pakistan mating is thought to occur in October, and the young are born in February (Roberts 1977). Newborn weight is 300–450 grams (Hayssen, Van Tienhoven, and Van Tienhoven 1993). According to Lekagul and McNeely (1977), pregnancy lasts 7–8 months, and usually two cubs are born in a cave or hollow tree in early winter. The young open their eyes after about 1 week and shortly thereafter begin to follow the female as she forages. They are weaned at about 3.5 months but remain with the mother until they are 2–3 years old. Females have been seen with two sets of cubs. Sexual maturity comes at about 3 years. An individual living at the zoo in Portland, Oregon, in December 1994 was approximately 36 years old (Marvin L. Jones, Zoological Society of San Diego, pers. comm., 1995).

The Asiatic black bear sometimes raids cornfields and attacks domestic livestock. Occasionally it has been reported to kill humans. Hayashi (1984), for example, reported that from 1970 to 1983 in Fukui Prefecture of Japan there were attacks on 16 people, one of whom was killed. For these reasons *U. thibetanus* is hunted by people, and it also has declined because of the destruction of its forest habitat (Cowan 1972). The subspecies *U. t. gedrosianus,* of southern Pakistan and possibly adjacent parts of Iran, is classified as critically endangered by the IUCN and as endangered by the USDI. There is no recent information on its status, but any remaining population would be very small and isolated. In neighboring Afghanistan *U. thibetanus* evidently has been completely extirpated, and populations in Bangladesh and Korea may be on the verge of disappearing. The subspecies *U. t. japonicus* is threatened by deforestation and subsistence and commercial hunting. It was wiped out on Kyushu in the 1950s, has been reduced to fewer than 100 individuals on Shikoku, and survives in viable numbers only in western Honshu (Servheen 1989). The same problems jeopardize *U. t. formosanus,* which is endemic to Taiwan, where probably fewer than 200 bears remain, and Hainan, where there are fewer than 50 (Garshelis 1995). Wild populations of *U. thibetanus* in mainland China are estimated to number about 12,000–18,000 and are generally declining (Guo 1995).

The species *U. thibetanus* has been given an overall classification of vulnerable by the IUCN and also is now on appendix 1 of the CITES. This species is probably the one most severely affected by the commercial trade in bear parts. Although bears long have been used in traditional Oriental foods and medicines, only recently has the extent of the resulting international trade and the devastating impact on bear populations been generally recogized (Rose and Gaski 1995). Nearly all parts of a bear's body are utilized (Highley and Highley 1995; Mills 1993, 1994, 1995). The meat, fat, and paws are popular foods, and their consumption is believed to give strength and ward off colds and other illnesses. With rising affluence in such countries as Singapore, Hong Kong, and South Korea, bear-paw dishes have become a status symbol and can cost hundreds of U.S. dollars per serving at some restaurants. The bones, brain, blood, and spinal cord also are reputed to have medicinal uses, and the whole head and skin are prized decorative items.

By far the most valued part, and the one that decidedly has made bears an international commodity, is the gallbladder. The bile salts found within that organ are used as a medicine to treat diseases of the liver, heart, and digestive system and for many other purposes, such as relieving pain, improving vision, and cleaning toxins from the blood.

There apparently is some scientific basis for these uses, as bears are the only mammals that manufacture the bile salt ursodeoxycholic acid. This substance has been shown in Western laboratory tests to be effective in treating some liver diseases, and a synthesized form is widely used to dissolve gallstones without surgery. A recent survey of traditional doctors in South Korea indicated that most sold bear bile for at least U.S. $37.50 per gram and would pay more than $1,000 for a gallbladder from a wild bear. Customs records show that the amount of bear bile imported by South Korea alone from 1970 to 1993 represented 2,867 bears annually; by far the greatest supplier was Japan. Bear populations in the Russian Far East reportedly also are being decimated by this trade. In China more than 10,000 bears are maintained in captivity for purposes of bile production. The animals are kept in small cages and "milked" of their bile by a tube surgically implanted in the gallbladder. Although this activity has been extolled as a conservation mechanism, wild bears still are being killed and sold regularly in China and their gallbladders are considered superior to those of captives for medicinal purposes.

Ursus americanus (American black bear)

Head and body length is 1,500–1,800 mm, tail length is about 120 mm, and shoulder height is up to 910 mm. Banfield (1974) listed weights of 92–140 kg for females and 115–270 kg for males. The most common color phases are black, chocolate brown, and cinnamon brown. Different colors may occur in the same litter. A white phase is generally rare and never in the majority but seems to be most common on the Pacific coast of central British Columbia. A blue black phase is also generally rare but occurs frequently in the St. Elias Range of southeastern Alaska. Rounds (1987) reported that coloration varies geographically, with nearly all bears in eastern North America being black but most in some southwestern populations being non-black. Compared with *U. arctos, U. americanus* has a shorter and more uniform pelage, shorter claws, and shorter hind feet. Females have three pairs of mammae (Banfield 1974).

The American black bear occurs mainly in forested areas. It may originally have avoided open country because of the lack of trees in which to escape *U. arctos.* The latter species is known to be a competitor with, and sometimes a predator upon, *U. americanus* (Jonkel 1978). The black bear now appears to have extended its range northward onto the tundra, possibly in response to the decline of the barren ground grizzly (Jonkel and Miller 1970; Veitch 1993). Following the extermination of the grizzly in the mountains of southern California, *U. americanus* moved into the area (E. R. Hall 1981).

The usual locomotion is a lumbering walk, but *U. americanus* can be quick when the need arises. It swims and climbs well. It may move about at any hour but is most active at night (Banfield 1974). Like other bears that sleep through the winter, it becomes fat with the approach of cold weather, finally ceases eating, and goes into a den in a protected location. The shelter may be under a fallen tree, in a hollow tree or log, or in a burrow. In the Hudson Bay area individuals may burrow into the snow. During hibernation body temperature drops from 38° C to 31°–34° C, the respiration slows, and the metabolic rate is depressed (Banfield 1974). The winter sleep is interrupted by excursions outside during periods of relatively warm weather. Such emergences are more numerous at southern latitudes. Hibernation begins as early as late September and may last until May. When dens are entered may depend on how much fat has been accumulated through feeding, and even after emergence a bear may remain lethargic and in the vicinity of its den until food again is plentiful. In Washington the average period of hibernation is 126 days, but three

American black bear *(Ursus americanus)*, photo from San Diego Zoological Garden.

American black bear *(Ursus americanus)*, photo by J. Perley Fitzgerald.

Louisiana bears each slept for 74–124 days (Lindzey and Meslow 1976; Lowery 1974; Rogers 1987). At least 75 percent of the diet consists of vegetable matter, especially fruits, berries, nuts, acorns, grass, and roots. In some areas sapwood is important; to reach it the bear peels bark from trees, thereby causing forest damage (Poelker and Hartwell 1973). The diet also includes insects, fish, rodents, carrion, and occasionally large mammals.

Banfield (1974) suggested an overall population density of about 1/14.5 sq km. Field studies in Alberta, Washington, and Montana, however, indicate a usual density of 1/2.6 sq km (Jonkel and Cowan 1971; Kemp 1976; Poelker and Hartwell 1973). Still higher densities have been reported: 1/1.3 sq km in southern California (Piekielek and Burton 1975) and 1/0.67 sq km on Long Island off southwestern Washington (Lindzey and Meslow 1977*b*). In the latter area home range was found to average 505 ha. for adult males and 235 ha. for adult females (Lindzey and Meslow 1977*a*). Farther north in Washington, however, Poelker and Hartwell (1973) determined home range to average about 5,200 ha. for males and 520 ha. for females. The ranges of males did not overlap one another, but the ranges of females overlapped with those of males and occasionally with those of other females. On the tundra of the Ungava Peninsula adult males commonly range over 50,000–100,000 ha., females over 5,000–20,000 ha. (Veitch 1993). In Idaho, Amstrup and Beecham (1976) found home ranges to vary from 1,660 to 13,030 ha., to remain stable from year to year, and to overlap extensively. Despite such overlap, individuals tend to avoid one another and to defend the space being used at a given time. A number of bears sometimes congregate at a large food source, such as a garbage dump, but they try to keep out of one another's way. Dominance hierarchies may be formed in such situations, but more tolerance is shown to familiar individuals than to strangers (Banfield 1974; Jonkel 1978; Jonkel and Cowan 1971; Rogers 1987). There are a variety of vocalizations. When startled, the ordinary sound is a "woof." When cubs are lonely or frightened they utter shrill howls.

Perhaps the most intensive study of *U. americanus* was carried out from 1969 to 1985 in northeastern Minnesota by Rogers (1987), who made use of live capture and radio tracking. The annual cycle of social behavior was found to be closely tied to plant growth, fruiting, and availability of food. The population density of bears, including cubs, was calculated at 1/4.1–6.3 sq km. Mature males used overlapping home ranges that averaged 75 sq km and appeared to be arranged so as to allow access to the maximum number of potentially estrous females. Females occupied territories averaging 9.6 sq km and vigorously chased intruders. However, after fruit and nuts disappeared in late summer, 67 percent of the males and 40 percent of the females foraged beyond their ranges or territories. Mothers tolerated their independent offspring within their territories and eventually allowed their daughters to take over a portion. Young males dispersed an average distance of 61 km from their natal areas.

The sexes come together briefly during the mating season, which generally peaks from June to mid-July. During this period individual females apparently are in estrus only 1–3 days. They usually give birth every other year but sometimes wait 3–4 years. Pregnancy generally lasts about 220 days, but there is delayed implantation. The fertilized eggs are not implanted in the uterus until the autumn, and embryonic development occurs only in the last 10 weeks of pregnancy. Births occur mainly in January and February, commonly while the female is hibernating. The number of young per litter ranges from one to five and is usually two or three. At birth the young weigh 225–330 grams and their eyes are tightly closed. They may appear naked but have a coat of short grayish hair (Frederick A. Ulmer, Jr., Academy of Natural Sciences, Philadelphia, pers. comm., 1994). They usually are weaned at around 6–8 months but remain with the mother and den with her during their second winter. Upon emergence in the spring they usually depart in order to avoid the aggression of the adult males in the breeding season. Females reach sexual maturity at 4–5 years, and males about a year later. One female is known to have lived 26 years and to have been in estrus at that age (Banfield 1974; Jonkel 1978; Poelker and Hartwell 1973; Rogers 1987). A captive reportedly lived for 31 years (Marvin L. Jones, Zoological Society of San Diego, pers. comm., 1995).

Except when wounded or attempting to protect its young, the black bear is generally harmless to people. In areas of total protection, such as national parks, the species has become accustomed to humans. It thus can be easily seen and is a popular attraction but is sometimes a nuisance, raiding campsites or begging for food along roads. Physical attacks are rare but occur with some regularity, often because the persons involved disregard safety regulations (Cole 1976; Jonkel 1978; Pelton, Scott, and Burghardt 1976). Black bears have killed people on occasion, most recently an adult woman in Alaska in July 1992 (*Washington Post*, 8 September 1992) and a four-year-old boy in British Columbia in September 1994 (*International Bear News* 41[1] [1995]: 19).

People have intensively killed *U. americanus* because of fear, to prevent depredations on domestic animals and crops, for sport, and to obtain fur and meat. According to Lowery (1974), attacks on livestock are negligible, but the bear does serious damage to cornfields and honey production. The economic loss caused to beekeepers in the Peace River Valley of Alberta was estimated at $200,000 in 1973, and a government control program is directed against the bear in that area (Gilbert and Roy 1977). In most of the states and provinces occupied by the black bear it is treated as a game animal, subject to regulated hunting. Approximately 40,000 individuals are killed legally each year in North America (McCracken, Rose, and Johnson 1995). Relatively few skins go to market now, as regulations sometimes forbid commerce and there is no great demand. The average price per pelt in the 1976–77 season was about $44 (Deems and Pursley 1978), but it had fallen to less than $20 by 1983–84 (Novak, Obbard, et al. 1987).

A more significant factor in recent years has been the killing of *U. americanus* and the exportation of its body parts for medicinal use or food in the Orient, as described above in the account of *U. thibetanus*. This problem may have intensified in association with a decline in the native bears of Asia. Indeed, the United States now is the second largest supplier of bear gallbladders to South Korea (Mills 1995). Other markets include China, Japan, Hong Kong, Taiwan, and Asian communities in the United States and Canada. In 1992 *U. americanus* was placed on appendix 2 of the CITES. According to McCracken, Rose, and Johnson (1995), this measure was taken primarily to prevent illegal commerce in the parts of Asian bears, which were being falsely labeled as those of American black bears. However, investigations have revealed an extensive trade in gallbladders and other parts of *U. americanus* itself, together with poaching operations in the United States and Canada. This activity appears to be intensifying, and CITES controls have thus far had only a very limited effect.

The distribution of the black bear has declined substantially, but the species is still common in Alaska, Canada, the western conterminous United States, the upper Great Lakes region, northern New England and New York, and parts of the Appalachians. Small native populations also survive in coastal lowlands from the Dismal Swamp of Virginia to the Okefenokee of Georgia and in the bottom land forests of southern Alabama and southeastern Arkansas (Wooding, Cox, and Pelton 1994). *U. americanus* evidently had been totally extirpated in Texas by the 1940s but recently has been reestablished through the apparent movement of animals from the Sierra del Carmen of northern Mexico into Big Bend National Park (Hellgren 1993). Data compiled by Cowan (1972) indicated the presence of about 170,000 black bears in the conterminous United States, and Raybourne (1987) gave an estimate of 400,000–500,000 for all of North America. A new survey (McCracken, Rose, and Johnson 1995) developed estimates of approximately 200,000 for the contiguous United States, 150,000 for Alaska, and 330,000 for Canada. Leopold (1959) thought the species was still widespread in Mexico, but more recent information indicates that it is critically endangered there (Ceballos and Navarro L. 1991).

The subspecies *U. a. floridanus*, of Florida and adjacent areas, is considered to be threatened through habitat loss, fragmentation of remnant populations, and persecution by beekeepers (Brady and Maehr 1985; Layne 1978). About 500–1,000 individuals are thought to survive (Maehr *in* Humphrey 1992). The subspecies *U. a. luteolus*, formerly found from eastern Texas to Mississippi, was by the mid–twentieth century reduced to a few individuals along the Mississippi and lower Atchafalaya rivers in eastern Louisiana and possibly neighboring Mississippi. In 1992 it was classified as threatened by the USDI, a measure that provided a basis for protection of substantial portions of its habitat (Neal 1993). During the 1960s the wildlife agencies of both Louisiana and Arkansas imported a number of bears from Minnesota (within the range of the subspecies *U. a. americanus*) to their respective states (Lowery 1974; Sealander 1979), thus further jeopardizing the genetic viability of the native populations. The Arkansas introduction has been spectacularly successful from a management standpoint, there now being about 2,100 bears in the Ozark and Ouachita highlands of the state; this population has expanded into adjacent parts of Missouri and Oklahoma, where there now may be an additional 300 bears (Smith and Clark 1994; Smith, Clark, and Shull 1993).

Ursus arctos (brown or grizzly bear)

Head and body length is 1,700–2,800 mm, tail length is 60–210 mm, and shoulder height is 900–1,500 mm. In any given population adult males are larger on average than adult females. The largest individuals—indeed, the largest of living carnivores—are found along the coast of southern Alaska and on nearby islands, such as Kodiak and Admiralty. In this area weight is as great as 780 kg. Size rapidly declines to the north and east. In southwestern Yukon, for example, Pearson (1975) found average weights of 139 kg for males and 95 kg for females. In the Yellowstone region, Knight, Blanchard, and Kendall (1981) found weights of full-grown animals to average 181 kg and to range from 102 to 324 kg. In Siberia and northern Europe weight is

Alaskan brown bear *(Ursus arctos)*, photo by Ernest P. Walker.

usually 150–250 kg. In parts of southern Europe average weight is only 70 kg (Grzimek 1975). Coloration is usually dark brown but varies from cream to almost black. In the Rocky Mountains the long hairs of the shoulders and back are often frosted with white, thus giving a grizzled appearance and the common name "grizzly" or "silvertip." From *U. americanus, U. arctos* is distinguished in having a prominent hump on the shoulders, a snout that rises more abruptly into the forehead, longer pelage, and longer claws.

The brown bear has one of the greatest natural distributions of any mammal. It occupies a variety of habitats but in the New World seems to prefer open areas such as tundra, alpine meadows, and coastlines. It was apparently common on the Great Plains prior to the arrival of European settlers. In Siberia the brown bear occurs primarily in forests (Stroganov 1969). Surviving European populations are restricted mainly to mountain woodlands (Van Den Brink 1968). Even when living in generally open regions *U. arctos* needs some areas with dense cover (Jonkel 1978). It shelters in such places by day, sometimes in a shallow excavation, and moves and feeds mainly during the cool of the evening and early morning. Egbert and Stokes (1976) noted that activity in coastal Alaska occurs throughout the day but peaks from 1800 to 1900 hours. Seasonal movements are primarily toward major food sources, such as salmon streams and areas of high berry production (Jonkel 1978). In Siberia individuals may travel hundreds of kilometers during the autumn to reach areas of favorable food supplies (Stroganov 1969).

According to Banfield (1974), the usual gait is a slow walk. *U. arctos* is capable of moving very quickly, however, and can easily catch a black bear. Its long foreclaws are not adapted for climbing trees. It has excellent senses of hearing and smell but relatively poor eyesight. The brown bear has great strength. Banfield saw one drag a carcass of a horse about 90 meters. In another case, a 360-kg grizzly killed and dragged a 450-kg bison.

Hibernation begins in October–December and ends in March–May. The exact period depends on the location, weather, and condition of the animal. In certain southerly areas hibernation is very brief or does not take place at all. In most cases the brown bear digs its own den and makes a bed of dry vegetation. The burrow is often located on a sheltered slope, either under a large stone or among the roots of a mature tree. The bed chamber has an average volume of around two cubic meters. A den is sometimes used year after year. During winter sleep there is a marked depression in heart rate and respiration but only a slight drop in body temperature. The animal can be aroused rather easily and can make a quick escape, if necessary (Craighead and Craighead 1972; Grzimek 1975; Stroganov 1969; Slobodyan 1976; Ustinov 1976).

The diet consists mainly of vegetation (Jonkel 1978). Early spring foods include grasses, sedges, roots, moss, and bulbs. In late spring succulent, perennial forbs become important. During the summer and early autumn berries are essential and bulbs and tubers are also taken. Banfield (1974) wrote that *U. arctos* consumes insects, fungi, and roots at all times of the year and also digs mice, ground squirrels, and marmots out of their burrows. In the Canadian Rockies the grizzly is quite carnivorous, hunting moose, elk, mountain sheep and goats, and even black bears. In Mount McKinley National Park, Alaska, Murie (1981) found *U. arctos* to feed mostly on vegetation but also to eat carrion whenever available and occasionally to capture young calves of caribou and moose. During the summer, when salmon are moving upstream along the Pacific coasts of Canada, southern Alaska, and northeastern Siberia, brown bears gather to feed on the vulnerable fish (Banfield 1974; Egbert and Stokes 1976; Kistchinski 1972). Perhaps because of this abundant food supply the bears of these areas are larger and are found at greater densities than anywhere else.

Some approximate reported population densities are: Carpathian Mountains, one bear per 20 sq km; Lake Baikal area, one per 60 sq km; coast of Sea of Okhotsk, one per 10 sq km; Kodiak Island, one per 1.5 sq km; Mount McKinley National Park, one per 30 sq km; northern parts of Alaska and Northwest Territories, one per 150 sq km; and Glacier National Park, Montana, one per 21 sq km (Dean 1976; Harding 1976; Kistchinski 1972; Martinka 1974, 1976; Slobodyan 1976; Ustinov 1976). In the Yellowstone region of the western United States overall average density is about one bear per 88 sq km. In summer, however, individuals have concentrated by night at feeding sites, so densities have reached about one per 0.05 sq km. Daytime dispersal has reduced density to about one per 0.36 sq km. In the Yellowstone ecosystem individual home range averages about 80 sq km and varies from about 20 to 600 sq km, with respect to the area used in the course of a year (Craighead 1976; Knight, Blanchard, and Kendall 1981). The home ranges of males are generally substantially larger than those of females. Lifetime individual ranges in Yellowstone have averaged 3,757 sq km for males and 884 sq km for females (Blanchard and Knight 1991). In the northern Yukon, Pearson (1976) found averages of 414 sq km for males and 73 sq km for females.

Home ranges overlap extensively and there is no evidence of territorial defense (Craighead 1976; Murie 1981). Although generally solitary, the grizzly is the most social of North American bears, occasionally gathering in large numbers at major food sources and often forming family foraging groups with more than one age class of young (Jonkel 1978). At a salmon stream in southern Alaska, Egbert and Stokes (1976) sometimes observed more than 30 bears at one time. Considerable intraspecific tolerance was demonstrated in such aggregations, but dominance hierarchies were enforced by aggression. The highest-ranking animals were the large adult males, which most other bears attempted to avoid. The most aggressive animals were females with young, and the least aggressive were adolescents. Overt fighting was usually brief, and no infliction of serious wounds was observed, but the researchers suspected that killing of young individuals by adult males was a factor in population regulation.

Except as noted, the following life history information was taken from Craighead, Craighead, and Sumner (1976), Egbert and Stokes (1976), Glenn et al. (1976), Jonkel (1978), Murie (1981), Pearson (1976), and Slobodyan (1976). The only lasting social bonds are those between females and young. During the breeding season males may fight over females. Successful males attend one or two females for 1–3 weeks. Mating takes place from May to July, implantation of the fertilized eggs in the uterus is usually delayed until October or November, and births generally occur from January to March, while the mother is in hibernation. The total period of pregnancy may last 180–266 days. Females have been reported to remain in estrus throughout the breeding season until mating, but data listed by Hayssen, Van Tienhoven, and Van Tienhoven (1993) suggest an estrus period of less than a month. After mating, females do not again enter estrus for at least 2, usually 3 or 4, years. The number of young in a litter averages about two and ranges from one to four. They weigh 340–680 grams at birth and their eyes are closed. They may appear nearly naked but are actually covered with short grayish brown hair (Frederick A. Ulmer, Jr., Academy of Natural Sciences, Philadelphia, pers. comm., 1994). They are weaned at about 5 months. They remain with the mother at least until their second spring of life and usually until their third or fourth. Litter mates sometimes maintain an association for 2–3 years after leaving the mother. They reach puberty at around 4–6 years but continue to grow. Males in southern Alaska may not reach full size until they are 10–11 years old. Females in the Yellowstone region are known to have lived 25 years and still be capable of reproduction. A female at the Leipzig Zoo bore 54 young in 19 litters and lived to the age of 39 years and 4 months, the known record life span for the species (Marvin L. Jones, Zoological Society of San Diego, pers. comm., 1995). Potential longevity in captivity may be as great as 50 years (Stroganov 1969).

The brown bear is reputedly the most dangerous animal in North America. If we disregard venomous insects, disease-spreading rodents, domestic animals, and people themselves, this may be true. Three persons were killed by grizzlies in Glacier National Park, Montana, in 1980 (*Washington Post*, 25 July 1980, A-14; 9 October 1980, A-54; 20 October 1980, A-5). Another was killed in Canada (J. R. Gunson, Alberta Fish and Wildlife Division, pers. comm., 1981). That was an unusually tragic year, but injuries and an occasional death have been reported from the western national parks since around 1900 (Cole 1976; Herrero 1970, 1976). During the nineteenth century, attacks on people apparently occurred with some regularity in California (Storer and Tevis 1955). Two people were killed in a single attack in Alaska in July 1995 (*Washington Post*, 3 July 1995), and another died in August 1996 when he inadvertently disturbed a mother bear with a young cub (ibid., 27 August 1996). Three men recently were killed by the bears they were photographing, one in Yellowstone in October 1986, one in Glacier Park in April 1987, and one on the Kamchatka Peninsula in August 1996 (*Audubon* 89[4]: 16–17; Mitchell 1987; *New York Times*, 22 September 1996). According to Ustinov (1976), more than 70 attacks and 17 deaths have been attributed to *U. arctos* in the Lake Baikal area of Siberia. Cicnjak and Ruff (1990) reported 4 deaths in Bosnia, Croatia, and Slovenia from 1986 to 1988. Many such incidents have probably been provoked by an effort to shoot or harass the animal, as the brown bear normally tries to avoid humans. It is unpredictable, however, if startled at close quarters, especially when accompanied by young or engrossed in a search for food. Jonkel (1978) cautioned that there may be more difficulties as recreational and commercial activity increases in areas occupied by the grizzly. He suggested that problems could be reduced by improved education and planning and by not locating campsites, trails, and residential facilities in places regularly used by bears.

The brown bear long has been persecuted as a predator of domestic livestock, especially cattle and sheep. Those parts of North America from which it has been eliminated correspond closely to areas of intensive ranching and grazing. In the nineteenth and early twentieth centuries there apparently were some remarkably destructive bears (Hubbard and Harris 1960; Storer and Tevis 1955), and their activities earned the entire species the lasting enmity of cattle ranchers and sheepherders. The brown bear also has been widely sought as a big-game trophy, and it is currently subject to regulated sport hunting in most of its range. During the 1983–84 season 1,441 brown bears were killed legally in the United States and Canada, and pelts sold in Canada for an average price of $162.05 (Novak, Obbard, et al. 1987).

The original eastern limits of *U. arctos* in North America are not certain. A skull found in a Labrador Eskimo midden dating from the late eighteenth century supports earlier stories that the grizzly used to occur to the east of Hudson Bay (Spiess 1976). Pelts, reportedly those of *U. arc-*

tos, were taken on the Ungava Peninsula as late as 1927 (Veitch 1993). The decline of the species on the Great Plains may have begun when the Indians of that region obtained the horse and hence an improved hunting capability. A precipitous drop in grizzly numbers came in the nineteenth century as settlers and livestock filled the West, setting up confrontations that usually ended to the detriment of the bear. This process was intensified by logging, mining, and road construction, which increased human presence in remote areas. The distinctive subspecies of California and northern Baja California, *U. a. californicus* (Hall 1984), evidently had disappeared by the 1920s.

In the early nineteenth century there may have been 100,000 grizzlies in the western conterminous United States. Now there are probably fewer than 1,000. Of these, about 200 are in Glacier National Park (Martinka 1974), 300 more are in nearby parts of northwestern Montana, and perhaps another 100 occupy isolated mountain ecosystems westward into northern Idaho and Washington. The population in Yellowstone National Park and vicinity has been estimated at 136 (Craighead, Varney, and Craighead 1974), 247 (Knight, Blanchard, and Kendall 1981), and 329 (Mattson et al. 1995). Although numbers have appeared to be stable or increasing for the last two decades and research and conservation efforts have intensified, this isolated population may not be able to survive anticipated long-term environmental perturbations and levels of mortality resulting from increased human habituation (Mattson, Blanchard, and Knight 1992; Mattson and Reid 1991). Some authorities have warned that the Yellowstone population may not be reproductively viable and that current management practices in and around the park are leading to conflicts between human interests and bears, often resulting in the death of the latter (Chase 1986). Such problems probably caused the recent disappearance of very small remnant groups in south-central Colorado and northwestern Mexico. The grizzly has been extirpated from the Great Plains of Canada, except for an isolated group in west-central Alberta (Banfield 1974). The species also has declined on the barrens of the Northwest Territories (Macpherson 1965). In the mountainous regions of western Canada and in Alaska *U. arctos* is still relatively common, perhaps numbering about 55,000 individuals (Cowan 1972; Jonkel 1987; Schoen, Miller, and Reynolds 1987; Servheen 1989).

Chestin et al. (1992) estimated that the number of brown bears in the countries of the old Soviet Union had increased from about 105,000 in the early 1960s to 131,000 in 1989. The annual legal kill was approximately 3,500 and poaching was thought to account for several thousand more. It must be noted that these figures were derived prior to a reported relaxation of controls and intensification of illegal killing that followed the breakup of the Soviet Union and is associated in part with the Far Eastern commerce in bear parts (see account of *Ursus thibetanus*). The bears of the Kamchatka Peninsula, comparable in size to those of southern Alaska, are said to have been especially hard hit (*Oryx* 28 [1994]: 224). The great majority of the bears estimated to be present in 1989 were in European Russia and Siberia, 3,500 were in the Caucasus, and 3,500 were in the Central Asian republics.

Vereschagin (1976) estimated that there were about 30,000 brown bears in Eurasia outside of the Soviet Union, though there may have been a reduction since then. There now are 4,500–7,600 in China (Guo 1995) and possibly some viable populations in Mongolia, Hokkaido, the Himalayas, and Turkey, but all appear to be declining in the face of uncontrolled hunting and habitat disruption (Servheen 1989). To the west of Russia there are about 700 in northern Scandinavia, 700 in southern Poland, 700 in Slovakia, 6,000 in Romania, 700 in Bulgaria, and as many as 2,000 in the countries originally comprised by Yugoslavia. There are small, isolated groups in Albania, Greece, Austria, Italy, southern France, northern Spain, and southern Norway (Elgmork 1978; Pasitschniak-Arts 1993; Servheen 1989; Smit and Van Wijngaarden 1981). A recent analysis of mitochondrial DNA, suggesting a phylogenetic divergence between the bears of Russia and northern Scandinavia, on the one hand, and those surviving in southern Scandinavia and southern Europe, on the other, may have effects on reintroduction planning (Taberlet and Bouvet 1994). *U. arctos* apparently disappeared from Great Britain in the tenth century (Pasitschniak-Arts 1993) and from northwestern Africa around the mid–nineteenth century (Harper 1945).

Appendix 2 of the CITES includes all North American populations of *U. arctos* except *U. a. nelsoni*, the Mexican grizzly bear (treated as a synonym of *U. a. arctos* by Hall 1984), and now also includes all Eurasian populations except those on appendix 1. On appendix 1 are *U. a. nelsoni*, *U. a. pruinosus* of Tibet and Mongolia, and *U. a. isabellinus* of the mountains of Central Asia. Schaller, Tulgat, and Navantsatsvalt (1993) considered the range of *isabellinus* to extend as far as Great Gobi National Park in western Mongolia, where there is a unique, isolated, and arid-adapted population of only about 30 bears. The IUCN classifies *U. a. nelsoni* as extinct. The USDI lists *U. a. nelsoni*, *U. a. pruinosus*, and the Italian populations of *U. arctos* as endangered. Those populations of *U. arctos* in the conterminous United States are listed as threatened by the USDI.

Ursus maritimus (polar bear)

Head and body length is 2,000–2,500 mm, tail length is 76–127 mm, and shoulder height is up to 1,600 mm. DeMaster and Stirling (1981) gave the weight as 150–300 kg for females and 300–800 kg for males. Banfield (1974), however, wrote that males usually weigh 420–500 kg. The color is often pure white following the molt but may become yellowish in the summer, probably because of oxidation by the sun. The pelage also sometimes appears gray or almost brown depending on the season and light conditions. The neck is longer than that of other bears, and the head is relatively small and flat. The forefeet are well adapted for swimming, being large and oarlike. The soles are haired, probably for insulation from the cold and traction on the ice. Females have four functional mammae (DeMaster and Stirling 1981).

The polar bear is often considered to be a marine mammal. It is distributed mainly in arctic regions around the North Pole. The southern limits of its range are determined by distribution of the pack ice. It has been recorded as far north as 88° N and as far south as the Pribilof Islands in the Bering Sea, the island of Newfoundland, the southern tip of Greenland, and Iceland. There also are permanent populations in James Bay and the southern part of Hudson Bay. Although found generally in coastal areas or on ice hundreds of kilometers from shore, individuals have wandered up to 200 km inland (Stroganov 1969).

According to DeMaster and Stirling (1981), the preferred habitat is pack ice that is subject to periodic fracturing by wind and sea currents. The refreezing of such fractures provides places where hunting by the bear is most successful. Some animals spend both winter and summer along the lower edge of the pack ice, perhaps undergoing extensive north-south migrations as this edge shifts. Others move onto land for the summer and disperse across the ice as it forms along the coast and between islands during winter. The bears of the Labrador coast sometimes move north to Baffin Island, and some individuals have traveled

Polar bears *(Ursus maritimus):* Top, photo from New York Zoological Society. Insets: A. Forefoot; B. Hind foot; photos from *Proc. Zool. Soc. London;* C. Young, 24 hours old, photo by Ernest P. Walker. Bottom, photo by Sue Ford, Washington Park Zoo, Portland, Oregon.

as far as 1,050 km, to the islands of northern Hudson Bay (Stirling and Kiliaan 1980). Individuals in the population of the Beaufort Sea off northern Alaska move several thousand kilometers annually and may use an area exceeding 500,000 sq km over a period of years (Amstrup, Garner, and Durner 1995). The population that summers along the southern shore of Hudson Bay spreads all across the partly ice-covered bay in November and returns to shore in July or August (Stirling et al. 1977). During the latter, ice-free period adult males occupy coastal areas and family groups and pregnant females move farther inland (Derocher and Stirling 1990*a*). Despite such movements, the polar bear is not a true nomad. There are a number of discrete populations, each with its own consistently used areas for feeding and breeding (Stirling, Calvert, and Andriashek 1980).

The polar bear can outrun a reindeer for short distances on land and can attain a swimming speed of about 6.5 km/hr. It swims rather high, with head and shoulders above the water. If killed in the water, it will not immediately sink. According to DeMaster and Stirling (1981), it has been reported to swim for at least 65 km across open water. It is capable of diving under the ice and surfacing in holes utilized by seals. It seems to be most active during the first third of the day and least active in the final third. From July to December in the James Bay region, when a lack of ice prevents seal hunting, *U. maritimus* spends about 87 percent of its time resting, apparently living off of stored fat (Knudsen 1978). The bears sometimes excavate depressions or complete earthen burrows on land during the summer in order to avoid the sun and keep cool (Jonkel et al. 1976).

Any individual bear may make a winter den for temporary shelter during severe weather, but only females, especially those that are pregnant, generally hibernate for lengthy periods. As with other bears, winter sleep involves a depressed respiratory rate and a slightly lowered body temperature but not deep torpor. Most pregnant females evidently do not spend the winter along the pack ice but hibernate on land from October or November to March or April. Maternal dens are usually found within 8 km of the coast, but in the southern Hudson Bay region they are concentrated 30–60 km inland. They are excavated in the snow to depths of 1–3 meters, often on a steep slope. They usually consist of a tunnel several meters long that leads to an oval chamber of about 3 cubic meters. Some dens have several rooms and corridors (DeMaster and Stirling 1981; Harington 1968; Larsen 1975; Stirling, Calvert, and Andriashek 1980; Stirling et al. 1977; Uspenski and Belikov 1976).

The polar bear feeds primarily on the ringed seal *(Phoca hispida)* (DeMaster and Stirling 1981). The bear either remains still until a seal emerges from the water or stealthily stalks its prey on the ice (Stirling 1974). It may also dig out the subnivean dens of seals to obtain the young (Stirling, Calvert, and Andriashek 1980). During summer and autumn in the southern Hudson Bay region *U. maritimus* often swims among sea birds and catches them as they sit on the water (Russell 1975). The diet also includes the carcasses of stranded marine mammals, small land mammals, reindeer, fish, and vegetation. Berries become important for some individuals during summer and autumn (Jonkel 1978).

Reported population densities range from 1/37 sq km to 1/139 sq km (DeMaster and Stirling 1981). Home ranges are not well defined but are thought to vary from 150 to 300 km in diameter and to overlap extensively (Kolenosky 1987). Although *U. maritimus* is generally solitary, large aggregations may form around a major source of food (Jonkel 1978). As many as 40 individuals have been seen at one time in the vicinity of the Churchill garbage dump, on the southern shore of Hudson Bay (Stirling et al. 1977). Of the adult males in the population that inhabits that region, more than half occur in aggregations during the ice-free season. These groups, with an average size of 4 bears, tend to occur at environmentally favorable sites and may help develop familiarity that will avoid future conflicts for resources (Derocher and Stirling 1990*b*). Wintering females evidently tolerate one another well, as dens on Wrangel Island are sometimes found at densities of one per 50 sq meters (Uspenski and Belikov 1976). High concentrations of summer dens also have been reported (Jonkel et al. 1976). Adult females with young are not subordinate to any other age or sex class but tend to avoid interaction with adult males, presumably because the latter are potential predators of the cubs (DeMaster and Stirling 1981).

Estrus lasts 3 days (Hayssen, Van Tienhoven, and Van Tienhoven 1993). The sexes usually come together only briefly during the mating season, March–June. Delayed implantation apparently extends the period of pregnancy to 195–265 days. The young are born from November to January, while the mother is in her winter den. Females give birth every 2–4 years. The number of young per litter averages about two and ranges from one to four. They weigh about 600 grams at birth and are well covered with short white fur, but their eyes are tightly closed. Upon emergence from the den in March or April the cubs weigh 10–15 kg. They usually leave the mother at 24–28 months. The age of sexual maturity averages about 5–6 years. Adult weight is attained at about 5 years by females but not until 10–11 years by males. Wild females apparently have a reduced natality rate after the age of 20. Annual adult mortality in a population is about 8–16 percent. Potential longevity in the wild is estimated at 25–30 years (DeMaster and Stirling 1981; Ramsay and Stirling 1988; Stirling, Calvert, and Andriashek 1980; Ulmer 1966; Uspenski and Belikov 1976). A female at the Detroit Zoo gave birth at 36 years and 11 months (Latinen 1987) and was still living at 45 years (Marvin L. Jones, Zoological Society of San Diego, pers. comm., 1995).

The polar bear often is considered to be dangerous to people, though usually the two species are not found in close proximity. An exception developed during the 1960s in the vicinity of the town of Churchill, on the southern shore of Hudson Bay (Stirling et al. 1977). Bears apparently increased in this area because of a decline in hunting. At the same time, more people moved in and several large garbage dumps were established. A number of persons were attacked and one was killed. Many bears were shot or translocated by government personnel. There also have been a number of confrontations in Svalbard, and one person was killed by a bear there in 1977 (Gjertz and Persen 1987).

The native peoples of the Arctic have long hunted the polar bear for its fat and fur. Sport and commercial hunting increased in the twentieth century. Of the regularly marketed North American mammal pelts, that of *U. maritimus* is the most valuable. During the 1976–77 season 530 skins from Canada were sold at an average price of $585.22 (Deems and Pursley 1978). Some individual prime pelts have brought more than $3,000 each (Smith and Jonkel 1975). There subsequently appears to have been a decline in demand, and other parts of the polar bear, particularly the gallbladder, seem to lack the market value of those of other bears (Frampton 1995*a*).

The use of aircraft to locate polar bears and to land trophy hunters in their vicinity developed in Alaska in the late 1940s. The annual kill by such means reached about 260 bears by 1972. In that year, however, the killing of *U. mar-*

itimus, except for native subsistence, was prohibited by the United States Marine Mammal Protection Act. Canada and Denmark (for Greenland) also limit hunting to resident natives, and Russia and Norway (for Svalbard) provide complete protection. In 1973 these five nations drafted an agreement calling for the restriction of hunting, the protection of habitat, and cooperative research on polar bears. The agreement was ratified by the United States in 1976. The yearly worldwide kill is now estimated at around 1,000 animals. The total number of polar bears in the wild in 1993 was estimated at 21,470–28,370, and populations are generally thought to be stable or increasing. *U. maritimus,* however, may be threatened by the exploitation of oil and gas reserves in the Arctic, especially by development in the limited areas suitable for denning by pregnant females (DeMaster and Stirling 1981; Stirling and Kiliaan 1980; U.S. Fish and Wildlife Service 1980; Wiig, Born, and Garner 1995).

There now also is concern that the influx of cash from oil and gas development will stimulate increased hunting by native peoples and that such hunting, not being restricted to adult males, could damage the relatively small and vulnerable polar bear populations off northern Alaska and northwestern Canada (Amstrup, Stirling, and Lentfer 1986). Another long-term concern is that global warming, resulting from the greenhouse effect of atmospheric polluting gases, will reduce the southerly extent of sea ice and thereby deny accessibility to seals. Even now the polar bear population of southwestern Hudson Bay is showing signs of nutritional stress (Stirling and Derocher 1993). The species is classified as conservation dependent by the IUCN and is on appendix 2 of the CITES. The provisions of both CITES and the United States Marine Mammal Protection Act do allow for limited importation under certain circumstances. The U.S. Fish and Wildlife Service recently proposed allowing importation of polar bear trophies taken in accordance with carefully regulated sport-hunting programs in the Northwest Territories of Canada (Frampton 1995*a*).

Ursus malayanus (Malayan sun bear)

This is the smallest bear. Head and body length is 1,000–1,400 mm, tail length is 30–70 mm, shoulder height is about 700 mm, and weight is 27–65 kg. The general coloration is black. There is a whitish or orange breast mark and a grayish or orange muzzle, and occasionally the feet are pale in color. The breast mark is often U-shaped but is variable and sometimes wholly lacking. The body is stocky, the muzzle is short, the paws are large, and the claws are strongly curved and pointed. The soles are naked.

The sun bear inhabits dense forests at all elevations (Lekagul and McNeely 1977). It is active at night, usually sleeping and sunbathing by day in a tree, two to seven meters above the ground. Tree branches are broken or bent to form a nest and lookout post. *U. malayanus* has a curious gait in that all the legs are turned inward while walking. The species is usually shy and retiring and does not hibernate. An expert tree climber, it is cautious, wary, and intelligent. A young captive observed the way in which a cupboard containing a sugar pot was locked with a key. It then later opened the cupboard by inserting a claw into the eye of the key and turning it. Another captive scattered rice from its feeding bowl in the vicinity of its cage, thus attracting chickens, which it then captured and ate.

The diet is omnivorous, and the front paws are used for most of the feeding activity. Trees are torn open in search of nests of wild bees and for insects and their larvae. The soft growing point of the coconut palm, known as "palmite," is ripped apart and consumed. After digging up termite colonies, the animal places its forepaws alternately in the nest and licks the termites off. Jungle fowl, small rodents, and fruit juices also are included in the diet.

Births may occur at any time of the year. In the East Berlin Zoo a female produced one litter on 4 April 1961 and another on 30 August 1961. The gestation period for six births at that zoo was 95–96 days (Dathe 1970). At the Fort Worth Zoo, however, three pregnancies lasted 174, 228, and 240 days, evidently because of delayed fertilization or im-

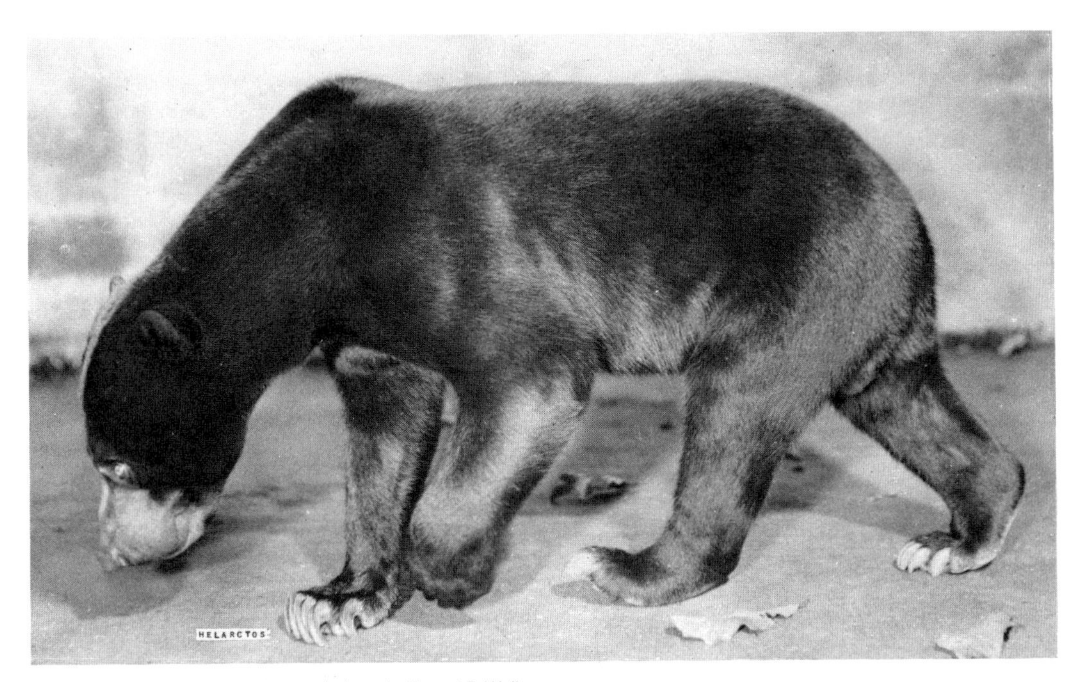

Malayan sun bear *(Ursus malayanus)*, photo by Ernest P. Walker.

plantation (McCusker 1974). Litters usually contain one or two young weighing about 325 grams each. They remain with the mother until nearly full grown (Lekagul and McNeely 1977). Two captives lived to an age of approximately 31 years (Marvin L. Jones, Zoological Society of San Diego, pers. comm., 1995).

In the wild, *U. malayanus* sometimes does considerable damage to coconut plantations and is said to be one of the most dangerous animals within its range (Lekagul and McNeely 1977). Young individuals make interesting pets but become unruly within a few years. According to Mills (1993, 1994, 1995), sun bear cubs are highly popular as pets in the Orient, but when they reach adult size they commonly are sold for their parts, which are used in the medicinal trade. International commerce involving such utilization also is adversely affecting wild populations (see account of *Ursus thibetanus*). Servheen (1989) reported a general decline because of hunting for gallbladders and other body parts, lack of effective regulation, and logging and conversion of vast areas of forest to cropland and rubber plantations. *U. malayanus* may now have been extirpated in India and is very rare in Bangladesh and China. It is on appendix 1 of the CITES.

Ursus ursinus (sloth bear)

Head and body length is 1,400–1,800 mm, tail length is 100–125 mm, shoulder height is 610–915 mm, and weight is 55–145 kg. The shaggy black hairs are longest between the shoulders. The overall black coloration is often mixed with brown and gray, but cinnamon and red individuals also have been noted. The chest mark, typically shaped like a V or Y, varies from white or yellow to chestnut brown.

The sloth bear has a number of structural modifications associated with an unusual method of feeding. The lips are protrusible, mobile, and naked; the snout is mobile; the nostrils can be closed at will; the inner pair of upper incisors is absent, thus forming a gap in the front teeth; and the palate is hollowed. These features enable the bear to feed on termites (white ants) in the following manner: the nest is dug up, the dust and dirt blown off, and the occupants sucked up in a "vacuum cleaner" action. The resulting noises can be heard more than 185 meters away and often lead to the bear's detection by hunters.

The sloth bear inhabits moist and dry forests, especially in areas of rocky outcrops. It may be active at any hour but is mainly nocturnal. During cool weather it spends the day in dense vegetation or shallow caves. The sense of smell is well developed, but sight and hearing are relatively poor. Hibernation is not known to occur. Termites are the most important food for most of the year, but the diet also includes other insects, grubs, honey, eggs, carrion, grass, flowers, and fruit.

In the Royal Chitawan National Park of Nepal, Sunquist (1982) found an adult male to move over an area of at least 10 sq km during a period of about 2 years. In the same area, Laurie and Seidensticker (1977) found a minimum density of about 0.1/sq km. Most observations were of lone bears or of females with cubs. Vocalizations, heard mainly in association with intraspecific agonistic encounters, included roars, howls, screams, and squeals. Births apparently occurred mostly from September to January.

Previous observations indicate that breeding takes place mainly in June in India and during most of the year in Sri Lanka. Pregnancy lasts about 6–7 months. The young, usually one or two and rarely three, are born in a ground shelter. They leave the den at 2–3 months and often ride on the mother's back. They remain with the mother until they are almost full grown, possibly 2 or 3 years. Captives have lived for 40 years.

The sloth bear normally is not aggressive but is held in great respect by some of the people that inhabit its range. Apparently because of its poor eyesight and hearing, it is sometimes closely approached by humans. It may then attack in what it considers to be self-defense and inflict severe wounds. Since it is thought to be dangerous, and since it sometimes damages crops, it has been extensively hunted. It also seems not to tolerate regular human disturbance and is losing habitat to agriculture, logging, settlement, and

Sloth bear *(Ursus ursinus)*: A. Photo by Hans-Jürg Kuhn. B. Photo from New York Zoological Society.

hydroelectric projects (IUCN 1978). Total numbers surviving throughout the range of the species are estimated at 7,500–10,000 individuals (Nobbe and Garshelis 1994). A major problem that only recently has received general recognition is the killing of *U. ursinus* for its body parts, which then are sold into international commerce and eventually used for food, alleged medicines, and decorative purposes, much the same as described above in the account of *Ursus thibetanus.* Bile from the gallbladder is widely used in the Orient to treat diseases of the liver, heart, and stomach, and bear paws are prized for consumption (Rose and Gaski 1995). Based on data showing the number of bear gallbladders exported from India to Japan from 1978 to 1988, Servheen (1989) calculated an annual kill of 728–1,548 sloth bears, probably a substantial part of the remaining populations. Such information contributed to the placing of *U. ursinus* on appendix 1 of the CITES in 1988, and the species now is also classified as vulnerable by the IUCN.

CARNIVORA; URSIDAE; **Genus AILUROPODA**
Milne-Edwards, 1870

Giant Panda

The single species, *A. melanoleuca,* now is known from the central Chinese provinces of Gansu, Shaanxi, and Sichuan; there also are historical records from much of eastern China to the south of the Huang River (Schaller et al. 1985).

There is much controversy regarding the systematic position of this genus. It once was considered to be a close relative of *Ailurus* and was placed with that genus in the family Procyonidae, but most authorities eventually came to treat *Ailuropoda* as a bear. E. R. Hall (1981) put it in the ursid subfamily Ailuropodinae, but Chorn and Hoffmann (1978) referred it to the otherwise extinct ursid subfamily Agriotheriinae. Thenius (1979) recognized the giant panda as an offshoot of the Ursidae but suggested that it represents a distinct family, the Ailuropodidae. A recent trend has been to again treat *Ailuropoda* and *Ailurus* as close relatives. Schaller et al. (1985) stated that on the basis of anatomical, biochemical, and paleontological evidence the giant panda's position remains equivocal but that certain characters of morphology, reproduction, and behavior indicate that *Ailuropoda* and *Ailurus* are related to both bears and raccoons, but more closely to the former, and that the two genera belong either in separate but closely related families of their own or together in their own family, the Ailuridae. Wozencraft (*in* Wilson and Reeder 1993) combined *Ailuropoda* and *Ailurus* in the subfamily Ailurinae of the family Ursidae.

Head and body length is 1,200–1,500 mm, tail length is about 127 mm, and weight is 75–160 kg. The coat is thick and woolly. The eye patches, ears, legs, and band across the shoulders are black, sometimes with a brownish tinge. The remainder of the body is white but may become soiled with age. There are scent glands under the tail.

Ailuropoda resembles other bears in general appearance but is distinguished by its striking coloration and certain characters associated with its diet. The head is relatively

Giant panda *(Ailuropoda melanoleuca),* photo from New York Zoological Society. Insets: A. Right hind foot; B. Right forefoot; photos from *Proc. Zool. Soc. London.* C. Skull showing dentition, photo by P. F. Wright of specimen in U.S. National Museum of Natural History.

massive because of the expanded zygomatic arches of the skull and the well-developed muscles of mastication. The second and third premolar teeth and the molars are relatively larger and broader than those of other bears. The forefoot has an unusual modification that is thought to aid in the grasping of bamboo stems. The pad on the sole of each forepaw has an accessory lobe, and the pad of the first digit—and to a lesser extent the pad of the second digit—can be flexed onto the summit of this accessory lobe and its supporting bone.

The giant panda is found in montane forests with dense stands of bamboo. Its usual elevational range is 2,700–3,900 meters, but it may descend to as low as 800 meters in the winter. It does not make a permanent den but takes shelter in hollow trees, rock crevices, and caves. It lives mainly on the ground but evidently can climb trees well. Activity is largely crepuscular and nocturnal. Schaller et al. (1985) found animals to be active about 14.2 hours per day. *Ailuropoda* does not hibernate but descends to lower elevations in the winter and spring (Chorn and Hoffmann 1978). Average daily movement has been reported to cover 596 meters (Johnson, Schaller, and Hu 1988).

The diet consists mainly of bamboo shoots, up to 13 mm in diameter, and bamboo roots. *Ailuropoda* spends 10–12 hours a day feeding, usually in a sitting position with the forepaws free to manipulate the bamboo. Average daily consumption is about 12.5 kg of bamboo (Johnson, Schaller, and Hu 1988). One radio-tracked individual used an area of 3.8 sq km during a 9-month period, except for a 15-day foray during the spring bamboo shooting season, which increased the overall area to 6.8 sq km (Johnson, Schaller, and Hu 1988). Other plants, such as gentians, irises, crocuses, and tufted grasses, are also taken. *Ailuropoda* occasionally hunts for fish, pikas, and small rodents.

Schaller et al. (1985) studied wild individuals by radio tracking in the Wolong Natural Reserve of China. About 145 pandas occurred in this 2,000 sq km area. Within the study area, which measured 35 sq km, there were 7 adult males, 5 or 6 adult females, 4 independent subadults, and 2 infants. Home range size varied from 3.9 to 6.4 sq km, with male ranges being only as large as or slightly larger than those of females. However, males were found to occupy greatly overlapping ranges lacking well-defined core areas. They shift frequently within their ranges, show no evidence of territorial behavior, and spend considerable time within the core areas of females and subadults. Females also may have overlapping ranges, but they spend most of their time within a discrete core area of only 30–40 ha. They evidently do not tolerate other females and subadults within their core areas. As the animals move about, they mark their routes by spraying urine, clawing tree trunks, and rubbing against objects. Captives are known to scent-mark with secretions from glands in the genital region (Kleiman 1983). The vocal repertoire consists of bleats, honks, squeals, growls, moans, barks, and chirps (Peters 1982).

Except as noted, the following information on reproduction and life history was taken from Schaller et al. (1985). *Ailuropoda* is generally solitary, but during the breeding season several males may compete for access to a female. Mating generally occurs from March to May. Females have a single estrous period of 12–25 days, but peak receptivity lasts only 1–5 days. Births usually take place during August or September in a cave or hollow tree. The overall period of pregnancy is 97–163 days; the variation evidently results from a delay in implantation of 45–120 days. The number of young per litter is usually one or two, and occasionally three, but normally only a single cub is raised. The neonatal/maternal weight ratio may be the smallest among the eutherian mammals (Kleiman 1983). At birth the offspring weighs 90–130 grams, is covered with sparse white fur, and has a tail that is about one-third as long as the head and body; adult coloration is attained by the end of the first month, and the eyes open after 40–60 days (Chorn and Hoffmann 1978). The young begin to walk at 3–4 months and to eat bamboo at 5–6 months. They may be fully weaned at 8–9 months, usually leave their mothers at about 18 months, and attain sexual maturity after 5.5–6.5 years. A captive specimen lived to an age of approximately 34 years (Marvin L. Jones, Zoological Society of San Diego, pers. comm., 1995).

The range of the giant panda began to decline in the late Pleistocene because of both climatic changes and the spread of people (Wang 1974). In the last 2,000 years the species has disappeared from Henan, Hubei, Hunan, Guizhou, and Yunnan provinces (Schaller et al. 1985). At present about 1,000 individuals are thought to survive, apparently divided into three isolated groups. In the mid-1970s about 100 pandas starved when an important food plant died over a large area. The species receives complete legal protection, and cooperative field investigations have been carried out by the Chinese government and the World Wildlife Fund (Schaller 1981, 1993). From 1985 to 1991, Chinese courts convicted 278 persons for panda poaching or pelt smuggling, with 16 sentenced to life imprisonment and 3 to death! Nonetheless, poaching to obtain the valuable skin continues to be a major threat (Schaller 1993). The giant panda is classified as endangered by the IUCN and the USDI and is on appendix 1 of the CITES. It is among the most popular of zoo animals but has been extremely difficult to breed. As of June 1993 there were 98 giant pandas in captivity in China and 15 in other countries (Frampton 1995*b*). A Chinese program of loaning pandas for exhibit in Western zoos became subject to intense controversy in the late 1980s, with supporting parties arguing that the fees charged would be used for panda conservation and opponents claiming that the animals should be kept together in China for breeding purposes (Drew 1989).

The West's foremost authority on the giant panda, George B. Schaller (1993), wrote that he initially had favored strictly regulated loans to foster cooperative research and conservation but had changed his mind after observing the resulting "undisciplined scramble for pandas." In response to political and financial interests, the involved countries ignored guidelines and CITES provisions designed to ensure that exportation would not be detrimental to maintenance of viable wild and captive breeding populations in China. The U.S. Fish and Wildlife Service was said to have buckled under political pressure, though more recently that agency proposed a new policy by which all imports would have to be part of an international effort to benefit panda conservation and not interfere with China's breeding and research program (Frampton 1995*b*). However, Schaller, like some other zoologists whose research has helped to publicize and stimulate the extensive capturing of an endangered species, wondered whether the giant panda would have been best left in "obscurity."

CARNIVORA; **Family PROCYONIDAE**

Raccoons and Relatives

This family contains 7 Recent genera and 19 species. Two subfamilies are traditionally recognized: Ailurinae, with the single genus *Ailurus* (lesser panda), found in the Himalayas and adjacent parts of eastern Asia; and Procyoninae, with the other 6 genera, which occur in temperate and tropical areas of the Western Hemisphere. Such an arrangement is supported by recent molecular and biochemical analyses (Wayne et al. 1989). However, Corbet and Hill (1991), Glatston (1994), and Roberts and Gittleman (1984) elevated the Ailurinae to full familial rank. Wozencraft (*in* Wilson and Reeder 1993) considered the Ailurinae to also include *Ailuropoda* (giant panda) and placed the group in the family Ursidae. On the basis of morphology, the remaining procyonids were divided into two subfamilies: the Procyoninae, with *Procyon, Bassariscus, Nasua,* and *Nasuella;* and the Potosinae, with *Potos* and *Bassaricyon* (Decker and Wozencraft 1991; Wozencraft *in* Wilson and Reeder 1993).

Head and body length is 305–670 mm, tail length is 200–690 mm, and weight is about 0.8–12.0 kg. Males are about one-fifth larger and heavier than females. The pelage varies from gray to rich reddish brown. Facial markings are often present, and the tail is usually ringed with light and dark bands. The face is short and broad. The ears are short, furred, erect, and rounded or pointed. The tail is prehensile in the arboreal kinkajou *(Potos)* and is used as a balancing and semiprehensile organ in the coatis *(Nasua).* Each limb bears five digits, the third being the longest. The claws are short, compressed, recurved, and in some genera semiretractile. The soles are haired in several genera. Males have a baculum.

The dental formula is usually: (i 3/3, c 1/1, pm 4/4, m 2/2) $\times$ 2 = 40. The premolars, however, number 3/3 in *Potos* and 3/4 in *Ailurus* (Stains 1984). The incisors are not specialized, the canines are elongate, the premolars are small and sharp, and the molars are broad and low-crowned. The carnassials are developed only in *Bassariscus.*

Procyonids walk on the sole of the foot with the heel touching the ground or partly on the sole and partly on the digits. The gait is usually bearlike. They are good climbers, and one genus *(Potos)* spends nearly its entire life in trees. Most procyonids shelter in hollow trees, on large branches, or in rock crevices. Most become active in the evening, but *Nasua* may be primarily diurnal. The diet is omnivorous, though *Potos* and *Bassaricyon* seem to depend largely on fruit and *Ailurus* feeds mainly on bamboo. Most genera travel in pairs or family groups and give birth in the spring.

The geological range of the Procyonidae is early Oligocene to Recent in North America, late Miocene to Recent in South America, late Eocene to early Pleistocene in Europe, and early Miocene to Recent in Asia (Stains 1984). Although *Ailurus* is now geographically far removed from the other procyonids, a related Pliocene genus, *Parailurus,* occurred in both Europe and North America (L. D. Martin 1989).

CARNIVORA; PROCYONIDAE; **Genus AILURUS**
F. Cuvier, 1825

Lesser Panda

The single species, *A. fulgens,* is known to occur in Nepal, Sikkim, Bhutan, northern Burma, and the provinces of Yunnan and Sichuan in south-central China and probably also exists along the border of Tibet and Assam (Ellerman and Morrison-Scott 1966; Roberts and Gittleman 1984).

Head and body length is 510–635 mm, tail length is 280–485 mm, and weight is usually 3–6 kg. The coat is long and soft and the tail is bushy. The upper parts are rusty to deep chestnut, being darkest along the middle of the back. The tail is inconspicuously ringed. Small, dark-colored eye patches are present, and the muzzle, lips, cheeks, and edges of the ears are white. The back of the ears, the limbs, and the underparts are dark reddish brown to black. The head is rather round, the ears are large and pointed, the feet have hairy soles, and the claws are semiretractile. The nonprehensile tail is about two-thirds as long as the head and body. There are glandular sacs in the anal region. Females have four mammae.

The lesser panda inhabits mountain forests and bamboo thickets at elevations of 1,800–4,000 meters. It seems to prefer colder temperatures than does the giant panda *(Ailuropoda).* It is usually said to be nocturnal and crepuscular (Roberts and Gittleman 1984) and to sleep by day in a tree.

Lesser panda *(Ailurus fulgens)*, photo by Arthur Ellis, *Washington Post.*

However, Reid, Hu, and Huang (1991) found it to be much more active in daylight, especially during summer coincident with arboreal foraging; it apparently rested in direct sunlight during winter to minimize heat loss. When sleeping, it generally curls up like a cat or dog, with its tail over its head, but it may also sleep while sitting on top of a limb, its head tucked under the chest and between the forelegs, as the American raccoon *(Procyon)* does at times. Although *Ailurus* is a capable climber, it seems to do most of its feeding on the ground. The diet consists mostly of bamboo sprouts, grasses, roots, fruits, and acorns. It also occasionally takes insects, eggs, young birds, and small rodents. A female radio tracked by Johnson, Schaller, and Hu (1988) for 9 months had an average daily movement of 481 meters and a home range of 3.4 sq km. Reid, Hu, and Huang (1991) estimated two home ranges to be 0.9 and 1.1 sq km and population density to be 1/2–3 sq km.

In the wild the lesser panda sometimes travels in pairs or small family groups. Such groups probably represent a consorting male and female or a mother with cubs. The young seem to stay with the mother for about a year, or until the next litter is about to be born. The disposition of *Ailurus* is mild; when captured, it does not fight, tames readily, and is gentle, curious, and generally quiet. The usual call is a series of short whistles or squeaking notes; when provoked, it utters a sharp, spitting hiss or a series of snorts while standing on its hind legs. A musky odor is emitted from the anus when the animal is excited. According to Roberts and Gittleman (1984), *Ailurus* scent-marks its territory by urine, feces, and secretions from anal and circumanal glands. In the wild, births take place in the spring and summer, mainly in June, in a hollow tree or rock crevice.

Studies at the U.S. National Zoo in Washington, D.C. (Roberts 1975, 1980), indicate that adult males can be kept with females and young but that adult females are not tolerant of one another. Reproduction is most successful when a single adult male and female are placed together, though they will sleep and rest apart except during the breeding season. Females apparently have only one estrus annually and are then receptive for just 18–24 hours. They may begin to build a nest of sticks and leaves several weeks before giving birth. Mating occurs from mid-January to early March, and births from mid-June to late July. Recorded gestation periods at the National Zoo are 114–45 days; however, there also is a record of only 90 days at the San Diego Zoo. It thus may be that there is delayed implantation in temperate zones but not in subtropical zones. Litters contain one to four, usually two, young. The cubs weigh about 200 grams at 1 week, open their eyes after 17–18 days, attain full adult coloration by 90 days, and take their first solid food at 125–35 days. The young are removed from the parents at 6–7 months and placed in small peer groups. Both sexes attain sexual maturity at about 18 months. According to Marvin L. Jones (Zoological Society of San Diego, pers. comm., 1995), a captive in China was still living at 17 years and 6 months.

The lesser panda is a very popular zoo animal and is frequently involved in the animal trade. It now is on appendix 1 of the CITES and is classified as endangered by the IUCN, which estimates that fewer than 2,500 mature individuals survive. There recently has been increasing evidence that the species is rare and continuing to decline because of the destruction of its forest habitat, killing for its pelt, and illegal trade in live animals (Glatston 1994). The species was found to be scarce even in a national park in Nepal; fecundity there was low, the habitat was deteriorating through intensive cattle grazing, and most young were killed as a result of human disturbance or attack by domestic dogs (Yonzon and Hunter 1991*a*, 1991*b*).

CARNIVORA; PROCYONIDAE; **Genus BASSARISCUS**
Coues, 1887

Ringtails, or Cacomistles

There are two species (E. R. Hall 1981):

B. astutus, southwestern Oregon and eastern Kansas to Baja California andsouthern Mexico;
B. sumichrasti, southern Mexico to western Panama.

The latter species formerly was often placed in a separate genus, *Jentinkia* Trouessart, 1904.

In *B. astutus* head and body length is 305–420 mm, tail length is 310–441 mm, and shoulder height is about 160

Ringtail *(Bassariscus astutus)*, photo by Woodrow Goodpaster.

Central American cacomistle *(Bassariscus sumichrasti)*, photo by I. Poglayen-Neuwall.

mm. Armstrong, Jones, and Birney (1972) listed weights of 824–1,338 grams. The upper parts are buffy with a black or dark brown wash, and the underparts are white or white washed with buff. The eye is ringed by black or dark brown, and the head has white to pinkish buff patches. The tail is bushy, longer than the head and body, and banded with black and white for its entire length. Females have four mammae.

In *B. sumichrasti* head and body length is 380–470 mm and tail length is 390–530 mm. One individual weighed 900 grams. The color is usually buffy gray to brownish, and the tail is ringed with buff and black. From *B. astutus, B. sumichrasti* is distinguished in having pointed (rather than rounded) ears, a longer tail, naked (rather than hairy) soles, nonretractile (rather than semiretractile) claws, and low (rather than high) ridges connecting the cusps of the molariform teeth.

B. astutus utilizes a varied habitat but seems to prefer rocky, broken areas, often near water. It dens in rock crevices, hollow trees, the ruins of old Indian dwellings, and the upper parts of cabins. Activity occurs mainly at night. *B. sumichrasti* is found in tropical forests and appears to be more arboreal than *B. astutus*, denning exclusively in trees and almost never coming to the ground (Poglayen-Neuwall and Poglayen-Neuwall 1995). The latter, however, is a good climber and travels quickly and agilely among cliffs and along ledges. In a study of captive *B. astutus*, Trapp (1972) determined that the hind foot can rotate at least 180°, permitting a rapid, headfirst descent and great dexterity. One individual traveled upside down along a cord 5 mm in diameter. Ringtails sometimes climb in a crevice by pressing all four feet on one wall and the back against the other. They also maneuver by ricocheting off of smooth surfaces to gain momentum to continue to an objective. The diet includes insects, rodents, birds, fruit, and other vegetable matter.

According to Poglayen-Neuwall (*in* Grzimek 1990), population density of *B. astutus* reaches 20/sq km in central California. Territories cover up to 136 ha., and those of animals of the same sex do not overlap. *Bassariscus* is generally solitary except during the mating season, and individuals may be aggressive toward one another. Nonetheless, aggregations of 5–9 *B. sumichrasti* have been observed in favored fruit trees (Poglayen-Neuwall and Poglayen-Neuwall 1995). *B. astutus* scent-marks its territory by regularly urinating at certain sites. *B. sumichrasti*, however, eliminates at random (Poglayen-Neuwall 1973). Both species have a variety of vocalizations. Adult *B. astutus* may emit an explosive bark, a piercing scream, and a long, plaintive, high-pitched call.

Female *B. sumichrasti* may enter estrus at any time but usually do so from February to June. Average length of the estrous cycle is 44 days, receptivity lasts 1 day, the gestation period is 63–66 days, and there commonly is a single young (Poglayen-Neuwall 1991, 1993; Poglayen-Neuwall and Poglayen-Neuwall 1995). *B. astutus* is seasonally monestrous, usually mating from February to May and giving birth from April to July. Heat usually does not last more than 24 hours, and the gestation period is about 51–54 days, the shortest among procyonids. The young number one to four, usually two or three, and are born in a nest or den. The female is mainly responsible for care, though the father may be tolerated in the vicinity and may play with the young as they grow older. They weigh about 25 grams at birth, open their eyes fully by 34 days, begin taking some solid food after 6–7 weeks, begin to forage with the adults at 2 months, and usually are completely weaned at 3 months. Both sexes attain sexual maturity and disperse at approximately 10 months (Poglayen-Neuwall and

Central American cacomistle *(Bassariscus sumichrasti)*, immature, photo from Jorge A. Ibarra through Museo Nacional de Historia Natural, Guatemala City.

Poglayen-Neuwall 1980, 1993). A captive *B. astutus* lived for more than 16 years (Poglayen-Neuwall *in* Grzimek 1990). A captive *B. sumichrasti* lived for 23 years and 5 months (Marvin L. Jones, Zoological Society of San Diego, pers. comm., 1995).

Ringtails, especially females obtained when young, make charming pets. They were sometimes kept about the homes of early settlers as companions and to catch mice. The coat is not of particularly high quality. It is known commercially as "California mink" or "civet cat," but there is no scientific basis for either name. In the 1976–77 trapping season 88,329 pelts of *B. astutus* were reported taken in the United States and sold for an average price of $5.50 (Deems and Pursley 1978). The harvest peaked at about 135,000 in 1978–79 but subsequently declined (Kaufmann 1987). In the 1991–92 season only 5,638 skins were taken in the United States and the average price was $3.62 (Linscombe 1994). *B. astutus* apparently extended its range into Kansas, Arkansas, and Louisiana in the twentieth century and has even been reported from Alabama and Ohio (E. R. Hall 1981). These occurrences might result partly from the habit of boarding railroad cars (Sealander 1979). *B. sumichrasti* now is classified as near threatened by the IUCN. It is completely dependent on forests, yet its entire range is in a region subject to intensive deforestation. It also is hunted for its fur and meat. The species has become rare or has disappeared over large areas (Glatston 1994).

CARNIVORA; PROCYONIDAE; **Genus PROCYON**
Storr, 1780

Raccoons

Two subgenera and seven species are currently recognized (Cabrera 1957; Gardner 1976; E. R. Hall 1981; Redford and Eisenberg 1992):

subgenus *Procyon* Storr, 1780

P. lotor, southern Canada to Panama;
P. insularis, Tres Marías Islands off western Mexico;
P. maynardi, New Providence Island (Bahamas);
P. pygmaeus, Cozumel Island off northeastern Yucatan;
P. minor, Guadeloupe Island (Lesser Antilles);
P. gloveralleni, Barbados (Lesser Antilles);

subgenus *Euprocyon* Gray, 1865

P. cancrivorus (crab-eating raccoon), eastern Costa Rica to eastern Peru, northern Argentina, andUruguay.

Lotze and Anderson (1979) suggested that several of the designated species of the subgenus *Procyon* might be conspecific with *P. lotor.* It was Olson and Pregill's (1982) view that *P. maynardi* probably is not a valid species and represents an introduced population of *P. lotor.* The latter species also now is established on Grand Bahama Island (Buden 1986). Based on lack of fossil evidence, Morgan and Woods (1986) suggested that all *Procyon* in the West Indies represent human introductions.

Head and body length is 415–600 mm, tail length is 200–405 mm, shoulder height is 228–304 mm, and weight

Raccoon *(Procyon lotor)*, photo from Zoological Society of Philadelphia.

is usually 2–12 kg. Generally, males are larger than females and northern animals are larger than southern ones. Five adult males in the Florida Keys averaged 2.4 kg. Mean weights in Alabama were 4.31 kg for males and 3.67 kg for females. Means in Missouri were 6.76 kg for males and 5.94 kg for females (Johnson 1970; Lotze and Anderson 1979). In Wisconsin the normal weight range is about 6–11 kg, but there is one record of a male weighing 28.3 kg (Jackson 1961).

The general coloration is gray to almost black, sometimes with a brown or red tinge. There are 5–10 black rings on the rather well-furred tail and a black "bandit" mask across the face. The head is broad posteriorly and has a pointed muzzle. The toes are not webbed, and the claws are not retractile. The front toes are rather long and can be widely spread. The footprints resemble those of people. Females have four pairs of mammae (Banfield 1974).

Raccoons frequent timbered and brushy areas, usually near water. They are more nocturnal than diurnal and are good climbers and swimmers. The den is usually in a hollow tree, with an entrance more than 3 meters above the ground (Banfield 1974). The den may also be in a rock crevice, an overturned stump, a burrow made by another animal, or a human building. Urban (1970) found most raccoons in a marsh to den in muskrat houses. Except when sequestered during severe winter weather or in cases of females with newborn, each den is usually occupied for only one or two days. The average distance between dens has been reported as 436 meters, and the general movements of *Procyon* are not extensive, but one individual was found to have traveled 266 km (Lotze and Anderson 1979).

Raccoons do not hibernate. In the southern parts of their range they are active throughout the year. In northern areas they may remain in a den for much of the winter but will emerge during intervals of relatively warm weather. While they are in winter sleep their heartbeat does not decline, their body temperature stays above 35° C, and their metabolic rate remains high. They do, however, live mostly off fat reserves accumulated the previous summer and autumn and may lose up to 50 percent of their weight (Lotze and Anderson 1979). The remarkable success of *P. lotor*, compared with other procyonids, in occupying a broad range of habitats and climates has been attributed to its well-defined cyclic changes in fat content and thermal conductance, high capacity for evaporative cooling, high level of heat tolerance, high basal metabolic rate, extraordinarily diverse diet, and high reproductive potential (Mugaas, Seidensticker, and Mahlke-Johnson 1993).

Procyon has a well-developed sense of touch, especially in the nose and forepaws. The hands are regularly used almost as skillfully as monkeys use theirs. Food is generally picked up with the hands and then placed in the mouth. Although raccoons have sometimes been observed to dip food in water, especially under captive conditions, the legend that they actually wash their food is without foundation (Lowery 1974). The omnivorous diet consists mainly of crayfish, crabs, other arthropods, frogs, fish, nuts, seeds, acorns, and berries.

As many as 167 raccoons have been found in an area of 41 ha., but more typical population densities are 1/5–43 ha. (Lotze and Anderson 1979). Reported home range size varies from 0.2 to 4,946 ha. but seems to be typified by the situation found by Lotze (1979) on St. Catherine's Island, Georgia: he reported an annual average of 65 ha. for males and 39 ha. for females but indicated that there was much variation and that different study methods might give different results. About the smallest population density (0.5–1.0/100 ha.) and the largest home range (means of 1,139 ha. for males and 806 ha. for females) were reported by Fritzell (1978) for the prairies of North Dakota. His study indicated that the ranges of adult males are largely exclusive of one another but do commonly overlap the ranges of 1–3 adult females and up to 4 yearlings. This and other studies (Lotze and Anderson 1979; Schneider, Mech, and Tester 1971) suggest that female ranges often are not exclusive and that territorial defense is not well developed in *Procyon* but that unrelated animals tend to avoid one another. Nonetheless, as many as 23 individuals have been found in the same winter den, and about the same number have congregated around artificial feeding sites (Lotze and Anderson 1979; Lowery 1974). Raccoons have a variety of vocalizations, most with little carrying power.

In the United States the reproductive season extends from December to August. Mating peaks in February and March, and births from April to June (Johnson 1970; Lotze and Anderson 1979). The breeding season of *P. cancrivorus* is July–September (Grzimek 1975). If a female *P. lotor* loses a newborn litter, she may ovulate a second time during the season (Sanderson and Nalbandov 1973). The estrous cycle has been reported to last 80–140 days (Hayssen, Van Tienhoven, and Van Tienhoven 1993). The gestation period averages 63 days and ranges from 60 to 73. The number of young per litter is one to seven, usually three or four. Captives in New York weighed 71 grams at birth. The eyes open after about 3 weeks (Banfield 1974), and weaning takes place at from 7 weeks to 4 months (Lotze and Anderson 1979). In Minnesota, Schneider, Mech, and Tester (1971) found that the young were kept in a den in a hollow tree until they were 7–9 weeks old and then were moved to one or a series of ground beds. At 10–11 weeks they were taken on short trips by the mother, and after another week the family began to move together. In November the members denned either together in one hollow tree or individually in nearby trees. The young usually separate from the mother at the end of winter. They may attain sexual maturity at about 1 year, but most do not mate until the following year. Few wild raccoons live more than 5 years, but some are estimated to have survived for 13–16 years (Lotze and Anderson 1979). One captive was still living after 20 years and 7 months (Jones 1982).

Raccoons sometimes damage corn and other crops but usually not to a serious extent (Jackson 1961). They make good pets and are interesting to observe in the wild; however, they carry pathogens known to cause such human diseases as leptospirosis, tularemia, and rabies. In 1992, 4,311 rabid *P. lotor* were reported in the United States, considerably more than any other animal, and there were expanding epizootics in the northeast and southeast (Krebs, Strine, and Childs 1993. This same species is currently the most valuable wild fur bearer in the United States, though prices have fallen since the peak in the late 1970s. The harvest in 44 states during the 1976–77 season was 3,832,802 skins, which sold at an average price of $26.00 (Deems and Pursley 1978). In the 1983–84 season the take in the United States and Canada was 3,410,548 pelts with an average value of $5.54 (Novak, Obbard, et al. 1987). During the 1991–92 season, 1,417,198 skins were taken in the United States and sold for an average price of $5.82 (Linscombe 1994). Because of its commercial value, *P. lotor* was introduced in France, the Netherlands, Germany, and various parts of the old Soviet Union, but now it is sometimes considered a nuisance in those areas (Corbet 1978; Glatston 1994). *P. lotor* seems to have extended its range and increased in numbers in certain parts of North America since the nineteenth century (Lotze and Anderson 1979).

In contrast, the IUCN now classifies the insular species *P. insularis*, *P. maynardi*, and *P. minor* as endangered. According to Glatston (1994), little is known about those three, but all are apparently rare and subject to hunting and habitat disruption. *P. gloveralleni,* the last living specimen of which was seen in 1964, is classed as extinct by the IUCN. Wilson (1991) indicated that the subspecies *P. insularis vicinus* of María Magdalena Island in the Tres Marías also may be extinct. Navarro L. and Suarez (1989) observed that *P. pygmaeus* is being killed for alleged depredations on fruit and also is jeopardized by construction of tourist facilities; it now is designated as endangered by the IUCN.

CARNIVORA; PROCYONIDAE; **Genus NASUA**
Storr, 1780

Coatis, or Coatimundis

There are two species (Decker 1991):

N. narica, Arizona to Gulf of Uraba in northwestern Colombia;
N. nasua, Colombia to northern Argentina and Uruguay.

E. R. Hall (1981) and many other authorities have recognized another species, *N. nelsoni,* on Cozumel Island off northeastern Yucatan, but Decker (1991) considered it a subspecies of *N. narica.* Glatston (1994) accepted *N. nelsoni* as a species but noted suggestions that it had been introduced to Cozumel by the Mayans.

Head and body length is 410–670 mm, tail length is 320–690 mm, and shoulder height is up to 305 mm. Poglayen-Neuwall (*in* Grzimek 1990) gave the weight as 3.5–6.0 kg. Males are usually larger than females. *N. narica nelsoni* has short, fairly soft, silky hair, but in other populations of both species the fur is longer and somewhat harsh. *Nelsoni* also is relatively small, but its size has been found to overlap that of mainland populations. The two recognized species can be distinguished consistently by a suite of cranial characters, the most easily discernible of which is that the palate is flat in *N. nasua* and depressed along the midline in *N. narica.* In both species the general color is reddish brown to black above and yellowish to dark brown below. Usually the muzzle, chin, and throat are whitish and the feet are blackish. Black and gray markings are present on the face, and the tail is banded. The muzzle is long and pointed, and the tip is very mobile. The forelegs are short, the hind legs are long, and the tapering tail is longer than the head and body.

Coatis are found mainly in wooded areas. They forage in trees as well as on the ground using the tail as a balancing and semiprehensile organ. While moving along the ground the animals usually carry the tail erect, except for the curled tip. The long, highly mobile snout is well adapted for investigating crevices and holes. Adult males are often active at night, but coatis are primarily diurnal. They move about 1,500–2,000 meters a day in their search for food and usually retire to a roost tree at night (Kaufmann 1962). The diet includes both plant and animal matter. When fruit is abundant coatis are almost exclusively frugivorous. At other times females and young forage for invertebrates on the

Coatimundi *(Nasua nasua),* photo from New York Zoological Society.

forest floor and adult males tend to prey on large rodents (Smythe 1970).

Reported population densities are 26–42/100 ha. on Barro Colorado Island in the Panama Canal Zone, 15–20/ha. in Guatemala, and 1.2–2.0/100 ha. in Arizona (Gompper 1995; Lanning 1976). Home ranges of 4 solitary males in Arizona were 70–270 ha. each (Kaufmann, Lanning, and Poole 1976). On Barro Colorado Island groups had overlapping, undefended home ranges of 35–45 ha. Each group, however, spent about 80 percent of its time in an exclusive core area within its home range (Kaufmann 1962).

The social system of *N. narica,* as studied on Barro Colorado Island, is an interesting example of the interrelationship of ecology and behavior (Kaufmann 1962; Russell 1981; Smythe 1970). All females, as well as males up to two years old, are found in loosely organized bands usually comprising 4–20 individuals. Males more than two years old commonly become solitary except during the breeding season. They are usually excluded from membership in the bands by the collective aggression of the adult females, sometimes supported by the juveniles. Several cases of adult males associating with female bands outside of the breeding season and also of two males in nonagonistic association were reported by Gompper and Krinsley (1992). In the breeding season an adult male is accepted into each of the female groups, but he is completely subordinate to the females. The breeding season corresponds with the period of maximum abundance of fruit. At this time there is thus a minimum of competition for food between the large males and the other animals. Moreover, at other times of the year, when the males become carnivorous, they may attempt to prey on young coatis and thus threaten the survival of the group. According to Gompper (1995), *N. narica* is highly vocal, with a repertoire of sounds associated with aggression, appeasement, alarm, and contact maintenance.

In the populations studied mating took place synchronously once a year during a period of 2–4 weeks. Births occurred in Arizona in June and in Panama in April or early May, at the beginning of the rains. The gestation period lasts 10–11 weeks. Pregnant females separate from the group and construct a tree nest, where they give birth to a litter of two to seven young. When the young are 5 weeks old, they leave the nest, and they and the mother join the group. The young weigh 100–180 grams at birth, open their eyes after 11 days, are weaned at 4 months, reach adult size at 15 months, and attain sexual maturity at 2 years (Gompper 1995; Kaufmann 1962; Poglayen-Neuwall *in* Grzimek 1990). A captive coati was still living after 17 years and 8 months (Jones 1982).

Coatis seem to have extended their range northward in the twentieth century. Numbers reached a peak in Arizona in the late 1950s, crashed in the early 1960s, and then began a slow recovery (Kaufmann, Lanning, and Poole 1976). According to Glatston (1994), however, *N. narica* now is uncommon in the United States and is subject to year-round hunting in Arizona, where the only substantial populations remain. Moreover, suitable habitat is being disrupted throughout its range, especially in northern Mexico, and there is a possibility that the U.S. populations will become isolated from those farther south. *N. narica nelsoni,* which is restricted to rapidly developing Cozumel Island, is classified as endangered by the IUCN. Coatis rarely damage crops and only infrequently take chickens. They are hunted for their meat by natives, who sometimes have dogs trained for this purpose. If cornered by a dog on the ground, a coati can inflict serious wounds with its large, sharp canine teeth. Coatis can be tamed and make interesting and inquisitive pets, but collection for such purposes can be damaging to wild populations.

CARNIVORA; PROCYONIDAE; **Genus NASUELLA**
Hollister, 1915

Mountain Coati

The single species, *N. olivacea,* is known from the Andes of western Venezuela, Colombia, and Ecuador (Cabrera 1957). Glatston (1994) indicated that the species also may occur in extreme northern Peru and noted that some authorities include it in the genus *Nasua,* to which it originally was assigned.

Nasuella resembles *Nasua* but is usually smaller and has a shorter tail. A specimen of a male from Colombia in the U.S. National Museum of Natural History has a head and body length of 383 mm and a tail length of 242 mm. Respective measurements for a female are 394 mm and 201 mm (John Miles, U.S. National Museum of Natural History, pers. comm., 1980). The general color is grayish sooty brown. The tail is ringed with alternating yellowish gray and dark brown bands. The skull of *Nasuella* is smaller and more slender than that of *Nasua,* the middle part of the facial portion is greatly constricted laterally, and the palate extends farther posteriorly.

Handley (1976) collected seven specimens in Venezuela at elevations of 2,000–3,020 meters. All were taken on the ground, four at dry and three at moist sites, and four were taken in cloud forest and three in paramo. Like *Nasua, Nasuella* probably feeds on insects, small vertebrates, and fruit. Glatston (1994) observed that the genus appears to be uncommon and that its upland forest habitat is being converted to agriculture and pine plantations.

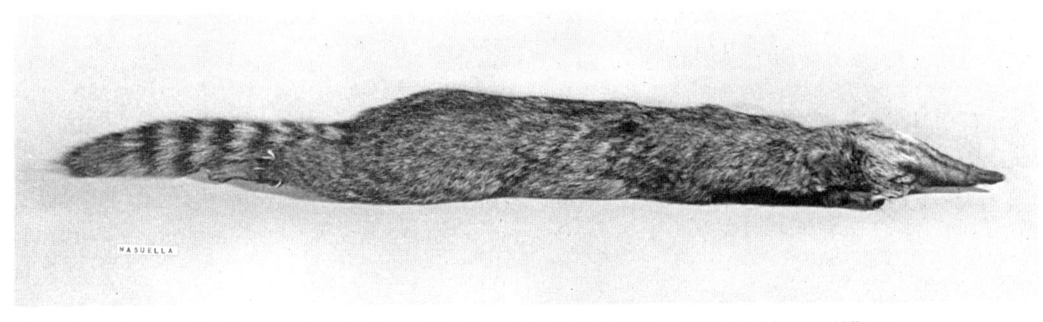

Mountain coati *(Nasuella olivacea),* photo by Howard E. Uible of skin in U.S. National Museum of Natural History.

Kinkajou *(Potos flavus)*, photo from New York Zoological Society.

CARNIVORA; PROCYONIDAE; **Genus POTOS**
E. Geoffroy St.-Hilaire and G. Cuvier, 1795

Kinkajou

The single species, *P. flavus,* is found from southern Tamaulipas in eastern Mexico to the Mato Grosso of central Brazil (Cabrera 1957; E. R. Hall 1981). Confusion sometimes results from the application of the vernacular name "potto" to this species because the same term is used for the African primate *Perodicticus.*

Head and body length is 405 to about 760 mm, tail length is 392–570 mm, shoulder height is as much as 254 mm, and weight is 1.4–4.6 kg. Males are generally larger than females (Kortlucke 1973). The upper parts and upper surface of the tail are tawny olive, yellow tawny, or brownish; some individuals have a black middorsal line. The underparts and undersurface of the tail are tawny yellow, buff, or brownish yellow, and the muzzle is dark brown to blackish. The hair is soft and woolly.

The kinkajou has a rounded head, a short face, a long, prehensile tail, and short, sharp claws. The hind feet are longer than the forefeet. The tongue is narrow and greatly extensible. *Potos* is similar to *Bassaricyon* but differs in its round, tapering, short-haired, prehensile tail; its stockier body form; and its face, which is not grayish. Females have two mammae (Grzimek 1975).

The kinkajou inhabits forests and is almost entirely arboreal. It spends the day in a hollow tree, sometimes emerging on hot, humid days to lie out on a limb or in a tangle of vines. By night it forages among the branches. Although it moves rapidly through a single tree, progress from one tree to another is made cautiously and relatively slowly. An individual probably returns to the same trees night after night. The long tongue of *Potos* is an adaptation for a frugivorous diet. It eats mainly fruit but also takes honey, flowers, insects, and small vertebrates (Husson 1978).

Population densities of 12.5–74.0/sq km have been reported in various parts of the range of *Potos,* and an individual home range of 8 ha. was estimated in Veracruz (Ford and Hoffmann 1988). Studies in French Guiana (Julien-Laferriere 1993) indicated that a male moved an average of 2,540 meters per night and a female averaged 1,495 meters; respective estimated home range sizes were 26.6–39.5 ha. and 15.7–17.6 ha. Individuals were considered to be territorial and solitary except during the mating season. Male territories evidently overlapped those of several females, and both sexes defended their areas against other animals of the same sex. Scent marking probably is used in territorial behavior or for intersexual communication. Small groups may form, but usually only temporarily in a fruit-bearing tree. *Potos* barks when disturbed and emits a variety of other vocalizations, but the usual call, given while feeding during the night, seems to be "a rather shrill, quavering scream that may be heard for nearly a mile" (Dalquest 1953:182).

According to Poglayen-Neuwall (*in* Grzimek 1990), females enter estrus every three months, but Husson (1978) wrote that in Surinam births reportedly take place in April and May. Hayssen, Van Tienhoven, and Van Tienhoven (1993) listed an estrous cycle range of 46–92 days and an estrus range of 12–24 days. The gestation period is 112–20 days. The number of young is usually one, rarely two. Born in a hollow tree, they weigh 150–200 grams at birth, open their eyes at 7–19 days, and begin to take solid food, and can hang by the tail, at 7 weeks. Sexual maturity comes at 1.5 years for males and 2.25 years for females. One pair lived together for 9 years and had their first litter at 12.5 years of age. One individual lived to an age of 32 years at the Bronx Zoo (Marvin L. Jones, Zoological Society of San Diego, pers. comm., 1995).

If captured when young and treated kindly, the kinkajou becomes a good pet. It is often sold under the name "honey bear." Its meat is said to be excellent (Husson 1978), and its pelt is used in making wallets and belts. It does little damage to cultivated fruit.

CARNIVORA; PROCYONIDAE; **Genus BASSARICYON**
J. A. Allen, 1876

Olingos

Five species are currently recognized (Cabrera 1957; Glatston 1994; E. R. Hall 1981; Handley 1976):

B. gabbii, Nicaragua to Ecuador and Venezuela;
B. pauli, known only from the type locality in western Panama;
B. lasius, known only from the type locality in central Costa Rica;
B. beddardi, Guyana and possibly adjacent parts of Venezuela and Brazil;
B. alleni, Ecuador, Peru, western Bolivia, possibly Venezuela.

All of the above species may be no more than subspecies of *B. gabbii* (Cabrera 1957; Grimwood 1969; E. R. Hall 1981).

Head and body length is 350–475 mm and tail length is 400–480 mm. Grzimek (1975) gave the weight as 970–1,500 grams. The fur is thick and soft. The upper parts are pinkish buff to golden mixed with black or grayish above, and the underparts are pale yellowish. The tail is somewhat flattened and more or less distinctly annulated along the median portion. The general body form is elongate, the head is flattened, the snout is pointed, and the ears are small and rounded. The limbs are short, the soles are partly furred, and the claws are sharply curved. The tail of *Bassaricyon,* unlike that of *Potos,* is long-haired and not prehensile. Females have a single pair of inguinal mammae.

Bassaricyon is found in tropical forests from sea level to 2,000 meters. It is primarily arboreal and nocturnal. It is thought to spend the day in a nest of dry leaves in a hollow tree. According to Grzimek (1975), it is an excellent climber and jumper and can leap 3 meters, from limb to limb, with-

Olingo *(Bassaricyon gabbii),* photo from New York Zoological Society. Inset: olingo *(B. pauli),* photo by Hans-Jürg Kuhn.

out difficulty. It feeds mainly on fruit but also hunts for insects and warm-blooded animals. Poglayen-Neuwall (1966) found that in captivity *Bassaricyon* required considerably more meat than *Potos*.

In the wild, *Bassaricyon* lives alone or in pairs. It is sometimes found in association with *Potos* as well as with opossums and douroucoulis *(Aotus)*. It scent-marks with urine, perhaps for its own orientation or to attract the opposite sex (Poglayen-Neuwall *in* Grzimek 1990). Studies of captives by Poglayen-Neuwall (1966, 1989) found that *Bassaricyon* is less social than *Potos* and that two males could not be kept together. There was no definite breeding season, gestation lasted 73–74 days, and each of 16 births yielded a single offspring. The young weighed about 55 grams at birth, opened their eyes after 27 days, began taking solid food after 2 months, and attained sexual maturity at 21–24 months. One female remained reproductively active until 12 years and ultimately lived to be more than 25 years old.

The IUCN now classifies *B. pauli* and *B. lasius* as endangered, estimating that each numbers fewer than 250 mature individuals, and *B. gabbii* and *B. beddardi* as near threatened. Glatston (1994) stated that those species, as well as *B. alleni*, are highly dependent on intact tropical forest and do not adapt readily to disturbed areas or secondary forest. They generally are found in a region where deforestation is rampant and seem to have become rare over much of their range.

CARNIVORA; **Family MUSTELIDAE**

Weasels, Badgers, Skunks, and Otters

This family of 25 Recent genera and 67 species occurs in all land areas of the world except the West Indies, Madagascar, Sulawesi and islands to the east, most of the Philippines, New Guinea, Australia, New Zealand, Antarctica, and most oceanic islands. One genus, *Enhydra*, inhabits coastal waters of the North Pacific. The sequence of genera presented here follows basically that of Simpson (1945), who recognized five subfamilies: the Mustelinae (weasels), with *Mustela, Vormela, Martes, Eira, Galictis, Lyncodon, Ictonyx, Poecilictis, Poecilogale,* and *Gulo;* the Mellivorinae (honey badger), with *Mellivora;* the Melinae (badgers), with *Meles, Arctonyx, Mydaus, Taxidea,* and *Melogale;* the Mephitinae (skunks), with *Mephitis, Spilogale,* and *Conepatus;* and the Lutrinae (otters), with *Lutra, Lutrogale, Lontra, Pteronura, Aonyx,* and *Enhydra.* A cladistic analysis of 46 morphological characters (Bryant, Russell, and Fitch 1993) generally supported these basic groupings but found the Melinae to be polyphyletic, with *Melogale* and perhaps *Taxidea* showing affinity with the Mustelinae and *Mydaus* showing some approach to the Mephitinae. Wozencraft (1989*a*) pointed out that *Taxidea* differs from

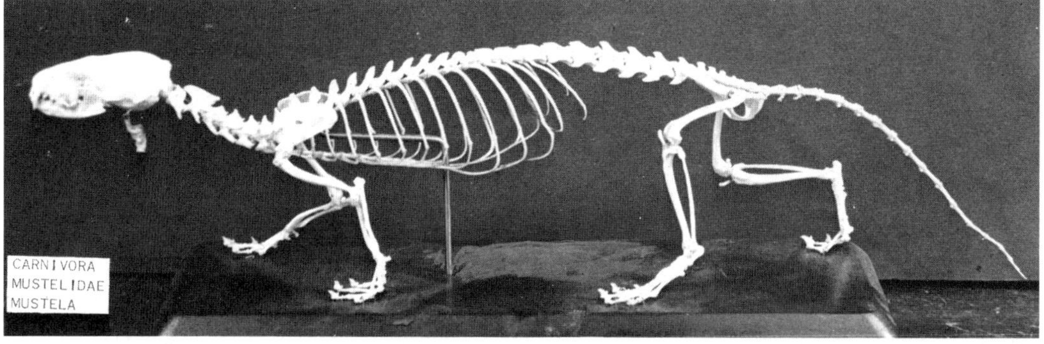

Giant otter *(Pteronura brasiliensis)*, photo by Bernhard Grzimek. Skeleton of mink *(Mustela vison)*, photo by P. F. Wright of specimen in U.S. National Museum of Natural History.

the other badgers in critical basicranial characters, and subsequently (*in* Wilson and Reeder 1993) he placed that genus in a separate subfamily, the Taxidiinae.

The smallest member of this family is the least weasel (*Mustela nivalis*), which has a head and body length of 114–260 mm and a weight as low as 25 grams. The largest members are the otters *Pteronura* and *Enhydra*, which have a head and body length of around 1 meter or more and a weight of 22–45 kg. Male mustelids are about one-fourth larger than females. The pelage is uniformly colored, spotted, or striped. Some species of *Mustela* turn white in winter in the northern parts of their range. The body is long and slender in most genera but stocky in the wolverine (*Gulo*) and in badgers. The short ears are either rounded or pointed. The limbs are short and bear five digits each. The claws are compressed, curved, and nonretractile. The claws of badgers are large and heavy for burrowing. The digits of otters are usually webbed for swimming. Well-developed anal scent glands are present in most genera. Males have a baculum.

The skull is usually sturdy and has a short facial region. The dental formula is: (i 3/2–3, c 1/1, pm 2–4/2–4, m 1/1–2) × 2 = 28–38. *Enhydra* is the only genus with 2 lower incisors. The incisors of mustelids are not specialized, the canines are elongate, the premolars are small and sometimes reduced in number, and a constriction is usually present between the lateral and medial halves of the upper molar. The carnassials are developed. The second lower molar, if present, is reduced to a simple peg.

Mustelids are either nocturnal or diurnal and often shelter in crevices, burrows, and trees. Badgers usually dig elaborate burrows. Mustelids move about on their digits or partly on their digits and partly on their soles. The smaller, slender forms usually travel by means of a scampering gait interspersed with a series of bounds. The larger, stocky forms proceed in a slow, rolling, bearlike shuffle. Mustelids often sit on their haunches to look around. Many genera are agile climbers, and otters and minks are skillful swimmers. Mustelids are mainly flesh eaters. They hunt by scent, though the senses of hearing and sight are also well developed. Some species occasionally feed on plant material, a few are omnivorous, and otters subsist mainly on aquatic animals. *Mustela, Poecilogale, Gulo,* and certain badgers reportedly store food.

Many mustelids use the secretions of their anal glands as a defensive measure. Some genera have a contrasting pattern of body colors; for example, skunks have black and white stripes. This pattern is thought to be a form of warning coloration associated with the fetid anal gland secretion and a reminder that the animal is better left alone. Some of the forms with contrasting body colors, such as the marbled polecat (*Vormela*) and certain skunks, expose and emphasize this contrasting pattern by means of bodily movements when they are alarmed.

Delayed implantation of the fertilized eggs in the uterus occurs in many genera. The actual period of developmental gestation is about 30–65 days, but with delayed implantation the total period of pregnancy is as long as 12.5 months in *Lutra canadensis*. There is usually a single litter per year. The offspring are usually tiny and blind at birth. The young of *Enhydra*, however, are born with their eyes open and in a more advanced stage than the young of other mustelids. The young of most genera can care for themselves at about 2 months and are sexually mature after a year or 2. The potential longevity in the wild is generally 5–20 years.

Mustelids sometimes kill poultry, but they also help to keep rodents in check. Many species are widely hunted for their fur.

The geological range of this family is early Oligocene to Recent in North America, Eocene to Recent in Europe, middle Oligocene to Recent in Asia; early Pliocene to Recent in Africa; and late Pliocene to Recent in South America (Stains 1984).

CARNIVORA; MUSTELIDAE; **Genus MUSTELA**
Linnaeus, 1758

Weasels, Ermines, Stoats, Minks, Ferrets, and Polecats

There are 5 subgenera and 17 species (Alberico 1994; Coetzee *in* Meester and Setzer 1977; Corbet 1978; Corbet and Hill 1992; Ellerman and Morrison-Scott 1966; E. R. Hall 1951, 1981; Izor and de la Torre 1978; Izor and Peterson 1985; Lekagul and McNeely 1977; Schreiber et al. 1989; Van Bree and Boeadi 1978; Youngman 1982):

subgenus *Grammogale* Cabrera, 1940

M. felipei (Colombian weasel), known by five specimens from western Colombia and northern Ecudaor;
M. africana (tropical weasel), Amazon Basin of Brazil, eastern Ecuador, and northeastern Peru;

subgenus *Mustela* Linnaeus, 1758

M. erminea (ermine, or stoat), Scandinavia and Ireland to northeastern Siberia and the western Himalayan region, Japan, Alaska and northern Greenland to northern New Mexico and Maryland;
M. nivalis (least weasel), most of the Palaearctic region from western Europe and Asia Minor to northeastern Siberia and Korea, parts of China and northern Viet Nam, Great Britain, several Mediterranean islands, Japan, northwestern Africa, Egypt, Alaska, Canada, north-central conterminous United States, Appalachian region;
M. frenata (long-tailed weasel), southern Canada to Guyana and Bolivia;
M. altaica (mountain weasel), southern Siberia to the Himalayan region and Korea;
M. kathiah (yellow-bellied weasel), Himalayan region to southern China and northern Viet Nam, Hainan;

subgenus *Lutreola* Wagner, 1841

M. lutreola (European mink), France to western Siberia and the Caucasus;
M. sibirica (Siberian weasel), eastern European Russia to eastern Siberia and Thailand, Japan, Taiwan;
M. lutreolina (Indonesian mountain weasel), southern Sumatra, Java;
M. nudipes (Malaysian weasel), Malay Peninsula, Sumatra, Borneo;
M. strigidorsa (back-striped weasel), Nepal to northern Viet Nam;

subgenus *Vison* Gray, 1843

M. vison (American mink), Alaska, Canada, conterminous United States except parts of Southwest;
M. macrodon (sea mink), formerly on Atlantic coast from New Brunswick to Massachusetts;

Weasel *(Mustela* sp.), photo by Ernest P. Walker.

subgenus *Putorius* Cuvier, 1817

M. putorius (European polecat), western Europe to Ural Mountains;
M. eversmanni (steppe polecat), steppe zone from Austria to Manchuria andTibet;
M. nigripes (black-footed ferret), plains region from Alberta and Saskatchewan to northeastern Arizona and Texas.

Grammogale was considered to be a full genus by Cabrera (1957) and Stains (1967) but was given only subgeneric rank by Hall (1951) and Izor and de la Torre (1978). Although E. R. Hall (1981) also recognized *Lutreola* and *Putorius* as subgenera, he noted that some members of these two taxa closely resemble some members of the subgenus *Mustela* and that distinction from the latter may not be warranted. Corbet (1978) did not use any subgeneric designations. Anderson (1989) and Youngman (1982) recognized the five subgenera with content as given above. A morphometric analysis by Van Zyll de Jong (1992) suggested that *M. subpalmata* of Egypt is a species distinct from *M. nivalis. M. macrodon* was considered a subspecies of *M. vison* by Manville (1966) but was given specific rank by E. R. Hall (1981). Several authorities have suggested that *M. nigripes* of North America is conspecific with *M. eversmanni* of the Old World (Anderson 1973; Corbet 1978), but recent electrophoretic evidence indicates that the two are distinct species (Miller et al. 1988); late Pleistocene remains indicate that the range of each extended into Alaska and Yukon about 35,000 years ago (Youngman 1994). Both *M. erminea* and *M. putorius* have been reported, with much doubt, from northwestern Africa (Coetzee *in* Meester and Setzer 1977).

There is considerable variation in size, but *Mustela* includes the smallest species in the Mustelidae. The body is usually long, lithe, and slender. The tail is shorter than the head and body, often less than half as long. The legs are short, and the ears are small and rounded. The skull has a relatively shorter facial region than that of *Martes*. The bullae are long and inflated. Additional information is given separately for each species.

Mustela felipei (Colombian weasel)

In the type specimen, an adult male, head and body length is 217 mm and tail length is 111 mm. In a female head and body length is 225 mm, tail length is 122 mm, and weight is 138 grams (Alberico 1994). The fur is relatively long, soft, and dense. The upper parts and the entire tail are uniformly blackish brown, and the lower parts are light orange buff. All four plantar surfaces are naked, and there is extensive interdigital webbing. These features are thought to be adaptations for semiaquatic life. Most of the known specimens were collected in riparian areas at elevations of 1,750 and 2,700 meters (Izor and de la Torre 1978). However, the most recent collection was at a rugged site not near riparian habitat (Alberico 1994). The species occurs in a limited area where deforestation is rampant. Schreiber et al. (1989) noted that it probably is the rarest carnivore in South America. It now is classified as endangered by the IUCN.

Mustela africana (tropical weasel)

Head and body length is 240–380 mm and tail length is about 160–210 mm. The upper parts are reddish to chocolate; the underparts are pale and have a longitudinal median stripe of the same color as the upper parts. All four plantar surfaces are nearly naked, and there is extensive interdigital webbing. The tropical weasel has been reported mainly from humid riparian forests (Izor and de la Torre 1978) and is reported to be a good swimmer and climber. However, an observation of a group of four by Ferrari and Lopes (1992) indicates that the species may also occur on high ground and be primarily terrestrial. Schreiber et al. (1989) were aware of only 30 specimens collected in the 170 years since the discovery of the species and noted that it may be of the highest priority for conservation-related research on carnivores in South America.

Mustela erminea (ermine, or stoat)

Except as noted, the information for the account of this species was taken from Banfield (1974), Jackson (1961), C. M. King (1983), Novikov (1962), and Stroganov (1969). Head and body length is 170–325 mm, tail length is 42–120 mm, and weight is 42–365 grams. Old World animals average larger than those of North America, and males average larger than females. Except in certain southern parts of its range the ermine changes color during three- to five-week molts in April–May and October–November. In summer the back, flanks, and outer sides of the limbs are rich chocolate brown, the tip of the tail is black, and the underparts are white. In winter the coat is white except for the tip of

Ermine, or stoat *(Mustela erminea)* in winter coat, photo by Bernhard Grzimek.

the tail, which remains black. The winter pelage is longer and denser than the summer pelage. Females have eight mammae.

The ermine is found in many habitats, from open tundra to deep forest, but seems to prefer areas with vegetative or rocky cover. It makes its den in a crevice, among tree roots, in a hollow log, or in a burrow taken over from a rodent. It maintains several nests within its range, which are lined with dry vegetation or the fur and feathers of its prey. It is primarily terrestrial but climbs and swims well. It generally hunts in a zigzag pattern, progressing by a series of leaps of up to 50 cm each. It can easily run over the snow, and if pursued, it may move under the snow. It may travel 10–15 km in a night, though the average hunt covers 1.3 km. Activity takes place at any hour but is primarily nocturnal. The ermine is swift, agile, and strong and has keen senses of smell and hearing. Its slender body allows it to enter and move quickly through the burrows of its prey. It generally kills by biting at the base of the skull. It sometimes attacks animals considerably larger than itself, such as adult hares. The diet consists mainly of small rodents and also includes birds, eggs, frogs, and insects. Food may be stored underground for the winter.

Population density fluctuates with prey abundance. Under good conditions there may be an ermine for every 10 ha. Individual home range is up to about 200 ha., usually 10–40 ha., and is generally larger for males than for females. There are several vocalizations, including a loud and shrill squeaking. In a radio-tracking study in southern Sweden, Erlinge (1977) found home range sizes of 2–3 ha. for females and 8–13 ha. for males. The ranges of the males included portions of those of the females. Resident animals of both sexes maintained exclusive territories. Boundaries were regularly patrolled and scent-marked, and neighbors usually avoided one another. Adult males were dominant over females and young. Females usually spent their lives in the vicinity of their birthplace, but juvenile males wandered extensively in the spring to find a territory.

Females are polyestrous but produce only one litter per year. The estrous cycle is 4 weeks (Hayssen, Van Tienhoven, and Van Tienhoven 1993). Mating occurs in late spring or early summer, but implantation of the fertilized eggs in the uterus is delayed until around the following March and birth takes place in April or May. Pregnancy thus lasts about 10 months, but embryonic development only a little more than 1 month. Litter size is 3–18 young, averaging about 6 in the New World and 8–9 in the Old. The young weigh about 1.5–3.0 grams at birth, are blind and helpless, and are covered with fine white hair. They grow rapidly and by 8 weeks are able to hunt with the mother. Females attain sexual maturity at 2–3 months and sometimes can mate in their first summer of life. Males do not reach full size and sexual maturity until 1 year.

The ermine rarely molests poultry and is valuable to human interests because it destroys mice and rats. Its white winter fur has long been used in trimming coats and making stoles. The reported number of ermine pelts taken in eight Canadian provinces during the 1976–77 trapping season was 55,216, with an average price of $1.03 (Deems and Pursley 1978).

Mustela nivalis (least weasel)

Except as noted, the information for the account of this species was taken from Banfield (1974), Jackson (1961), and Stroganov (1969). The least weasel is the smallest carnivore, though there is considerable variation in size across its vast range. Head and body length is 114–260 mm, tail length is 17–78 mm, and weight is 25–250 grams. Old World animals average larger than those of North America, and males average larger than females. Except in certain southern parts of its range the least weasel changes color during the spring and autumn. In summer the upper parts are brown and the underparts are white. In winter the entire coat is white, though there may be a few black hairs at the tip of the tail.

The least weasel is much like *M. erminea* in details of habitat, nest construction, and movements. It is, however, even more agile and does not travel over such large areas.

A. Least weasel *(Mustela nivalis).* B. Least weasel changing into winter coat. Photos by Ernest P. Walker.

It feeds almost entirely on small rodents and may store food for the winter. In an area of 27 ha. in England, King (1975) determined that not more than 4 adult males were resident at one time. Home range size was 7–15 ha. for males and 1–4 ha. for females. For the species as a whole Sheffield and King (1994) reported population densities of 1/1–100ha., male home ranges of 0.6–26.2 ha., and female ranges of 0.2–7.0 ha. Male ranges commonly include those of 1 or more females but do not overlap with those of other males. Both sexes scent-mark their ranges and defend them vigorously against individuals of the same sex.

According to Sheffield and King (1994), breeding may continue throughout the year but is concentrated in spring and late summer. Even in the Arctic, when small rodents are abundant *M. nivalis* breeds during winter under the snow. Females experience a postpartum estrus and can bear more than one litter annually. Delayed implantation does not occur, and the gestation period is 34–37 days. The number of young per litter averages about 5 and ranges from 3 to 10. The offspring weigh about 1.5 grams at birth, open their eyes at 26–30 days, are weaned at 42–56 days, leave their mother at 9–12 weeks, and attain adult size at 12–15 weeks. Females born in the spring are sexually mature at 3 months and may produce a litter in their first summer. Most wild individuals do not survive even 1 year, but captives have lived up to 10 years.

The least weasel is rare and of little or no commercial value. It is not known to prey on domestic animals and is beneficial to people through its destruction of mice and rats. It evidently has been introduced by human agency on Malta and Crete in the Mediterranean, as well as in the Azores, New Zealand, and Sao Tomé off west-central Africa (Corbet 1978; Sheffield and King 1994).

Mustela frenata (long-tailed weasel)

In Canada and the United States females have a head and body length of 203–28 mm, a tail length of 76–127 mm, and a weight of 85–198 grams; males have a head and body length of 228–60 mm, a tail length of 102–52 mm, and a weight of 198–340 grams (Burt and Grossenheider 1976). In Mexico the head and body length is 250–300 mm, tail length is 140–205 mm, and one male weighed 365 grams (Leopold 1959). Except as noted, the information for the remainder of this account was taken from Banfield (1974), Jackson (1961), and Lowery (1974). In Canada and the northern United States a color change occurs in the course of 25- to 30-day molts from early October to early December and from late February to late April. During the summer in these regions, and throughout the year farther south, the upper parts are brown, the underparts are ochraceous or buff, and the tip of the tail is black. During the winter in these regions the entire coat is pure white, except for

Immature long-tailed weasel *(Mustela frenata)*, photo by Ernest P. Walker.

the terminal quarter of the tail, which is black. Females have eight mammae.

The long-tailed weasel occurs in a variety of habitats but shows preference for open, brushy or grassy areas near water. It dens in a hollow log or stump, among rocks, or in a burrow taken over from a rodent. Its den is lined with the fur of its victims. It is primarily nocturnal but is frequently active by day. It can climb and swim, but apparently not as well as *M. erminea.* Its long and slender shape allows it to follow a mouse to the end of a burrow or to enter a chicken coop through a knothole. This shape also prevents curling into a spherical resting posture to conserve heat, and thus the metabolic rate of *M. frenata* (and presumably of other weasels) is 50–100 percent higher than that of "normally" shaped mammals of the same weight (Brown and Lasiewski 1972). It thus has a voracious appetite but also speed, agility, and determination. *M. frenata* seizes its prey with its claws and teeth and usually kills by a bite to the back of the neck. It may kill animals larger than itself and has even been known to attack humans who get between it and its prey. The diet, however, consists mainly of rodents and other small mammals. Although weasels are sometimes said to suck blood, this behavior has not been scientifically documented.

Population density varies from about 1/2.6 ha. to 1/260 ha., and home range from 4 to 120 ha. Home ranges may overlap, but individuals seldom meet except during the reproductive season. The voice of *M. frenata* has been reported to consist of a "trill, screech, and squeal" (Svendsen 1976).

Females are monestrous. Mating occurs in July and August, but implantation of the fertilized eggs in the uterus is delayed until around the following March. Embryonic development then proceeds for approximately 27 days until birth in April or May. The total period of pregnancy is 205–337 days. The mean number of young per litter is six, and the range is three to nine. The young weigh about 3.1 grams at birth, open their eyes after 35–37 days, and are weaned at 3.5 weeks. The father has been reported to assist in the care of the offspring. Females attain sexual maturity at 3–4 months, but males do not mate until the year following their birth.

The long-tailed weasel is more prone to raid henhouses than are other species of *Mustela* but is generally beneficial in the vicinity of poultry farms because it destroys the rats that prey on young chickens. The white winter pelage, known collectively with that of *M. erminea* as "ermine," has varied in value, with prime pelts sometimes bringing as much as $3.50 each. The reported number of pelts taken in the United States and Canada during the 1976–77 trapping season was 61,175, with an average price of about $1.00 (Deems and Pursley 1978). The reported take in the United States declined to 18,218 pelts in 1987–88 and to only 3,957 in 1991–92, but prices have held steady at around $1.00 (Linscombe 1994).

Mustela altaica (mountain weasel)

The information for the account of this species was taken from Stroganov (1969). In males head and body length is 224–87 mm, tail length is 108–45 mm, and weight is 217–350 grams. In females head and body length is 217–49 mm, tail length is 90–117 mm, and weight is 122–220 grams. *M. altaica* resembles *M. sibirica* but is smaller and has shorter fur and a less luxuriant tail. There are spring and autumn molts. The winter pelage is yellowish brown above and pale yellow below. In summer the coat is gray to grayish brown.

This weasel occurs in highland steppes and forests at elevations up to 3,500 meters. It nests in rock crevices, among tree roots, or in expropriated rodent burrows. It is quick, agile, and chiefly nocturnal or crepuscular. It feeds mainly on rodents, pikas, and small birds. In Kazakh mating occurs in February and March. The gestation period is 40 days, litters contain two to eight young, and lactation lasts 2 months. Following independence, litter mates remain together until autumn. Of little importance in the fur trade, *M. altaica* is considered to be beneficial to agricultural interests.

Mustela kathiah (yellow-bellied weasel)

The information for the account of this species was taken from Mitchell (1977). Head and body length is 250–70 mm and tail length is 125–50 mm. The tail is more than half and sometimes nearly two-thirds as long as the head and body. The upper parts are dark brownish and the underparts are deep yellow. This weasel inhabits pine forests and also occurs above the timber line. The elevational range is 1,800–4,000 meters. The diet includes birds, rodents, and other small mammals.

Mustela lutreola (European mink)

Except as noted, the information for the account of this species was taken from Novikov (1962) and Stroganov (1969). In males head and body length is 280–430 mm, tail length is 124–90 mm, and weight is up to 739 grams. In females head and body length is 320–400 mm, tail length is 120–80 mm, and weight is up to 440 grams. The general coloration is reddish brown to dark cinnamon. The underparts are somewhat paler than the back, and there may be some white on the chin, chest, and throat. The dense pelage is short even in winter.

The European mink inhabits the densely vegetated banks of creeks, rivers, and lakes. It is rarely found more than 100 meters from fresh water. It may excavate its own burrow, take one from a water vole *(Arvicola)*, or den in a crevice, among tree roots, or in some other sheltered spot. It swims and dives well. Activity is mainly nocturnal and crepuscular. The summer is generally spent in an area of 15–20 ha., but there may be extensive autumn and winter movements to locate swift, nonfrozen streams. The chief prey is the water vole. The diet also includes other small rodents, amphibians, mollusks, crabs, and insects. Food is often stored.

In Russia reported population densities are 2–12 individuals per 10 km of shoreline and average territory size is 32 ha. for males and 26 ha. for females (Youngman 1990). Mating occurs from February to March, with births in April and May. Pregnancy lasts 35–72 days, the variation probably resulting from delayed implantation in some females. The number of young per litter is two to seven, usually four or five. The young open their eyes after 4 weeks, are weaned at 10 weeks, disperse in the autumn, and attain sexual maturity the following year. Longevity is 7–10 years.

The European mink appears to be in a precipitous decline, and it now is classified as endangered by the IUCN. Although its fur is not as valuable as that of *M. vison*, it has been widely trapped for commercial purposes. An annual average of 49,850 pelts was taken from 1922 to 1924 in the old Soviet Union (Youngman 1990). It also has been killed as a predator, has lost much habitat through hydroelectric developments and water pollution, and has suffered badly from competition with the introduced *M. vison*. Tumanov and Zverev (1986) reported that in the old Soviet Union its numbers had declined to 40,000–45,000 individuals. However, according to Maran (1992, 1994), recent surveys show that there are only about 25,000 in that region, the great majority in Russia, and that the estimated total for the world is less than 30,000. Introductions have been carried out in the Kuril Islands and Tajikistan, far from the natural range of the species. Outside of Russia, there are very small and declining populations in Georgia, Ukraine, Estonia, Belarus, Moldava, and possibly Finland and the Danube Delta of Romania. There also are about 2,000 in western France and 1,000 in northern Spain. Interestingly, the species may have spread through France as recently as the eighteenth and nineteenth centuries during a period of climatic amelioration at the same time that it was vanishing from central Europe, and it appears to have entered Spain only about 1950 (Youngman 1982).

Mustela sibirica (Siberian weasel)

Except as noted, the information for the account of this species was taken from Stroganov (1969). In males head and body length is 280–390 mm, tail length is 155–210 mm, and weight is 650–820 grams. In females head and body length is 250–305 mm, tail length is 133–64 mm, and weight is 360–430 grams. In winter the upper parts are bright ochre to straw yellow and the flanks and underparts are somewhat paler. The summer pelage is darker, shorter, coarser, and sparser. Females have four pairs of mammae.

The Siberian weasel dwells mainly in forests, especially along streams, but sometimes enters towns and cities. It dens in tree hollows, under roots or logs, between stones, in modified rodent burrows, or in buildings. It lines its nest with fur, feathers, and dried vegetation. It is swift and agile, has good senses of smell and hearing, and can climb and swim well. It is mainly nocturnal and crepuscular and has been observed to cover a distance of 8 km in one night. In the autumn it may move from upland areas to valleys. There are reports of mass migrations associated with food shortages. The diet consists mainly of small rodents but also includes pikas, birds, eggs, frogs, and fish. Food may be stored for winter use.

Several males may pursue and fight over a single female. Mating occurs in late winter and early spring, and births take place from April to June. The gestation period is 28–30 days, litter size is 2–12 young, the offspring open their eyes after 1 month, and lactation lasts 2 months. The young leave their mother by the end of August, but litter mates may travel together through the autumn. A captive lived 8 years and 10 months (Jones 1982).

The Siberian weasel is important in the fur trade. It occasionally attacks domestic fowl but is generally considered beneficial because of its destruction of noxious rodents. It has been introduced on Sakhalin and Iriomote islands (Corbet 1978).

Mustela lutreolina (Indonesian mountain weasel)

The information for the account of this species was taken from Van Bree and Boeadi (1978). Based on 6 male specimens, head and body length is 297–321 mm, tail length is 136–70 mm, and weight is 295–340 grams. The overall color is glossy dark russet, and there is no mask or other facial markings. *M. lutreolina* bears a striking resemblance to *M. lutreola* in size and color but resembles *M. sibirica* in characters of the skull. The 11 known specimens have been collected at elevations of 1,000–2,200 meters in the mountains of Java and southern Sumatra. *M. lutreolina* is a close relative of *M. sibirica* and sometimes is considered conspecific with the latter. Its lifestyle probably is much the same. Apparently, during a glacial phase of the Pleistocene a cooler climate and lower sea level allowed the ancestral stock of *M. lutreolina* to spread southward. Changing conditions subsequently isolated it in its present mountain habitat. This area now is again declining because of human disruption, and *M. lutreolina* is classified as endangered by the IUCN.

Mustela nudipes (Malaysian weasel)

The information for the account of this species was taken from Lekagul and McNeely (1977). Head and body length is 300–360 mm and tail length is 240–60 mm. The body coloration varies from pale grayish white to reddish brown, and the head is much paler than the body. The soles of the feet are naked around the pads. Females have two pairs of mammae. This species is not well known, but its habits are thought to be like those of other weasels. A litter of four young has been recorded.

Mustela strigidorsa (back-striped weasel)

The information for the account of this species was taken from Lekagul and McNeely (1977) and Mitchell (1977). Head and body length is 250–325 mm and tail length is 130–205 mm. The general coloration is dark brown, but the upper lips, cheeks, chin, and throat are pale yellow. There is a narrow whitish stripe down the middle of the back and another along the venter. As in *M. nudipes,* the area around the foot pads is entirely naked. Females have two pairs of mammae.

The back-striped weasel inhabits evergreen forest at elevations of 1,200–2,200 meters. One was taken in a tree hole 3–4 meters above the ground. An individual was observed to attack a bandicoot rat three times its own size. According to Schreiber et al. (1989), this species is probably rare, though it now is known from at least 31 specimens. The IUCN classifies it as vulnerable, noting that fewer than 10,000 mature individuals survive and a decline is continuing.

Mustela vison (American mink)

Except as noted, the information for the account of this species was taken from Banfield (1974), Burt and Grossenheider (1976), Jackson (1961), and Lowery (1974). In males head and body length is 330–430 mm, tail length is 158–230 mm, and weight is 681–2,310 grams. In females head and body length is 300–400 mm, tail length is 128–200 mm, and weight is 790–1,089 grams. The pelage is soft and luxurious. Its general color varies from rich brown to almost black, but the ventral surface is paler and may have some white spotting. Captive breeding has produced a number of color variants. Anal scent glands emit a strong, musky odor that some persons consider to be more obnoxious than that of skunks. Females have three pairs of mammae.

The mink is found along streams and lakes as well as in swamps and marshes. It prefers densely vegetated areas. It dens under stones or the roots of trees, in expropriated beaver or muskrat houses, or in self-excavated burrows. Such burrows may be about 3 meters long and 1 meter beneath the surface and have one or more entrances just above water level. The mink is an excellent swimmer, can dive to depths of 5–6 meters, and can swim underwater for about 30 meters. It is primarily nocturnal and crepuscular but is sometimes active by day. It normally is not wide-ranging but may travel up to about 25 km in a night during times of food shortage. The most important dietary components are small mammals, fish, frogs, and crayfish; other foods include insects, worms, and birds.

Population densities of about 1–8/sq km have been recorded. Females have home ranges of about 8–20 ha. The ranges of males are larger, sometimes up to 800 ha. Individuals are generally solitary and hostile to one another except when opposite sexes come together for breeding. Females are polyestrous but have only one litter per year. Mating occurs from February to April, with births in late April and early May. Because of a varying period of delay in implantation of the fertilized eggs, pregnancy may last from 39 to 78 days. Actual embryonic development takes 30–32 days. The number of young per litter averages 5 and ranges from 2 to 10. The young are born in a nest lined with fur, feathers, and dry vegetation. They are blind and naked at birth, open their eyes after 5 weeks, are weaned at 5–6 weeks, leave the nest and begin to hunt at 7–8 weeks, and separate from the mother in the autumn. Females reach adult weight at 4 months and sexual maturity at 12 months; males reach adult weight at 9–11 months and sexual maturity at 18 months. Potential longevity is 10 years.

Most of the mink fur used in commerce is produced on farms. The preferred breeding stock results from crossing the large Alaskan and dark Labrador forms. Selective breeding has led to development of strains that regularly yield such colors as black, white, platinum, and blue (Grzimek 1975). In Canada during the 1971–72 season 72,674 wild and 1,155,020 farm-raised mink pelts were sold (Banfield 1974). During the 1976–77 season the reported number of wild mink taken and the average price per pelt were 116,537 and $19.67 in Canada and 320,823 and $14.00 in the United States (Deems and Pursley 1978). In 1983–84 the total harvest of wild mink was 392,122, with an average value of $9.71 (Novak, Obbard, et al. 1987). In the 1991–92 season 129,106 skins were taken in the United States and sold for an average price of $18.47 (Linscombe 1994). *M. vison* was introduced deliberately in many parts of the old Soviet Union, and escaped animals have established populations in Iceland, Ireland, Great Britain, France, Spain, Portugal, Scandinavia, Germany, and Poland (Braun 1990; Corbet 1978; Romanowski 1990; Ruiz-Olmo and Palazon 1991; Smit and Van Wijngaarden 1981). The subspecies *M. v. evergladensis,* of southern Florida, appears to be rare and may be jeopardized by human water diversion projects (Smith and Cary 1982). Humphrey and Setzer (1989) have suggested that *evergladensis* actually is a disjunct population of the subspecies *M. vison mink,* which otherwise occurs no farther south than central Georgia. Populations of other subspecies are found in coastal areas of northwestern and extreme northeastern Florida.

Mustela macrodon (sea mink)

The information for the account of this extinct species was taken from Allen (1942), Campbell (1988), and Manville (1966). The sea mink was said to resemble *M. vison* but to

American mink *(Mustela vison)*, photo by Ernest P. Walker.

European polecat *(Mustela putorius)*, photo from Zoological Society of London.

be much larger, to have a coarser and more reddish fur, and to have an entirely different odor. Head and body length has been estimated at 660 mm and tail length at 254 mm. No complete specimen is known to exist, and descriptions are based only on recorded observations and numerous bone fragments and teeth found at Indian middens along the New England coast.

The sea mink reportedly made its home among the rocks along the ocean. Its den had two entrances. It is thought to have been nocturnal and solitary. The diet consisted mainly of fish and probably also included mollusks. An adult and four young estimated to be three or four weeks old were seen along a beach in August. Because of the large size of *M. macrodon*, its pelt brought a higher price than that of *M. vison* and was persistently sought. Some persons pursued the species from island to island, using dogs to locate individuals on ledges and in rock crevices and then digging or smoking them out. The sea mink apparently had been exterminated by about 1880, and its range seems subsequently to have been occupied by *M. vison*.

Mustela putorius (European polecat)

Except as noted, the information for the account of this species was taken from Blandford (1987), Grzimek (1975), Novikov (1962), and Stroganov (1969). Males have a head and body length of 295–460 mm, a tail length of 105–90 mm, and a weight of 405–1,710 grams. Females have a head and body length of 205–385 mm, a tail length of 70–140 mm, and a weight of 205–915 grams. The general coloration is dark brown to black; the underfur is pale yellow and is clearly seen through the guard hairs; and the area between the eye and the ear is silvery white.

The European polecat is most common in open forests and meadows. It dens in such places as crevices, hollow logs, and burrows made by other animals. It sometimes enters settled areas and buildings occupied by people. It is nocturnal and terrestrial but is capable of climbing. A radio-tracking study by Brzezinski, Jedrzejewski, and Jedrzejewska (1992) showed individuals to occupy stretches of about 0.65–3.05 km along a stream, to have an average daily movement of 1.1 km, and to use several dens each. The diet consists of small mammals, birds, frogs, fish, and invertebrates.

Population density has been calculated to be 1/1,000 ha., and home range about 100–150 ha. *M. putorius* is usually solitary and silent but has a variety of squeals, screams, and other sounds. Mating occurs from March to June, heat lasts 3–5 days, and the gestation period is about 42 days. The number of young per litter is 2–12, usually 3–7. The young weigh 9–10 grams at birth, open their eyes and are weaned after about 1 month, and become independent at around 3 months. Sexual maturity may come in the first or second year of life. Longevity in the wild is usually as high as 5–6 years, but captives have lived as long as 14 years.

The domestic ferret sometimes is given the subspecific name *M. putorius furo*. It generally is thought to be a descendant of the European polecat, but Lynch (1995) indicated that *M. eversmanni* may also be involved in its ancestry. It was bred in captivity as early as the fourth century B.C. It is usually tame and playful and is used to control rodents and to drive rabbits from their burrows. It is now found in captivity in much of the world. An estimated 1 million are kept as pets in the United States, though there have been a few reported cases of rabid individuals and savage attacks on children (*Washington Post*, 5 April 1988). Unlike the wild polecat, it is generally white or pale yellow in color. According to Asdell (1964), females may have two or three litters annually. Both the domestic ferret and the polecat apparently were introduced in New Zealand, and large feral populations are now established there. These animals are trapped for their fur in New Zealand as well as in their original range. The pelt is sometimes called "fitch." The wild polecat generally has been persecuted as a predator of small domestic mammals and birds. By the early twentieth century it had been eliminated throughout Great Britain, except in Wales, but there has been some recovery in recent years. Recently, Lynch (1995) expressed concern about possible hybridization between wild and domestic polecats in Britain.

Mustela eversmanni (steppe polecat)

The information for the account of this species was taken from Stroganov (1969). Males have a head and body length of 370–562 mm, a tail length of 80–183 mm, and a weight of up to 2,050 grams. Females have a head and body length of 290–520 mm, a tail length of 70–180 mm, and a weight of up to 1,350 grams. There is much variation in color pattern, but generally the body is straw yellow or pale brown, somewhat darker above than below. There is a dark mask across the face. The chest, limbs, groin area, and terminal third of the tail are dark brown to black. Some individuals bear a striking resemblance to the North American black-footed ferret *(M. nigripes)*.

The steppe polecat is found in open grassland and semidesert. It usually expropriates the burrow of a ground squirrel or some other animal and modifies the home for its own use. Some burrow systems, especially those of females, may be occupied for several years and become rather

Steppe polecat *(Mustela eversmanni)*, photo by Ernest P. Walker.

complex. *M. eversmanni* is quick and agile and has keen senses, especially of smell and hearing. It moves by leaps of up to 1 meter and constantly changes direction during hunts. It is nocturnal and has been known to cover up to 18 km in the course of a winter night. Local migrations may occur in response to extreme snow depth or food shortage. The diet consists mainly of pikas, voles, marmots, hamsters, and other rodents. Food is sometimes stored for later use.

Mating usually occurs from February to March, with births from April to May. If a litter is lost, however, the female may produce a second later in the year. The gestation period is 38–41 days. The average number of young per litter is about 8–10 and the range is 4–18. The young weigh 4–6 grams at birth, open their eyes after 1 month, are weaned and start hunting with the mother at 1.5 months, disperse at 3 months, and attain sexual maturity at 9 months.

The steppe polecat is considered to be beneficial to agriculture because of its destruction of rodents. Its fur has commercial importance but is not as valuable as that of *M. putorius.* The subspecies *M. e. amurensis,* of southeastern Siberia and Manchuria, is classified as vulnerable by the IUCN because of excessive human exploitation and habitat disruption.

Mustela nigripes (black-footed ferret)

Head and body length is 380–500 mm and tail length is 114–50 mm. The linear measurements of males are about 10 percent greater than those of females. Two males weighed 964 and 1,078 grams, and two females weighed 764 and 854 grams. The body color is generally yellow buff, being palest on the underparts. The forehead, muzzle, and throat are nearly white. The top of the head and middle of the back are brown. The face mask, feet, and terminal fourth of the tail are black. Females have three pairs of mammae (Burt and Grossenheider 1976; Henderson, Springer, and Adrian 1969; Hillman and Clark 1980).

The black-footed ferret is found mainly on short and midgrass prairies. It is closely associated with prairie dogs *(Cynomys)* and utilizes their burrows for shelter and travel; it may modify such burrows for its own use. It is primarily nocturnal and is thought to have keen senses of hearing, smell, and sight. It depends largely on prairie dogs for food, but captives have readily accepted other small

Black-footed ferret *(Mustela nigripes)*, photo by Luther C. Goldman.

mammals. Studies of a recently discovered population near Meeteetse, Wyoming (Richardson et al. 1987), indicate that the ferret confines its activity primarily to prairie dog colonies and seldom moves from one colony to another. Average nightly movement during winter was 1,406 meters, and areas of activity for periods of 3–8 nights were 0.4–98.1 ha. Some food was stored in many small caches.

Studies in South Dakota (Hillman, Linder, and Dahlgren 1979) indicate that *M. nigripes* kills only enough to eat. A prairie dog town of 14 ha. was occupied by a single ferret for six months, but the prey population was not severely reduced. Females raising litters require relatively large prairie dog towns. Whereas overall average town size was found to be 8 ha., the average town size occupied by females with young was 36 (10–120) ha. The mean distance between a town and the nearest neighboring town was 2.4 km. The mean distance between two towns occupied by ferrets was 5.4 km. Investigation of the wild population near Meeteetse, Wyoming (Clark 1987*b*), found a density of about 1 ferret per 50 ha. of prairie dog colonies. This population consisted of 67 percent juveniles and 33 percent adults each August.

M. nigripes is solitary except during the breeding season, and males apparently do not assist in the rearing of young (Henderson, Springer, and Adrian 1969; Hillman and Clark 1980). Individuals kept in a captive colony in Wyoming from 1986 to 1990 mated mostly in March and April and gave birth in May and June. Estrus lasted 32–42 days, the gestation period was 42–45 days, and litter size averaged 3.0 kits and ranged from 1 to 6 (Williams et al. 1991). In the wild the number of young per litter has been found to average 3.3. The young emerge from the burrow in early July and separate from the mother in September or early October. Young males then disperse for a considerable distance, but young females often remain in the vicinity of their mother's territory. Sexual maturity is attained by 1 year (Forrest et al. 1988; Hillman and Clark 1980; Miller et al. 1988). Captives are estimated to have lived up to about 12 years (Hillman and Carpenter 1980).

The black-footed ferret may once have been common on the Great Plains, but its subterranean and nocturnal habits made it difficult to locate and observe. According to Clark (1976), there had been about 1,000 reports of the species since 1851, and there are about 100 specimens in museums. In 1920 ferret numbers were estimated at more than 500,000 (Clark 1987*a*). During the twentieth century *M. nigripes* apparently declined in association with the extermination of prairie dogs by human agency. In Kansas, for example, the area occupied by prairie dog towns has been reduced by 98.6 percent since 1905. The ferret became so rare that some persons considered it to be extinct. Reports continued in several states, however, and in 1964 a population was discovered in South Dakota. This group was regularly observed and studied until 1974, when confirmed records ceased. Several captives were taken, but eventually all died without leaving surviving offspring. There again was fear that the species was extinct, though Clark (1978) gathered a number of reliable reports in Wyoming, and Boggess, Henderson, and Choate (1980) found the skull of a ferret that may have died as recently as 1977 in Kansas.

Then on 26 September 1981 a ferret was found that had been killed by dogs on a ranch near Meeteetse, Wyoming. Agents of the U.S. Fish and Wildlife Service quickly live-captured, radio-collared, and released another individual in the same area and found evidence of several additional animals. It soon was realized that a substantial ferret population, indeed the largest ever scientifically observed, was present. Intensive field studies were initiated, and by July 1984 the population was estimated to contain 129 individuals. The following year, however, the number of ferrets began to fall, apparently as a result of a decline in their major prey, white-tailed prairie dogs, caused by sylvatic plague. Amidst efforts to control the plague and live-capture a portion of the remaining ferrets, canine distemper somehow was introduced into the wild ferret population, which promptly collapsed (the first 6 ferrets to be caught also died of the deadly disease). From late 1985 to early 1987 all of the ferrets known to survive, a total of 18 individuals, were brought into captivity and used as the basis of a breeding program. By 1989 there were 58 ferrets in a captive facility operated by the Wyoming Game and Fish Department and another 12 in the National Zoological Park at Front Royal, Virginia (Bender 1988; Clark 1987*b;* Collins 1989; Forrest et al. 1988; Miller et al. 1988; Solt 1981). There has been considerable controversy regarding the manner in which the ferret study and conservation program was handled, especially with respect to the length of time taken to establish a proper captive breeding program (Carr 1986; Clark 1987*a*, 1987*b;* May 1986; T. Williams 1986).

By the start of the 1991 breeding season there were 180 individuals in captivity, and during the season about 150 surviving kits were produced. It was then decided that the population was large enough to sustain removal of some animals for reintroduction. In the autumn of 1991, 49 ferrets were released in the Shirley Basin of southeastern Wyoming; 91 more were released there in 1992, and another 48 in 1993. Most of these animals are thought to have died, but some are known to have survived and reproduced; at least six litters were born in the wild in 1993 (Russell et al. 1994; Thorne and Russell *in* Bowdoin et al. 1994). By 1996 there were about 400 individuals in captivity at seven separate facilities, and the U.S. Fish and Wildlife Service had carried out additional reintroductions in north-central Montana, southwestern South Dakota, and northwestern Arizona (Christopherson and Torbit 1993; Reading et al. 1996; Searls and Torbit 1993; Wada 1995). At present there are no known nonintroduced wild populations, though reports continue to come from various areas, and two skulls were found in Montana in 1984 that evidently were from animals that had died not more than 10 years previously (Clark, Anderson, et al. 1987). The species is thought to be present in Canada, beyond the range of prairie dogs but in areas of native grasslands with high densities of ground squirrels (Laing and Holroyd 1989). *M. nigripes* is classified as extinct in the wild by the IUCN (because there has been no confirmation of a surviving nonintroduced population) and as endangered by the USDI and is on appendix 1 of the CITES. Prospects for the long-term survival of the species in the context of human manipulation and restricted size and genetic viability is covered in detail in Seal et al. (1989).

CARNIVORA; MUSTELIDAE; **Genus VORMELA**
Blasius, 1884

Marbled Polecat

The single species, *V. peregusna*, occurs in the steppe and subdesert zones from the Balkans and Palestine to Inner Mongolia and Pakistan (Corbet 1978).

Head and body length is 290–380 mm, tail length is 150–218 mm, and weight is 370–715 grams (Grzimek 1975; Stroganov 1969). *Vormela* resembles *Mustela putorius* but differs in its broken and mottled color pattern on the upper parts and in its long claws. The mottling on the back is red-

Marbled polecat *(Vormela peregusna)*, photo by Bernhard Grzimek.

dish brown and white or yellowish, and the tail is usually whitish with a dark tip. The underparts are dark brown or blackish, and the facial mask is dark brown. Females have five pairs of mammae (Roberts 1977).

Like most mustelids, *Vormela* possesses anal scent glands, from which a noxious-smelling substance is emitted. When this animal is threatened, it throws its head back, bares its teeth, erects its body hairs, and bristles and curls its tail over its back. This behavior results in the fullest display of the contrasting body colors, and the pattern thus exposed is thought to be a warning associated with the fetid anal gland secretion.

The marbled polecat seems to prefer steppes and foothills. With its strong paws and long claws it excavates deep, roomy burrows. It may also shelter in the burrows of other animals. It is chiefly nocturnal and crepuscular but is sometimes active by day. A radio-tracking study (Ben-David, Hellwing, and Mendelssohn 1988*a*) determined that *Vormela* moved up to 1 km per night, seldom used the same path repeatedly, and changed its den and activity area every 2–3 days. It is a good climber but feeds mainly on the ground. It preys on rodents, birds, reptiles, and other animals.

Vormela is solitary except during the breeding season. Investigations in Israel found each animal to occupy a home range of 0.5–0.6 sq km; there was some overlap of ranges and there were some encounters between animals, but each foraged and rested alone (Ben-David, Hellwing, and Mendelssohn 1988*a*). Observations in captivity suggest that males form a dominance hierarchy. In March, just prior to the mating season, all males that were housed alone underwent a conspicuous color change, the yellow spots of the fur changing to bright orange patches. However, of the males that were housed in groups only dominant individuals changed color (Ben-David, Mendelssohn, and Hellwing 1989).

The mating season in Israel extends from mid-April to early June, and some wild males undergo the color change described above at that time. Females kept in captivity gave birth 8–11 months after mating, meaning that delayed implantation is involved. Litter size was 1–8 young. They did not open their eyes until they were 40 days old, but they began eating solid food at 30 days. Females attained adult size and sexual maturity at only 3 months of age, but males continued to grow for another 2 months and reached sexual maturity at 1 year (Ben-David, Hellwing, and Mendelssohn 1988*b*). Only the mother cares for the young, which are reared in a nest of grass and leaves within a burrow. A captive specimen was still living after 8 years and 11 months (Jones 1982).

The fur of the marbled polecat has been sought at certain times but is not of major commercial importance (Grzimek 1975; Stroganov 1969). *Vormela* sometimes preys on poultry and has been eliminated in parts of its range. The subspecies *V. p. peregusna,* of Europe and Asia Minor, is classified as vulnerable by the IUCN and Russia. Schreiber et al. (1989) suggested that the major decline of this subspecies on the steppes of the Balkans and Ukraine, like the disappearance of the North American *Mustela nigripes,* has resulted from usurpation of grassland habitat by agriculture and the elimination of rodent prey.

CARNIVORA; MUSTELIDAE; **Genus MARTES**
Pinel, 1792

Martens, Fisher, and Sable

There are three subgenera and eight living species (Anderson 1970; Chasen 1940; Cholley 1982; Corbet 1978; Ellerman and Morrison-Scott 1966; Graham and Graham 1994; E. R. Hall 1981; Hsu and Wu 1981; Lekagul and McNeely 1977; Pilgrim 1980; Schreiber et al. 1989):

subgenus *Martes* Pinel, 1792

M. foina (beech marten, or stone marten), Denmark and Spain to Mongolia and the Himalayas, Crete, Rhodes, Corfu;
M. martes (European pine marten), western Europe to western Siberia and the Caucasus, Ireland, Great Britain, Balearic Islands, Corsica, Sardinia, Elba, Sicily;
M. zibellina (sable), originally the entire taiga zone from Scandinavia to eastern Siberia and North Korea, Sakhalin, Hokkaido;
M. melampus (Japanese marten), South Korea, Honshu, Kyushu, Shikoku, Tsushima;
M. americana (American pine marten), Alaska to Newfoundland, south in mountainous areas to central California and northern New Mexico, Great Lakes region, New England, historically as far south as Iowa and southern Ohio;

subgenus *Pekania* Gray, 1865

M. pennanti (fisher, or pekan), southern Yukon to Labrador, south in mountainous areas to central California and Utah, Great Lakes region, New England, Appalachian region, historically as far south as northeastern Alabama;

subgenus *Charronia* Gray, 1865

M. flavigula (yellow-throated marten), southeastern Siberia to Malay Peninsula, Himalayan region, Hainan, Taiwan, Sumatra, Bangka Island, Java, Borneo;
M. gwatkinsi (Nilgiri marten), Nilgiri Hills of extreme southern India.

An additional species, *M. nobilis* (noble marten), has been described from subfossil and fossil specimens from the Yukon, Idaho, Wyoming, Colorado, Nevada, and California and may have survived until about 1,200 years ago (Graham and Graham 1994). Youngman and Schueler (1991) concluded that *M. nobilis* is a synonym of *M. americana*, but Anderson (1994) questioned that view and treated *nobilis* as a subspecies of *M. americana*. In any event, *nobilis* was a large form, related to modern *M. americana* and evidently adapted to a wider variety of environments; the reason for its extinction is unknown but could have involved human impacts (Graham and Graham 1994; Grayson 1984). Anderson (1970, 1994) stated that additional study might show that *M. martes, M. zibellina, M. melampus*, and *M. americana* are conspecific. Corbet (1978) suggested that *M. gwatkinsi* is conspecific with *M. flavigula*, and Corbet (1984) cited information suggesting that *M. latinorum* of Sardinia is a species distinct from *M. martes*.

From *Mustela, Martes* is generally distinguished by a larger and heavier body, a longer and more pointed nose, larger ears, longer limbs, and a bushier tail. *Martes* has typical mustelid anal sacs, but they are poorly developed compared with those of other genera (Buskirk 1994). Additional information is provided separately for each species.

Martes foina (beech marten, or stone marten)

Except as noted, the information for the account of this species was taken from Anderson (1970), Grzimek (1975), Stroganov (1969), and Waechter (1975). Head and body length is about 400–540 mm, tail length is 220–300 mm, and weight is 1.1–2.3 kg. The general coloration is pale grayish brown to dark brown, and most specimens have a prominent white or pale yellow neck patch. The fur is coarser than that of *M. martes*, and the tail is relatively longer. The soles are covered with sparse hairs, through which the pads stand out markedly.

The stone marten is less dependent on forests than *M. martes* and prefers rocky and open areas. It is found in mountains at elevations of up to 4,000 meters. It often enters towns and may occupy buildings. Natural nest sites include rocky crevices, stone heaps, abandoned burrows of other animals, and hollow trees. *M. foina* is a good climber but rarely goes high in trees. It is nocturnal and crepuscular. The diet consists of rodents, birds, eggs, and berries. Vegetable matter forms a major part of the summer food in some areas.

American pine marten *(Martes americana)*, photo by Howard E. Uible.

A radio-tracking study in western Germany determined seasonal home range size to vary widely, from 12 to 211 ha. In general, ranges were largest during summer and smallest in winter, the smallest ranges were in areas of highest-quality habitat, male ranges were significantly larger than those of females, and adult ranges were larger than those of immature animals, especially during the mating season (Herrmann 1994). Mating occurs in midsummer, but because of delayed implantation of the fertilized eggs in the uterus, births do not occur until the following spring. The total period of pregnancy is 230–75 days, though only about a month is true gestation. Litters usually contain three to four young but may have as many as eight. Canivenc et al. (1981) reported that in southwestern France, where births take place in late March and early April, lactation lasts until mid-May. Sexual maturity has been reported to come at 15–27 months for both sexes (Mead 1994). A captive was still living after 18 years and 1 month (Jones 1982).

The stone marten is considered common in most of Eurasia. Its pelt is hunted but never reaches the quality of that of *M. martes* (Kruska *in* Grzimek 1990). A distinctive but undescribed form of *M. foina* found on Ibiza in the Balearic Islands apparently had been hunted to extinction by about 1960 (Clevenger 1993*a*, 1993*b*; Schreiber et al. 1989). An apparently breeding population of *M. foina* has been established in southeastern Wisconsin for at least 25 years (C. A. Long 1995).

Martes martes (European pine marten)

Except as noted, the information for the account of this species was taken from Grzimek (1975) and Stroganov (1969). Head and body length is 450–580 mm, tail length is 160–280 mm, and weight is 800–1,800 grams. The general coloration is chestnut to gray brown. There is a light yellow patch on the chest and lower neck. The winter pelage is luxuriant and silky, the summer pelage shorter and coarser. The paws are covered by dense hair. Females have four mammae.

The pine marten dwells in forests, both coniferous and deciduous. It is better adapted than *M. zibellina* for an aboreal life. An individual has several nests, located preferably in hollow trees. Activity is mainly nocturnal, and 20–30 km may be covered in a night's hunting. The diet consists of murids, sciurids, other small mammals, honey, fruit, and berries. There are several food storage sites within the home range, which may be 5 km in diameter.

Average home range size is 23 sq km for males and 6.5 sq km for females; there is little or no overlap between the ranges of individuals of the same sex, but the ranges of males greatly overlap those of one or more females (Powell 1994). Independent subadults are tolerated within the otherwise exclusive ranges of adult animals of the same sex, suggesting that the social system of this species (as well as that of many others in the Mustelinae) is determined not strictly by availability of resources but also by an inherent intolerance (Balharry 1993). Mating occurs in midsummer, but because of delayed implantation births do not take place until March or April. Captive females reportedly exhibit one to four periods of sexual receptivity, which usually last 1–4 days and recur at intervals of 6–17 days during the breeding season (Mead 1994). Pregnancy lasts 230–75 days. The number of young per litter is two to eight, usually three to five. The young weigh about 30 grams at birth, open their eyes after 32–38 days, are weaned at 6–7 weeks, separate from the mother in the autumn, and usually attain sexual maturity at 2 years. Maximum known longevity is 17 years.

The fur of the pine marten is more valuable than that of *M. foina*. Wild populations have been excessively trapped and greatly reduced during the twentieth century, but the species is not in immediate danger of extinction. Efforts at captive breeding have had only limited success. *M. martes* occurs in the Balearic Islands, perhaps through human introduction, but a large subspecies, *M. m. minoricensis*, has been described from Menorca. Trappers had nearly exterminated it by 1970, but subsequent legal protection has allowed recovery (Clevenger 1993*a*, 1993*b*). Populations on Mallorca, Elba, Sardinia, and Corsica also are of conservation concern (Clevenger 1993*a*, 1993*b*; Schreiber et al. 1989).

Martes zibellina (sable)

Except as noted, the information for the account of this species was taken from Grzimek (1975), Novikov (1962), and Stroganov (1969). Males have a head and body length of 380–560 mm, a tail length of 120–90 mm, and a weight of 880–1,800 grams. Females have a head and body length of 350–510 mm, a tail length of 115–72 mm, and a weight of 700–1,560 grams. The winter pelage is long, silky, and luxurious. Coloration varies but is generally pale gray brown to dark black brown. The summer pelage is shorter, coarser, duller, and darker. The soles are covered with extremely dense, stiff hairs. The body is very slender, long, and supple.

The sable dwells in both coniferous and deciduous forests, sometimes high in the mountains and preferably near streams. It is mainly terrestrial but can climb. An individual may have several permanent and temporary dens located in holes among or under rocks, logs, or roots. A burrow several meters long may lead to the enlarged nest chamber, which is lined with dry vegetation and fur. The sable hunts either by day or by night. It tends to remain in one part of its home range for several days and then move on. It sometimes stays in its nest for several days during severe winter weather. There may be migrations to higher country in summer and also large-scale movements associated with food shortage. The diet consists mostly of rodents but also includes pikas, birds, fish, honey, nuts, and berries.

Reported population densities vary from 1/1.5 sq km in some pine forests to 1/25 sq km in larch forests. Individual home range is usually several hundred hectares but may be as great as 3,000 ha. in more desolate parts of Siberia. At least part of the home range is defended against intruders, but a male may sometimes share its territory with a female. Mating occurs from June to August, with births usually in April or May. In contrast to *M. martes*, *M. zibellina* usually has a single estrus period of 2–8 days (Mead 1994). Actual embryonic development takes perhaps 25–40 days, but because of delayed implantation the total period of pregnancy is 250–300 days. The number of young per litter ranges from one to five and is usually three or four. The young weigh 30–35 grams at birth, open their eyes after 30–36 days, emerge from the den at 38 days, are weaned at about 7 weeks, and attain sexual maturity at 15–16 months. Maximum known longevity is 15 years.

The sable is one of the most valuable of fur bearers. Its pelt has been avidly sought since ancient times. Several hundred thousand skins were traded annually during the late eighteenth century in the western Siberian city of Irbit. Because of excessive trapping, the take dropped to 20,000–25,000/year by 1910–13, and the sable then had disappeared from much of its range. Subsequently, through programs of protection and reintroduction, the species increased in numbers and distribution. Total numbers in the wild in Russia are now estimated at 1,000,000–1,300,000, and 300,000–350,000 are taken each year (Bakeyev and

Sinitsyn 1994); approximately 27,000 captive-bred sables also are harvested annually. About half of the pelts obtained are released to the world market, and these may sell for more than $500 each (Daniloff 1986). China also was once a major source of sable fur, but the species is now rare in that country (Ma and Li 1994).

Martes melampus (Japanese marten)

Head and body length is 470–545 mm, tail length is 170–223 mm, and weight averaged 1,563 grams for nine males and 1,011 grams for four females (Anderson 1970; Tatara 1994). The general coloration is yellowish brown to dark brown, and there is a whitish neck patch. Little is known about the habits of the Japanese marten. It reportedly is decreasing in numbers through excessive hunting for its fur and because of the harmful effects of agricultural insecticides. Tatara (1994) noted that the hunting season is 1 December–31 January but that a possibly introduced population in southern Hokkaido is fully protected.

The subspecies *M. m. tsuensis*, of Tsushima Island, is classified as vulnerable by the IUCN. Schreiber et al. (1989) noted that this subspecies now is legally protected. According to Tatara (1994), it is found mainly in broad-leaved forests, dens in trees or ground burrows, and is nocturnal. The diet consists largely of fruit and insects but also includes frogs, birds, and small mammals. Both sexes occupy home ranges of about 0.5–1.0 sq km and are territorial. Predation by feral dogs and highway mortality appear to be major threats.

Martes americana (American pine marten)

Males have a head and body length of 360–450 mm, a tail length of 150–230 mm, and a weight of 470–1,300 grams. Females have a head and body length of 320–400 mm, a tail length of 135–200 mm, and a weight of 280–850 grams. The pelage is long and lustrous. The upper parts vary in color from dark brown to pale buff, the legs and tail are almost black, the head is pale gray, and the underparts are pale brown with irregular cream or orange spots. The body is slender, the ears and eyes are relatively large, and the claws are sharp and recurved. Females have eight mammae (Banfield 1974; Burt and Grossenheider 1976; Clark, Grensten, et al. 1987).

The American pine marten is found mainly in coniferous forest. It dens in hollow trees or logs, in rock crevices, or in burrows. The natal nest is lined with dry vegetation. The marten is primarily nocturnal and partly arboreal but spends considerable time on the ground. It can swim and dive well. It is active all winter but may descend to lower elevations if living in a mountainous area. The diet consists mostly of rodents and other small mammals and also includes birds, insects, fruit, and carrion.

Population density varies from about 0.5/sq km to 1.7/sq km of good habitat (Banfield 1974). In Maine, Soutiere (1979) found densities of 1.2 resident martens per sq km in undisturbed and partly harvested forests but only 0.4/sq km in commercially clear-cut forests. Average home range size throughout North America is 8.1 sq km for males and 2.3 sq km for females, and degree of overlap varies (Powell 1994). In a radio-tracking study in northeastern Minnesota, Mech and Rogers (1977) determined home ranges to be 10.5, 16.6, and 19.9 sq km for 3 males and 4.3 sq km for 1 female. There was considerable overlap between the ranges of 2 of the males. The marten is primarily a solitary species, but Herman and Fuller (1974) regularly observed what seemed to be an adult male and female together, sometimes apparently with two of their offspring, even though it was not the breeding season.

Mating occurs from June to August. Captive females reportedly exhibit one to four periods of sexual receptivity, which last 1–4 days and recur at intervals of 6–17 days during the breeding season, though possibly wild individuals have a longer estrus (Mead 1994). Implantation of the fertilized eggs in the uterus is delayed until February, and embryonic development then proceeds for about 28 days. The total period of pregnancy is 220–75 days. The average number of young per litter is 2.6 (1–5). The young weigh 28 grams at birth, open their eyes after 39 days, are weaned at 6 weeks, reach adult size at 3.5 months, and attain sexual maturity at 15–24 months. Wild females have still been capable of reproduction at 12 years (Banfield 1974; Clark, Grensten, et al. 1987). Captives have lived up to 17 years.

The fur of the marten is valuable and is sometimes referred to as "American sable." In the 1940s marten pelts were worth as much as $100 each. In the mid–nineteenth century the Hudson's Bay Company traded as many as 180,000 Canadian skins each year (Banfield 1974). By the early twentieth century excessive trapping had severely depleted *M. americana* in Alaska, Canada, and the western conterminous United States. Protective regulations subsequently allowed the species to make a comeback in some areas, but in the eastern United States the marten survives only in small parts of Minnesota, New York, and Maine (Blanchard 1974; Mech 1961; Mech and Rogers 1977; Yocom 1974). Reintroduction projects have been carried out in northern Michigan and Wisconsin (Knap 1975), and it appears that a self-sustaining population has been restored in that region (Slough 1994). Reintroduction also has been attempted in New Hampshire (Strickland and Douglas 1987) and in various other parts of the northeastern United States and southeastern Canada. However, continued absence or only very low populations in most of this region seem attributable to forestry practices that leave relatively poor habitat for the species (Thompson 1991). During the 1976–77 trapping season 27,898 marten skins were taken in the United States, mostly in Alaska, and sold for an average price of $14.00. The respective figures for Canada were 102,632 skins and $19.92 (Deems and Pursley 1978). In 1983–84 the total kill was 188,647 and the average price was $18.61 (Novak, Obbard, et al. 1987). In 1991–92 the kill in the United States was 22,827 and the average price was $44.06 (Linscombe 1994).

Martes pennanti (fisher, or pekan)

Head and body length is 490–630 mm and tail length is 253–425 mm. Males weigh 2.6–5.5 kg and females weigh 1.3–3.2 kg. The head, neck, shoulders, and upper back are dark brown to black. The underparts are brown, sometimes with small white spots. The thick pelage is coarser than that of *M. americana*. The body is slender but rather stocky for a weasel. Females have eight mammae.

Except as noted, the information for the remainder of this account was taken from Powell (1981, 1982). The fisher inhabits dense forests with an extensive overhead canopy and avoids open areas. It generally lacks a permanent den but seeks temporary shelter in hollow trees and logs, brush piles, abandoned beaver lodges, holes in the ground, and snow dens. All dens known to have been used for raising young were located high up in hollow trees. Activity may take place at any hour. The fisher is adapted for climbing but is primarily terrestrial. It is capable of traveling long distances; one individual covered 90 km in three days. Usual daily movement in New Hampshire, however, was found to be 1.5–3.0 km. The fisher generally forages in a zigzag pattern, constantly investigating places where prey might be concealed.

The diet consists mainly of small to medium-sized birds and mammals, and carrion. It has been calculated that a

Fisher *(Martes pennanti)*, photo from San Diego Zoological Garden.

fisher requires either 1 porcupine every 10–35 days, 1 snowshoe hare every 2.5–8.0 days, 1 kg of deer carrion every 2.5–8.0 days, 1–2 squirrels per day, or 7–22 mice per day. Hares are killed with a quick rush and a bite to the back of the neck. Porcupines are taken only on the ground and are killed by repeatedly circling and biting at the face.

Population density in preferred habitat is 1/2.6–7.5 sq km, but in other areas it may be as low as 1/200 sq km. Individual home range size averages 38 sq km for males and 15 sq km for females (Powell 1994). There is little overlap between the ranges of animals of the same sex but extensive overlap between the ranges of opposite sexes. *M. pennanti* is solitary except during the breeding season, at which time, according to Leonard (1986), normal spacing mechanisms seem to break down, with males leaving their territories to seek as many females as possible and perhaps coming into physical conflict with one another.

Mating occurs from March to May, but implantation of the fertilized eggs in the uterus is delayed until the following January to early April, and births occur from February to May. Postimplantation embryonic development lasts about 30 days, but the total period of pregnancy is nearly 1 year. Females probably mate within 10 days of giving birth and thus are pregnant almost continually (estrus is the same as described for *M. americana*). Litters contain an average of about three young, but there may be as many as six. The newborn weigh less than 40 grams (LaBarge, Baker, and Moore 1990). They are blind and only partly covered with fine hair. They do not open their eyes for about 7 weeks, and they do not walk until about 8–9 weeks. Weaning begins at 8–10 weeks, and separation from the mother occurs in the fifth month. Females reach adult weight after 6 months, and males after 1 year. Sexual maturity comes at 1–2 years. Longevities of about 10 years have been recorded for both wild and captive animals.

In the nineteenth and early twentieth centuries the fisher declined over most of its range because of excessive fur trapping and habitat destruction through logging. The species was almost totally eliminated in the United States and greatly reduced in eastern Canada. As a result, porcupines increased in numbers and began to do considerable forest damage. Widespread closed seasons and other protective regulations were initiated in the 1930s, and reintroductions were made in various areas in the 1950s and 1960s. To the south of Canada, populations of *M. pennanti* are now present in Washington, Oregon, northern California, Idaho, Montana, Minnesota, Wisconsin, Michigan, New York, Vermont, New Hampshire, Maine, Massachusetts, and West Virginia (Cottrell 1978; Coulter 1974; Handley 1980; Mohler 1974; Mussehl and Howell 1971; Olterman and Verts 1972; Penrod 1976; Petersen, Martin, and Pils 1977; Schempf and White 1977; Yocom and McCollum 1973). There also have been several recent records in North and South Dakota (Gibilisco 1994). In some of these states during the past two decades the fisher has been taken legally for the fur market. The average price of a fisher pelt was $100 in the 1920s. The price subsequently declined but then again increased. During the 1976–77 trapping season 12,557 skins were taken in the United States and Canada, selling for an average of about $95 each (Deems and Pursley 1978). In 1983–84 the take was 20,248 and the average price was $49.45 (Novak, Obbard, et al. 1987), but in 1986 in Ontario the highest-quality pelts were selling for $450 (Douglas and Strickland 1987). According to Strickland (1994), there was a reduction in the popularity of furs in the 1990s, and trapping pressure on the fisher eased. During the 1991–92 season 3,336 skins were taken in the United States and sold for an average price of $34.43 (Linscombe 1994).

Martes flavigula (yellow-throated marten)

The information for the account of this species was taken from Lekagul and McNeely (1977) and Stroganov (1969). Head and body length is 450–650 mm, tail length is

Yellow-throated marten *(Martes flavigula)*, photo from Zoological Society of London.

370–450 mm, and weight is 2–3 kg. The fur is short, sparse, and coarse. There is much variation in color, but generally the top of the head and neck, the tail, the lower limbs, and parts of the back are dark brown to black. The rest of the body is pale brown, except for a bright yellow patch from the chin to the chest. Females have four mammae.

The yellow-throated marten is generally found in forests. It climbs and maneuvers in trees with great agility but often comes to the ground to hunt. Activity is primarily diurnal. The diet includes rodents, pikas, eggs, frogs, insects, honey, and fruit. In the northern parts of its range *M. flavigula* evidently preys heavily on the musk deer *(Moschus)* and the young of other ungulates. Individuals often hunt in pairs or family groups, and lifelong pair bonding has been suggested. Births occur in April (Mead 1994). The number of young per litter is usually two or three and may be as many as five. Maximum known longevity is about 14 years. The pelt of this marten has little commercial value. However, the subspecies on Taiwan, *M. f. chrysospila*, has become very rare because of hunting for use of its inner organs for food and through deforestation (Schreiber et al. 1989); it is classified as endangered by the USDI. The subspecies *M. f. robinsoni* of Java has not been collected since 1959, though it was sighted as late as 1979 (Schreiber et al. 1989); it is classified as endangered by the IUCN.

Martes gwatkinsi (Nilgiri marten)

According to Wirth and Van Rompaey (1991), a male had a head and body length of 515 mm, a tail length of 419 mm, and a weight of about 2 kg. As in *M. flavigula*, the general coloration is dark brown and there is a yellowish patch from the chin to the chest. The dorsal profile of the skull of *M. gwatkinsi* is flat, not convex as in *M. flavigula*. The Nilgiri marten is found only in a small isolated tract of hill forest, seldom occurring below 900 meters. It long has been considered rare, only 5–10 specimens are known, and it sometimes is persecuted by beekeepers (Schreiber et al. 1989). It is classified as vulnerable by the IUCN. Madhusudan (1995) reported a recent observation of an individual with an estimated total length of about 1,200 mm sleeping in and moving through the tree canopy in a small patch of forest.

CARNIVORA; MUSTELIDAE; **Genus EIRA**
Hamilton-Smith, 1842

Tayra

The single species, *E. barbara*, is found from southern Sinaloa (west-central Mexico) and southern Tamaulipas (east-central Mexico) to northern Argentina and on the island of Trinidad (Cabrera 1957; Goodwin and Greenhall 1961; E. R. Hall 1981).

Head and body length is 560–680 mm, tail length is 375–470 mm, and weight is 4–5 kg (Grzimek 1975; Leopold 1959). The short, coarse pelage is gray, brown, or black on the head and neck, has a yellow or white spot on the chest, and is black or dark brown on the body. There also is a rare, light-colored form, which is pale buffy and has a darker head. The tayra has a long and slender body, short limbs, and a long tail. The head is broad, the ears are short and rounded, and the neck is long. The soles are naked, and the strong claws are nonretractile.

The tayra is a forest dweller. It nests in a hollow tree or log, a burrow made by another animal, or tall grass. It can climb, run, and swim well. When pursued by dogs it may run on the ground for some distance, then climb a tree and leap through the trees for about 100 meters before descending to the ground again. It thus gains time while the dogs are trying to pick up the trail. Leopold (1959) observed a group of 4 animals springing swiftly through the trees with incredible agility. *Eira* is active both at night and, especially when there is cloud cover, in the morning. The diet seems to consist mostly of rodents but also includes rabbits, birds, small deer *(Mazama)*, honey, and fruit.

A radio-collared female on the llanos of Venezuela was found to use a home range of 9 sq km (Eisenberg 1989). *Eira* is often seen alone, in pairs, or in small family groups. Hall and Dalquest (1963) wrote that the genus is not social to any extent, but Leopold (1959) referred to an old record of hunting troops made up of 15–20 individuals. He also cited a report that 2 young are born in February, that they open their eyes after 2 weeks, and that they forage with their mother by the time they are 2 months old. Other reports suggest that the young may be born in any season.

Tayra *(Eira barbara)*, photo by Ernest P. Walker.

Poglayen-Neuwall (1992) and Poglayen-Neuwall et al. (1989) reported that *Eira* is a nonseasonal, polyestrous species, with an estrous cycle averaging 17 days and receptivity lasting only 2–3 days. Poglayen-Neuwall (1975) and Poglayen-Neuwall and Poglayen-Neuwall (1976) determined that six gestation periods lasted 63–65 days, that newborn weigh about 74–92 grams, that they do not open their eyes until they are 35–58 days old, and that they suckle for 2–3 months. According to Vaughn (1974), litters of 2 young each were produced in captivity on 4 March and 22 July, following gestation periods of about 67–70 days.

The tayra can live 18 years in captivity, loves to play, and can be tamed. It reportedly was used long ago by the Indians to control rodents (Grzimek 1975). It is of no particular importance as a fur bearer or predator of game. It may occasionally eat poultry but does no substantial damage (Leopold 1959). It also has been accused of raiding corn and sugar cane fields (Hall and Dalquest 1963). Schreiber et al. (1989) reported that the range of the tayra has been greatly reduced in Mexico because of the destruction of tropical forests and spread of agriculture. Remaining populations are small and threatened by habitat loss and hunting. The subspecies *E. b. senex*, of Mexico, Guatemala, Belize, and northern Honduras, is now classified as vulnerable by the IUCN.

CARNIVORA; MUSTELIDAE; **Genus GALICTIS**
Bell, 1826

Grisóns

There are two species (Cabrera 1957; E. R. Hall 1981):

G. vittata (greater grisón), southern Mexico to central Peru and southeastern Brazil;
G. cuja (little grisón), central and southern South America.

Galictis has often been referred to as *Grison* Oken, 1816. Cabrera placed *G. cuja* in the subgenus *Grisonella* Thomas, 1912.

In *G. vittata* the head and body length is 475–550 mm, tail length is about 160 mm, and weight is 1.4–3.3 kg. In *G. cuja* the head and body length is 280–508 mm, tail length is 120–93 mm, and weight is 1.0–2.5 kg (Redford and Eisenberg 1992). The color pattern is striking. In both species the black face, sides, and underparts, including the legs and feet, are sharply set off from the back. The back is smoky gray in *G. vittata* and yellow gray or brownish in *G. cuja*. A white stripe extends across the forehead and down the sides of the neck, separating the black of the face from the gray

Greater grisón *(Galictis vittata)*, photo by Ernest P. Walker.

or brown of the back. Short legs and slender bodies give the animals of this genus an appearance somewhat like that of *Mustela,* but the color pattern immediately distinguishes them.

Grisóns are found in forests and open country from sea level to 1,200 meters. They live under tree roots or rocks, in hollow logs, or in burrows made by other animals, such as the viscachas of South America. They are probably capable burrowers themselves. Quick and agile, they are are good climbers and swimmers and are active both by day and by night. The diet includes small mammals, birds and their eggs, cold-blooded vertebrates, invertebrates, and fruit.

A radio-collared female on the llanos of Venezuela was found to use a home range of 4.2 sq km (Eisenberg 1989). Grisóns are sometimes seen in pairs or groups and playing together. They have a number of vocalizations, including sharp, growling barks when threatened. Various reports indicate that offspring have been produced in March, August, September, and October (Grzimek 1975; Leopold 1959). The gestation period is 39 days (Eisenberg 1989). Litter size is two to four young. A captive *G. vittata* was still living after 10 years and 6 months (Jones 1982).

Young grisóns tame readily and make affectionate pets. In early-nineteenth-century Chile grisóns reportedly were domesticated by natives and used in the same manner as ferrets to enter the crevices and holes of chinchillas to drive the latter out (Osgood 1943).

CARNIVORA; MUSTELIDAE; **Genus LYNCODON**
Gervais, 1844

Patagonian Weasel

The single species, *L. patagonicus,* is found in Argentina and southern Chile (Cabrera 1957).

Head and body length is usually 300–350 mm and tail length is 60–90 mm. The coloration on the back is grayish brown with a whitish tinge. The top of the head is creamy or white, and this color extends as a broad stripe on either side to each shoulder. The nape, throat, chest, and limbs are dark brown, and the rest of the lower surface is lighter brown varied with gray. The color pattern is characteristic and quite attractive. *Lyncodon* is somewhat similar externally to *Galictis* but differs in such features as coloration and the shorter tail. Internally *Lyncodon* has fewer teeth than *Galictis,* there being only two upper and two lower premolars, and one upper and one lower molar, on each side. Like most mustelids, *Lyncodon* has a slender body and short legs.

The Patagonian weasel inhabits the pampas. Its habits are little known, but Ewer (1973:177) wrote: "The reduced molars and cutting carnassials of *Lyncodon* strongly suggest that it is a highly carnivorous species." Redford and Eisenberg (1992:163)) added: "It is reported to be nocturnal or crepuscular and to enter the burrows of *Ctenomys* and *Microcavia* in pursuit of prey." This weasel reportedly was sometimes kept in the houses of ranchers for the purpose of destroying rats. Miller et al. (1983) considered it to be rare in Chile.

Patagonian weasel *(Lyncodon patagonicus),* photo by Tom Scott of mounted specimen in Royal Scottish Museum, Edinburgh.

Zorilla *(Ictonyx striatus)*, photo from Zoological Society of London.

CARNIVORA; MUSTELIDAE; **Genus ICTONYX**
Kaup, 1835

Zorilla, or Striped Polecat

The single species, *I. striatus,* occurs from Mauritania to Sudan and south to South Africa (Coetzee *in* Meester and Setzer 1977).

Head and body length is 280–385 mm, tail length is 200–305 mm, and weight is 420–1,400 grams. Males are generally larger than females. The body is black with white dorsal stripes, the tail is more or less white, and the face has white markings. The appearance is somewhat like that of the spotted skunks *(Spilogale)* of North America, and early writers sometimes confused the two genera. *Ictonyx* also bears some resemblance to the African genera *Poecilictis* and *Poecilogale* but may be recognized by its color pattern, long hair, and bushy tail. Females have two pairs of mammae (Rowe-Rowe 1978*a*).

The zorilla is found in a variety of habitats but avoids dense forest. It is mainly nocturnal, resting during the day in rock crevices or in burrows excavated by itself or some other animal. It occasionally shelters under buildings and in outhouses in farming areas. It is terrestrial but can climb and swim well. The usual pace is an easy trot, slower than that of a mongoose, with the back slightly hunched. The diet consists mainly of small rodents and large insects but also includes eggs, snakes, and other kinds of animals. The zorilla may take a chicken on occasion but is often useful in eliminating rodents from houses and stables.

At the sight of an enemy, such as a dog, a zorilla may erect its hair and tail and perhaps emit its anal gland secretion. Such behavior would seem to make the zorilla more formidable than it really is. When actually attacked, it usually emits fluid into the face of the enemy and then feigns death. The ejected fluid may vary in potency with the individual animal and perhaps with age and the time of year. Some writers have remarked that the fluid is much less pungent than that of the American skunks, but others have stated that it is most repulsive, acrid, and persistent.

Ictonyx is solitary. Captive males are totally intolerant of one another, and even adults of opposite sexes are amicable only at the time of mating (Rowe-Rowe 1978*b*). There are several adult vocalizations associated with threat, defense, and greeting (Channing and Rowe-Rowe 1977). In a study of captives, Rowe-Rowe (1978*a*) determined the reproductive season to extend from early spring to late summer. All births occurred from September to December. There was usually a single annual litter, but if all young died at an early stage, the female sometimes mated and gave birth a second time. Gestation periods of 36 days were recorded. Litter size was one to three young. They weighed about 15 grams at birth, began to take solid food after about

32 days, opened their eyes at 40 days, were completely weaned at 18 weeks, and were almost full grown at 20 weeks. A male first mated at 22 months, and a female bore its first litter at 10 months. According to Jones (1982), a captive zorilla lived 13 years and 4 months.

CARNIVORA; MUSTELIDAE; **Genus POECILICTIS**
Thomas and Hinton, 1920

North African Striped Weasel

The single species, *P. libyca,* occurs from Morocco and Senegal to the Red Sea (Rosevear 1974). There is some question whether *Poecilictis* warrants generic distinction from *Ictonyx.* The two taxa are sometimes confused in northern Nigeria and Sudan, where their ranges overlap. Wozencraft (*in* Wilson and Reeder 1993) chose to place *Poecilictis* in the synonymy of *Ictonyx,* but the two were treated as distinct genera by Corbet and Hill (1991).

Head and body length is 200–285 mm and tail length is 100–180 mm. Three males weighed 200–250 grams each, though Sitek (1995) reported weights of 600 grams for males and 500 grams for females. The snout is black, the forehead is white, and the top of the head is black. The back is white with a variable pattern of black bands. The tail is white but becomes darker toward the tip. The underparts and limbs are black. From *Ictonyx, Poecilictis* differs in color pattern, in smaller size, in larger bullae, and in having hairy soles except for the pads. The well-developed anal glands are capable of ejecting a malodorous fluid (Rosevear 1974).

According to Rosevear (1974), this weasel is restricted to the edges of the Sahara and the contiguous arid zones. It is nocturnal, sheltering throughout the daylight hours in single subterranean burrows. These it digs for itself, either down into level surfaces or in the sides of dunes. Maternal dens consist of a single gallery ending in an unlined chamber. The diet apparently consists of rodents, young ground birds, eggs, lizards, and insects.

Births reportedly occur in the wild from January to March. Rosevear (1974) wrote that gestation may be as short as 37 days or as long as 77. Litters usually contain two or three young. At birth they are blind and covered with short hair. Sitek (1995) reported that cubs born in captivity weighed 5 grams at birth, took some solid food after 5 weeks, weighed 250 grams at 2 months, and were separated from their mother at 3 months. Some captives have lived for 5 years or more. *Poecilictis* probably does not make a good pet, as it constantly has a disagreeable smell and is aggressive toward people.

CARNIVORA; MUSTELIDAE; **Genus POECILOGALE**
Thomas, 1883

African Striped Weasel

The single species, *P. albinucha,* is found from Zaire and Uganda to South Africa (Coetzee *in* Meester and Setzer 1977).

Head and body length is 250–360 mm and tail length is 130–230 mm. Rowe-Rowe (1978*b*) listed weights of 283–380 grams for males and 230–90 grams for females. From the white on the head and nape four whitish to orange yellow stripes and three black stripes extend on the black back toward the tail, which is white. The legs and underparts are black. There is little variation in pattern.

Poecilogale is smaller and more slender than *Ictonyx* and has narrower back stripes. It resembles *Mustela* in its remarkably slender and elongate body and short legs. Like other weasels, it is able to enter any burrow that it can get its head into. Females have two pairs of mammae (Rowe-Rowe 1978*a*).

This weasel is found in a variety of habitats, including forest edge, grassland, and marsh. It is almost entirely nocturnal, usually spending the day in a burrow that is either self-excavated or taken over from another animal (Kingdon 1977). It can climb well but spends most of its time on the ground. It cannot run rapidly, and patiently trails prey by scent. The diet consists mainly of small mammals and birds but also includes snakes and insects.

Like most weasels, *Poecilogale* attacks by grabbing the throat or neck and hanging on and chewing until the victim is dead. Fair-sized prey, such as springhare, may run some distance before dropping from exhaustion, with the weasel retaining its hold. *Poecilogale* kills venomous snakes in much the same manner as the mongoose *Her-*

North African striped weasel *(Poecilictis libyca)*, photo by Robert E. Kuntz.

African striped weasel *(Poecilogale albinucha):* Top, photo of mounted specimen from Field Museum of Natural History; Bottom, photo by D. T. Rowe-Rowe.

pestes. It repeatedly provokes the snake to strike until the reptile is tired and slower in recovery and then seizes the snake by the back of the head.

Poecilogale is generally solitary, but groups of two to four individuals—apparently family parties—have been observed on occasion. Adults of opposite sexes lived together quite amicably in captivity, but adult males fought each other at every encounter (Rowe-Rowe 1978*b*). The area around the den is marked by defecation (Kingdon 1977).

When attacked or under stress *Poecilogale* emits a noxious odor from its anal glands. Although nauseating, this odor is not as strong or persistent as that of the American skunks or the African *Ictonyx.* Normally *Poecilogale* is silent, but when alarmed it utters a loud sound described as being between a growl and a shriek. Several adult vocalizations, associated with threat, defense, and greeting, were recorded by Channing and Rowe-Rowe (1977).

In studies of captives and wild-caught animals in South Africa, Rowe-Rowe (1978*a*) determined that births occurred from September to April. Females were polyestrous and would mate a second time in the season if their first litter was lost. Gestation periods of 31–33 days were recorded. Litters contained one to three young. They weighed four grams at birth, started taking solid food after 35 days, opened their eyes at 51–54 days, were completely weaned at about 11 weeks, and were nearly full grown at 20 weeks. A male first mated at 33 months, and a female had her first litter at 19 months. According to Smithers (1971), one individual lived for 5 years and 2 months after capture.

Meester (1976) considered this weasel to be rare but not endangered in South Africa. However, Rowe-Rowe (1990) indicated that it is at risk because of loss of habitat to agriculture and overgrazing and through predation by and competition from increasing numbers of domestic dogs. Although accused of killing poultry on occasion, *Poecilogale* is generally beneficial to human interests because of its destruction of rats, mice, and springhares. It also eats quantities of locusts when available and digs their larvae out of the ground. Some African tribes use the skins of *Poecilo-*

gale in ceremonial costumes or as ornaments, and the Zulu people reportedly use its skin and other parts for medicinal purposes.

CARNIVORA; MUSTELIDAE; **Genus GULO**
Pallas, 1780

Wolverine

The single species, *G. gulo,* originally occurred from Scandinavia and Germany to northeastern Siberia, throughout Alaska and Canada, and as far south as central California, southern Colorado, Indiana, and Pennsylvania (Corbet 1978; E. R. Hall 1981). Although Hall treated the North American populations as a separate species, *G. luscus,* he indicated that there was evidence supporting their recognition as a subspecies of *G. gulo.*

Head and body length is 650–1,050 mm, tail length is 170–260 mm, shoulder height is 355–432 mm, and weight is 7–32 kg. Females average 10 percent less than males in linear measurements and 30 percent less in weight (E. R. Hall 1981). The fur is long and dense. The general coloration is blackish brown. A light brown band extends along each side of the body from shoulder to rump and joins its opposite over and across the base of the tail. *Gulo* resembles a giant marten, with a heavy build, a large head, relatively small and rounded ears, a short tail, and massive limbs (Stroganov 1969). Females have four abdominal and four inguinal mammae (Pasitschniak-Arts and Larivière 1995).

The wolverine occurs in the tundra and taiga zones. It may be found in forests, mountains, or open plains. For shelter it may construct a rough bed of grass or leaves in a cave or rock crevice, in a burrow made by another animal, or under a fallen tree (Banfield 1974; Stroganov 1969). Most maternal dens in Finland were found in holes under the snow (Pulliainen 1968). *Gulo* is mainly terrestrial, the usual gait being a sort of loping gallop, but it can climb trees with considerable speed and is an excellent swimmer. It has a keen sense of smell but apparently poor eyesight and indifferent hearing. It seems to be unexcelled in strength among mammals of its size and has been reported to drive bears and cougars from their kills. It is largely nocturnal but is occasionally active in daylight. In the far north, where there are periods of extended light or darkness, the wolverine reportedly alternates 3- to 4-hour periods of activity and sleep. In winter it has been known to cover up to 45 km in a day. It can maintain a gallop for a lengthy period, sometimes moving 10–15 km without rest (Stroganov 1969). In mountainous areas it moves to lower elevations during winter (Pasitschniak-Arts and Larivière 1995). Juveniles normally disperse 30–100 km from their natal range, and an individual marked on 6 March 1981 was found 378 km away on 29 November 1982 (Gardner, Ballard, and Jessup 1986).

The diet includes carrion, the eggs of ground-nesting birds, lemmings, and berries. Large mammals such as reindeer, roe deer, and wild sheep are taken mainly in winter, when the snow cover allows *Gulo* to travel faster than its prey. Most large mammals, however, are obtained in the form of carrion. Small rodents may be chased, pounced upon, or dug out of the ground (Pasitschniak-Arts and Larivière 1995). Caches of prey or carrion are covered with earth or snow or, sometimes, wedged in the forks of trees.

Gulo occurs at relatively low population densities (Van Zyll de Jong 1975). In Scandinavia the estimates vary from 1/200 sq km to 1/500 sq km. An adult male there has a territory that may be as large as 2,000 sq km in winter, and individuals of the same sex are not tolerated within it. Territories are regularly marked, mainly by secretions from anal

Wolverine *(Gulo gulo)*, photo by James K. Drake.

scent glands but also with urine. Within the territory of each male are the territories of three or four adult females and their young (Kruska *in* Grzimek 1990; Pulliainen and Ovaskainen 1975).

Somewhat different information was derived from studies in Montana (Hornocker 1983; Hornocker and Hash 1981; Koehler, Hornocker, and Hash 1980). A study area of 1,300 sq km there contained a minimum population of 20 wolverines. Average yearly home range was 422 sq km for males and 388 sq km for females; however, individuals sometimes left their ranges, and females occupied much smaller areas while they were nursing young. There was extensive overlap between the ranges of both the same and the opposite sexes, and no territorial defense was observed. The extensive marking was considered not to define a territory but rather to identify the area being utilized at a given time.

Investigations in arctic Alaska indicated that females generally exclude one another from their home ranges but males tolerate the presence of females (Gipson 1985). In south-central Alaska, Whitman, Ballard, and Gardner (1986) determined annual home range to average 535 sq km for males and 105 sq km for females with young. Annual ranges in Yukon were found to be 209–69 sq km for males and 76–269 sq km for females (Banci and Harestad 1990).

The wolverine is solitary except during the breeding season. Females are monestrous and apparently give birth about every two years. Mating usually occurs from late April to July, but implantation of the fertilized eggs in the uterus is delayed until the following November to March, with births from January to April. The usual number of young per litter is two to four, and the range is one to five. Active gestation lasts 30–40 days. The young weigh 90–100 grams at birth, nurse for 8–10 weeks, separate from the mother in the autumn, and attain adult size after 1 year. Sexual maturity comes in the second or third year of life (Banci and Harestad 1988; Banfield 1974; Grzimek 1975; Pulliainen 1968; Rausch and Pearson 1972; Stroganov 1969). A captive female approximately 10 years of age gave birth after a pregnancy of 272 days (Mehrer 1976). Another captive wolverine lived 17 years and 4 months (Jones 1982).

The pelt of the wolverine is not used widely in commerce but is valued for parkas by persons living in the Arctic because it accumulates less frost than other kinds of fur. During the 1976–77 trapping season 1,922 skins were reported taken in Canada, Alaska, and Montana. The average selling price was about $182 (Deems and Pursley 1978). In 1983–84 the take was 1,377, with an average value of $77.16 (Novak, Obbard, et al. 1987). In 1991–92, 591 skins were taken in Alaska and sold at an average price of $235 (Linscombe 1994). Fur trapping has contributed to a decline in the numbers and distribution of the wolverine, but a more important factor may be human consideration of the genus as a nuisance. In Scandinavia it was intensively hunted, often for bounty, because of alleged predation on domestic reindeer. Throughout its range the wolverine came into conflict with people by following traplines and devouring the fur bearers it found and by breaking into cabins and food caches and spraying the contents with its strong scent.

In Europe *Gulo* is now found only in parts of Scandinavia, where it is very rare, and northern Russia (Pasitschniak-Arts and Larivière 1995; Smit and Van Wijngaarden 1981). Only 120–50 individuals remain in Norway; although legally protected, they are regularly killed in response to claims of attacks on the increasing numbers of sheep being allowed to graze in mountainous areas (Bevanger 1992). The genus has disappeared over most of eastern and south-central Canada (Van Zyll de Jong 1975). It is classified as vulnerable by the IUCN and has been designated as endangered in eastern Canada. By the early twentieth century the wolverine also had been nearly eliminated in the conterminous United States. One population, however, held out and subsequently increased in the mountains of northern and eastern California, and another population reestablished itself in western Montana. Since 1960 there have been numerous reliable reports from Washington, Oregon, Idaho, Wyoming, and Colorado and a few questionable records in Minnesota, Iowa, and South Dakota (Birney 1974; Blus, Fitzner, and Fitzner 1993; Field and Feltner 1974; Johnson 1977; Nead, Halfpenny, and Bissell 1985; Nowak 1973; Schempf and White 1977). The U.S. Fish and Wildlife Service recently was petitioned to add the wolverine in the conterminous United States to the List of Endangered and Threatened Wildlife but decided against such a measure (Nordstrom 1995), seemingly in contradiction of the IUCN classification and the whole history of the species. The subspecies *G. g. katschemakensis* of Alaska's Kenai Peninsula now numbers only about 50 individuals and reportedly continues to decline because of an excessively long hunting season (Schreiber et al. 1989).

CARNIVORA; MUSTELIDAE; **Genus MELLIVORA**
Storr, 1780

Honey Badger, or Ratel

The single species, *M. capensis,* originally occurred from Palestine and the Arabian Peninsula to Turkmenistan and eastern India and from Morocco and lower Egypt to South Africa (Coetzee *in* Meester and Setzer 1977; Corbet 1978).

Head and body length is 600–770 mm, tail length is usually 200–300 mm, and shoulder height is usually 250–300 mm. Kingdon (1977) listed weight as 7–13 kg. The upper parts, from the top of the head to the base of the tail, vary from gray to pale yellow or whitish and contrast sharply with the dark brown or black of the underparts. Completely black individuals, however, have been found in Africa, particularly in the Ituri Forest of northern Zaire. The color pattern of the honey badger has been interpreted as being a warning coloration because it makes the animal easily recognizable. Females have two pairs of mammae.

The body is heavily built, the legs and tail are relatively short, the ears are small, and the muzzle is blunt. The large forefeet are armed with very large, strong claws. The hair is coarse and quite scant on the underparts. The skin is exceedingly loose on the body and very tough. The skull is massive and the teeth are robust. Anal glands secrete a vile-smelling liquid. This combination of characters provides an effective system of deterrence and defense.

Mellivora is very difficult to kill. Except on the belly the skin is so tough that a dog can make little impression. The ratel can twist about in its skin, so it can even bite an adversary that has seized it by the back of the neck. Porcupine quills and bee stings have little effect, and snake fangs are rarely able to penetrate. *Mellivora* seems to be devoid of fear, and it is doubtful that any animal of equivalent size can regularly kill it. It may rush out from its burrow and charge an intruder, especially in the breeding season. Horses, antelopes, cattle, and even buffalo have been attacked and severely wounded in this manner.

The honey badger occupies a variety of habitats, mainly in dry areas but also in forests and wet grasslands. It lives among rocks, in hollow logs or trees, and in burrows. Its

Honey badger, or ratel *(Mellivora capensis)*, photo by Bernhard Grzimek.

powerful limbs and large claws make it a capable and rapid digger. It is primarily terrestrial but can climb, especially when attracted by honey. It travels by a jog trot but is tireless and trails its prey until the prey is run to the ground. The diet includes small mammals, the young of large mammals, birds, reptiles, arthropods, carrion, and vegetation. Honey and bees are important foods at certain times of the year.

A remarkable association has developed, at least in tropical Africa, between a bird—the honey guide *(Indicator indicator)*—and *Mellivora*. The association is mutually beneficial in the common exploitation of the nests of wild bees. In the presence of any mammal, even people, the honey guide has the unusual habit of uttering a series of characteristic calls. If a honey badger hears these calls, it follows the bird, which invariably leads it to the vicinity of a beehive. The badger breaks open the nest, and the bird obtains enough of the honey and insects to pay for its work.

Mellivora may travel alone or in pairs. It is generally silent but may utter a very harsh, grating growl when annoyed. The sparse data on reproduction were summarized by Hancox (1992*b*), Kingdon (1977), Rosevear (1974), and Smithers (1971). Mating has been noted in South Africa in February, June, and December. A lactating female was found in Botswana in November, and newborn were recorded in Zambia in December. Seasonal breeding has been reported in Turkmenistan, with mating occurring in autumn and births in the spring. The gestation period is thought to be about 5–6 months. The number of young in a litter is commonly two, and the range is one to four. The young are born in a grass-lined chamber and evidently remain close to the burrow for a long time. *Mellivora* appears to thrive in captivity: one specimen lived 26 years and 5 months (Jones 1982).

If captured before it is half grown, the honey badger can become a satisfactory pet, as it is docile, affectionate, and active. It is, however, incredibly strong and energetic and can wreck cages and damage property in its explorations. Wild individuals sometimes prey on poultry, tearing through wire netting to effect entry (Smithers 1971). Destruction of commercial beehives in East Africa seems to be a significant problem, and *Mellivora* has thus been intensively poisoned, trapped, and hunted (Kingdon 1977). The overall range of the genus has declined, probably through human persecution. The ratel is classified as vulnerable in South Africa (Smithers 1986) and as rare by Russia.

CARNIVORA; MUSTELIDAE; **Genus MELES**
Storr, 1780

Old World Badger

The single species, *Meles meles*, originally occurred throughout Europe, including the British Isles and several Mediterranean islands (Sicily, Crete, Rhodes), and in Asia as far east as Japan and as far south as Palestine, Iran, Tibet, and southern China (Corbet 1978; Long and Killingley 1983). The authority for the generic name *Meles* sometimes is given as Brisson, 1762.

Head and body length is 560–900 mm, tail length is 115–202 mm, and weight is usually 10–16 kg; Novikov (1962), however, wrote that old males attain weights of 30–34 kg in the late autumn. The upper parts are grayish, and the underparts and limbs are black. On each side of the face is a dark stripe that extends from the tip of the snout to the ear and encloses the eye; white stripes border the dark stripe. Like other badgers, *Meles* has a stocky body, short limbs, and a short tail. It is distinguished from *Arctonyx* by its black throat and shorter tail, which is the same color as the back. Females have three pairs of mammae (Grzimek 1975).

The Old World badger is found mainly in forests and densely vegetated areas. It usually lives in a large communal burrow system that covers about 0.25 ha. There are numerous entrances, passages, and chambers. Nests may be located 10 meters from an entrance and 2–3 meters below the surface of the ground and have a diameter of 1.5 meters. A burrow system may be used for decades or centuries, by one generation of badgers after another; it continually increases in complexity and may eventually cover several hectares (Grzimek 1975; Novikov 1962). The animals occupying a system may utilize one nest for several months and then suddenly move to another part of the burrow. The living quarters are kept quite clean. Bedding material, in the form of dry grass, brackens, moss, or leaves, is dragged backward into the den. Occasionally this bedding is brought up and strewn around the entrance to air for an hour or so in the early morning. Around the burrows are dung pits, sunning grounds, and areas for play (badgers play all sorts of games, including leapfrog). Well-defined foraging paths may extend outward for 2–3 km (Novikov 1962). Kruuk (1978*b*) distinguished two kinds of burrows:

Old World badger *(Meles meles)*, photo from Paignton Zoo, Great Britain.

"main setts," with an average of 10.5 entrances; and small "outliers," usually having only a single entrance.

Meles usually does not emerge from its burrow until after sundown. During periods of very cold weather and high snow it may spend days or weeks in the den. Such intervals of winter sleep extend to several months in northern Europe and up to seven months in Siberia. There is no substantial drop in body temperature, and the badger can be aroused easily, but the animal lives off of fat reserves accumulated in the summer and autumn. The omnivorous diet includes almost any available food—small mammals, birds, reptiles, frogs, mollusks, insects, larvae of bees and wasps, carrion, nuts, acorns, berries, fruits, seeds, tubers, rhizomes, and mushrooms. Earthworms have been found to be of major dietary importance in some areas (Kruuk 1978*a;* Skoog 1970).

In a study in England, Kruuk (1978*b*) found badgers to be organized into clans of as many as 12 individuals. The minimum distance between the main burrows of clans was 300 meters. Most clans used ranges of 50–150 ha., and there was little overlap. The ranges were marked by defecation and secretions from subcaudal glands. Several fights were observed along territorial boundaries. Most clans had more females than males, but one, which used a range of only 21 ha., consisted solely of males (it was suggested that in other parts of the range of *Meles* the bachelors may be nomadic). Individuals moved around alone within the clan range. Adult males always slept in the main setts, but females sometimes slept in the outliers, especially in the summer.

Mating is possible throughout the year but occurs mostly from late winter to midsummer. Pregnancy sometimes proceeds without interruption, with birth 2 months after mating. However, development of the fertilized eggs commonly stops at the blastocyst stage, and implantation in the uterus is delayed for about 10 months. The time of implantation seems to be controlled by conditions of light and temperature. Following implantation, embryonic development proceeds for 6–8 weeks, and births occur mainly from February to March. The total period of pregnancy may thus be about 9–12 months. Females may experience a postpartum estrus. The number of young in a litter is two to six, usually three or four. The young weigh 75 grams at birth, open their eyes after about 1 month, nurse for 2.5 months, and usually separate from the female in the autumn. Both males and females apparently attain sexual maturity at the age of 1 year (Ahnlund 1980; Canivenc and Bonnin 1979; Hancox 1992*b;* Kruska *in* Grzimek 1990; Novikov 1962). One specimen lived in captivity for 16 years and 2 months (Jones 1982).

Meles sometimes damages ripening grapes, corn, and oats. It also has been widely accused of killing small game. Its hair is used to make various kinds of brushes, and its skin has been used in northern China to make rugs. For these reasons and because of habitat disruption *Meles* had declined in much of its European range by the late nineteenth century, and the decline continues in some areas. However, a recent survey of the continent west of Russia, Belarus, and Ukraine (Griffiths and Thomas 1993) indicates that populations are stable or increasing in most countries. About half of the countries provide year-round legal protection, though many animals are killed illegally and by vehicles. The minimum number of badgers in the region surveyed was estimated at 1,220,000 and there was an annual legal kill of 118,000. Sweden, with 350,000–400,000, had the largest number of any country. Previous surveys in Great Britain (Clements, Neal, and Yalden 1988; Cresswell et al. 1989) had shown a general recovery from a low point before World War I, when the species was intensively persecuted by gamekeepers. The species now occupies most of the island and has been estimated to number about 250,000 in 43,000 social groups. About 50,000 are killed by vehicles annually, however, and there are concerns about the effects of future urbanization. Evidence of transmission of bovine tuberculosis to cattle from badgers led to efforts to reduce numbers of the latter in parts of Britain and Ireland during recent decades, but such programs are now thought to be scientifically and economically unjustified (Hancox 1992*a*).

Hog badger *(Arctonyx collaris)*, photo by Ernest P. Walker.

CARNIVORA; MUSTELIDAE; **Genus ARCTONYX**
F. Cuvier, 1825

Hog Badger

The single species, *A. collaris,* occurs from Sikkim and northeastern China to peninsular Thailand and on the island of Sumatra (Corbet 1978; Lekagul and McNeely 1977).

Head and body length is 550–700 mm, tail length is 120–70 mm, and weight is usually 7–14 kg. The back is yellowish, grayish, or blackish, and there is a pattern of white and black stripes on the head. The dark stripes run through the eyes and are bordered by white stripes that merge with the nape and with the white of the throat. The ears and tail also are white, and the feet and belly are black. The body form is stocky. This badger is distinguished from *Meles* by its white, rather than black, throat and by its long and mostly white tail, as distinct from the short tail, colored the same as the back, in *Meles.* Another external difference is that in *Arctonyx* the claws are pale in color, whereas in *Meles* they are dark. The common name "hog badger" refers to the long, truncate, mobile, and naked snout, which is often compared to that of a pig *(Sus).*

The general habits probably resemble those of *Meles.* As with *Meles* and *Mellivora,* the color pattern has been interpreted as a means of warning potential enemies that the animal so marked is best left alone. Like these other genera, *Arctonyx* is a savage and formidable antagonist. It has thick and loose skin, powerful jaws, fairly strong teeth, well-developed claws, and a potent anal gland secretion. The snout is thought to be used in rooting for the various plants and small animals that compose the diet.

According to Lekagul and McNeely (1977), the hog badger is usually found in forested areas at elevations of up to 3,500 meters. It is nocturnal, spending the day in natural shelters, such as rock crevices, or in self-excavated burrows. A study in Shaanxi Province, central China (Zheng et al. 1988), found *Arctonyx* to inhabit agricultural areas and mountain grassland as well as forests. There were two peaks in activity, 0300–0500 and 1900–2100, and earthworms were the favorite food. The genus hibernated from November to February or March and was normally solitary. Mating occurred in May and parturition in the following February or March, suggesting a period of delayed implantation as in *Meles.* Females with litters of three to five young were captured from March to July. The young were weaned at about 4 months and then became independent.

Parker (1979) reported that a captive pair from China was received at the Toronto Zoo in July 1976. The female gave birth to two cubs in February 1977; one cub survived and reached approximate adult size at 7.5 months. In February 1978 the same female gave birth to four young. Matings had been observed from April to September 1977, but delayed implantation was suspected, and true gestation was postulated at less than 6 weeks. According to Jones (1982), a captive lived 13 years and 11 months.

CARNIVORA; MUSTELIDAE; **Genus MYDAUS**
F. Cuvier, 1821

Stink Badgers

Long (1978) recognized two subgenera and two species:

subgenus *Mydaus* F. Cuvier, 1821

M. javanensis, Sumatra, Java, Borneo, Natuna Islands;

subgenus *Suillotaxus* Lawrence, 1939

M. marchei, Palawan and Calamian Islands (Philippines).

Suillotaxus often has been given generic rank.

In *M. javanensis* the head and body length is 375–510 mm, tail length is usually 50–75 mm, and weight is usually 1.4–3.6 kg. The coloration is blackish except for a white crown and a complete or partial narrow white stripe down

Stink badger *(Mydaus javanensis)*, photo from Michael Riffel.

the back onto the tail. In *M. marchei* the head and body length is 320–460 mm, tail length is 15–45 mm, and weight is about 2.5 kg. The upper parts are brown to black, with a scattering of white or silvery hairs on the back and sometimes on the head, and the underparts are brown.

Both species have a pointed face, a somewhat elongate and mobile snout, short and stout limbs, and well-developed anal scent glands. Compared with *M. javanensis, M. marchei* has smaller ears, a shorter tail, and larger teeth.

M. javanensis reportedly is a montane species, often being found at elevations over 2,100 meters (Long and Killingley 1983). It is nocturnal, residing by day in holes in the ground dug either by itself or by the porcupines with which it sometimes lives. The burrows usually are not more than 60 cm deep. On Borneo this species reportedly inhabits caves. Captives have consumed worms, insects, and the entrails of chickens.

M. marchei inhabits grassland-thicket and cultivated areas (Long and Killingley 1983). Grimwood (1976) wrote that it is active both by day and by night. It is common, leaving its tracks and scent along roads and paths. It moves with a rather ponderous, fussy walk. One individual shammed death when first touched and allowed itself to be carried but finally squirted a jet of yellowish fluid from its anal glands into the lens of a camera about 1 meter away.

M. javanensis may growl and attempt to bite when handled. If molested or threatened, it raises its tail and ejects a pale greenish fluid. This vile-smelling secretion is reported by natives sometimes to asphyxiate dogs or even to blind them if they are struck in the eye. The old Javanese sultans used this fluid, in suitable dilution, in the manufacture of perfumes.

Some natives eat the flesh of *Mydaus*, removing the scent glands immediately after the animals are killed. Others mix shavings of the skin with water and drink the mixture as a cure for fever or rheumatism. The IUCN classifies *M. marchei* as vulnerable because its restricted habitat is being lost to human encroachment.

CARNIVORA; MUSTELIDAE; **Genus TAXIDEA**
Waterhouse, 1839

American Badger

The single species, *T. taxus*, is found from northern Alberta and southern British Columbia to Ohio, central Mexico, and Baja California (E. R. Hall 1981).

Head and body length is 420–720 mm, tail length is 100–155 mm, and weight is 4–12 kg. The upper parts are grayish to reddish, and a dorsal white stripe extends rearward from the nose. In the north this stripe usually reaches only to the neck or shoulders, but in the south it usually extends to the rump (Long 1972). Black patches are present on the face and cheeks; the chin, throat, and midventral region are whitish; the underparts are buffy; and the feet are dark brown to black. The hairs are longest on the sides. *Taxidea* can be recognized by its flattened and stocky form, large foreclaws, distinctive black and white head pattern, long fur, and short, bushy tail. There are anal scent glands. Females have eight mammae (Jackson 1961).

The American badger is usually found in relatively dry, open country. It is a remarkable burrower and can quickly dig itself out of sight. The usual signs of its presence are the large holes that it digs in pursuit of rodents. For shelter it either excavates a burrow or modifies one initially made by another animal. The burrow can be as long as 10 meters and can extend as far as 3 meters below the surface. A bulky nest of grass is located in an enlarged chamber, and the entrances are marked by mounds of earth (Banfield 1974).

Taxidea may be active at any hour but is mainly nocturnal. Its movements are restricted, especially in winter, and it shows a strong attachment to a home area. In the summer, however, the young animals disperse over a considerable distance; one traveled 110 km. The badger is active all year, but it may sleep in its den for several days or weeks during severe winter weather. One female was found to have emerged only once during a 72-day period (Messick and Hornocker 1981).

American badger *(Taxidea taxus)*, photo by E. P. Haddon from U.S. Fish and Wildlife Service.

Most food is obtained by excavating the burrows of fossorial rodents, such as ground squirrels. Also eaten are other small mammals, birds, reptiles, and arthropods (Banfield 1974; Messick and Hornocker 1981). Food is sometimes buried and used later. If a sizable meal, such as a rabbit, is obtained, the badger may dig a hole, carry in the prey, and remain below ground with it for several days. There are reports that a badger sometimes forms a "hunting partnership" with a coyote (see account of *Canis latrans*).

On the basis of a radio-tracking study in southwestern Idaho, Messick and Hornocker (1981) estimated a population density of up to 5/sq km and found average home ranges of 2.4 sq km for males and 1.6 sq km for females. The ranges overlapped, but individuals were solitary except during the reproductive season. A female radio-tracked in Minnesota used an area of 752 ha. during the summer. She had 50 dens within this area and was never found in the same den on two consecutive days. In the autumn she shifted to an adjacent area of 52 ha. and often reused dens; in the winter she used a single den and traveled only infrequently within an area of 2 ha. (Sargeant and Warner 1972).

Mating occurs in summer and early autumn, but implantation of the fertilized eggs in the uterus is delayed until December–February and births take place in March and early April (Long 1973). The total period of pregnancy is thus about 7 months, but actual embryonic development lasts only about 6 weeks (Ewer 1973). The litter of one to five young, usually two, is born and raised in a nest of dry grass within a burrow. Newborn weight is about 94 grams (Hayssen, Van Tienhoven, and Van Tienhoven 1993). The young are weaned at about 6 weeks and disperse shortly thereafter. The study by Messick and Hornocker (1981) indicated that 30 percent of the young females mate in the first breeding season following birth, when they are about 4 months old, but that males wait until the following year. The oldest wild badger caught in this study had attained an age of 14 years. A captive badger lived 26 years (Jones 1982).

Taxidea is generally beneficial to human interests as it destroys many rodents and its burrows provide shelter for many kinds of wildlife, including cottontail rabbits. Because the badger's burrows and holes may constitute a hazard to cattle, horses, and riders, ranchers have often killed badgers. Many badgers also have been killed by poison put out for coyotes (Hall 1955). Although *Taxidea* has declined in numbers in some areas, it has extended its range eastward in the twentieth century. It now occupies most of Ohio and is invading southeastern Ontario (E. R. Hall 1981; Long 1978). There also are recent records from New York, New England, and southern Yukon, but it is not known whether these represent natural occurrences (Messick 1987).

For many years badger fur was used to make shaving brushes, but it has now been largely replaced by synthetic materials. The fur is still used for trimming garments. In the 1976–77 trapping season 49,807 pelts were reported taken in the United States and Canada, selling at an average price of about $38 (Deems and Pursley 1978). By 1983–84 the take had fallen to only about 20,000 and the average price to $10.00 (Messick 1987). In the 1991–92 season 10,825 skins were taken in the United States and sold for an average price of $6.45 (Linscombe 1994).

CARNIVORA; MUSTELIDAE; **Genus MELOGALE**
I. Geoffroy St.-Hilaire, 1831

Ferret Badgers

There are four species (Corbet and Hill 1992; Long 1978; Riffel 1991; Zheng and Xu 1983):

M. moschata, Assam to central China and northern Indochina, Taiwan, Hainan;
M. personata, Nepal to Indochina;
M. orientalis, Java, Bali;
M. everetti, Borneo.

Long (1978) noted that *M. moschata sorella*, of southeastern China, may be a full species.

Ferret badger *(Melogale moschata)*, photo by Gwilym S. Jones.

Head and body length is 330–430 mm, tail length is 145–230 mm, and weight is 1–3 kg. The general coloration of the upper parts is gray brown to brown black; the underparts are somewhat paler. A white or reddish dorsal stripe is usually present. *Melogale* is distinguished by the striking coloration of the head, which combines black with patches of white or yellow. The tail is bushy, the limbs are short, and the feet are broad and have long, strong claws for digging (Lekagul and McNeely 1977).

Ferret badgers are found in wooded country and grassland. They reside in burrows and natural shelters during the day and are active at dusk and during the night. They climb on occasion. *M. moschata* on Taiwan is reported to be a good climber and often to sleep on the branches of trees. Ferret badgers are savage and fearless when provoked or pressed and have an offensive odor. The conspicuous markings on the head have been interpreted as a warning signal. The omnivorous diet is known to include small vertebrates, insects, earthworms, and fruit. A ferret badger is sometimes welcome to enter a native hut because of its destruction of insect pests.

The young, usually one to three per litter, are born in a burrow in May and June. They apparently are dependent on the milk of the mother for some time, as two nearly full-grown suckling animals and their mother were once found in a burrow. According to Marvin L. Jones (Zoological Society of San Diego, pers. comm., 1995), a specimen of *M. moschata* lived for more than 17 years in captivity.

The species *M. everetti* is classified as vulnerable by the IUCN because its restricted habitat is being lost to human encroachment. *M. orientalis* (treated by the IUCN as a subspecies of *M. personata*) is designated as near threatened.

CARNIVORA; MUSTELIDAE; **Genus SPILOGALE**
Gray, 1865

Spotted Skunks

There are three species (Dragoo et al. 1993; E. R. Hall 1981):

S. pygmaea, Pacific coast of Mexico from southern Sinaloa to Oaxaca;
S. gracilis, southern British Columbia and central Wyoming to Costa Rica and central Texas, Baja California, Channel Islands;
S. putorius, eastern Wyoming, Minnesota, and southern Pennsylvania to northeastern Mexico and Florida.

E. R. Hall (1981) and Wozencraft (*in* Wilson and Reeder 1993) included *S. gracilis* in *S. putorius,* but the two were regarded as distinct species by Dragoo et al. (1993) and Jones et al. (1992). The association of these two taxa is unusual. They are entirely allopatric, their ranges meeting all along a line through Wyoming, Colorado, Texas, and northeastern Mexico. On a morphological basis they do not appear to be more than subspecifically separable, but molecular analysis indicates a greater distinction. These data support earlier investigations showing a profound difference between the breeding patterns and seasons of the two species, effectively making them reproductively isolated from one another (see below).

Head and body length is 115–345 mm, tail length is 70–220 mm, and weight is usually 200–1,000 grams. Of the three genera of skunks *Spilogale* has the finest fur. The hairs are longest on the tail and shortest on the face. The basic color pattern consists of six white stripes extending along the back and sides; these are broken into smaller stripes and spots on the rump. There is a triangular patch in the middle of the forehead, and the tail is usually tipped with white. The variations are infinite as no two individuals have been found to have exactly the same pattern. This genus may be distinguished from the other two genera by its small size, forehead patch, and pattern of stripes and spots, the white never being massed. There is a pair of scent glands under the base of the tail, from which a jet of strong-

Spotted skunk *(Spilogale putorius)*, photo by Ernest P. Walker.

smelling fluid can be emitted through the anus. Females have 10 mammae (Lowery 1974).

Spotted skunks occur in a variety of brushy, rocky, and wooded habitats but avoid dense forests and wetlands. They generally remain under cover more than striped skunks *(Mephitis)*. They usually den underground but can climb well and sometimes shelter in trees. The dens are lined with dry vegetation. Spotted skunks are largely nocturnal and are active all year. The omnivorous diet consists mainly of vegetation and insects in the summer and of rodents and other small animals in the winter.

The white-plumed tail is used to warn other animals that spotted skunks should not be molested. If, however, the sudden erection of the tail is not sufficient deterrence, *Spilogale* may stand on its forefeet and sometimes even advance toward its adversary. Finally, the fluid from the anal glands is discharged at the enemy, usually after the skunk has returned its forefeet to the ground and assumed a horseshoe position (Lowery 1974).

According to Banfield (1974), population densities reach 5/sq km in good agricultural land and winter home range is approximately 64 ha. In spring the males wander over an area of about 5–10 sq km, but females have smaller ranges. Spotted skunks are very playful with one another. As many as eight individuals sometimes share a den.

The reproductive pattern is not the same in all parts of the range (Foresman and Mead 1973; Mead 1968*a*, 1968*b*). Populations in South Dakota and Florida *(S. putorius)* apparently mate mainly in March and April. Implantation of the fertilized eggs in the uterus occurs only 14–16 days later, and births take place in late May and June. Pregnancy is estimated to last 50–65 days. Populations farther to the west *(S. gracilis)* mate in September and October, but implantation is delayed until the following March or April, with births from April to June. The total period of pregnancy thus lasts 230–50 days, but actual embryonic development takes only 28–31 days. The number of young per litter is two to nine, usually three to six. The young weigh about 22.5 grams at birth, have adult coloration after 21 days, open their eyes at 32 days, can spray musk at 46 days, are weaned at about 54 days, and attain adult size at about 15 weeks. A captive specimen lived 9 years and 10 months (Egoscue, Bittmenn, and Petrovich 1970).

Spotted skunks have been reported to carry rabies and occasionally to take poultry and eggs but generally benefit people through their destruction of rodents and insects. The pelts are very attractive and durable, but they generally sold for well under $1.00 each until about 1970 (Jackson 1961; Lowery 1974). In the 1976–77 trapping season the reported harvest in the United States was 41,952 skins, with an average selling price of $4.00 (Deems and Pursley 1978). The subspecies *S. putorius interrupta*, which apparently expanded into Minnesota and Wisconsin only in the early twentieth century in response to favorable agricultural development, now is in decline throughout the northern part of its range as its habitat undergoes further human modification (Boppel and Long 1994). The species *S. pygmaea* occupies a restricted area of generally deteriorating habitat and apparently has become rare (Schreiber et al. 1989). The subspecies *S. gracilis amphiala*, found only on

Santa Cruz and Santa Rosa in the Channel Islands off southern California, also appears to have become very rare (Crooks 1994).

CARNIVORA; MUSTELIDAE; **Genus MEPHITIS**
E. Geoffroy St.-Hilaire and G. Cuvier, 1795

Striped Skunk and Hooded Skunk

There are two species (E. R. Hall 1981; Janzen and Hallwachs 1982):

M. mephitis (striped skunk), southern Canada to northern Mexico;
M. macroura (hooded skunk), Arizona and southwestern Texas to Costa Rica.

Head and body length is 280–380 mm, tail length is 185–435 mm, and weight is 700–2,500 grams. Both species have black and white color patterns, but with considerable variation. *M. mephitis* usually has white on top of the head and on the nape extended posteriorly and separated into stripes. In some individuals of this species the top and sides of the tail are white, whereas in others the white is limited to a small spot on the forehead. The white areas are composed entirely of white hairs, with no black hairs intermixed. *M. macroura* has a white-backed color phase and a black-headed color phase. In the former there are some black hairs mixed with the white hairs of the back; in the latter the two white stripes are widely separated and are situated on the sides of the animal instead of being narrowly separated and situated on the back as in *M. mephitis.* Female *Mephitis* have 10–14 mammae (Jackson 1961; Leopold 1959).

According to Lowery (1974), the well-known scent of *Mephitis* is expelled from two tiny nipples located just inside the anus, which mark the outlets of the two ducts leading from glands lying adjacent to the anus. This musk is discharged either as an atomized spray or as a short stream of rain-sized drops. The skunk usually employs this weapon only after much provocation. When confronted by an antagonist, it arches its back, elevates its tail, erects the hairs thereon, and sometimes stamps its feet on the ground. Finally, it makes its body into a U, with the head and tail facing the intruder. The musk usually travels 2–3 meters, but the smell can be detected up to 2.5 km downwind. Lowery observed that one squirt is sufficient to send the most ferocious dog yelping in agony from burning eyes and nostrils and retching with nausea.

These skunks are found in a variety of habitats, includ-

Striped skunk *(Mephitis mephitis):* A. Facing danger that does not appear to be imminent. B. Aimed toward the enemy ready to spray its scent. Photos by Ernest P. Walker.

ing woods, grasslands, and deserts. They generally are active at dusk and through the night and spend the day in a burrow, under a building, or in any dry, sheltered spot. In Minnesota, Houseknecht and Tester (1978) found a general shift from underground, upland dens in winter to aboveground, lowland dens in summer. Although skunks tended to remain at a single den for a long time in winter, females with young able to travel changed dens every one or two days. Bjorge, Gunson, and Samuel (1981) reported that females moved a minimum daily distance of 1.5 km between dens. Juveniles were found to disperse up to 22 km. in summer.

In northern parts of its range *M. mephitis* stays in one den and sleeps through much of the winter. Males tend to sleep for shorter periods than females and to become active more readily during intervals of mild weather (Banfield 1974). The degree of lethargy achieved during the winter is not well understood. It does not appear to be deep torpor, but some skunks are known to have remained underground for more than 100 consecutive days (Sunquist 1974). In Alberta the overall period of female hibernation is 120–50 days, whereas in Illinois it is 62–87 days (Gunson and Bjorge 1979). A striped skunk may become very fat in the autumn before hibernation. The diet is omnivorous and includes rodents, other small vertebrates, insects, fruit, grains, and green vegetation.

Density estimates for striped skunk populations have ranged from 0.7/sq km to 18.5/sq km, but most are 1.8–4.8/sq km (Wade-Smith and Verts 1982). The home ranges of 6 females radio-tracked for 45–105 days each in Alberta (Bjorge, Gunson, and Samuel 1981) averaged 208 ha. and varied from 110 to 370 ha. Males wandered over larger areas, most notably in the autumn. *Mephitis* is generally solitary, but there is a tendency for individuals to den together, especially in the north, as a means of optimizing winter survival and reproductive success. In Alberta, Gunson and Bjorge (1979) found that only males, both adults and juveniles, denned alone during the winter. Communal winter dens contained an average of 6.7 (2–19) individuals. Usually there was only a single adult male per den and an average of 5.8 females. A male apparently wanders in search of a group of females during the autumn and then keeps other males away. *Mephitis* is usually silent but makes several sounds, such as low churrings, shrill screeches, and birdlike twitters (Lowery 1974).

Mating takes place from mid-February to mid-April, with births in May and early June. The period of pregnancy is 59–77 days, and delayed implantation may be involved. Females are usually monestrous but sometimes have a second estrus and parturition subsequent to the normal period if their first pregnancy is not successful. Litters contain 1–10 young, usually about 4–5. The young weigh about 30 grams at birth, open their eyes after 3 weeks, are weaned at 8–10 weeks, and separate from the mother by the autumn. Females may bear their first litter at 1 year (Lowery 1974; Wade-Smith and Richmond 1975, 1978). The average longevity in captivity is about 6 years, but one individual was still living after 12 years and 11 months (Jones 1982).

Striped skunks are generally beneficial to human interests because of their destruction of rodents and insects. *Mephitis,* however, sometimes attacks poultry and is reportedly the principal carrier of rabies among North American wildlife (Wade-Smith and Richmond 1975). The fur is durable and of good texture, but demand and value have varied widely (Lowery 1974; Jackson 1961). During the 1976–77 season 175,884 skins were reported taken in the United States and Canada, selling at an average price of about $2.25 (Deems and Pursley 1978). In 1983–84 the take was about the same, but the price fell to less than $1.00 (Rosatte 1987). In the 1991–92 season 45,148 skins were taken in the United States and sold at an average price of about $2.00 (Linscombe 1994).

CARNIVORA; MUSTELIDAE; **Genus CONEPATUS**
Gray, 1837

Hog-nosed Skunks

Five species are now recognized (Cabrera 1957; Ewer 1973; E. R. Hall 1981; Kipp 1965; Manning, Jones, and Hollander 1986; Pine, Miller, and Schamberger 1979; Redford and Eisenberg 1992):

C. mesoleucus, southern Colorado and eastern Texas to Nicaragua;
C. leuconotus, southern Texas, eastern Mexico;
C. semistriatus, southern Mexico to northern Peru and eastern Brazil;
C. chinga, central and southern Peru, Bolivia, southern Brazil, Chile, northern and western Argentina, Uruguay;
C. humboldti, southern Chile and Argentina.

Statements by the various authorities cited above suggest that some, perhaps all, of the listed species are conspecific.

Head and body length is 300–490 mm, tail length is 160–410 mm, and weight is usually 2.3–4.5 kg. *Conepatus* has the coarsest fur of all skunks. There are two main color patterns, with variations. In one the top of the head, the back, and the tail are white, and the remainder of the animal is black. This coloration occurs most commonly in areas where the ranges of *Conepatus* and *Mephitis* overlap. In the other pattern the pelage is black except for two white stripes, beginning at the nape and extending on the hips, and a mostly white tail. This coloration resembles that of *Mephitis mephitis* and seems to be most common in areas where *Conepatus* is the only kind of skunk present. In all cases hog-nosed skunks lack the thin white stripe down the center of the face that is present in *Mephitis. Conepatus* may be distinguished from the other two genera of skunks by its nose, which is bare, broad, and projecting. Females have three pairs of mammae (Leopold 1959).

Hog-nosed skunks are found in both open and wooded areas but avoid dense forests. They occur at all elevations up to at least 4,100 meters (Grimwood 1969). Dens are located in rocky places, hollow logs, or burrows made by other animals. Like other skunks, *Conepatus* is mainly nocturnal, is generally slow-moving, does not ordinarily climb, and defends itself by expelling musk from anal scent glands.

The diet may consist principally of insects and other invertebrates, though fruit and small vertebrates, including snakes, probably are also eaten. Hog-nosed skunks may turn over the soil in a considerable area with their bare snout and their claws when in search of food. Like *Mephitis,* they also pounce on insects. At least in the Andes hog-nosed skunks are resistant to the venom of pit vipers. There is some evidence that the spotted skunks *(Spilogale)* also are resistant to rattlesnake venom. Since the musk of skunks produces an alarm reaction in rattlers (the same reaction that they exhibit in the presence of king snakes, which prey on them), it may be that skunks feed on rattlesnakes quite extensively.

In southern Chile Fuller et al. (1987) found *C. humboldti* to be solitary but to occupy overlapping home ranges of

Hog-nosed skunk *(Conepatus mesoleucus)*, photo by Lloyd G. Ingles.

7–16 ha. According to Davis (1966), *C. mesoleucus* is not as social as *Mephitis*, and usually only one individual lives in a den. The breeding season in Texas begins in February, most mature females are pregnant by March, and births occur in late April or early May. Gestation lasts approximately 2 months. Of six pregnant females on record three contained three embryos each and three contained two each. By August most of the young are weaned and foraging for themselves. Available evidence indicates that in Mexico the young are also born in the spring (Hall and Dalquest 1963; Leopold 1959). The gestation period of one South American species is 42 days, and litter size is usually two to five young. Sexual maturity in *C. mesoleucus* has been reported to come at 10–11 months (Hayssen, Van Tienhoven, and Van Tienhoven 1993). A captive of that species lived 8 years and 8 months (Marvin L. Jones, Zoological Society of San Diego, pers. comm., 1995).

The pelt of *Conepatus* is inferior in quality to that of *Mephitis*, but large numbers have been marketed from Texas (Davis 1966). Perhaps because of this commerce, the subspecies *C. mesoleucus telmalestes*, isolated in the Big Thicket area of southeastern Texas, has declined to near the point of extinction (Schmidly 1983). It now is classified as extinct by the IUCN. *C. chinga rex* of northern Chile also is hunted for its pelt and now seems to be rare (Miller et al. 1983). Some natives use the skins for capes or blankets, and others consider the meat to have curative properties. *C. humboldti* is on appendix 2 of the CITES. Records of the CITES indicate that during the 1970s about 155,000 skins of *Conepatus* were exported annually, each with a value of about U.S. $8.00; the trade may subsequently have declined (Broad, Luxmoore, and Jenkins 1988).

CARNIVORA; MUSTELIDAE; **Genus LUTRA**
Brünnich, 1771

Old World River Otters

There are two subgenera and three species (Chasen 1940; Coetzee *in* Meester and Setzer 1977; Corbet 1978; Corbet and Hill 1992; Ellerman and Morrison-Scott 1966; Lekagul and McNeely 1977; Rosevear 1974; Van Zyll de Jong 1972, 1987):

subgenus *Lutra* Brünnich, 1771

L. lutra, western Europe to northeastern Siberia and Korea, Asia Minor and certain other parts of southwestern Asia, Himalayan region, southern India, China, Burma, Thailand, Indochina, northwestern Africa, British Isles, Sri Lanka, Sakhalin, Japan, Taiwan, Hainan, Sumatra, doubtfully Java;

L. sumatrana (hairy-nosed otter), Indochina, Thailand, Malay Peninsula, Sumatra, Bangka, Java, Borneo;

subgenus *Hydrictis* Pocock, 1921

L. maculicollis (spotted-necked otter), Sierra Leone to Ethiopia, and south to South Africa.

The New World river otters (genus *Lontra*) and the smooth-coated otter (genus *Lutrogale*) of southern Asia (see accounts thereof) often are considered only subgenerically distinct from *Lutra*, but Van Zyll de Jong (1972, 1987) considered all three to be full genera. However, he did not agree with Rosevear (1974), who stated that there is a strong case for regarding *Hydrictis* as a full genus. With respect to the possibly extinct otter populations of Japan, Imaizumi and Yoshiyuki (1989) restricted *L. lutra* to Hok-

European river otter *(Lutra lutra)*, photo by Annelise Jensen.

kaido and described a separate species, *L. nippon*, from Shikoku, Honshu, and Kyushu.

Head and body length is 500–820 mm, tail length is 330–500 mm, and weight is 5–14 kg. Males average larger than females (Van Zyll de Jong 1972). The upper parts are brownish and the underparts are paler; the lower jaw and throat may be whitish. The fur is short and dense. The head is flattened and rounded; the neck is short and about as wide as the skull; the trunk is cylindrical; the tail is thick at the base, muscular, flexible, and tapering; the legs are short; and the digits are webbed. The small ears and the nostrils can be closed when the animal is in the water.

These aquatic mammals inhabit all types of inland waterways as well as estuaries and marine coves. Otters are excellent swimmers and divers and usually are found no more than a few hundred meters from water. They may shelter temporarily in shallow burrows or in piles of rocks or driftwood, but they also have at least one permanent burrow beside the water (Stroganov 1969). The main entrance may open underwater and then slope upward into the bank to a nest chamber that is above the high-water level. Erlinge (1967) found *L. lutra* to utilize the following types of facilities in southern Sweden: "dens," generally with several passages and a chamber lined with dry leaves and grass; "rolling places," bare spots near water where the otters roll and groom themselves; "slides," either on slopes or in level places but most common on winter snow; "feeding places," including holes kept open through winter ice; "runways," well-defined paths on land that connect waterways and other facilities; and "sprainting spots and sign heaps," prominent points of land where the animals mark by scratching and elimination.

Otters swim by movements of the hind legs and tail and usually dive for one or two minutes, five at the most (Kruska *in* Grzimek 1990). When traveling on ground, snow, or ice they may use a combination of running and sliding. Although normally closely associated with water, river otters sometimes move many kilometers overland to reach different river basins and to find ice-free water in winter (Stroganov 1969). They may be either diurnal or nocturnal but are generally more active at night. With the possible exception of the Old World badger *(Meles)*, river otters are the most playful of the Mustelidae. Some species engage in the year-round activity of sliding down mud and snow banks, and individuals of all ages participate. Sometimes they tunnel under snow to emerge some distance beyond. The diet consists largely of fish, frogs, crayfish, crabs, and other aquatic invertebrates. Birds and land mammals, such as rodents and rabbits, are also taken. Studies indicate that the fish consumed are mainly nongame species. River otters capture their prey with the mouth, not the hands (Rowe-Rowe 1977).

Investigations by Erlinge (1967, 1968) in southern Sweden indicate that *L. lutra* occurs at population densities of 0.7–1.0/sq km of water area, or 1 for every 2–3 km of lakeshore or 5 km of stream. The straight-line length of a home range, including land area, was found to average about 15 km for adult males and 7 km for females with young. The ranges of males constitute territories, which may overlap the ranges of 1 or more adult females and from which other males are excluded. Females also defend their ranges against individuals of the same sex. Territories are marked with scent, and fights occasionally take place. Apparently, males form a dominance hierarchy, with the highest-ranking animals occupying the most favorable ranges. Erlinge noted that the males generally are solitary and ignore the females and young. Studies in Scotland (Mason and Macdonald 1986) suggest a somewhat less rigid social system, with females occupying overlapping home ranges, though again the dominant males have relatively exclusive territories. Still another variation evidently exists on Shetland Island, where up to five reproductively active females were found to share a group range of about 5–14 km of coastline; male ranges were larger and overlapped those of two or more female groups (Kruuk and Moorhouse 1991). On Lake Victoria *L. maculicollis* may undergo a regular cycle of aggregation and dispersal, with males and females each forming their own groups. The males' groups grow larger after the mating season, when males are not tolerated by the females with young. These groups may contain 8–20 individuals from January to May but then become smaller from June to August, when the older males leave to pair with the females (Kingdon 1977).

Within the wide geographic range of *Lutra* there is considerable variation in reproductive pattern (Duplaix-Hall 1975; Ewer 1973; Kingdon 1977; Liers 1966; Mason and Macdonald 1986; Stroganov 1969). Female *L. lutra* are polyestrous, with the cycle lasting about 4–6 weeks and estrus about 2 weeks. In some areas (e.g., most of England) mating and birth may occur at any time of the year. In areas with a more severe climate (e.g., Sweden and Siberia) mating takes place in late winter or early spring, with births in April or May. The gestation period of this species is 60–63 days. The length of pregnancy is about the same in *L. maculicollis*. Litter size in the genus is one to five young, usually two or three. The young weigh about 130 grams at birth, open their eyes after 1 month, emerge from the den and begin to swim at 2 months, nurse for 3–4 months, separate from the mother at about 1 year, and attain sexual maturity in the second or third year of life. Captives have lived up to 22 years (Kruska *in* Grzimek 1990).

River otters have been intensively hunted for their excellent fur and also have proved highly vulnerable to human environmental disruption and pollution of their aquatic habitat. The major problem in Europe is thought to be contamination of fish prey by bioaccumulating organochlorine pollutants, both pesticides and polychlorinated biphenyls (Macdonald and Mason 1994). The Eurasian *L. lutra* has declined drastically in such diverse places as Great Britain (Chanin and Jefferies 1978), Germany (Roben 1974), southeastern Siberia (Kucherenko 1976), and Japan (Mikuriya 1976). Since the 1980s there have been signs of recovery in England following a crash caused by hunting and insecticide pollution, and populations in Scotland and Ireland are now widespread (Foster-Turley, Macdonald, and Mason 1990; Jefferies 1989). There also is a thriving population in Latvia, Lithuania, and Belarus (Baranauskas et al. 1994). Extensive recent surveys of *L. lutra* in the Mediterranean region and southern Europe have found the species to be still relatively common in parts of Spain, Portugal, Greece, various other parts of the Balkans, the Jordan River Valley, and North Africa but rare or absent in Italy, Austria, Switzerland, and most of France (Macdonald and Mason 1994; Mason and Macdonald 1986). It has nearly or completely disappeared in Belgium, the Netherlands, Switzerland, and most of Germany and Poland. It is still common in Finland and northern Norway but rare in the rest of Scandinavia. It is rare in all parts of Asia where its status is known, has not been sighted in Japan since 1986, and may also have been extirpated in Thailand (Foster-Turley, Macdonald, and Mason 1990; Macdonald and Mason 1994).

L. lutra is on appendix 1 of the CITES, and the other species of *Lutra* are on appendix 2. *L. maculicollis* has become very rare in most of West Africa because of destruction of riparian forests and contamination of waterways (Foster-Turley, Macdonald, and Mason 1990) and is considered endangered in South Africa (Stuart 1985). The IUCN classifies *L. sumatrana* as vulnerable; it also is subject to habitat disruption and is still regularly hunted for its pelt. It appears to be rare throughout its range and may have disappeared completely from Thailand and mainland Malaysia (Foster-Turley, Macdonald, and Mason 1990). Neither it nor *L. lutra* was found during recent surveys on Java (Melisch, Asmoro, and Kusumawardhani 1994).

CARNIVORA; MUSTELIDAE; **Genus LUTROGALE**
Gray, 1865

Smooth-coated Otter

The single species, *L. perspicillata*, is found in southern Iraq, from Pakistan to Indochina and the Malay Peninsula, and on the islands of Sumatra, Java, and Borneo. *Lutrogale* often has been considered a subgenus or synonym of *Lutra* but was treated as a distinct genus by Corbet and Hill (1992), Van Zyll de Jong (1972, 1987), and Wozencraft (*in* Wilson and Reeder 1993).

Head and body length is 650–790 mm, tail length is 400–505 mm, and weight is 7–11 kg. The upper parts are raw umber to smoky gray-brown; the underparts are a lighter drab color. The cheeks, upper lip, throat, neck, and upper chest are whitish. The coat is short and very smooth and sleek rather than coarse, and the large footpads are smooth rather than granular. The general external form is similar to that of *Lutra* except that the dorsoventral flattening of the tail is much more pronounced and there are distinct lateral keels distally. Compared with that of *Lutra*, the skull of *Lutrogale* is less depressed, its muzzle shorter with larger orbits set lower and farther forward, its braincase deeper and more inflated, and the teeth larger and flatter (Foster-Turley, Macdonald, and Mason 1990; Harrison and Bates 1991; Lekagul and McNeely 1977).

According to Lekagul and McNeely (1977), *Lutrogale* is a plains otter, inhabiting mostly areas of low elevation. It is found in lakes, streams, reservoirs, canals, and flooded fields and will enter the open sea. It is quite active on land, often traveling long distances in search of suitable streams, and during the dry season may become a jungle hunter. It is a capable burrower and may dig its own breeding den. Family groups may combine when fishing, swimming in a semicircle and driving fish before them. Villagers in India have taken advantage of this habit by using *Lutrogale* to drive fish into nets. On the Malay Peninsula this genus typically occurs in groups consisting of a mated adult pair and as many as four young, which have a territory of 7–12 km of river. Breeding may occur in the early part of the year, and delayed implantation may be a factor. Both sexes bring nesting material into the den and carry food to the cubs. The true gestation period is 63 days, and a captive female first mated at 3 years of age. Chakrabarti (1993) reported that a captive male lived to an age of about 20 years.

The smooth-coated otter is classified as vulnerable by the IUCN (under the genus *Lutra*) and is on appendix 2 of the CITES. It still is generally common but is declining because of destruction of riparian habitat, deforestation, water pollution by industrial wastes and agricultural pesticides, damming and impoundment of streams, and killing by people who seek its pelt or consider it a threat to fisheries. The population on Java may already be extinct (Foster-Turley, Macdonald, and Mason 1990). At the opposite end of the range is the isolated subspecies *L. p. maxwelli*, known only from a few sites in the southern marshes of Iraq (Harrison and Bates 1991). A captive individual of this subspecies was featured in Gavin Maxwell's book *Ring of Bright Water*, published in 1960. Unfortunately, no information on its status has since become available. Its restricted habitat has been the scene of much recent military activity and also probably is being severely affected by current projects to drain the marshes (*Oryx* 28 [1944]:8), reportedly as a means of extending government control over the indigenous human population.

CARNIVORA; MUSTELIDAE; **Genus LONTRA**
Gray, 1843

New World River Otters

There are four species (E. R. Hall 1981; Redford and Eisenberg 1992; Van Zyll de Jong 1972, 1987):

L. canadensis, Alaska, Canada, conterminous United States;
L. longicaudis (neotropical river otter), northwestern Mexico to Uruguay and Buenos Aires Province of Argentina;
L. provocax, central and southern Chile, southern Argentina, Tierra del Fuego;
L. felina (marine otter), Pacific coast from northern Peru to Tierra del Fuego.

Lontra was considered a subgenus or synonym of *Lutra* by E. R. Hall (1981) and Jones et al. (1992) but was treated as a distinct genus by Van Zyll de Jong (1972, 1987) and Wozencraft (*in* Wilson and Reeder 1993). The basis on which various authors rejected generic status for *Lontra* was questioned by Kellnhauser (1983).

Head and body length is 460–820 mm, tail length is 300–570 mm, and weight is 3–15 kg. Males average larger than females (Van Zyll de Jong 1972). The upper parts are various shades of brown, the underparts are light brown or grayish, and the muzzle and throat may be whitish or silvery gray. The fur is short and sleek, with dense underfur overlaid by glossy guard hairs. The general morphology is much like that of *Lutra,* but in the skull the posterior palatine foramina are located more posteriorly, the vomer-ethmoid partition of the nasal cavity extends posteriorly to or behind the first upper molar tooth (in *Lutra* it extends only to between the third and fourth premolars), the first upper molar has a prominent cingulum and expanded talon (in *Lutra* the cingulum is little developed and the talon is small), and the sectorial fourth upper premolar has a talon extending more than two-thirds the length of the tooth (in *Lutra* the talon extends less than two-thirds the length). In both genera the toes are webbed to the terminal digit pads or beyond, the proximal part of the tail is broad and moderately dorsoventrally flattened, and females have four mammae, except in *Lontra provocax,* which has more than four (Van Zyll de Jong 1987).

The natural history of New World river otters, includ-

Canadian river otters *(Lontra canadensis)*, photos by Ernest P. Walker.

ing their aquatic habits and playful behavior, is much like that of *Lutra* (see account thereof). However, there are some particularly different aspects of the ecology of *L. felina* (Estes 1986; Redford and Eisenberg 1992). That species is found largely or exclusively along the exposed seashore, though it may enter freshwater estuaries and large rivers. It stays within about 500 meters of the coast, mainly in areas characterized by rocky outcroppings, heavy seas, and strong winds. It shelters in caves that open at water level, is active mostly in the afternoon, and makes food dives of 15–45 seconds. It feeds mostly on crustaceans, mollusks, and fish. *L. provocax* also sometimes is found along rocky coastlines, whereas *L. longicaudis* depends on permanent streams or lakes with ample riparian vegetation, and shelters in a self-excavated burrow.

According to Melquist and Dronkert (1987), *L. canadensis* is found in both marine and freshwater environments and from coastal areas to high mountains. Densities appear highest in food-rich habitats such as estuaries, lower stream drainages, coastal marshes, and interconnected small lakes and swamps. Favored locations include those with riparian vegetation or rocks that can be used for dens. An individual may use numerous dens and temporary shelters in the course of a year. Beaver lodges are frequently occupied, sometimes simultaneously with the builder. Otters are mainly nocturnal and crepuscular, but daytime activity is not unusual. The diet consists primarily of fish and also includes crustaceans, reptiles, amphibians, and occasionally birds and mammals. Most reported population densities have been about 1/1–10 km of shoreline or waterway length, and home range lengths have been 4–78 km.

In Idaho, Melquist and Hornocker (1983) found *L. canadensis* to have an overall population density of 1/3.9 km of waterway. Seasonal home range length was 8–78 km, with males generally having larger ranges than females; however, there was extensive overlap between the ranges of both the same and opposite sexes. There was some mutual avoidance and defense of personal space but no strong territorial behavior. The basic social group consisted of an adult female and her juvenile offspring. Such families broke up before the female again gave birth, though yearlings occasionally associated with the group. Fully adult males were not observed to accompany family groups. Observations in other areas, however, suggest that although the male is excluded from the vicinity of the female when the latter has small young, he joins the family when the cubs are about 6 months old (Banfield 1974; Jackson 1961). *Lontra* has a variety of vocalizations, and like *Lutra*, it communicates through scent marking with urine, feces, and anal gland secretions (Melquist and Dronkert 1987).

In the southern part of the range of *L. felina* there seems to be a birth peak in September and October, the gestation period is 60–65 days, and litter size is two young; they are born in an earthen den or rocky crevice and stay with the female for approximately 10 months (Redford and Eisenberg 1992). Little is known about reproduction in *L. longicaudis*, but recently a captive pair in southern Brazil was observed to produce litters on 1 April 1992, 21 July 1992, and 14 February 1993; the young of the first two litters died shortly after parturition (Blacher 1994). In most populations of *L. canadensis*, of North America, there is delayed implantation of the fertilized eggs in the uterus. Mating occurs in the winter or spring, and births take place the following year, usually from January to May. The total period of pregnancy has been reported to vary from 290 to 380 days, though actual embryonic development is about 60–63 days, the same as in other kinds of river otters. Populations in southern Florida may not experience delayed implantation. The female does not excavate her own den but uses that of another animal or some natural shelter. The male does not assist in rearing the young. Litter size is one to six cubs, usually two or three. They weigh about 120–60 grams at birth, emerge from the den at about 2 months, are weaned at 5–6 months, and usually leave their mother just before she again gives birth. Both females and males attain sexual maturity at about 2 years, but males generally cannot successfully breed until 5–7 years of age. Wild individuals up to 14 years old have been taken, and captives have lived about 25 years.

As a group, river otters have suffered severely through habitat destruction, water pollution, misuse of pesticides, excessive fur trapping, and persecution as supposed predators of game and commercial fish. *L. canadensis* has disappeared or become rare throughout the conterminous United States except in the Northwest, the upper Great Lakes region, New York, New England, and the states along the Atlantic and Gulf coasts. The southwestern subspecies *L. c. sonora* has nearly disappeared, though there have been several recent reports in Arizona (Polechla *in* Foster-Turley, Macdonald, and Mason 1990). Otters of another subspecies were released in Arizona in 1981, perhaps inadvisably considering the possibility of genetic modification of the native population. Since 1976 there also have been efforts to reintroduce otters in Colorado, Iowa, Kansas, Kentucky, Missouri, Nebraska, Oklahoma, Pennsylvania, and West Virginia (Melquist and Dronkert 1987).

The IUCN classifies *L. felina* as endangered, noting that it has declined by at least 50 percent over the past decade, and *L. provocax* as vulnerable (both are included in the genus *Lutra*). *L. longicaudis, L. provocax*, and *L. felina* are listed as endangered by the USDI and are on appendix 1 of the CITES, and *L. canadensis* is on appendix 2. There may now be fewer than 1,000 individuals of *L. felina* (Estes 1986); it has declined in Chile because of excessive hunting for its fur and in Peru because of persecution for alleged damage to prawn fisheries (Thornback and Jenkins 1982). *L. provocax* remains common at a few isolated sites in extreme southern Chile and Argentina but has disappeared from most of its range because of overhunting and habitat alteration. *L. longicaudis* is still widespread but has disappeared from the highlands of Mexico and is threatened in the rest of that country by habitat destruction and fragmentation (Foster-Turley, Macdonald, and Mason 1990).

The beautiful and durable fur of river otters is used for coat collars and trimming. During the 1976–77 trapping season 32,846 pelts of *L. canadensis* were reported taken in the United States, and the average selling price was $53.00. Respective figures for Canada that season were 19,932 pelts and $69.04 (Deems and Pursley 1978). In 1983–84 the total take was 33,135, and the average selling price was $18.71 (Novak, Obbard, et al. 1987). In the 1991–92 season 10,916 pelts were taken in the United States and sold at an average price of $22.34 (Linscombe 1994). *L. longicaudis* also was taken in large numbers for its valuable skin, with probably about 30,000 killed annually during the early 1970s in Colombia and Peru alone. Continued illegal hunting, along with habitat loss and water pollution, jeopardizes the survival of this species and *L. provocax* (Mason and Macdonald 1986; Thornback and Jenkins 1982).

Giant otter *(Pteronura brasiliensis brasiliensis)*, photo from New York Zoological Society.

CARNIVORA; MUSTELIDAE; **Genus PTERONURA**
Gray, 1867

Giant Otter

The single species, *P. brasiliensis*, originally was found in Colombia, Venezuela, the Guianas, eastern Ecuador and Peru, Brazil, Bolivia, Paraguay, Uruguay, and northeastern Argentina (Cabrera 1957; Thornback and Jenkins 1982).

The remainder of this account is based largely on Duplaix's (1980) study of *Pteronura* in Surinam. Head and body length is 864–1,400 mm and tail length is 330–1,000 mm. Males weigh 26–34 kg and females, 22–26 kg. The short fur generally appears brown and velvetlike when dry and shiny black chocolate when wet. On the lips, chin, throat, and chest there are often creamy white to buff splotches, which may unite to form a large white "bib." The feet are large, and thick webbing extends to the ends of the five clawed digits. The tail is thick and muscular at the base but becomes dorsoventrally flattened with a noticeable bilateral flange. There are subcaudal anal glands for secretion of musk.

The giant otter is found mainly in slow-moving rivers and creeks within forests, swamps, and marshes. It prefers waterways with gently sloping banks that have good cover. At certain points along a stream, areas of about 50 sq meters are cleared and used for rest and grooming. Some of these sites have dens, which consist of one or more short tunnels leading to a chamber about 1.2–1.8 meters wide. *Pteronura* seems clumsy on land but may move a considerable distance between waterways. When swimming slowly or remaining stationary in the water it paddles with all four feet. When swimming at top speed it depends largely on undulations of the tail and uses the feet for steering. It is entirely diurnal. During the dry season, when cubs are being reared, activity is generally restricted to one portion of a waterway. In the wet season movements are far more extensive, and spawning fish are followed into the flooded forest. Prey is caught with the mouth and then may be held in the forepaws while being consumed. Small fish may be eaten in the water, but larger prey is taken to shore. The diet consists mainly of fish and crabs.

Group home range, including land area, measures about 12 km in both length and width. During the dry season, at least, several kilometers of stream form a defended territory. Both sexes regularly patrol and mark the area, but groups tend to avoid one another, and fighting is evidently rare. *Pteronura* is more social than *Lutra*. A population includes both resident groups and solitary transients. As many as 20 individuals have reportedly been seen together, but groups of 4–8 are usually observed. A group consists of a mated adult pair, 1 or more subadults, and 1 or more young of the year. There is a high degree of pair bonding and group cohesiveness. A male and female stay close together and share the same den even when cubs are present. *Pteronura* is much noisier than *Lutra.* Nine vocalizations have been distinguished, including screams of excitement, often given while swimming with the forepart of the body steeply out of the water, and coos upon close intraspecific contact.

Although data are scanty, in the wild the young apparently are born from late August to early October, at the start of the dry season. If the first litter is lost, a second is sometimes produced from December to April. The gestation period is 65–70 days. The number of young per litter is one to five, usually one to three. The cubs weigh about 200 grams at birth and are able to eat solid food by 3–4 months. They remain with the parents at least until the birth of the next litter and probably for some time afterward. In a captive colony in Germany births occurred throughout the year, sexual maturity was attained at about 2 years, and one individual lived at least 14 years and 6 months (Hagenbeck and Wunnemann 1992).

The giant otter is classified as vulnerable by the IUCN and as endangered by the USDI and is on appendix 1 of the CITES. It has become very rare or has entirely disappeared over vast parts of its range. The main factor in its decline is excessive hunting by people for its large and valuable pelt. Because of its noise, diurnal habits, and tendency to approach intruders, it is relatively easy to locate and kill. In the early 1980s a skin could be sold by a hunter for the equivalent of U.S. $50 and on the market in Europe for $250. Such trade is now prohibited, but illegal hunting continues, facilitated by the opening of wilderness habitat (Thornback and Jenkins 1982). Recent surveys indicate that *Pteronura* survives in viable numbers in several parts of South America but has nearly or completely disappeared from Argentina, Uruguay, and southeastern Brazil (Foster-Turley, Macdonald, and Mason 1990).

CARNIVORA; MUSTELIDAE; **Genus AONYX**
Lesson, 1827

Clawless Otters

There are three subgenera and three species (Chasen 1940; Coetzee *in* Meester and Setzer 1977; Corbet and Hill 1992; Ellerman and Morrison-Scott 1966; Lekagul and McNeely 1977; Rosevear 1974):

subgenus *Aonyx* Lesson, 1827

A. capensis, Senegal to Ethiopia, and south to South Africa;

subgenus *Paraonyx* Hinton, 1921

A. congica, southeastern Nigeria and Gabon to Uganda and Burundi;

subgenus *Amblonyx* Rafinesque, 1832

A. cinerea, northwestern India to southeastern China and Malay Peninsula, southern India, Hainan, Sumatra, Java, Borneo, Riau Archipelago, Palawan.

Amblonyx was treated as a separate genus by Wozencraft (*in* Wilson and Reeder 1993) but not by Corbet and Hill (1991, 1992).

In the African species, *A. capensis* and *A. congica,* head and body length is 600–1,000 mm, tail length is 400–710 mm, and weight is 13–34 kg. In the smaller *A. cinerea,* of Asia, head and body length is 450–610 mm, tail length is 250–350 mm, and weight is 1–5 kg. The general coloration is brown, with paler underparts and sometimes white markings on the face, throat, and chest.

Aonyx differs from *Lutra* and *Pteronura* in having webbing that either does not extend to the ends of the digits or is entirely lacking and in having much smaller claws. In *A. congica* all the toes bear small, blunt claws; in *A. cinerea* the claws of adults are only minute spikes that do not project beyond the ends of the digital pads; and in *A. capensis* the only claws are tiny ones on the third and fourth toes of the hind feet. In association with these adaptations, *Aonyx* has developed very sensitive forepaws and considerable digital movement. *A. capensis* and *A. cinerea* have relatively large, broad cheek teeth, apparently for purposes of crushing the shells of crabs and mollusks. *A. congica* has lighter and sharper dentition, more adapted to cutting flesh.

The general habitat of *A. capensis* varies from dense rainforest to open coastal plain and semiarid country. The species is usually found near water, preferring quiet ponds and sluggish streams, but may sometimes wander a considerable distance overland. In coastal areas it has been seen to forage both in the sea and in adjoining freshwater streams and marshes (Verwoerd 1987). It is mainly nocturnal but may be active by day in areas remote from human disturbance. It dens under boulders or driftwood, in crannies under ledges, or in tangles of vegetation. It apparently does not dig its own burrow. *A. cinerea* occurs in rivers, creeks, estuaries, and coastal waters (Lekagul and McNeely 1977). *A. congica* is seemingly found only in small, torrential mountain streams within heavy rainforest.

These otters use their sensitive and dexterous forepaws to locate prey in mud or under stones. Captive *A. capensis* usually take food with the forepaws and do not eat directly off the ground. In the wild this species also catches most of its food with the forefeet, not with the mouth as do *Lutra* and *Pteronura* (Rowe-Rowe 1978*b*). The diet of both *A. capensis* and *A. cinerea* seems to consist mainly of crabs, other crustaceans, mollusks, and frogs; fish are relatively unimportant. There is thus apparently little competition for food between *Aonyx* and the fish-eating *Lutra* where both genera occur together. Piles of cracked crab and mollusk shells are signs of the presence of *A. capensis.* Donnelly and Grobler (1976) observed this species to use hard objects as anvils on which to break open mussel shells.

Little has been recorded about the habits of *A. congica.* Because of its scanty hair, weakly developed facial vibrissae, digital structure, and dental features, there has been speculation that it is more terrestrial than other otters. It is thought to feed mainly on relatively soft matter, such as small land vertebrates, eggs, and frogs. If this supposition is correct, there would be little competition for food between *A. congica* and the other kinds of otters that may occur in the same region.

Arden-Clarke (1986) studied a population of *A. capensis* inhabiting a rugged, densely vegetated environment along the coast of South Africa. The mean population density there was 1/1.9 km of coast, and dens were spaced at intervals of 470 meters. A radio-tracked adult male had a minimum home range of 19.5 km of coast, with a core area of 12.0 km, where it spent most of its time. An adult female had a 14.3-km home range with a 7.5-km core area. There apparently was a clan-type social organization, with groups of related animals defending joint territories. The home ranges of 4 adult males overlapped completely, and some of these animals were seen foraging together. In another

African clawless otter *(Aonyx capensis)*, photo from Zoological Society of London. Insets: A. Forefoot; B. Hind foot; photos by U. Rahm.

coastal area, Verwoerd (1987) found that an adult male might maintain a loose association with a female and cubs. *A. capensis* emits powerful, high-pitched shrieks when disturbed or trying to attract attention. According to Timmis (1971), *A. cinerea* lives in loose family groups of about 12 individuals and has a vocabulary of 12 or more calls, not including basic instinctive cries.

Births of *A. capensis* have been recorded in July and August in Zambia, and young have been found in March and April in Uganda (Kingdon 1977). There is probably no set breeding season in West Africa (Rosevear 1974). Most births in a coastal area of South Africa occurred in December and January (Verwoerd 1987). This species has a gestation period of 63 days and a litter size of two young. The

Oriental small-clawed otter *(Aonyx cinerea)*, photo by Lim Boo Liat.

young remain with the parents for at least 1 year. Female *A. cinerea* have an estrous cycle of 24–30 days, with an estrus of 3 days. They may produce two litters annually. The gestation period is 60–64 days, and litters contain one to six young, usually one or two. They open their eyes after 40 days, first swim at 9 weeks, and take solid food after 80 days (Duplaix-Hall 1975; Lekagul and McNeely 1977; Leslie 1971; Timmis 1971). Little is known of the reproduction of *A. congica*, but it probably has a gestation period of about 2 months, gives birth to two or three young, and attains sexual maturity at about 1 year. A captive specimen of *A. capensis* lived 14 years and a captive *A. cinerea* lived about 16 years (Marvin L. Jones, Zoological Society of San Diego, pers. comm., 1995).

If captured when young, these otters make intelligent and charming pets, though such activity may be illegal and detrimental to wild populations. *A. cinerea* has been trained to catch fish by Malay fishermen. The fur of *Aonyx* is not as good as that of *Lutra;* nonetheless, the subspecies *A. congica microdon* of Nigeria and Cameroon has declined seriously through uncontrolled commercial hunting. It is on appendix 1 of the CITES (other subspecies are on appendix 2) and is listed as endangered by the USDI. Data compiled by Foster-Turley, Macdonald, and Mason (1990) indicate that it is extremely rare. These data also show that the species *A. capensis* has become rare throughout West Africa, mainly in association with deforestation. Both *A. congica* and *A. cinerea* are designated as near threatened by the IUCN, and both *A. capensis* and *A. cinerea* are on appendix 2 of the CITES. *A. cinerea* apparently remains more common in southeastern Asia than the species of *Lutra* that occur there, but it is declining because of habitat loss and pollution and seems rare in most mainland parts of its range (Foster-Turley, Macdonald, and Mason 1990).

CARNIVORA; MUSTELIDAE; **Genus ENHYDRA**
Fleming, 1822

Sea Otter

The single species, *E. lutris*, was originally found in coastal waters off Hokkaido, Sakhalin, Kamchatka, the Commander Islands, the Pribilof Islands, the Aleutians, southern Alaska, British Columbia, Washington, Oregon, California, and western Baja California (Estes 1980). There are three subspecies: *E. l. lutris*, Hokkaido to the Commander Islands; *E. l. kenyoni*, Aleutians to Oregon; and *E. l. nereis*, California and Baja California (Wilson et al. 1991). The recent confirmation of the validity of these subspecies is important, both for consideration in reintroduction planning and to dispel claims that certain populations of bioconservation concern are of no systematic significance.

Head and body length is usually 1,000–1,200 mm and tail length is 250–370 mm. Males weigh 22–45 kg and females, 15–32 kg (Estes 1980). The color varies from reddish brown to dark brown, almost black, except for the gray or creamy head, throat, and chest. Albinistic individuals are rare. The head is large and blunt, the neck is short and thick, and the legs and tail are short. The ears are short, thickened, pointed, and valvelike. The hind feet are webbed and flattened into broad flippers; the forefeet are small and have retractile claws. *Enhydra* is the only carnivore with only four incisor teeth in the lower jaw. The molars are broad, flat, and well adapted to crushing the shells of such prey as crustaceans, snails, mussels, and sea urchins. Unlike most mustelids, the sea otter lacks anal scent glands. Females have two abdominal mammae (Estes 1980).

The sea otter differs from most marine mammals in that it lacks an insulating subcutaneous layer of fat. For protection against the cold water it depends entirely on a layer of air trapped among its long, soft fibers of hair. If the hair becomes soiled, as if by oil, the insulating qualities are lost and the otter may perish. The underfur, about 25 mm long, is the densest mammalian fur, averaging about 100,000 hairs per sq cm (Rotterman and Simon-Jackson 1988). It is protected by a scant coat of guard hairs.

Although the sea otter is a marine mammal, it rarely ventures more than 1 km from shore. According to Estes (1980), it forages in both rocky and soft-sediment communities on or near the ocean floor. Off California *Enhydra* seldom enters water of greater depth than 20 meters, but in the Aleutians it commonly forages at depths of 40 meters or more; the maximum confirmed depth of a dive was 97 meters. The usual period of submergence is 52–90 seconds, and the longest on record is 4 minutes and 25 seconds. The sea otter is capable of spending its entire life at sea but sometimes rests on rocks near the water. Such hauling-out behavior is more common in the Alaskan population than in that of California. *Enhydra* walks awkwardly on land. When supine on the surface of the water, it moves by paddling with the hind limbs and sculling with the tail. For rapid swimming and diving it uses dorsoventral undulations of the body. It can attain velocities of up to 1.5 km/hr on the surface and 9 km/hr for short distances underwater. The sea otter is generally diurnal, with crepuscular peaks and a midday period of rest (Riedman and Estes 1990). It often spends the night in a kelp bed, lying under strands of kelp to avoid drifting while sleeping. It sometimes sleeps with a forepaw over the eyes. Daily movements usually extend over a few kilometers, and there may be local seasonal movements but no extensive migrations (Riedman and Estes 1990).

The diet consists mainly of slow-moving fish and marine invertebrates, such as sea urchins, abalone, crabs, and mollusks (Estes 1980). Prey is usually captured with the forepaws, not the jaws. *Enhydra* floats on its back while eating and uses its chest as a "lunch counter." It is one of the few mammals known to use a tool. While floating on its back, it places a rock on its chest and then employs the rock as an anvil for breaking the shells of mussels, clams, and large sea snails in order to obtain the soft internal parts. This activity is most frequent in the population off California, and recent research there has shown considerable variation, such as using the rock as a hammer or using one rock as a hammer and another as an anvil (Riedman and Estes 1990). The sea otter requires a great deal of food: it must eat 20–25 percent of its body weight every day. It obtains about 23 percent of its water needs from drinking sea water and most of the rest from its food.

According to Estes (1980), *Enhydra* is basically solitary but sometimes rests in concentrations of as many as 2,000 individuals. Males and females usually come together only briefly for courtship and mating. At most times there is sexual segregation, with males and females occupying separate sections of coastline. Males usually occur at higher densities. Recent studies (Garshelis and Garshelis 1984; Garshelis, Johnson, and Garshelis 1984; Jameson 1989; Loughlin 1980; Rotterman and Simon-Jackson 1988) indicate that during the breeding season (which may be for most of the year) some males move into the areas occupied by females and establish territories. Such behavior has been documented in both Alaska and California, and males have been observed to return to the same place for up to seven years. The most favorable territories—those that seem to attract the most females—are characterized by availability of food; density of canopy-forming kelp, to which the ani-

Sea otters *(Enhydra lutris):* A. Juvenile walking. B. *E. lutris* floating on its back eating the head of a large codfish. C. Mother floating on her back with newborn pup. Photos by Karl W. Kenyon.

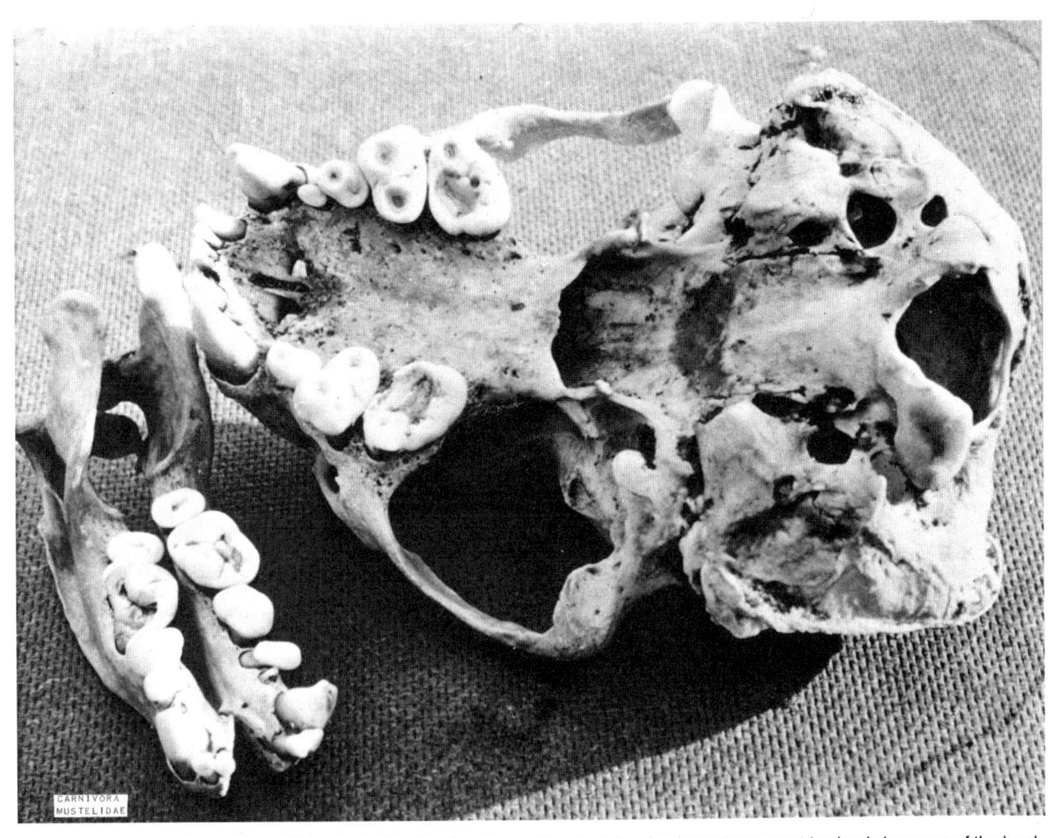

Skull and lower jaw of a sea otter *(Enhydra lutris)*, showing the cavities that develop in the teeth of old animals because of the hard, rough materials that they eat, photo by H. Robert Krear.

mals can attach themselves for secure resting; and associated shoreline features that provide a degree of shelter from the open sea (Riedman and Estes 1990). The boundaries of these territories are vigorously patrolled and intruding males are repulsed, but serious fighting is rare. The owner seeks to mate with any female that enters, though sometimes a pair bond is formed for a few days or weeks. Male territories are usually about 20–50 ha. and are smaller than female home ranges in the same area. Annual movements of both sexes frequently cover 50–100 km, considering foraging, breeding, and the passage of males between their territories and the all-male areas. McShane et al. (1995) described 10 vocalizations of *Enhydra,* including screams of distress, heard especially when mothers and young were separated, and coos, heard mostly when individuals were content or in familiar company.

Reproductive data have been summarized by Estes (1980) and Riedman and Estes (1990). Breeding occurs throughout the year, but births peak in late May and June in the Aleutians and from January to March off California; in the latter area there apparently is a secondary peak in late summer and early autumn. Males may mate with more than one female during the season. Pair bonding, and presumably estrus, lasts about 3 days. Females are capable of giving birth every year but usually do so at greater intervals. If a litter does not survive, the female may experience a postpartum estrus. Females are known to adopt and nurse orphaned pups. Reports of the period of pregnancy range from 4 to 12 months, and delayed implantation is probably involved. Estimates of the period of actual implanted gestation vary from about 4.0 to 5.5 months. Births probably occur most often in the water. There is normally a single offspring. About 2 percent of births are multiple, but only one young can be successfully reared. The pup weighs 1.4–2.3 kg at birth. While still small it is carried, nursed, and groomed on the mother's chest as the mother swims on her back. The pup begins to dive in the second month of life. It may take some solid food shortly after birth but may nurse almost until it attains adult size. The period of dependency on the mother is thought to be about 5–8 months. Females become sexually mature at about 4 years. Males are capable of mating at 5–6 years but usually do not become active breeders until several more years have passed. Wendell, Ames, and Hardy (1984) concluded that the reproductive cycle of the California population is shorter than elsewhere, with some females giving birth each year. It now is known that Alaskan females also are capable of annual reproduction (Garshelis, Johnson, and Garshelis 1984). According to Rotterman and Simon-Jackson (1988), a captive male fathered young when at least 19 years old, and maximum estimated longevity for wild females is 23 years. Data from Alaska indicate that sea otter populations have the potential to increase by about 20 percent annually, but the population off California has tended to increase by only 5 percent a year, probably because of high preweaning mortality of young that may be associated with pollutants imparted through lactation (Estes 1990; Riedman et al. 1994).

The fur of the sea otter may be the most valuable of any mammal's. During the 1880s prices on the London market ranged from $105 to $165 per skin. By 1903, when the species had become scarce, large, high-quality skins sold for up to $1,125. Pelts taken in Alaska in the late 1960s, during a brief reopening of commercial activity, sold for an average of $280 each (Kenyon 1969).

Estimates of the original numbers of *Enhydra* are 150,000–300,000 (Riedman and Estes 1990). Intensive exploitation of the genus was begun by the Russians in 1741. Hunting was uncontrolled until 1799, when some conservation measures were established. Unregulated killing resumed in 1867, when Alaska was purchased by the United States. By 1911, when the sea otter was protected by a treaty among the United States, Russia, Japan, and Great Britain, probably only 1,000–2,000 of the animals survived worldwide (Kenyon 1969). Under protection of the treaty, state and national laws, and finally the United States Marine Mammal Protection Act of 1972, the sea otter has steadily increased in numbers and distribution. There are now probably 100,000–150,000 individuals in the major populations off southwestern and south-central Alaska and another 17,000 off Kamchatka and the Kuril and Commander islands in Russia. Alaskan populations are subject to limited killing for native subsistence purposes, may come into conflict with shellfisheries, and are potentially jeopardized by oil spills. Reintroduced populations (from Alaskan stock) apparently have been established off southeastern Alaska (now numbering 4,500 animals), Vancouver Island (about 350), and Washington (280). Reintroduced groups off Oregon and in the Pribilof Islands do not seem to have done well and have all but disappeared (Estes 1980; Jameson 1993; Jameson et al. 1982; Riedman and Estes 1990; Rotterman and Simon-Jackson 1988).

The magnitude of the threat posed by oil spills from damaged tanker ships was tragically demonstrated by the *Exxon Valdez* disaster in 1989. As many as 5,000 sea otters are estimated to have been killed directly (*Oryx* 26 [1992]: 195). Hence, a significant part of the entire world's population of *Enhydra* was destroyed in this single incident. More than 1,000 dead or dying sea otters were actually recovered. However, an intensive effort was made by the U.S. Fish and Wildlife Service and cooperating organizations to save as many of the affected animals as possible. Several hundred were rescued and treated, and 197 were released back into the wild (Bayha and Kormendy 1990).

The southern sea otter (subspecies *E. l. nereis*), which originally ranged from Baja California to at least Oregon was generally considered extinct by 1920. Apparently, however, a group of 50–100 individuals survived off central California in the vicinity of Monterey. In 1938 the presence of this population became generally known. By the 1970s it had grown to include about 1,800 animals, but subsequently numbers stabilized or even declined. The sea otter now regularly occurs along about 350 km of the central California coast, and there have been scattered reports of individuals from southern California and northern Baja California. As this population increased, there was concern that stocks of abalone and other shellfish were being depleted. Some parties with a commercial or recreational interest in these stocks have advocated control of the sea otter population, and there have been cases of illegal killing. Some otters also are being drowned accidentally in fishing nets. Another fear is that an oil spill, associated with either the extensive tanker traffic in the area or offshore drilling, could devastate the population (Armstrong 1979; Carey 1987; Estes and VanBlaricom 1985; Leatherwood, Harrington-Coulombe, and Hubbs 1978; U.S. Fish and Wildlife Service 1980).

There has been concern that the genetic viability of *E. l. nereis* was severely reduced when it approached extinction earlier in the century, but Ralls, Ballou, and Brownell (1983) calculated that the existing population should theoretically retain about 77 percent of the original diversity and that transplanted colonies should also be viable. An effort to establish such a colony was started in 1987 when 63 otters taken from the main California population were released around San Nicolas Island (Brownell and Rathbun 1988). By June of 1990, 137 animals had been brought there, but only 15 were known to have remained in the area (Riedman and Estes 1990) and the experiment was called unsuccessful (*Oryx* 28 [1994]: 95). The southern sea otter is listed as threatened by the USDI and is on appendix 1 of the CITES (other subspecies of *E. lutris* are on appendix 2).

CARNIVORA; **Family VIVERRIDAE**

Civets, Genets, and Linsangs

This family of 19 Recent genera and 35 species is found in southwestern Europe, southern Asia, the East Indies, Africa, and Madagascar. Certain genera have been introduced to areas in which the family does not naturally occur. The following sequence of genera, grouped in 5 subfamilies, is based on the classifications of Coetzee (*in* Meester and Setzer 1977), Corbet and Hill (1992), Ellerman and Morrison-Scott (1966), and Ewer (1973):

Subfamily Viverrinae (civets, genets, linsangs), genera *Viverra, Civettictis, Viverricula, Genetta, Osbornictis, Poiana, Prionodon;*

Subfamily Nandiniinae (African palm civet), genus *Nandinia;*

Subfamily Paradoxurinae (Asian palm civets), genera *Arctogalidia, Paradoxurus, Paguma, Macrogalidia, Arctictis;*

Subfamily Hemigalinae (banded palm and otter civets), genera *Hemigalus, Diplogale, Chrotogale, Cynogale;*

Subfamily Euplerinae (Malagasy civets), genera *Fossa, Eupleres.*

The Viverridae often have been considered to include the family Herpestidae (see account thereof). Wozencraft (*in* Wilson and Reeder 1993) placed the subfamily Cryptoproctinae in the Viverridae, but it is here included in the Herpestidae. Wozencraft is followed here in the use of the name Euplerinae rather than Fossinae. Wozencraft also is followed in placing *Nandinia* in its own subfamily. This is essentially a compromise position between that of Meester et al. (1986) and numerous other authorities who treated *Nandinia* as a member of the Paradoxurinae and that of Hunt and Tedford (1993), who suggested that this genus is so strikingly different from other viverrids that it actually represents a separate family. The latter view is based largely on the structure of the basicranium, a highly complex region of the skull thought to express phylogenetic relationships. In particular, *Nandinia* has a primitive auditory bulla, differing from that of other viverrids and herpestids in that it does not inflate during ontogeny and has a single (rather than double) chamber, no septum bullae, and a cartilaginous (rather than ossified) caudal entotympanic that uniquely intervenes between the ectotympanic and rostral entotympanic.

Viverrids are characteristic small and medium-sized carnivores of the Old World. Head and body length is 350–950 mm, tail length is 130–900 mm, and adult weight is 0.6–20.0 kg. There are various striped, spotted, and uniform color patterns. In some genera the tail is banded or ringed. The body is long and sinewy with short legs and generally a long, bushy tail. One genus *(Arctictis)* has a truly prehensile tail. The head is elongate and the muzzle is pointed. Most genera have five toes on each foot. The claws are retractile or semiretractile in some genera. Female viverrids usually have two or three pairs of abdominal mammae. Males have a baculum.

Most viverrids have scent glands in the anal region that secrete a nauseous-smelling fluid as a defensive measure. The conspicuous pattern of pelage in some genera has been interpreted as being a warning that the fetid secretion is present. Such a color pattern is also found in skunks and certain other members of the family Mustelidae. Rubbed on various objects, the secretion of the scent glands is recognized by other individuals of the same species and is probably used to communicate various information.

The skull is usually long and flattened. The dental formula is: (i 3/3, c 1/1, pm 3–4/3–4, m 1–2/1–2) × 2 = 32–40. The second lower incisor is raised above the level of the first and third, the canines are elongate, and the carnassials are developed.

Viverrids are essentially forest inhabitants, but they also live in dense brush and thick grass. They are either diurnal or nocturnal and shelter in any convenient retreat, usually a hole in a tree, a tangle of vines, ground cover, a cave, a crevice, or a burrow. A few species dig their own burrows. Those species living near people sometimes seek refuge under rafters or in the drains of houses. Those viverrids that walk on their digits (such as *Genetta*) have a gait described as "a waltzing trot," whereas the members of the family that walk on their soles, with their heels touching the ground (such as *Arctictis*), have a bearlike shuffle. Many genera are agile and extremely graceful in their movements. A number of species are skillful climbers; some apparently spend most of their lives in trees. Some genera take to water readily and swim well; two, *Osbornictis* and *Cynogale*, are semiaquatic. Sight, hearing, and smell are acute.

Viverrids may fight when cornered. They seek their prey in trees and on the ground either by stalking or by pouncing from a hiding place. They eat small vertebrates and various invertebrates and occasionally consume vegetable matter such as fruit, bulbs, and nuts. Carrion is taken by some species. Viverrids are solitary or live in pairs or groups. Breeding may occur seasonally or throughout the year. A number of genera have two litters annually. The one to six offspring are born blind but haired. Most species probably have a potential longevity of 5–15 years.

The secretion of the scent glands, civet, is obtained from several genera *(Civettictis, Viverra,* and *Viverricula)* for both perfumery and medicinal purposes. Some viverrids are tamed and kept to extract the musk. They may also be kept as pets. Viverrids occasionally kill poultry but also prey on rodents.

The geological range of this family is late Eocene to Recent in Europe, early Oligocene to Recent in Asia, early Miocene to Recent in Africa, and Pleistocene to Recent in Madagascar (Stains 1984).

CARNIVORA; VIVERRIDAE; **Genus VIVERRA**
Linnaeus, 1758

Oriental Civets

There are two subgenera and three species (Chasen 1940; Corbet and Hill 1992; Ellerman and Morrison-Scott 1966; Medway 1978; Taylor 1934):

subgenus *Viverra* Linnaeus, 1758

V. zibetha, Nepal and Bangladesh to southeastern China and Malay Peninsula, Hainan;
V. tangalunga, Malay Peninsula, Riau Archipelago, Sumatra, Bangka, Borneo, Belitung and Karimata islands, Palawan and Calamian Islands (Philippines);

Oriental civet *(Viverra tangalunga)*, photo by Ernest P. Walker.

subgenus *Moschothera* Pocock, 1933

V. megaspila, extreme southwestern peninsular India, extreme southeastern China, Burma, Thailand, Indochina, Malay Peninsula and some nearby small islands, possibly Sumatra.

Civettictis sometimes is regarded as a third subgenus of *Viverra* (see Coetzee *in* Meester and Setzer 1977). Corbet and Hill (1992) followed Wozencraft (1989*a* and *in* Wilson and Reeder 1993) in listing *V. civettina* of southwestern India as distinct from *V. megaspila* but noted that specific separation is very doubtful based on examination of the characters of the few available specimens.

Head and body length is 585–950 mm, tail length is 300–482 mm, and weight is 5–11 kg. The fur, especially in winter, is long and loose. It is usually elongated in the median line of the body, forming a low crest or mane. The color pattern of the body is composed of black spots on a grayish or tawny ground color. The sides of the neck and throat are marked with black and white stripes—usually three black and two white collars. The crest is marked by a black spinal stripe that runs from the shoulders to the tail, and the tail is banded or ringed with black and white. The feet are black. In *V. zibetha* and *V. tangalunga* the third and fourth digits of the forefeet are provided with lobes of skin that act as protective sheaths for the retractile claws. *Viverra* is distinguished from *Viverricula* by its larger size, by the presence of a dorsal crest of erectile hairs, and by the insertion of the ears, the inner edges of which are set farther apart on the forehead.

Oriental civets occur in a wide variety of habitats in forest, brush, and grassland. They stay in dense cover by day and come out into the open at night. They are mainly terrestrial and often live in holes in the ground dug by other animals. They apparently can climb readily but seldom do so. Like *Viverricula*, they are often found near villages and are common over most of their range. Like most civets, they are easily trapped. They are vigorous hunters, killing small mammals, birds, snakes, frogs, and insects and taking eggs, fruit, and some roots. The species *V. zibetha* has been observed fishing in India, and the remains of crabs have been found in the stomachs of two individuals from China.

Viverra is generally solitary. An adult male *V. zibetha*, radio-tracked by Rabinowitz (1991) in Thailand, moved within an area of 12 sq km in a period of 7 months. Its average monthly range was 5.4 sq km and its average daily movement was 1.7 km. *V. zibetha* is said to breed all year and to bear two litters annually (Lekagul and McNeely 1977). The number of young per litter is one to four, usually two or three. The young are born in a hole in the ground or in dense vegetation. The young of *V. zibetha* open their eyes after 10 days, and weaning begins at 1 month (Medway 1978). There are captive specimens of *V. zibetha* in the Ahmedabad Zoo in India that are more than 20 years old (Smielowski 1986).

Viverra is one of the sources of civet, a substance used commercially in producing perfume. Because of this function, *V. tangalunga* has been introduced through much of the East Indies, including most of the Philippines, Sulawesi, the Sangihe Islands, and the Moluccan islands of Buru, Seram, Amboina, and Halmahera (Corbet and Hill 1992; Groves 1976; Laurie and Hill 1954). The Malabar civet, *V. civettina*, known only from an isolated belt of rainforests in the Western Ghats of southwestern India, is recognized by the IUCN as a species distinct from *V. megaspila* and classified as critically endangered; fewer than 250 mature individuals are thought to survive. This form also is listed as endangered by the USDI. It evidently has become very rare through hunting by people and loss of habitat to agriculture. No specimens had been known to science since the nineteenth century, but in 1987 three were captured at Elayur (Schreiber et al. 1989). A survey in 1990 determined that a number of other individuals had been killed or captured in the last few decades and that a small population survives but is immediately jeopardized by land clearing for rubber plantations (Ashraf, Kumar, and Johnsingh 1993).

CARNIVORA; VIVERRIDAE; **Genus CIVETTICTIS** *Pocock, 1915*

African Civet

The single species, *C. civetta*, is found from Senegal to Somalia and south to northern Namibia and northeastern South Africa. Recognition of *Civettictis* as a distinct genus is in keeping with Ewer (1973), Kingdon (1977), Ray (1995), and Rosevear (1974). Some other authorities, such as Coetzee (*in* Meester and Setzer 1977) and Rowe-Rowe (1978*b*), have included *C. civetta* in the genus *Viverra*.

Head and body length is 670–890 mm, tail length is 340–470 mm, and weight is 7–20 kg (Kingdon 1977; Ray 1995). The color is black with white or yellowish spots, stripes, and bands. There is much variation in the pattern of markings, and some individuals are melanistic. The long and coarse hair is thick on the tail. The perineal glands under the tail contain the oily scented matter used commercially in making perfume. All the feet have five claws and the soles are hairy. From *Viverra*, *Civettictis* is distinguished by much larger molar teeth and a far broader lower carnassial (Rosevear 1974).

The African civet is widely distributed in both forests and savannahs, wherever long grass or thickets are sufficient to provide daytime cover (Ewer and Wemmer 1974). It seems to use a permanent burrow or nest only to bear young. It is nocturnal and almost completely terrestrial but takes to water readily and swims well. The omnivorous diet includes carrion, rodents, birds, eggs, reptiles, frogs, crabs, insects, fruits, and other vegetation. Poultry and young lambs are sometimes taken (Rosevear 1974).

Civettictis is generally solitary but has a variety of visual, olfactory, and auditory means of communication. Individuals may have defined and well-marked territories. The scent glands have a major social role, leaving scent along a path to convey information, such as whether a female is in estrus (Kingdon 1977). There are three agonistic vocalizations—the growl, cough-spit, and scream—but the most commonly heard sound is the "ha-ha-ha" used in making contact (Ewer and Wemmer 1974).

Available data on reproduction (Ewer and Wemmer 1974; Hayssen, Van Tienhoven, and Van Tienhoven 1993; Kingdon 1977; Mallinson 1973, 1974; Ray 1995; Rosevear 1974) suggest that breeding occurs throughout the year in West Africa, from March to October in East Africa, and in the warm, wet summer from August to January in South Africa. Females are polyestrous and there may be two or even three litters annually. The gestation period is usually 60–72 days but is occasionally extended to as many as 81 days, perhaps because of delayed implantation. The number of young per litter is one to four, usually two or three. The young are born fully furred, weigh about 300 grams at birth, open their eyes within a few days, cease suckling at 14–20 weeks, and attain sexual maturity at about 1 year. According to Jones (1982), a captive lived for 28 years.

In Ethiopia, and to a lesser extent in other parts of Africa,

African civet *(Civettictis civetta)*, photo from New York Zoological Society.

the natives keep civets in captivity and remove the musk from them several times a week. An average animal yields 3–4 grams weekly. The natives do not raise the civets, however, but merely capture wild ones. In 1934 Africa produced about 2,475 kg of musk with a value of U.S. $200,000. In that same year the United States imported 200 kg of musk. The production of civet musk is an old industry; King Solomon's supply came from East Africa. Rosevear (1974) reported that the trade in civet musk now has diminished considerably. However, Schreiber et al. (1989) indicated that in 1988 there still were more than 2,700 captive civets in Ethiopia and that their musk, exported mainly to France, was selling for about $438 per kg.

CARNIVORA; VIVERRIDAE; **Genus VIVERRICULA**
Hodgson, 1838

Lesser Oriental Civet, or Rasse

The single species, *V. indica,* occurs naturally from Pakistan, through most of India, to southeastern China and the Malay Peninsula as well as in Sri Lanka, Taiwan, Hainan, Sumatra, Java, the Kangean Islands, and Bali (Corbet and Hill 1992; Ellerman and Morrison-Scott 1966; Roberts 1977). The name *V. malaccensis* was used for this species by Medway (1978) and Lekagul and McNeely (1977).

Head and body length is 450–630 mm, tail length is 300–430 mm, and weight is usually 2–4 kg. The fur is harsh, rather coarse, and loose. The body color is buffy, brownish, or grayish and the feet are black. Small spots are present on the forequarters, and larger spots, tending to run into longitudinal lines, are present on the flanks. There are six to eight dark stripes on the back, and the tail is ringed black and white by six to nine rings of each color.

Viverricula is distinguished from *Viverra* by its smaller size, the absence of a dorsal crest of erectile hairs, and the insertion of the ears, the inner edges of which are set closer together on the forehead than those of *Viverra.* The muzzle is also shorter and more pointed. Internally the two genera differ in a number of cranial and dental features.

The rasse inhabits grasslands or forests. It probably excavates its own burrow (Roberts 1977) but may also shelter in thick clumps of vegetation, buildings, or drains (Lekagul and McNeely 1977). It is generally nocturnal but may be seen hunting by day in areas not populated by humans. It is mainly terrestrial but is said to climb well. It usually tries to escape from dogs by dodging and twisting through the underbrush. The diet consists of small vertebrates, carrion, insects and their grubs, fruits, and roots.

An adult male radio-tracked for six months in Thailand by Rabinowitz (1991) moved within a total area of 3.1 sq km, had an average monthly range of 0.83 sq km, and had an average daily movement of 500 meters. *Viverricula* is usually solitary but occasionally associates in pairs. It breeds throughout the year in Sri Lanka. Captives in Shanghai mate mostly from February to April and to some extent in August–September (Xu and Sheng 1994). The two to five young are born in a shelter on the ground. They are weaned after 4–4.5 months (Hayssen, Van Tienhoven, and Van Tienhoven 1993). A captive lived 10 years and 6 months (Jones 1982).

The rasse is kept in captivity by natives for the purpose of extracting the civet that is secreted and retained in sacs close to the genitals in both sexes. The removal of this secretion is accomplished by scraping the inside of the sac with a spoonlike implement. In India this secretion is used as a perfume to flavor the tobacco that is smoked by some natives. *Viverricula* was introduced by people to Socotra, the Comoro Islands, Madagascar, and perhaps the Philippines, probably for the production of civet. Its presence on the islands of Lombok and Sumbawa, to the east of Bali, also is thought to have resulted from introduction (Corbet and Hill 1992; Laurie and Hill 1954).

CARNIVORA; VIVERRIDAE; **Genus GENETTA**
Oken, 1816

Genets

There are 3 subgenera and 10 species (Ansell 1978; Coetzee *in* Meester and Setzer 1977; Crawford-Cabral 1981; Harrison and Bates 1991; Lamotte and Tranier 1983;

Lesser oriental civet *(Viverricula indica)*, photo by Ernest P. Walker.

Meester et al. 1986; Schlawe 1980; Schreiber et al. 1989; Skinner and Smithers 1990):

subgenus *Pseudogenetta* Dekeyser, 1949

G. thierryi, savannah zone from Senegal to the area south of Lake Chad;
G. abyssinica, Ethiopia, possibly Djibouti and northern Somalia;

subgenus *Paragenetta* Kuhn, 1960

G. johnstoni, southern Guinea, Liberia, Ivory Coast, possibly Ghana;

subgenus *Genetta* Oken, 1816

G. servalina, southern Nigeria to western Kenya;
G. victoriae, northern and eastern Zaire, Uganda;
G. genetta, France, Spain, Portugal, Balearic Islands, southwestern Saudi Arabia, Yemen, southern Oman, northwestern Africa, savannah zone of Africa from Senegal to northeastern Sudan and south to Tanzania, savannah and desert zone from southwestern Angola to southern Mozambique and South Africa;
G. angolensis, southern Zaire, central and northeastern Angola, western Zambia, northern Mozambique, probably southern Tanzania, possibly northern Zimbabwe;
G. pardina, Gambia to Cameroon;
G. maculata, Senegal to Somalia, and south to Namibia and Natal (South Africa);
G. tigrina, Cape Province and Natal (South Africa), Lesotho.

Corbet (1978) suggested that the European populations of *G. genetta* are the result of introduction by human agency. Schlawe (1980) restricted the name *G. genetta* to Europe and northwestern Africa. He referred the populations to the south of the Sahara and on the southwestern Arabian Peninsula to a separate species, *G. felina*, and he showed that the reported presence of *Genetta* in Palestine is not correct. Later, Schlawe (1981) indicated that much work remains to be done before the systematics of this genus can be reasonably well understood. Corbet (1984) recognized the specific distinction of *G. felina*, but Corbet and Hill (1991), Meester et al. (1986), and Wozencraft (*in* Wilson and Reeder 1993) did not. Wozencraft included *G. pardina* within *G. maculata*, but the two were listed as separate species by Corbet and Hill (1991). Meester et al. (1986) considered *G. maculata* and *G. tigrina* to be conspecific, noting that the two intergraded over much of Natal. That same po-

Genets *(Genetta tigrina):* Top, photo by John Markham; Bottom, photo by John Visser.

sition was taken by Coetzee (*in* Meester and Setzer 1977), but the two taxa were treated as separate species by Corbet and Hill (1991), Crawford-Cabral (1981), Crawford-Cabral and Pacheco (1992), Schlawe (1981), and Wozencraft (1989*a* and *in* Wilson and Reeder 1993). Some of those authorities used the name *G. rubiginosa* in place of *G. maculata*. Heard and Van Rompaey (1990) suggested that *G. cristata* of southeastern Nigeria and southwestern Cameroon may be a species distinct from *G. servalina*.

Head and body length is usually 420–580 mm, tail length is 390–530 mm, and weight is 1–3 kg. Coloration is variable, but the body is generally grayish or yellowish with brown or black spots and blotches on the sides that tend to be arranged in rows. A row of black erectile hairs is usually present along the middle of the back. The tail has black and white rings. Melanistic individuals seem to be fairly common. Genets have a long body, short legs, a pointed snout, prominent and rounded ears, short and curved retractile claws, and soft and dense hair. They have the ability to emit a musky-smelling fluid from their anal glands. Females have two pairs of abdominal mammae.

Genets inhabit forests, savannahs, and grasslands. They are active at night, usually spending the day in rock crevices, in burrows excavated by other animals, in a hollow trees, or on large branches. They seem to return daily to the same shelter. They climb trees to prey on nesting and roosting birds, but much of their food is taken on the ground. They are silent and stealthy hunters; when stalking prey they crouch until their body and tail seem to glide along the ground. At the same time, the body seems to lengthen. The genet's slender and loosely jointed body allows it to go through any opening its head can enter. Nightly movements in Spain were found to average 2.78 km (Palomares and Delibes 1994). The diet consists of any small animals that can be captured, including rodents, birds, reptiles, and insects. Genets sometimes take game birds and poultry.

Radio-tracking studies of *G. genetta* in Spain over a period of more than two years indicated an average maximum home range size of 7.8 sq km (Palomares and Delibes 1994), though one adult male wandered across 50 sq km during about five months (Palomares and Delibes 1988). Home ranges of adult males and females overlapped greatly, but those of animals of the same sex were exclusive. Radio-tracking studies of *G. maculata* in Kenya during June–August found average home ranges of 5.9 sq km for three males and 2.8 sq km for two females (Fuller, Biknevicius, and Kat 1990). Genets travel alone or in pairs. They communicate with one another by a variety of vocal, olfactory, and visual signals. Breeding seems to correspond with the wet seasons in both West and East Africa. In Kenya, for example, pregnant and lactating females have been taken in

May and from September to December. A pair of *G. genetta* in the National Zoo in Washington, D.C., regularly produced two litters per year, one in April–May and another in July–August. Gestation periods ranging from 56 to 77 days have been reported. The number of young per litter is one to four, usually two or three. The young weigh 61–82 grams at birth, begin to take solid food at 2 months, and attain adult weight at 2 years. One female *G. genetta* became sexually mature at about 4 years and produced young regularly until she died at the age of 13 years (Kingdon 1977; Rosevear 1974; Wemmer 1977). Another captive *G. genetta* lived 21 years and 6 months (Jones 1982).

According to Smit and Van Wijngaarden (1981), the genet has declined in Europe because of persecution for alleged depredations on game birds and poultry. In addition, its winter pelt is highly esteemed. A recently described subspecies on Ibiza Island in the Balearics, *G. genetta isabelae,* is classified as vulnerable by the IUCN. Although sometimes considered to represent an introduction by people long ago, this subspecies is clearly distinguishable from the nearby mainland populations of Europe and Africa (Schreiber et al. 1989). *G. cristata,* restricted to Nigeria and Cameroon, is recognized by the IUCN as a species distinct from *G. servalina* and classified as endangered. Although its survival has recently been confirmed, its only known habitat is rapidly being degraded (Heard and Van Rompaey 1990). There also is concern for *G. abyssinica* and *G. johnstoni,* both of which are apparently very rare, have not been definitely recorded from the field for many years, and occur in a limited area of habitat that is increasingly subject to human disturbance (Schreiber et al. 1989).

CARNIVORA; VIVERRIDAE; **Genus OSBORNICTIS**
J. A. Allen, 1919

Aquatic Genet

The single species, *O. piscivora,* is known only by about 30 specimens taken in northeastern Zaire (Coetzee *in* Meester and Setzer 1977; Hart and Timm 1978; Van Rompaey 1988).

An adult male had a head and body length of 445 mm, a tail length of 340 mm, and a weight of 1,430 grams; an adult female weighed 1,500 grams (Hart and Timm 1978). The body is chestnut red to dull red and the tail is black. There is a pair of elongated white spots between the eyes. The front and sides of the muzzle and the sides of the head below the eyes are whitish. Black spots and bands are absent, and the tail is not ringed. The pelage is long and dense, especially on the tail. The palms and soles are bare, not furred as in *Genetta* and other related genera. The skull is long and lightly built, and the teeth are relatively small and weak.

Osbornictis is among the rarest genera of carnivores. All specimens probably originated in areas of dense forest at elevations of 500–1,500 meters (Hart and Timm 1978). The genus is generally thought to be semiaquatic, as several

Aquatic genet *(Osbornictis piscivora)*, photo of mounted specimen by M. Colyn.

specimens have been taken in or near streams, and available evidence suggests that fish constitute a major part of the diet. Hart and Timm, for example, noted the following: the stomach of one specimen contained the remains of fish; natives of the area indicated that fish is the favored prey; the dentition of *Osbornictis* seems to be adapted to deal with slippery vertebrate prey, such as fish and frogs; and the bare palms may be an adaptation allowing the genet to feel for fish in muddy holes and then handle the prey. *Osbornictis* is apparently solitary. A pregnant female with a single embryo 15 mm long was taken on 31 December.

CARNIVORA; VIVERRIDAE; **Genus POIANA**
Gray, 1864

African Linsang, or Oyan

The single species, *P. richardsoni*, occurs from Sierra Leone to northern Zaire and on the island of Bioko (Fernando Poo) (Coetzee *in* Meester and Setzer 1977); the generic name reflects the occurrence on this island. Rosevear (1974) recognized the populations in the western part of the range of *Poiana* as a distinct species, *P. leightoni.*

The average head and body length is 384 mm and the average tail length is 365 mm (Rosevear 1974). The general color effect is light brownish gray to rusty yellow; dark brown to black spots and rings are present. Some individuals have alternating broad and narrow black bands on the tail, whereas others have only the broad bands. This genus differs from the Asiatic linsangs *(Prionodon)* in that the spots are smaller and show no tendency to run into bands or stripes except in the region of the head and shoulder. It also differs from them, and resembles *Genetta*, in having a narrow bare line on the sole of each hind foot.

The oyan is a forest animal and is nocturnal. According to Hans-Jürg Kuhn (Anatomisches Institut der Universität Frankfurt am Mein, pers. comm.), *Poiana* builds a round nest of green material, in which several individuals sleep for a few days, and then moves on and builds a new nest. The nests are at least two meters above the ground, usually higher. Although the oyan has been reported to sleep in the abandoned nests of squirrels, reliable hunters say that the reverse is true: the squirrels sleep in abandoned nests of *Poiana*. The diet includes cola nuts, other plant material, insects, and young birds. In the Liberian hinterland natives make medicine bags from the skins of *Poiana*.

Observations by Charles-Dominique (1978) in northeastern Gabon indicate a population density of 1/sq km. A lactating female has been noted in October. As in some other genera of viverrids, there may be two litters per year. The number of young per birth is 2 or 3. A captive oyan lived five years and four months (Jones 1982). The subspecies *P. r. liberiensis*, of Liberia, Ivory Coast, and Sierra Leone, is rare, isolated from the more easterly populations, and known only from about 14 museum specimens, including 2 killed in 1987 (Schreiber et al. 1989; Taylor 1989).

African linsang *(Poiana richardsoni)*, photo from *Proc. Zool. Soc. London.*

Banded linsang *(Prionodon linsang)*, photo by Ernest P. Walker.

CARNIVORA; VIVERRIDAE; **Genus PRIONODON**
Horsfield, 1822

Oriental Linsangs

There are two subgenera and two species (Chasen 1940; Corbet and Hill 1992; Ellerman and Morrison-Scott 1966; Lekagul and McNeely 1977):

subgenus *Prionodon* Horsfield, 1822

P. linsang (banded linsang), western and southern Thailand, Tenasserim, Malay Peninsula, Sumatra, Bangka, Java, Borneo, Belitung Island;

subgenus *Pardictis* Thomas, 1925

P. pardicolor (spotted linsang), eastern Nepal to northern Indochina and nearby parts of southern China.

Head and body length is 310–450 mm and tail length is 304–420 mm. Medway (1978) listed the weight of *P. linsang* as 598–798 grams; on the average, *P. pardicolor* is slightly smaller. In *P. linsang* the ground color varies from whitish gray to brownish gray and becomes creamy on the underparts. The dark pattern consists of four or five broad, transverse black or dark brown bands across the back; there is one large stripe on each side of the neck. The sides of the body and legs are marked with dark spots, and the tail is banded. Some individuals of *P. pardicolor* have a ground color of orange buff, whereas others are pale brown. Black spots on the upper parts are arranged more or less in longitudinal rows, and the tail has 8–10 dark rings.

These animals are extremely slender, graceful, and beautiful. The fur is short, dense, and soft; it has the appearance and feel of velvet. The claws are retractile; claw sheaths are present on the forepaws, and protective lobes of skin are present on the hind paws. The skull is long, low, and narrow, and the muzzle is narrow and elongate. Unlike many viverrids, *Prionodon* seems to be free from odor.

Oriental linsangs dwell mainly in forests. They are nocturnal and generally arboreal but frequently move to the ground in search of food (Lekagul and McNeely 1977). *P. linsang* constructs a nest of sticks and leaves; in one case a nest was located in a burrow at the base of a palm. This species is also said to live in tree hollows. The diet includes small mammals, birds, eggs, and insects.

The limited data on reproduction suggest that *P. linsang* has no clear breeding season (Lekagul and McNeely 1977). Two pregnant females, one with two embryos and the other with three, were collected in May, and two lactating females were found in April and October. *P. pardicolor* is said to breed in February and August and to have litters of two young. Hayssen, Van Tienhoven, and Van Tienhoven (1993) listed a newborn weight of 40 grams and an estrus length of 11 days for *P. linsang*. A captive of that species lived 10 years and 8 months (Jones 1982).

The species *P. pardicolor* is listed as endangered by the USDI and is on appendix 1 of the CITES. *P. linsang* is on appendix 2 of the CITES. Schreiber et al. (1989) noted that the former seems to be very rare but the latter is still relatively numerous in certain areas.

CARNIVORA; VIVERRIDAE; **Genus NANDINIA**
Gray, 1843

African Palm Civet

The single species, *N. binotata*, occurs from Guinea-Bissau to southern Sudan and south to northern Angola and eastern Zimbabwe (Coetzee *in* Meester and Setzer 1977).

Head and body length is 440–580 mm and tail length is 460–620 mm. Kingdon (1977) listed weight as 1.7–2.1 kg, but Charles-Dominique (1978) reported that males weighed as much as 5 kg. Coloration is quite variable but is usually grayish or brownish tinged with buffy or chestnut. Often two creamy spots are present between the shoulders, and obscure dark brown spots are present on the lower back and top of the tail. The tail, which is somewhat darker than the body, is the same color above and below and has a variable pattern of black rings. The throat tends to be grayish,

African palm civet *(Nandinia binotata)*, photo by Ernest P. Walker.

and the underparts are grayish tinged with yellow. The pelage is short and woolly but coarse-tipped. The ears are short and rounded, the tail is fairly thick, the legs are short, and the claws are sharp and curved. There are scent glands on the palms, between the toes, on the lower abdomen, and possibly on the chin (Kingdon 1977). *Nandinia* has a number of unique cranial characters (see account of family Viverridae).

In a radio-tracking study in Gabon, Charles-Dominique (1978) found *Nandinia* to be largely arboreal and to occur mainly 10–30 meters above the ground in various types of forest. It was nocturnal, sleeping by day in a fork, on a large branch, or in a bundle of lianas. Stomach contents consisted of 80 percent fruit, on the average, but also included remains of rodents, bird eggs, large beetles, and caterpillars.

Charles-Dominique found a population density of 5/sq km in his study area. Adult females established territories averaging 45 ha. They allowed immature females on these areas but did not tolerate trespassing by other adult females. Large, dominant adult males had territories averaging about 100 ha., which overlapped a number of female territories. The large males drove away other animals of the same size and sex but allowed smaller adult males to remain; however, the small adult males were not permitted access to the females. Territories were marked with scent. Fighting was severe, sometimes resulting in death. Loud calls were exchanged during courtship.

In West Africa breeding apparently can occur during the wet or dry season (Rosevear 1974). Records from East Africa suggest that there are two birth peaks or seasons, May and October. The gestation period is 64 days. The number of young is usually two but up to four (Kingdon 1977). As soon as they are weaned, young males leave the territory of their mother. Sexual maturity is attained in the third year of life (Charles-Dominique 1978). One individual was still alive after 15 years and 10 months in captivity (Jones 1982).

The African palm civet is easily tamed and will drink milk in captivity. It is said to be quite clean and to keep houses free of rats, mice, and cockroaches.

CARNIVORA; VIVERRIDAE; **Genus ARCTOGALIDIA**
Merriam, 1897

Small-toothed or Three-striped Palm Civet

The single species, *A. trivirgata,* is found from Assam to Indochina and the Malay Peninsula and on Sumatra, Bangka, Java, Borneo, and numerous small nearby islands of the East Indies (Chasen 1940; Ellerman and Morrison-Scott 1966).

Head and body length is 432–532 mm, tail length is 510–660 mm, and weight is usually 2.0–2.5 kg. The color of the upper parts, proximal part of the tail, and outside of the limbs varies from dusky grayish tawny to bright orangish tawny. The head is usually darker and grayer, and the paws and distal part of the tail are brownish. There is a median white stripe on the muzzle, and there are three brown or black longitudinal stripes on the back. The median stripe is usually complete and distinct, whereas the laterals may be broken up into spots or almost absent. The undersides are grayish white or creamy buff with a whitish patch on the chest.

Only the females of this genus possess the civet gland, which is located near the opening of the urinogenital tract. *Arctogalidia* closely resembles *Paradoxurus* in external form as well as in the length of the legs and tail but differs externally in characters of the feet. Internally, the skull differs from that of *Paradoxurus,* and the back teeth are smaller.

Arctogalidia inhabits dense forests. In some areas it fre-

Small-toothed palm civet *(Arctogalidia trivirgata)*, photo by Lim Boo Liat.

quents coconut plantations, though Lekagul and McNeely (1977) reported that it avoids human settlements. It is nocturnal, resting by day in the upper branches of tall trees (Medway 1978). It is arboreal, climbing actively and leaping from branch to branch with considerable agility. The omnivorous diet includes squirrels, birds, frogs, insects, and fruit.

Three animals, representing both sexes, occupied an empty nest of *Ratufa bicolor* about 20 meters above the ground in a tree. Mewing calls and light snarls, accompanied by playful leaps and chases, have been noted for a male and female at night in the wild. The young are reared in hollow trees. According to Lekagul and McNeely (1977), breeding probably continues throughout the year, there may be two litters annually, and litter size is two or three young. Batten and Batten (1966) reported that a female about 2 weeks old was captured in Borneo in August 1961. It entered estrus for the first time in December 1962 and then again at intervals of 6 months. In August 1964 it gave birth to its first litter (three young) after a gestation period of approximately 45 days. The young opened their eyes at 11 days and were suckled for more than 2 months. The father was reintroduced to the family when the young were 2.5 months old and was soon accepted by the others. According to Marvin L. Jones (Zoological Society of San Diego, pers. comm., 1995), one individual lived in captivity for 16 years and 1 month.

The subspecies *A. t. trilineata*, of Java, is classified as endangered by the IUCN. Schreiber et al. (1989) indicated that it was already rare 50 years ago; the last confirmed record was in 1978.

CARNIVORA; VIVERRIDAE; **Genus PARADOXURUS**
F. Cuvier, 1821

Palm Civets, Musangs, or Toddy Cats

There are four species (Chasen 1940; Corbet and Hill 1992; Ellerman and Morrison-Scott 1966; Laurie and Hill 1954; Schreiber et al. 1989):

P. hermaphroditus, occurs naturally from Kashmir and peninsular India to southeastern China and the Malay Peninsula and in Sri Lanka, Hainan, Sumatra and the nearby Simeulue and Enggano islands (but not the intervening Mentawai Islands), Java, Kangean Islands, Borneo, Palawan, and many small nearby islands of the East Indies;
P. lignicolor, Siberut, Sipura, North Pagai, and South Pagai in the Mentawai Islands off western Sumatra;
P. zeylonensis (golden palm civet), Sri Lanka;
P. jerdoni, extreme southwestern India.

Taylor (1934) referred populations of *Paradoxurus* in the Philippines to three species—*P. philippinensis, P. torvus*, and *P. minax*—but Corbet and Hill (1992) indicated that these populations, other than that on Palawan, probably were introduced and that the only species in the Philippines is *P. hermaphroditus*. Wozencraft (*in* Wilson and Reeder 1993) included *P. lignicolor* in *P. hermaphroditus*, but Corbet and Hill (1992) found them to be distinct species.

Head and body length is 432–710 mm, tail length is 406–660 mm, and weight is 1.5–4.5 kg. The ground color is grayish to brownish but is often almost entirely masked by the black tips of the guard hairs. There is a definite pattern of dorsal stripes and lateral spots, at least in the new coat, but this is sometimes concealed by the long black hairs. The pattern is most plainly shown in the species *P. hermaphroditus*, where it consists of longitudinal stripes on the back and spots on the shoulders, sides, and thighs and sometimes on the base of the tail. A pattern of white patches and a white band across the forehead may be present on the head of this species.

The species *P. hermaphroditus* can always be distinguished from *P. zeylonensis* and *P. jerdoni* by the backward direction of the hairs on the neck. In the other species the hairs on the neck grow forward from the shoulders to the head.

Paradoxurus differs from *Arctogalidia* and *Paguma* in color pattern and in characters of the skull and teeth. According to Lekagul and McNeely (1977), the teeth of *Paradoxurus* are less specialized for eating meat than those of most viverrids, having low, rounded cusps on rather square molars. Both sexes have well-developed anal scent glands. Females have three pairs of mammae.

Palm civet *(Paradoxurus hermaphroditus)*, photo by Ernest P. Walker.

Musangs are nocturnal forest dwellers. They are expert climbers and spend most of their time in trees, where they utilize cavities or secluded nooks. They are often found about human habitations, probably because of the presence of rats and mice. Under such conditions they shelter in thatched roofs and in dry drain tiles and pipes. They eat small vertebrates, insects, fruits, and seeds. They are fond of the palm juice, or "toddy," collected by the natives—thus one of the vernacular names.

An older male radio-tracked for 12 months in Thailand by Rabinowitz (1991) moved within a total area of 17 sq km and had an average monthly range of 3.2 sq km and an average daily movement of 1 km; respective figures for a younger male tracked for 7 months were 4.25 sq km, 0.74 sq km, and 660 meters. A female with five kittens radio-tracked for about a month in Nepal used a home range of 1.2 ha. (Dhungel and Edge 1985). Hayssen, Van Tienhoven, and Van Tienhoven (1993) listed a gestation period of 60 days and newborn weights of 69–102 grams for *P. hermaphroditus*. Reproduction occurs throughout the year, though Lekagul and McNeely (1977) stated that the young of *P. hermaphroditus* seem to be seen most often from October to December. Litter size is two to five young. Sexual maturity is attained at 11–12 months in *P. hermaphroditus*. One individual was still living after 25 years in captivity (Marvin L. Jones, Zoological Society of San Diego, pers. comm., 1995).

Groves (1976) noted that *P. hermaphroditus* has been carried about from island to island by people for use as a rat catcher. Such activity is probably responsible for the presence of this species on most islands of the Philippines, Sulawesi, the Lesser Sundas as far east as Timor, Seram, the Sula and Kei islands, and Aru (Corbet and Hill 1992). *P. hermaphroditus* is generally widespread and often lives in close association with people; however, there is concern for the other three species, which have restricted ranges and depend on natural forest habitat. The IUCN classifies *P. lignicolor* (treated as a subspecies of *P. hermaphroditus*) and *P. jerdoni* as vulnerable. Schreiber et al. (1989) pointed out that the former is known only by four specimens and that its island habitat is being destroyed by commercial logging; the latter also occurs in a limited area and reportedly has been sighted only twice in the last 20 years. *P. zeylonensis* may still be common in some parts of Sri Lanka and is respected by older villagers, who realize that it spreads seeds of the valuable kitul palm, but it has lost much of its habitat and is trapped and eaten by younger people. The subspecies *P. hermaphroditus kangeanus* is restricted to the Kangean Islands off northeastern Java and may be jeopardized through interbreeding with introduced subspecies.

CARNIVORA; VIVERRIDAE; **Genus PAGUMA**
Gray, 1831

Masked Palm Civet

The single species, *P. larvata*, occurs from northern Pakistan and Kashmir to Indochina and the Malay Peninsula,

Masked palm civet *(Paguma larvata)*, photo from New York Zoological Society.

in much of eastern and southern China, and on the Andaman Islands, Taiwan, Hainan, Sumatra, and Borneo (Chasen 1940; Ellerman and Morrison-Scott 1966). In 1993 an individual was sighted in western Java, but this may have resulted from human introduction (Brooks and Dutson 1994).

Head and body length is 508–762 mm, tail length is usually 508–636 mm, and weight is usually 3.6–5.0 kg. In the facial region there is generally a mask, which consists of a median white stripe from the top of the head to the nose, a white mark below each eye, and a white mark above each eye extending to the base of the ear and below. The general color is gray, gray tinged with buff, orange, or yellowish red. There are no stripes or spots on the body and no spots or bands on the tail. The distal part of the tail may be darker than the basal part, and the feet are blackish. This genus differs externally from *Paradoxurus* and *Arctogalidia* in the absence of the striping and spotting. Like *Paradoxurus, Paguma* has a potent anal gland secretion, which it uses to ward off predators. The conspicuously marked head has been interpreted as a warning signal of the presence of the secretion. Females have two pairs of mammae (Lekagul and McNeely 1977).

The masked palm civet frequents forests and brush country. It reportedly raises its young in tree holes, at least in Nepal. It is arboreal and nocturnal (Roberts 1977). The omnivorous diet includes small vertebrates, insects, and fruits. An adult female radio-tracked for 12 months in Thailand by Rabinowitz (1991) moved within a total area of 3.7 sq km, had an average monthly range of 0.93 sq km, and moved an average of 620 meters per day.

Paguma is solitary, and apparently most young in the western parts of its range are born in spring and early summer (Roberts 1977). Births in Borneo have taken place in October (Banks 1978). Data on captives indicate that there may be two breeding seasons, in early spring and late autumn. Litters contain one to four young. They open their eyes at 9 days and are almost the size of adults by 3 months (Lekagul and McNeely 1977; Medway 1978). An individual lived at the London Zoo for more than 20 years (Marvin L. Jones, Zoological Society of San Diego, pers. comm., 1995).

In Tenasserim *Paguma* is reported to be a great ratter and not to destroy poultry. Medway (1978), however, wrote that it has been known to raid hen runs. The genus has been introduced on Honshu and Shikoku, Japan (Corbet and Hill 1992).

CARNIVORA; VIVERRIDAE; **Genus MACROGALIDIA**
Schwarz, 1910

Sulawesian Palm Civet

The single species, *M. musschenbroeki,* occurs only on Sulawesi (Laurie and Hill 1954).

Wemmer et al. (1983) reported that an adult male had a head and body length of 715 mm, a tail length of 540 mm, and a weight of 6.1 kg and that respective measurements for two adult females were 650 and 680 mm, 480 and 445 mm, and 3.8 and 4.5 kg. The upper parts are light brownish chestnut to dark brown. The underparts range from fulvous to whitish, with a reddish breast. The cheeks and a patch above the eye are usually buffy or grayish. Faint brown spots and bands are usually present on the sides and lower back, and the tail is ringed with dark and pale brown. The tail has more bands than that of *Arctogalidia* or *Paradoxurus.* Other distinguishing characters are the short, close fur and a whorl in the neck with the hairs directed forward. Both of the females noted above had two pairs of inguinal mammae.

According to Wemmer and Watling (1986), *Macrogalidia* occurs in both lowland and montane forests up to

Sulawesian palm civet *(Macrogalidia musschenbroeki)*, photo by Christen Wemmer.

about 2,600 meters. It seems to be dependent on primary forests but also moves onto adjacent grasslands and farms. It is a skillful climber and sometimes moves through trees but probably forages mainly on the ground. It feeds mainly on small mammals and fruits, especially palm fruit, and also raids domestic chickens and pigs. It is probably solitary, though a female and young may share a home range for a period after weaning. Although once thought to be extinct or restricted to the northern peninsula of Sulawesi and currently classified as vulnerable by the IUCN, recent observations indicate that *Macrogalidia* occurs in most parts of the island. It now seems to be neither abundant nor scarce but could be potentially jeopardized in some areas by logging and agricultural activity.

CARNIVORA; VIVERRIDAE; **Genus ARCTICTIS**
Temminck, 1824

Binturong

The single species, *A. binturong,* occurs from Sikkim, and probably originally from Nepal, to Indochina and the Malay Peninsula and on the Riau Archipelago, Sumatra, Bangka, Java, Borneo, and Palawan (Corbet and Hill 1992).

Head and body length is 610–965 mm and tail length is 560–890 mm. Weight is usually 9–14 kg, but a 19-kg individual reportedly was taken in Viet Nam (Rozhnov 1994). The fur is long and coarse, that on the tail being longer than that on the body. The lustrous black hairs often have a gray, fulvous, or buff tip. The head is finely speckled with gray and buff, and the edges of the ears and the whiskers are white. The ears have long hairs on the back that project beyond the tips and produce a fringed or tufted effect. The tail is particularly muscular at the base and prehensile at the tip. The only other carnivore with a truly prehensile tail is the kinkajou *(Potos),* which the binturong resembles in habits to some extent. Females have two pairs of mammae.

The binturong lives in dense forests and is nowhere abundant. It is mainly arboreal and nocturnal. It usually lies curled up, with the head tucked under the tail, when resting. It has never been observed to leap; rather, it progresses slowly but skillfully, using the tail as an extra hand. Its movements, at least during daylight hours, are rather slow and cautious, the tail slowly uncoiling from the last support as the animal moves carefully forward. According to Medway (1978), the binturong is reported to dive, swim, and catch fish. The diet also includes birds, carrion, fruit, leaves, and shoots.

Medway (1978) stated that *Arctictis* occurs either alone or in small groups of adults with immature offspring. Captives are very vocal, uttering high-pitched whines and howls, rasping growls, and, when excited, a variety of grunts and hisses. Lekagul and McNeely (1977) wrote that breeding seems to occur throughout the year. Numerous observations in captivity (Bulir 1972; Gensch 1963; Grzimek 1975; Kuschinski 1974; Xanten, Kafka, and Olds 1976) indicate that females are nonseasonally polyestrous and may give birth to two litters annually. Wemmer and Murtaugh (1981) reported the following data for reproduction in captivity: breeding occurs year-round, but there is a pronounced birth peak from January to March; the estrous cycle averages 81.8 days; the mean gestation period is 91.1

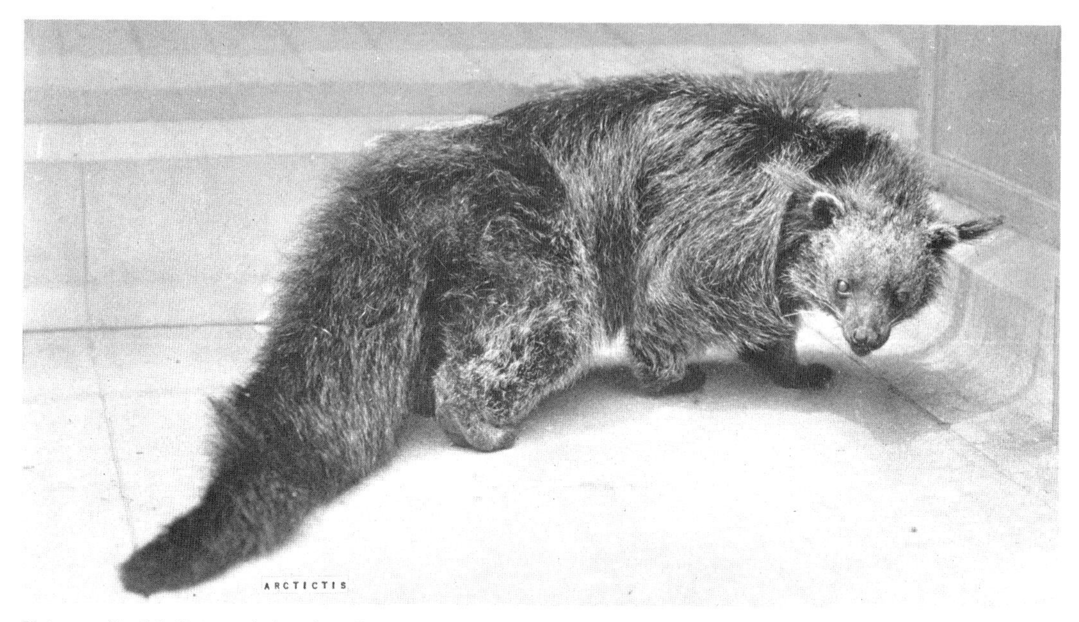

Binturong *(Arctictis binturong)*, photo from New York Zoological Society.

days (84–99); the number of young per litter averages 1.98 (1–6); the young weigh an average of 319 grams at birth and begin to take solid food at 6–8 weeks; the mean age of first mating is 30.4 months in females and 27.7 months in males; and both sexes can remain fertile until at least 15 years. According to Marvin L. Jones (Zoological Society of San Diego, pers. comm., 1995), a binturong lived at the Stuttgart Zoo for 25 years and 11 months.

Arctictis is sometimes kept as a pet. It is said to be easily domesticated, to become quite affectionate, and to follow its master like a dog. The subspecies *A. b. whitei* of Palawan is classified as vulnerable by the IUCN; it is declining because of habitat destruction by people.

CARNIVORA; VIVERRIDAE; **Genus HEMIGALUS**
Jourdan, 1837

Banded Palm Civet

The single species, *H. derbyanus*, occurs on the Malay Peninsula, including Tenasserim and southern Thailand, and on Sumatra, the Mentawai Islands (Siberut, Sipura, South Pagai), and Borneo (Chasen 1940; Corbet and Hill 1992; Ellerman and Morrison-Scott 1966; Lekagul and McNeely 1977; Medway 1977). The genera *Diplogale* and *Chrotogale* (see accounts thereof) sometimes are included in *Hemigalus*.

Head and body length is 410–620 mm, tail length is 255–383 mm, and weight is usually 1.75–3.0 kg. On the head a narrow, median dark streak extends from the nose to the nape; on either side of this streak a broader dark stripe encircles the eye and passes backward over the base of the ear. Two broad stripes, sometimes more or less broken into shorter stripes or spots, run backward from the neck and curve downward to the elbow. Behind these are two shorter stripes. The back behind the shoulders is marked with four or five broad transverse stripes separated by pale, usually narrower spaces, and there are two imperfect stripes at the base of the tail. The ground color is whitish to orange buff, usually lighter and more buffy below, and the tail is usually black. The tips of the hairs on the back of the neck point forward. The five-toed feet have strongly curved claws that are retractile like those of cats. Small scent glands are present.

On the Malay Peninsula *H. derbyanus* is restricted to tall forest and apparently is largely terrestrial (Medway 1978); however, it is at least partly arboreal and climbs well (Lekagul and McNeely 1977). All specimens of *Hemigalus* taken by Davis (1962) in Borneo were collected on the ground. Davis observed that the animals were exclusively nocturnal and apparently foraged on the forest floor, picking food up from the surface. Orthopterans and worms made up 80 percent of the contents of 12 stomachs, the remaining 20 percent consisted mostly of other invertebrates.

Gangloff (1975) reported that captive *H. derbyanus* were fond of fruit, did not construct nests, and marked with scent. A pregnant female *H. derbyanus* with one embryo was taken in Borneo in February. A captive female in the Wassenaar Zoo, Netherlands, had two young. They weighed 125 grams at birth, opened their eyes after 8–12 days, and first took solid food at about 70 days (Ewer 1973). One specimen lived in captivity for about 18 years (Marvin L. Jones, Zoological Society of San Diego, pers. comm., 1995).

Schreiber et al. (1989) expressed concern for the subspecies *H. derbyanus minor* and *H. d. sipora* of the Mentawai Islands. These are the only small islands on which *Hemigalus* is found, which suggests that its occurrence there is natural and that the genus was not widely carried about by people as some other viverrids of the East Indies were. The status of these subspecies is not well known, but they probably are being adversely affected by commercial logging and direct killing by people who consider them predators on domestic chickens. The species *H. derbyanus* is on appendix 2 of the CITES.

Banded palm civet *(Hemigalus derbyanus)*, photo from Duisburg Zoo.

CARNIVORA; VIVERRIDAE; **Genus DIPLOGALE**
Thomas, 1912

Hose's Palm Civet

The single species, *D. hosei*, is known only from the mountains of northern Borneo. *Diplogale* sometimes has been included in *Hemigalus* but was considered a separate genus by Corbet and Hill (1992), Schreiber et al. (1989), and Wozencraft (*in* Wilson and Reeder 1993).

Head and body length is about 470–540 mm and tail length is about 280–335 mm. Coloration is dark brown or black above and grayish, yellowish white, or slightly rufescent below. The ears are thinly haired and white inside. A buffy gray patch extends from above the eye to the cheek and terminates where it meets the white of the lips and throat. The inner side of the limb near the body is grayish, while the remainder of each limb is black. The tail is not banded but is dark throughout. From *Hemigalus, Diplogale* differs in more uniform coloration, lack of a median foramen between the paired incisive foramina of the skull, lack of strongly curved incisor tooth rows, and more prominent accessory cusps on the anterior premolars (Corbet and Hill 1992).

This genus is found mainly in montane forest and is largely terrestrial (Medway 1977). It probably is nocturnal, and its long facial whiskers and hairs between the footpads suggest that it may specialize in foraging for small animals among mossy boulders and in streams; one stomach contained various small insects (Payne, Francis, and Phillips 1985). It is known only from 15 museum specimens, the last of which was collected in 1955 (Schreiber et al. 1989). It is classified as vulnerable by the IUCN.

CARNIVORA; VIVERRIDAE; **Genus CHROTOGALE**
Thomas, 1912

Owston's Palm Civet

The single species, *C. owstoni*, now is known from about 40 specimens taken in Yunnan and Guangxi provinces in extreme southern China, Laos, and northern and central Viet Nam (Rozhnov, Kuznetzov, and Anh 1992; Schreiber et al. 1989). *Chrotogale* was included within *Hemigalus* by Corbet and Hill (1992) but not by Wozencraft (*in* Wilson and Reeder 1993).

Head and body length is 560–720 mm, tail length is 350–470 mm, and weight is 2.5–4.0 kg (Dang, Anh, and Huynh 1992). The body and base of the tail have alternating and sharply contrasting dark and light transverse bands, and longitudinal stripes are present on the neck. The pattern of stripes and bands resembles that of *Hemigalus derbyanus* but it is supplemented by black spots on the

Owston's palm civet *(Chrotogale owstoni)*, photo from Hanoi Zoo through Jorg Adler.

sides of the neck, the forelimbs and thighs, and the flanks. Four seems to be the maximum number of dorsal bands. It has been suggested that this striking pattern serves as a warning signal as *Chrotogale* is thought to possess a particularly foul-smelling anal gland secretion. The underparts are pale buffy, and a narrow orange midventral line runs from the chest to the inguinal region. The terminal two-thirds of the tail is completely black.

Form and markings are strikingly similar to those of *Hemigalus derbyanus*, but the hairs on the back of the neck of *Chrotogale* are not reversed in direction. *Chrotogale* is also distinguished by cranial and dental characters. The incisor teeth are remarkable in that they are broad, close-set, and arranged in practically a semicircle, a type unique among carnivores and only approached in certain of the marsupials. The other teeth and the skull are also peculiar, indicating habits and a mode of life different from that of other genera in the family Viverridae.

The natural history of this genus was almost unknown until considerable new information was provided by Dang, Anh, and Huynh (1992). *Chrotogale* prefers forests and other areas of dense vegetation in the vicinity of streams and lakes. It makes simple dens under large tree trunks or in thick brush and also frequently uses natural holes in trees, rocks, or the ground. It is terrestrial but can climb well and often enters trees in search of food. Activity is mainly nocturnal, usually commencing at dusk and ending early in the morning. The natural diet consists mostly of earthworms but also includes small vertebrates, insects, and fruit. Captives readily accept beef, chicken, and bananas.

Although the genus is reported to be solitary in the wild, several individuals of both sexes have lived together peacefully in captivity and have accepted new members to the group without any show of aggression. They are generally silent and mark their home area with secretions from the anal-genital region. Mating seems to occur mainly from January to March but may continue until November, and there are one or two litters annually. The gestation period is about 60 days, there are one to three young, and newborn weight is 75–88 grams.

Habitat loss and intensive hunting during the last few decades have greatly reduced the range and numbers of *Chrotogale*, and it now is legally protected as an endangered species in Viet Nam. Schreiber et al. (1989), however, indicated that it can survive close to villages and even approaches houses in search of kitchen wastes. It is classified as vulnerable by the IUCN.

CARNIVORA; VIVERRIDAE; **Genus CYNOGALE**
Gray, 1837

Otter Civets

Corbet and Hill (1992) and Schreiber et al. (1989) recognized two species:

C. bennettii, peninsular Malaysia and possibly adjacent southern Thailand, Sumatra, Borneo;
C. lowei, known with certainty only by a single specimen from Bac Can in northern Viet Nam, but likely records from southern Yunnan and northern Thailand.

Wozencraft (*in* Wilson and Reeder 1993) included *C. lowei* in *C. bennettii.*

Head and body length is 575–675 mm, tail length is 130–205 mm, and weight is usually 3–5 kg. The form is somewhat like that of the otters *(Lutra)*. The underfur is close, soft, and short. It is pale buff near the skin and shades to dark brown or almost black at the tip. The longer, coarser guard hairs are usually partially gray, which gives a frosted or speckled effect on the head and body. The lower side of the body is lighter brown and not speckled with gray. The whiskers are remarkably long and plentiful; those on the snout are fairly long, but those on a patch under the ear are the longest. The newly born young lack dorsal

Otter civet *(Cynogale bennettii)*, photo from San Diego Zoological Society.

speckling; they have some gray on the forehead and ears and two longitudinal stripes down the sides of the neck extending under the throat.

Because of the deepening and expansion of the upper lip, the rhinarium occupies a horizontal position, with the nostrils opening upward on top of the muzzle. The nostrils can be closed by flaps, an adaptation for aquatic life. The ears also can be closed. Although the webbing on the feet does not extend farther toward the tips of the digits than in such genera as *Paradoxurus,* it is quite broad, and the fingers are capable of considerable flexion. A glandular area, merely three pores in the skin, is located near the genitals and secretes a mild scent material. The premolar teeth are elongate and sharp, adapted for capturing and holding prey, and the molars are broad and flat, for crushing. Females have four mammae.

Cynogale is usually found near streams and swampy areas. It can climb well and often takes refuge in a tree when chased by dogs. While walking it usually carries its head and tail low and arches its back. Although it is partly adapted for an aquatic life, its tail is short and lacks special muscular power, and the webbing between the digits is only slightly developed. *Cynogale* thus is probably a slow swimmer and cannot turn quickly in the water. It probably captures aquatic animals only after they have taken shelter from the chase and catches some birds and mammals as they come to drink. It cannot be seen by its prey because it is submerged with only the tip of the nose exposed above the surface of the water. The diet includes crustaceans, perhaps mollusks, fish, birds, small mammals, and fruits. There are records of pregnant females with two and three embryos. Young still with the mother have been noted in May in Borneo. A captive specimen lived five years (Jones 1982).

Cynogale bennettii (including *C. lowei*) now is classified as endangered by the IUCN and is on appendix 2 of the CITES. It is thought to have declined by at least 50 percent in the past decade and to now number fewer than 2,500 mature individuals. Schreiber et al. (1989) noted that *C. bennettii* is probably jeopardized by expanding human settlement and agriculture along rivers. The single known specimen of *C. lowei* was collected in 1926; however, there have been a few recent reports.

CARNIVORA; VIVERRIDAE; **Genus FOSSA**
Gray, 1864

Malagasy Civet

The single species, *F. fossana,* originally was found throughout the rainforests of eastern and southern Madagascar (Schreiber et al. 1989). Although the generic name is the same as the vernacular term for another Malagasy carnivore *(Cryptoprocta),* the two animals are not closely related. The specific name sometimes is given as *fossa.*

Head and body length is 400–450 mm and tail length is 210–30 mm (Albignac 1972). Males weigh up to 2 kg and females, up to 1.5 kg (Coetzee *in* Meester and Setzer 1977). The ground color is grayish, washed with reddish. There are four rows of black spots on each side of the back and a few black spots on the backs of the thighs. These spots may merge to form stripes, and the gray tail is banded with brown. The underparts are grayish or whitish and more or less obscurely spotted. The limbs are slender, perhaps being adapted for running. There are no anal scent glands, but there probably are marking glands on the cheeks and neck (Albignac 1972).

The Malagasy civet inhabits evergreen forests and shelters in hollow trees or crevices. It is nocturnal and may occur in trees or on the ground. The preferred foods are crustaceans, worms, small eels, and frogs. Other kinds of animal matter and fruit are also taken (Coetzee *in* Meester and Setzer 1977).

Fossa lives in pairs, which share a territory. Vocalizations include cries, groans, and a characteristic "coq-coq," heard only in the presence of more than one individual. Births have been recorded from October to January. The gestation period is 3 months and a single offspring is born. It weighs 65–70 grams at birth, is weaned after 2 months, and prob-

Malagasy civets *(Fossa fossa)*, photo from U.S. National Zoological Park.

ably attains adult weight at 1 year (Albignac 1970*b*, 1972; Coetzee *in* Meester and Setzer 1977). A captive specimen lived 11 years (Jones 1982).

The Malagasy civet is classified as vulnerable by the IUCN and is on appendix 2 of the CITES. It still is not uncommon, but its habitat is rapidly declining because of human encroachment and now has been reduced to isolated patches; it also is subject to hunting and trapping (Schreiber et al. 1989).

CARNIVORA; VIVERRIDAE; **Genus EUPLERES**
Doyère, 1835

Falanouc

The single species, *E. goudotii*, is found in the coastal forests of Madagascar (Coetzee *in* Meester and Setzer 1977).

Head and body length is 450–650 mm, tail length is 220–50 mm, and weight is 2–4 kg (Albignac 1974). The subspecies *E. g. goudotii* is fawn-colored above and lighter below. In the subspecies *E. g. major* the males are brownish and the females are grayish. The pelage is woolly and soft, being made up of a dense underfur and longer guard hairs. The tail is covered by rather long hairs that give a bushy appearance.

Eupleres has a pointed muzzle, a narrow and elongate head, and short, conical teeth. It resembles the civets in some structural features and the mongooses in others. The small teeth are similar to each other and resemble those of insectivores rather than carnivores. Indeed, *Eupleres* was classified as an insectivore before its somewhat obscure relationship with the mongooses was detected. The claws are relatively long and not retractile, or are only imperfectly so. The feet are peculiar in the comparatively large size and low position of the great toe and thumb.

The falanouc inhabits humid, lowland forests. It is crepuscular and nocturnal, resting by day in crevices and burrows. It is terrestrial, and if threatened, it may either run or remain motionless. In autumn up to 800 grams of fat can be accumulated in the tail, and it has been suggested that *Eupleres* hibernates during winter; however, active individuals have been observed in winter. The diet consists mainly of earthworms and also includes other invertebrates and frogs but apparently not reptiles, birds, rodents, or fruit (Albignac 1974; Coetzee *in* Meester and Setzer 1977).

Eupleres may live alone or in small family groups. It has several vocalizations and other means of communication. Mating probably takes place in July or August (winter), and a birth was observed in November. Litters contain one or two young. Weight of the newborn is 150 grams, and weaning occurs at nine weeks (Albignac 1974; Coetzee *in* Meester and Setzer 1977).

The falanouc is classified as endangered by the IUCN and is on appendix 2 of the CITES. It has declined in numbers and distribution because of deforestation, drainage of marshes, excessive hunting for use as food, predation by domestic dogs, and possibly competition from the introduced *Viverricula indica*. It still occurs over a large area but is nowhere common (Schreiber et al. 1989).

CARNIVORA; **Family HERPESTIDAE**

Mongooses and Fossa

This family of 19 Recent genera and 39 species is found in southern Asia, the East Indies, Africa, and Madagascar. One

Falanoucs *(Eupleres goudotii),* photos by R. Albignac.

genus *(Herpestes)* occurs in Spain and Portugal, probably brought there by people in ancient times, and also has been introduced in other parts of Europe and on many islands around the world. The following sequence of genera, grouped in 3 subfamilies, is based on the classifications of Coetzee (*in* Meester and Setzer 1977), Ellerman and Morrison-Scott (1966), and Ewer (1973):

Subfamily Galidiinae (Malagasy mongooses), genera *Galidia, Galidictis, Mungotictis, Salanoia;*

Subfamily Herpestinae (mongooses), genera *Herpestes, Galerella, Mungos, Crossarchus, Liberiictis, Helogale, Dologale, Bdeogale, Rhynchogale, Ichneumia, Atilax, Cynictis, Paracynictis, Suricata;*

Subfamily Cryptoproctinae (fossa), genus *Cryptoprocta.*

The above three subfamilies often have been considered components of the family Viverridae, but there seems to be a growing consensus that the Herpestidae warrant full familial distinction (Corbet and Hill 1991, 1992; Hunt and Tedford 1993; Wozencraft (1989*a*, 1989*b*, and *in* Wilson and Reeder 1993). Some evidence even suggests that the Herpestidae are more closely related to the Hyaenidae than to the Viverridae, with the latter family being closest to the Felidae. There still is disagreement about the affinity of the Cryptoproctinae. Wozencraft (1989*a* and *in* Wilson and Reeder 1993) placed that subfamily in the Viverridae, but recent DNA analyses by Véron and Catzeflis (1993) indicate that it is more closely related to the Herpestidae.

The Herpestidae resemble the Viverridae in general appearance, size, distribution, and many aspects of natural history. They are on average somewhat smaller; excluding *Cryptoprocta,* head and body length is 180–710 mm, tail length is 150–530 mm, and weight is 230–5,200 grams. In *Cryptoprocta* head and body length is as much as 800 mm and weight is as much as 12 kg. The pelage of herpestids tends to be more uniform in color than that of viverrids, though some genera have stripes, bands, or contrasting patterns. The body is long, the limbs are rather short, and the tail usually is relatively shorter than in the Viverridae. Most genera have five toes on each foot; *Cynictis* has only four digits on the hind foot, and *Bdeogale, Paracynictis,* and *Suricata* have only four digits on all feet (Meester et al. 1986). The claws usually are not retractile. Female herpestids have two or three pairs of abdominal mammae. Males have a baculum. Like viverrids, herpestids have scent glands in the anal region. In herpestids, however, these glands open into a pouch or saclike depression outside the anus proper in which the secretion is stored. The skull is commonly shorter and broader than that of the Viverridae (Schliemann *in* Grzimek 1990). The dental formula is: (i 3/3, c 1/1, pm 3–4/3–4, m 2/2) × 2 = 36–40.

The presence of prominent anal scent glands and a large associated sac is one of the critical characters distinguishing the Herpestidae as a family from the Viverridae, in which the scent glands are simple. Other diagnostic characters are found in the skull, especially in the basicranial region (Hunt and Tedford 1993; Wozencraft 1989*b;* Wyss and Flynn 1993). In the Herpestidae the caudal entotympanic does not penetrate into the anterior chamber of the auditory bulla (as it does in the Viverridae), the ectotympanic contributes to an external auditory meatal tube (it does not in the Viverridae), and the internal carotid artery is enclosed in an osseous tube within the bony wall of the middle ear prior to entering the cranial cavity (in the Viverridae this artery is not enclosed in a bony tube and actually enters the internal space of the bulla). In these characters the Viverridae resemble the Felidae more than they do the Herpestidae.

Mongooses tend to live in more open country than do civets, to be more diurnal, and to form larger social groups. Therefore, some genera have been studied extensively. Several genera, including *Cynictis* and *Suricata,* live in colonies in ground burrows. Some genera of mongooses associate in bands and take refuge as a group in any convenient shelter. Mongooses, particularly of the genus *Herpestes,* have been introduced into several areas to check the numbers of rodents and venomous snakes. Such introductions, however, generally have not proven beneficial because the mongooses quickly multiply and destroy many desirable forms of mammals and birds.

The geological range of this family is Oligocene to Recent in Europe, Miocene to Recent in Asia and Africa, and Pleistocene to Recent in Madagascar.

CARNIVORA; HERPESTIDAE; **Genus GALIDIA**
I. Geoffroy St.-Hilaire, 1837

Malagasy Ring-tailed Mongoose

The single species, *G. elegans,* is found in eastern, west-central, and extreme northern Madagascar (Albignac 1969).

Head and body length is about 380 mm and tail length is about 305 mm. Weight is 700–900 grams (Coetzee *in* Meester and Setzer 1977). The general coloration is dark chestnut brown, and the tail is ringed with dark brown and black. *Galidia* has some of the structural features of civets and some of mongooses. The feet differ from those of *Galidictis* in having shorter digits, fuller webbing, shorter claws, and more hairy soles. The lower canine teeth are smaller. From *Salanoia, Galidia* is distinguished by its ringed tail and very small second upper premolar. Dissection of two individuals indicates the presence of a scent gland, closely associated with the external genitalia, in males but not in females.

Galidia occurs in humid forests. It shelters in burrows, which it digs very rapidly, and probably also in hollow trees. It is mainly diurnal but may also be active at night. More arboreal than most mongooses, it is able to climb and descend on vertical trunks only 4 cm in diameter. It can also swim. The diet consists mainly of small mammals, birds and their eggs, and frogs and also includes fruits, fish, reptiles, and invertebrates (Albignac 1972; Coetzee *in* Meester and Setzer 1977).

Galidia is less social than most mongooses, being found alone or in pairs. Mating in Madagascar occurs from April to November, with births from July to February. The gestation period is 79–92 days, and a single young weighing 50 grams is produced. Physical maturity is reportedly attained at 1 year, and sexual maturity at 2 years (Coetzee *in* Meester and Setzer 1977; Larkin and Roberts 1979). One animal lived in captivity for 24 years and 5 months (Marvin L. Jones, Zoological Society of San Diego, pers. comm., 1995).

Malagasy ring-tailed mongoose *(Galidia elegans)*, photo from U.S. National Zoological Park.

CARNIVORA; HERPESTIDAE; **Genus GALIDICTIS**
I. Geoffroy St.-Hilaire, 1839

Malagasy Broad-striped Mongooses

There are two species (Coetzee *in* Meester and Setzer 1977; Wozencraft 1986, 1987):

G. fasciata, eastern Madagascar;
G. grandidieri, southwestern Madagascar.

Head and body length is 320–40 mm and tail length is 280–300 mm (Albignac 1972). The general body color is pale brown or grayish. In the subspecies *G. fasciata striata* there are usually 5 longitudinal black bands or stripes on the back and sides, and the tail is whitish. In *G. fasciata fasciata* there are usually 8–10 stripes, and the tail is bay-colored and somewhat bushy. *G. grandidieri* is larger than *G. fasciata* and has 8 dark brown longitudinal stripes on the back and sides that are narrower than the intervening spaces.

Galidictis differs from other viverrids in color pattern and in cranial and dental characters. The feet differ from those of *Galidia* in having longer digits, less extensive webbing, and longer claws. A scent pouch is present in females.

The species *G. fasciata* occurs in forests and is nocturnal and crepuscular. The diet consists mainly of small vertebrates, especially rodents, and also includes invertebrates. The three known specimens of *G. grandidieri* were taken in an area of spiny desert vegetation. *Galidictis* is found in pairs or small social groups. It apparently produces one young per year, during the summer (Albignac 1972; Coetzee *in* Meester and Setzer 1977). Young individuals are said to tame readily, to follow their masters, and even to sleep in their masters' laps. *G. fasciata* is classified as vulnerable by the IUCN, *G. grandidieri* as endangered. Schreiber et al. (1989) reported the former to be locally common but threatened by habitat destruction. *G. grandidieri* had not been recorded since 1929, when a report of its possible survival in southwestern Madagascar was received from villagers in 1987.

CARNIVORA; HERPESTIDAE; **Genus MUNGOTICTIS**
Pocock, 1915

Malagasy Narrow-striped Mongoose

The single species, *M. decemlineata*, is found in western and southwestern Madagascar (Coetzee *in* Meester and Setzer 1977).

Head and body length is 250–350 mm, tail length is 230–70 mm, and weight is 600–700 grams. The fur is rather dense and generally gray beige in color. There are usually 8–10 dark stripes on the back and flanks. The underparts are pale beige. The soles of the feet are naked and the digits are partly webbed. Glands on the side of the head and on the neck appear to be for marking with scent. Females have a

Malagasy broad-striped mongoose *(Galidictis fasciata)*, photo by Peter Schachenmann through Steven M. Goodman.

single pair of inguinal mammae (Albignac 1971, 1972, 1976).

Mungotictis is found on sandy, open savannahs. It is diurnal and both arboreal and terrestrial. It spends the night in tree holes during the wet summer and in ground burrows during the dry winter. It can swim well. The diet consists mostly of insects but also includes small vertebrates, birds' eggs, and invertebrates. To break an egg or the shell of a snail, *Mungotictis* lies on one side, grasps the object with all four feet, and throws it abruptly until it breaks and the contents can be lapped up (Albignac 1971, 1972, 1976).

In a radio-tracking study Albignac (1976) determined that 22 individuals inhabited 300 ha. The animals were divided into two stable social units that sometimes engaged in agonistic encounters where their ranges met. Each group contained several adults of both sexes, as well as juveniles

Malagasy narrow-striped mongoose *(Mungotictis decemlineata)*. Inset: head *(M. decemlineata)*. Photos by Don Davis.

and young of the year. Group cohesiveness and intrarelationships varied. Generally, adult males and females came together in the summer. In the winter there was division into small units, such as temporary pairs, maternal family parties, all-male groups, and solitary males.

The breeding season extends from December to April, peaking in February and March (summer). The gestation period is 90–105 days. There is usually a single offspring weighing 50 grams at birth. Weaning occurs at 2 months, but the young remain with their mothers until they are 2 years old (Albignac 1972, 1976).

This mongoose is now classified as vulnerable by the IUCN. Schreiber et al. (1989) reported that it appears to be subject to very little direct human persecution but that its habitat is being burned and cleared at an alarming rate.

CARNIVORA; HERPESTIDAE; **Genus SALANOIA**
Gray, 1864

Malagasy Brown-tailed Mongoose, or Salano

The single species, *S. concolor,* is found in northeastern Madagascar (Coetzee *in* Meester and Setzer 1977).

Head and body length is 250–300 mm and tail length is 200–250 mm (Albignac 1972). The general coloration is brown, and there are either dark or pale spots. The tail is the same color as the body and is not ringed. The claws are not strongly curved, the ears are broad and short, and the muzzle is pointed.

The salano occurs only in the evergreen forests on the northeastern part of the central plateau of Madagascar. It is typically diurnal, sheltering at night in tree trunks or burrows. It feeds mainly on insects and fruits but also on frogs, small reptiles, and rodents. It occurs individually or in pairs depending on the season. The young are born mainly during the summer (Albignac 1972; Coetzee *in* Meester and Setzer 1977). A captive lived 4 years and 9 months (Jones 1982). *Salanoia* is declining because of habitat loss and now is classified as vulnerable by the IUCN.

CARNIVORA; HERPESTIDAE; **Genus HERPESTES**
Illiger, 1811

Mongooses

There are 2 subgenera and 10 species (Chasen 1940; Coetzee *in* Meester and Setzer 1977; Corbet 1978; Corbet and Hill 1992; Ellerman and Morrison-Scott 1966; Medway 1977; Sanborn 1952; D. R. Wells 1989; Wells and Francis 1988):

subgenus *Herpestes* Illiger, 1811

H. ichneumon, southern Spain and Portugal (probably through human introduction), Asia Minor to Palestine, Morocco to Tunisia, Egypt, possibly eastern Libya, most of Africa south of the Sahara;
H. javanicus, Iraq through northern India to extreme southern China and Indochina, Malay Peninsula, Hainan, Java;
H. edwardsii, eastern Arabian Peninsula through Afghanistan and India to Assam, Sri Lanka;
H. smithii, peninsular India, Sri Lanka;
H. fuscus, southwestern India, Sri Lanka;
H. vitticollis, southwestern India, Sri Lanka;
H. urva, Nepal to southeastern China and peninsular Malaysia, Taiwan, Hainan;
H. semitorquatus, Sumatra, Borneo;
H. brachyurus, Malay Peninsula, Sumatra, Borneo, Palawan;

Malagasy brown-tailed mongoose *(Salanoia concolor)*, photo of mounted specimen in Field Museum of Natural History.

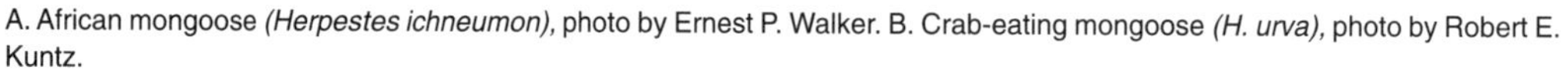
A. African mongoose *(Herpestes ichneumon)*, photo by Ernest P. Walker. B. Crab-eating mongoose *(H. urva)*, photo by Robert E. Kuntz.

subgenus *Xenogale* J. A. Allen, 1919

H. naso, southeastern Nigeria to eastern Zaire.

H. auropunctatus, found from Iraq to Southeast Asia, sometimes has been treated as a species distinct from *H. javanicus,* but the two were considered conspecific by Corbet and Hill (1992) and Lekagul and McNeely (1977). *H. palustris,* described from the vicinity of Calcutta in eastern India (Ghose 1965; Ghose and Chaturvedi 1972), was listed as a separate species by Wozencraft (*in* Wilson and Reeder 1993) but placed within *H. javanicus* by Corbet and Hill (1992). Wozencraft also included *H. fuscus* in *H. brachyurus,* but Corbet and Hill argued that such a procedure is unjustified. Rosevear (1974) treated *Xenogale* as a distinct genus, but it was regarded as a subgenus or synonym of *Herpestes* by Corbet and Hill (1991), Meester et al. (1986), and Wozencraft (*in* Wilson and Reeder 1993). *Galerella* (see account thereof) also sometimes is included in *Herpestes.*

Head and body length is 250–650 mm, tail length is 200–510 mm, and weight is 0.5–4.0 kg. Coloration varies considerably. Some forms are greenish gray, yellowish brown, or grayish brown. Others are finely speckled with white or buff. The underparts are generally lighter than the back and sides and are white in some species. The fur is short and soft in some species and rather long and coarse in others. The body is slender, the tail is long, there are five digits on each limb, the hind foot is naked to the heel, and the foreclaws are sharp and curved. Small scent glands are situated near the anus, and some species can eject a vile-smelling secretion. Females have four or six mammae.

Mongooses occupy a wide variety of habitats ranging from densely forested hills to open, arid plains. They shelter in hollow logs or trees, holes in the ground, or rock crevices. They may be either diurnal or nocturnal. They are basically terrestrial but are very agile, and some species can climb skillfully (Grzimek 1975). During the morning they frequently stretch out in an exposed area to sun themselves. The diet includes insects, crabs, fish, frogs, snakes, birds, small mammals, fruits, and other vegetable matter. Some species kill cobras and other venomous snakes. Contrary to popular belief, mongooses are not immune to the bites of these reptiles. Rather, they are so skillful and quick in their movements that they avoid being struck by the snake and almost invariably succeed in seizing it behind the head. The battle usually ends with the mammal's eating the snake.

Population densities of 1–14/sq km have been recorded for *H. javanicus* in the West Indies (Hoagland, Horst, and Kilpatrick 1989). Data cited by Ewer (1973) indicate that in Hawaii the individual home range diameter of *H. javanicus* is about 1.6 km for males and 0.8 km for females. Mongooses are found alone, in pairs, or in groups of up to 14 individuals. In Spain, Palomares and Delibes (1993) observed most *H. ichneumon* alone, but pairs consisting of an adult male and female also were sighted year-round. Males were territorial, and the core areas of their ranges overlapped with those of one or more females. Ben-Yaacov and Yom-Tov (1983) found that in Israel family groups of *H. ichneumon* occupied permanent home ranges and consisted of 1 adult male, 2–3 females, and young. The group produced a single annual litter in the spring, and the young remained with the family for a year or more.

Some species breed throughout the year, with females giving birth two or three times annually. One female *H. edwardsii* produced five litters in 18 months. In *H. javanicus* the estrous cycle lasts about 3 weeks, estrus 3–4 days, and gestation 42–49 days. In *H. ichneumon* gestation has been variously reported to last 60–84 days. Litters contain one to four young. The young of *H. javanicus* are weaned after 4–5 weeks and attain sexual maturity at 1 year. Captive *H. ichneumon* have lived more than 20 years (Ewer 1973; Grzimek 1975; Kingdon 1977; Lekagul and McNeely 1977; Nellis 1989; Roberts 1977).

Mongooses have been widely introduced by people to kill rats and snakes. The presence of *H. ichneumon* in Spain and Portugal probably results from introduction in ancient times. That species also has been introduced to Madagascar. *H. javanicus* was introduced in the West Indies during the early 1870s, and populations now are present on 29 islands there, including all of the Greater Antilles. *H. javanicus* also has been established on the northeastern coast of South America, Mafia Island off East Africa, Mauritius, islands in the Adriatic off Croatia, and the Hawaiian and Fiji islands. Several individuals of this species have been taken on the mainland of North America. *H. edwardsii* has been introduced in central Italy and on the Malay Peninsula (though the population there probably is now extinct), Mauritius, and the Ryukyu Islands (Carpaneto 1990; Corbet 1978, 1984; Corbet and Hill 1992; Gorman 1976; Krystufek and Tvrtkovic 1992; Nellis et al. 1978; Van Gelder 1979; D. R. Wells 1989). In addition to killing rats and snakes, mongooses have destroyed harmless birds and mammals and have contributed to the extinction or endangerment of many desirable species of wildlife. They also have become pests by preying on poultry. The importation or possession of mongooses is therefore now forbidden by law in some countries.

CARNIVORA; HERPESTIDAE; **Genus GALERELLA**
Gray, 1865

Gray and Slender Mongooses

There are four species (Skinner and Smithers 1990; Taylor and Goldman 1993; Watson 1990; Watson and Dippenaar 1987):

G. pulverulenta (Cape gray mongoose), South Africa to south of 28° S, Lesotho;
G. nigrita, southern Angola, northern and central Namibia;
G. sanguinea (slender mongoose), throughout Africa from Senegal and possibly Western Sahara to Sudan, and south to northern and eastern South Africa, but not the central forest zone of Gabon, Congo, and some adjacent areas.
G. ochracea, Somalia.

Galerella was treated as a subgenus of *Herpestes* by Taylor and Goldman (1993) but was considered a full genus by the other authorities cited above, as well as Corbet and Hill (1991), Meester et al. (1986), and Wozencraft (*in* Wilson and Reeder 1993). Watson and Dippenaar (1987) indicated that another named species, *G. swinnyi*, may have occurred in southeastern South Africa but that it could not be properly evaluated based on the limited available material. Wozencraft included *swinnyi* in *G. sanguinea* and noted that it is believed to have been extirpated from the type locality. Wozencraft followed Crawford-Cabral (1989) in using the name *G. flavescens* in place of *G. nigrita* and followed Watson (1990) in listing *G. swalius* of central and southern Namibia as a species distinct from *G. sanguinea*. Taylor and Goldman (1993) are followed here with regard to both those issues and also in accepting *G. ochracea* as a species separate from *G. sanguinea*.

Galerella generally is smaller than *Herpestes*. Head and body length is 268–425 mm, tail length is 205–340 mm, and weight is 373–1,250 grams. Males are about 9 percent larger than females. The pelage may be short or shaggy, and the color varies extensively. The upper parts are dark brown or even black, drab or pale brown, grayish, or grizzled reddish or yellowish. The hairs of the underparts lack annulations, being grayer in gray specimens, redder or yellow in the more brightly colored. The tip of the tail is dark or completely black. Females have been reported to have two or three pairs of mammae. From that of *Herpestes*, the skull of *Galerella* differs in being smaller, usually lacking the lower first premolar in adults, and in having the anterior chamber of the auditory bulla inflated and comparable in size to the posterior portion, rather than flattish and much smaller than the inflated posterior portion (Kingdon 1977; Meester et al. 1986; Skinner and Smithers 1990; Watson 1990; Wozencraft 1989*a*).

Skinner and Smithers (1990) wrote that these mongooses are found in savannah, woodland, arid country, and a variety of other wet and dry habitats but not in dense forests or deserts. The elevational range is sea level to 3,600 meters. They are diurnal and terrestrial but good climbers and will readily take to trees. The diet consists mostly of insects and also includes a substantial amount of small vertebrates and some plant material.

In a radio-tracking study of *G. pulverulenta* in Cape Province, Cavallini and Nel (1990) found home ranges of 21–63 ha., with much overlap both between and within sexes. Ewer (1973) noted that family parties of *G. pulverulenta* den together but forage individually. Taylor (1975) wrote that *G. sanguinea* travels alone or in pairs, that its home range may be as small as 1 sq km but is much larger in desert areas, that it may be territorial, and that it is generally silent. Skinner and Smithers (1990), however, reported a variety of vocalizations and indicated that animals usually are solitary. In southern Africa *G. sanguinea* apparently gives birth in the wet summer months, from about October to March. *G. pulverulenta* has been reported to give birth from about August through December in Cape Province. The young are born in holes in the ground, crevices in rocks, hollow logs, or other suitable shelters. Litter sizes of one to three young have been recorded (Cavallini 1992). The gestation period has been reported as 58–62 days, the lactation period as 50–65 days (Hayssen, Van Tienhoven, and Van Tienhoven 1993).

CARNIVORA; HERPESTIDAE; **Genus MUNGOS**
E. Geoffroy St.-Hilaire and G. Cuvier, 1795

Banded and Gambian Mongooses

There are two species (Coetzee *in* Meester and Setzer 1977):

M. gambianus (Gambian mongoose), savannah zone from Gambia to Nigeria;
M. mungo (banded mongoose), Gambia to northeastern Ethiopia and south to South Africa.

The name *Mungos* formerly was sometimes used for many other species of mongooses, including those now assigned to *Herpestes*.

Head and body length is 300–450 mm, tail length is 230–90 mm, and weight is 1.0–2.2 kg. *M. mungo* is brownish gray with dark brown and well-defined yellowish or whitish bands across the back. The banded pattern is produced by hair markings of the same type that produce the ground color of the remainder of the body. The hairs are alternately ringed with dark and light bands; the color rings of the individual hairs coincide with like colors on adjacent hairs. *M. gambianus* lacks the transverse bands but has a dark streak on the side of the neck (Coetzee *in* Meester and Setzer 1977).

The pelage is coarse; compared with other mongooses, *Mungos* has little underfur. Although the tail is not bushy, it is covered with coarse hair and is tapered toward the tip. The foreclaws are elongate, the soles are naked to the wrist and heel, and there are five digits on each limb. There is no naked grooved line from the tip of the nose to the upper lip. Females have six mammae.

The information in the remainder of this account applies to *M. mungo* and was taken from Kingdon (1977), Neal (1970), Rood (1974, 1975), Rosevear (1974), and Simpson (1964). The banded mongoose is found in grassland, brushland, woodland, and rocky, broken country. It dens mainly in old termite mounds but also in such places as erosion gullies, abandoned aardvark holes, and hollow logs. Dens are communal and consist of one to nine entrance holes, a central sleeping chamber of about 1–2 cubic meters, and perhaps several smaller chambers. Most dens are used only for a few days, but some favorite sites may be occupied for as long as 2 months. *M. mungo* is terrestrial and diurnal and has excellent senses of vision, hearing, and smell. A group generally emerges from the den around 0700–0800 hours, forages for several hours, rests in a shady spot during the hottest part of the day, forages again, and returns to a den before sunset. More time is spent in the vicinity of the den if young are present therein. A group generally covers 2–3 km per day, moving in a zigzag pattern and searching among rocks and vegetation for food. The diet consists largely of invertebrates, especially beetles and millipedes, and also includes small vertebrates. To break an egg and obtain the contents, *M. mungo* grasps the item with its forefeet and propels it backward between its hind feet and against a hard object.

In the Ruwenzori National Park of Uganda population density was found to be about 18/sq km, and group home range varied from about 38 to 130 ha. In the Serengeti

Banded mongoose *(Mungos mungo)*, photo from Zoological Garden Berlin-West through Ernst von Roy.

Banded mongooses *(Mungos mungo)*, photo by J. P. Rood.

home range may be more than 400 ha. Ranges overlap, but intergroup encounters are generally noisy and hostile and sometimes involve chasing and fighting. Groups contain as many as 40 individuals, usually about 10–20, including several adults of both sexes. Captive females have been seen to dominate males. The animals sleep in contact and forage in a fairly close, but not bunched, formation. Groups are cohesive, but some splitting off has been observed, and mating between members of different groups sometimes occurs. Individuals may mark one another with scent from anal glands. The most common vocalization is a continuous birdlike twittering that probably serves to keep the group together during foraging. There are also various agonistic growls and screams and an alarm chitter.

In East Africa, at least, reproduction continues throughout the year. Breeding is synchronized within a given group, with several females bearing litters at approximately the same time. Groups breed as many as four times per year, though it cannot be said that any one female has that many litters annually. A period of mating often begins within 1–2 weeks after the birth of young. The gestation period is about 2 months. The number of young per litter seems usually to be about two or three but may be as high as six. The young weigh about 20 grams at birth but grow rapidly. They apparently are kept together and raised commonly by the group. They suckle indiscriminately from any lactating female and are usually guarded by one or two adult males while the rest of the group forages. They begin to travel with the others at about 1 month. Females attain sexual maturity at 9–10 months. According to Van Rompaey (1978), a captive *M. mungo* lived to an age of approximately 12 years.

CARNIVORA; HERPESTIDAE; **Genus CROSSARCHUS**
F. Cuvier, 1825

Cusimanses

There are four species (Colyn and Van Rompaey 1990; Colyn et al. 1995; Goldman 1984; Schreiber et al. 1989):

C. obscurus, Sierra Leone to Ghana;
C. platycephalus, southern Benin to southeastern Cameroon and possibly Congo;
C. alexandri, Zaire, Uganda, possibly Central African Republic and Zambia;
C. ansorgei, central Zaire, one specimen from northwestern Angola.

C. platycephalus was considered conspecific with *C. obscurus* by Wozencraft (1989*a* and *in* Wilson and Reeder 1993) but not by Corbet and Hill (1991), Goldman (1984), or Van Rompaey and Colyn (1992).

Head and body length is 305–450 mm, tail length is 150–255 mm, and weight is 450–1,450 grams. The body is covered by relatively long, coarse hair that is a mixture of browns, grays, and yellows. The head is usually lighter-colored than the remainder of the body, while the feet and legs are usually the darkest. The legs are short, the tail is tapering, the ears are small, and the face is sharp.

Cusimanses live in forests and swampy areas. Various reports suggest that they may be active either by day or by night (Kingdon 1977). They travel about in groups, seldom remaining in any one locality longer than two days and taking temporary shelter in any convenient place. While seeking food they scratch and dig in dead vegetation and in the soil. The diet consists principally of insects, larvae, small reptiles, crabs, tender fruits, and berries. It is said that cusimanses crack the shells of snails and eggs by hurling them

Cusimanse *(Crossarchus obscurus)*, photo by Ernest P. Walker.

with the forepaws back between the hind feet and against some hard object.

Groups contain 10–24 individuals. These probably represent 1–3 family units, each with a mated pair and the surviving members of 2–3 litters. There is no evidence of seasonal breeding in either East or West Africa. Observations in captivity indicate that there may be several litters annually (Kingdon 1977; Rosevear 1974). According to Goldman (1987), the gestation period averages 58 days and litters contain two to four, usually four, young; the young open their eyes after 12 days, take solid food at 3 weeks, and attain sexual maturity at about 9 months. Captives have been estimated to live for 9 years.

Cusimanses tame easily and make good pets. They are affectionate, playful, clean, and readily housebroken, but they sometimes mark objects with their anal scent glands (Rosevear 1974). The subspecies *C. ansorgei ansorgei* is known only by a single specimen collected in 1908 in a relict forest in northern Angola (Schreiber et al. 1989).

CARNIVORA; HERPESTIDAE; **Genus LIBERIICTIS**
Hayman, 1958

Liberian Mongoose

The single species, *L. kuhni,* now is known by 27 specimens from northeastern Liberia (Carnio 1989; Carnio and Taylor 1988) and has been reported from southwestern Ivory Coast (Hoppe-Dominik 1990; M. E. Taylor 1992).

Head and body length of an adult male was 423 mm, tail length was 197 mm, and weight was 2.3 kg. In an adult female head and body length was 478 mm and tail length was

Liberian mongoose *(Liberiictis kuhni)*, photo by F. Faigal, Metropolitan Toronto Zoo.

Liberian mongoose *(Liberiictis kuhni)*, photo of dead specimen by Lynn Robbins.

205 mm (Goldman and Taylor 1990). The predominant color of the pelage is dark brown. A dark stripe bordered above and below by a pale stripe is present on the neck. The throat is pale, the tail is slightly bicolored, and the legs are dark. From *Crossarchus, Liberiictis* is distinguished externally by the presence of neck stripes, a more robust body, and apparently longer ears (Schlitter 1974). The skull of *Liberiictis* is larger than that of *Crossarchus,* the rostrum and nasals are more elongate, the teeth are proportionally smaller and weaker, and there is an additional premolar in both the upper and lower jaws.

Schlitter (1974) noted that the long claws of the front feet, the long mobile snout, and the weak dentition of *Liberiictis* indicate that it is a terrestrial animal with a primarily insectivorous diet. Carnio and Taylor (1988) stated that the diet also includes worms, eggs, and small vertebrates. Goldman and Taylor (1990) reported that a captive was fed ground meat, dog food, young chickens, and fish. Schlitter's two specimens were taken in a densely forested area traversed by numerous streams. People native to the area said that *Liberiictis* is diurnal and found only on the ground. An adult male was taken in a snare on the ground. A juvenile female was excavated from a burrow associated with a termite mound on 29 July. An adult of unknown sex was also in the burrow but was not preserved. Goldman and Taylor (1990) indicated that breeding probably coincides with the rainy season, May–September. Schlitter stated that *Liberiictis* is eaten by human hunters. In a letter of 26 June 1963 to Ernest P. Walker, Dr. Hans-Jürg Kuhn, who had traveled in Liberia, wrote that native people say that *Liberiictis* is usually found in tree holes and lives in groups of 3–5 individuals. Carnio and Taylor (1988) reported that a group of 15 had been seen foraging. They added, however, that the genus seems to be very rare and may be jeopardized by human hunting and habitat destruction. In January 1989 a live specimen was obtained and placed in the Metro Toronto Zoo (Carnio 1989). The IUCN now classifies *Liberiictis* as endangered.

CARNIVORA; HERPESTIDAE; **Genus HELOGALE**
Gray, 1861

Dwarf Mongooses

There are two species (Coetzee *in* Meester and Setzer 1977; Yalden, Largen, and Kock 1980):

H. parvula, Ethiopia to Angola and eastern South Africa;
H. hirtula, southern Ethiopia, southern Somalia, northern Kenya.

Head and body length is 180–260 mm, tail length is 120–200 mm, and weight is 230–680 grams (Kingdon 1977). Coloration is variable, but generally the upper parts are speckled brown to grayish. The lower parts are only slightly paler, and the tail and lower parts of the legs are dark. In some individuals there is a rufous patch on the throat and breast and the basal portion of the lower side of the tail is reddish brown. Other individuals are entirely black.

Dwarf mongooses are found in savannahs, woodlands, brush country, and mountain scrub, from sea level to elevations of about 1,800 meters. They are mainly terrestrial and diurnal. They seek shelter at dusk in deserted or active termite mounds, among gnarled roots of trees, and in crevices. Their slender bodies enable them to squeeze into small openings. On occasion they may excavate their own burrows. Dens are changed frequently, sometimes daily (Rood 1978). Most of the day is spent in an active and noisy search for food among brush, leaves, and rocks. The diet consists mainly of insects (Rasa 1977; Rood 1980) and also includes small vertebrates, eggs, and fruit.

Helogale is found in organized groups that are thought to use a definite home range, though there is conflicting information on group movements and relationships. According to Rasa (1977), a portion of the range is occupied for two or three months and then there is a shift to another part, presumably because of depleted food supplies. Kingdon (1977) wrote that a group observed for seven years stayed mainly within an area of 2 ha.; another group used 2 ha. during the dry season but moved away with the coming of the rains. In the Serengeti, Rood (1978) found group home ranges to average 30 ha., to overlap by 5–40 percent with the ranges of one to four neighboring packs, and to contain 10–20 dens each. Rasa (1973, 1977) reported that

Dwarf mongoose *(Helogale parvula)*, photo by Bernhard Grzimek.

captives have been observed to mark the vicinity of dens with secretions from cheek and anal glands and to show considerable intergroup aggression. Rasa (1986) stated that groups inhabit home ranges of 0.65–0.96 sq km that show little overlap with one another and are traversed every 20–26 days, the length of time it takes for the marking secretions to decay.

The social organization of *Helogale* is unique among mammals (Kingdon 1977; Rasa 1972, 1973, 1975, 1976, 1977, 1983; Rood 1978, 1980). There are as many as 40 individuals in a group but usually about 10–12. The groups are matriarchal families founded and led by an old female. She initiates movements and has priority regarding food. The second highest ranking member of the group is her mate, an old male. These two dominant animals are monogamous and usually the only members of the group to produce offspring; they suppress sexual activity in other group members. The latter form a hierarchy, with the *youngest* individuals ranking highest. This arrangement probably serves to allow the young animals to obtain sufficient food, without competition from older and stronger mongooses, during the unusually long period of growth in this genus. Within any age class in the group females are dominant over males.

Despite the rigid class structure, or perhaps because of it, intragroup relations are generally harmonious, and severe fights are rare. Subordinate adults clean, carry, warm, and bring food to helpless young and take turns "baby-sitting" while the rest of the group forages. Females in addition to the mother sometimes nurse the young. The youngest mobile animals seem to have the role of watching for danger and alerting the others by means of visual signals or a shrill alarm call. Often a single animal occupies an exposed position, where it serves as a group guard. One series of observations showed that when a low-ranking male became sick, it was allowed a higher than normal feeding priority and was also warmed by other group members. In another case, a group restricted its normal movements in order to provide care and food for an injured member. Individuals communicate by depositing scent from the cheek and anal glands, and they also sometimes mark one another as an apparent sign of acceptance. As surviving subordinate animals grow older, they seem not to leave the group, even though they are not allowed to mate. If the dominant female dies, however, the group may split up.

In the Serengeti National Park of Tanzania, births occur mainly in the rainy season, from November to May, and the alpha female usually has three litters per year (Rood 1978, 1980). In a captive colony in Europe the young are born regularly in spring and autumn (Rasa 1972). According to Rasa (1977), females there normally give birth twice a year, entering estrus 4–7 days after lactation ceases. If the newborn young die, however, females may quickly remate, and they thus have the potential of producing five litters annually. The gestation period is 49–56 days. The number of young per litter averages about four and ranges from one to seven. Nursing lasts for at least 45 days, but group members begin to bring solid food to the young before weaning is complete. The young start to forage with the group by the time they are 6 months old. Females may reach physiological sexual maturity at as early as 107 days (Zannier 1965), but social restrictions normally delay breeding for several years. Apparently, full physical maturity is not attained until 3 years (Rasa 1972). One dominant pair was observed to mate first when about 3 years old and then to continue to breed for 7 years (Kingdon 1977). An individual lived at the Basel Zoo for 12 years and 3 months (Marvin L. Jones, Zoological Society of San Diego, pers. comm., 1995).

CARNIVORA; HERPESTIDAE; **Genus DOLOGALE**
Thomas, 1926

African Tropical Savannah Mongoose

The single species, *D. dybowskii*, is found in the Central African Republic, northeastern Zaire, southern Sudan, and western Uganda (Coetzee *in* Meester and Setzer 1977). This species was originally assigned to *Crossarchus* and has sometimes been referred to *Helogale*.

Head and body length is about 250–330 mm and tail length is 160–230 mm. Kingdon (1977) listed the weight as approximately 300–400 grams. Stripes are lacking. The head and neck are black, grizzled with grayish white. The back, tail, and limbs are lighter in color and have brownish spots. The underparts are reddish gray. The fur is short, even, and fine, in contrast to the loose, coarse pelage of *Crossarchus*. The snout is not lengthened like that of *Crossarchus*. *Dologale* closely resembles *Helogale* but does not have a groove on the upper lip and has weaker teeth (Coetzee *in* Meester and Setzer 1977).

Kingdon (1977) stated that the few known specimens of

African tropical savannah mongoose *(Dologale dybowskii)*, photo by F. Petter of mounted specimen in Museum National d'Histoire Naturelle.

Dologale suggest adaptation to a variety of habitats—thick forest, savannah-forest, and montane forest grassland. Some evidence indicates that the genus is at least partly diurnal. It has robust claws, suggesting digging habits as in *Mungos*. Asdell (1964) wrote that a litter of four young had been noted in Zaire. Schreiber et al. (1989) noted that there had been no records or sightings of *Dologale* for at least 10 years.

CARNIVORA; HERPESTIDAE; **Genus BDEOGALE**
Peters, 1850

Black-legged Mongooses

There are two subgenera and three species (Coetzee *in* Meester and Setzer 1977; Nader and Al-Safadi 1991):

subgenus *Bdeogale* Peters, 1850

B. crassicauda, Yemen, southern Kenya to central Mozambique;

subgenus *Galeriscus* Thomas, 1894

B. nigripes, southeastern Nigeria to northern Zaire and northern Angola;
B. jacksoni, southeastern Uganda, central Kenya.

Rosevear (1974) considered *Galeriscus* to be a separate genus. Kingdon (1977) suggested that *B. jacksoni* is only a subspecies of *B. nigripes*.

Head and body length is 375 to at least 600 mm and tail length is 175–375 mm. Kingdon (1977) listed weight as 0.9–3.0 kg. There is considerable variation in color both within and between species. The predominant general coloration is some shade of gray or brown, and the legs are usually black. The fur of adults is rather close, dense, and short; that of the young is nearly twice as long and lighter in color.

Bdeogale resembles *Ichneumia* in having black feet, soft underfur, and long, coarse hair over the upper parts of the body. It differs in lacking the first or inner toe on each foot and in having larger premolar teeth. *Bdeogale* differs from *Rhynchogale* in having a naked groove from the nose to the upper lip. The foreparts of the feet of *Bdeogale* are naked, but the hind parts are well haired.

Taylor (1986, 1987) described *B. crassicauda* as rare, unspecialized, nocturnal, insectivorous, and solitary. According to Kingdon (1977), *B. crassicauda* inhabits woodland and moist savannah, and the other species live in tropical forest. *B. crassicauda* feeds almost entirely on insects, especially ants and termites, but may also take crabs and rodents. *B. nigripes* seems to prefer ants but also eats small vertebrates and carrion. These mongooses are frequently seen in pairs. A female and a quarter-grown young were taken in December on the coast of Kenya, a pregnant female with a large fetus was also taken in December, and a female with a newborn infant was found in southeastern Tanzania in late November. Information compiled by Rosevear (1974) suggests that adults are basically solitary in the wild but are not quarrelsome when kept together in captivity; that births in West Africa occur from November to January; and that litters normally contain a single young.

Black-legged mongoose *(Bdeogale* sp.), photo by Don Davis.

According to Jones (1982), a captive *B. nigripes* lived 15 years and 10 months.

The IUCN classifies *B. jacksoni* as vulnerable and the subspecies *B. crassicauda omnivora*, of coastal Kenya and Tanzania, as endangered, and *B. c. tenuis*, of Zanzibar, as indeterminate. Schreiber et al. (1989) noted that *B. c. omnivora* is very rare and that its restricted forest habitat is rapidly being cut over. In 1988 a specimen of *B. crassicauda* was collected near Sanaa in central Yemen, the first record of *Bdeogale* outside of Africa. If human introduction is not involved, this record would be one in a series of recent discoveries demonstrating the faunal affinity of East Africa and the Arabian Peninsula and suggesting that the southern part of the peninsula holds one of the world's largest reservoirs of mammalian surprises.

CARNIVORA; HERPESTIDAE; **Genus RHYNCHOGALE**
Thomas, 1894

Meller's Mongoose

The single species, *R. melleri*, occurs from southern Zaire and Tanzania to eastern South Africa and possibly northeastern Angola (Coetzee *in* Meester and Setzer 1977).

Head and body length is 440–85 mm and tail length is usually 300–400 mm. Kingdon (1977) listed weight as 1.7–3.0 kg. The general coloration is grayish or pale brown, the head and undersides are paler, and the feet are usually darker. *Rhynchogale* resembles *Ichneumia* in having coarse guard hairs protruding from the close underfur and the same dental formula. *Rhynchogale* differs from *Ichneumia* in the frequent reduction of the hallux and the lack of a naked crease from the nose to the upper lip. *Rhynchogale* has hind soles that are hairy to the roots of the toes. Females have two abdominal pairs of mammae.

According to Kingdon (1977), the habitat appears to be restricted to the woodland belt and possibly to moister and more heavily grassed or wooded areas, such as drainage lines and rock outcrops. Available information suggests that *Rhynchogale* is terrestrial, nocturnal, and solitary. The diet includes wild fruit, termites, and probably small vertebrates. A litter of two newborn young with eyes still unopened was found in a small cave on a rocky hill in Zambia in December. A pregnant female containing two embryos was found in the same area, also in December. In Zimbabwe births occur around November and litters contain as many as three young.

CARNIVORA; HERPESTIDAE; **Genus ICHNEUMIA**
I. Geoffroy St.-Hilaire, 1837

White-tailed Mongoose

The single species, *I. albicauda*, occurs all across the southern part of the Arabian Peninsula and in Africa from Senegal to southeastern Egypt and south to northeastern Namibia and eastern South Africa, but not in the dense forest zone of West and Central Africa or in the deserts of the north or southwest (Coetzee *in* Meester and Setzer 1977; Corbet 1978; Gallagher 1992; Harrison and Bates 1991; Nader 1979; Skinner and Smithers 1990).

Head and body length is 470–710 mm, tail length is 355–470 mm, and weight is 1.8–5.2 kg (Kingdon 1977; Taylor 1972). Long, coarse, black guard hairs protrude from a yellowish or whitish close, woolly underfur, producing a grayish general body color. The four extremities, from the elbows and knees, are black. The basal half of the tail is of the general body color. The terminal portion is usually white but occasionally black. *Ichneumia* is characterized by large size; its bushy, tapering tail; having the soles of the forelimbs naked to the wrist; and the division of the upper lip by a naked slit from the nose to the mouth. Females have four mammae.

The white-tailed mongoose is found mainly in savannahs and grassland. It prefers areas of thick cover, such as forest edge and bush-fringed streams. It is basically terrestrial and nocturnal, sheltering by day in porcupine or aardvark burrows, termite mounds, or cavities under roots or rocks. The diet consists mainly of insects and also includes snakes, other small vertebrates, and fruit. *Ichneumia* breaks eggs by hurling them back between its hind legs and against some hard object. It may defend itself by ejecting a particularly noxious secretion from its anal scent glands (Kingdon 1977).

In Kenya, Taylor (1972) found home range of one individual to be about 8 sq km. Detailed studies on the Serengeti of Tanzania (Waser and Waser 1985) revealed average home range to be 0.97 sq km for adult males and 0.64 sq km for females. The male ranges did not overlap, but there was complete overlap between the ranges of opposite

Meller's mongoose *(Rhynchogale melleri)*, photo from *Proc. Zool. Soc. London.*

White-tailed mongoose *(Ichneumia albicauda)*, photo by Ernest P. Walker.

sexes. Some female ranges were exclusive, but in other cases several females and their offspring used a common range, though they foraged separately. Such an arrangement was thought to represent a matrilineal clan of related individuals. Reported pairs and family groups probably reflect observations of consorting individuals and mothers with young, respectively. *Ichneumia* is highly vocal, the most unusual sound being a doglike yap that may be associated with sexual behavior. Kingdon (1977) wrote that the two to four young are born in a burrow. All litters observed by Waser and Waser (1985) on the Serengeti contained one or two young, seen most frequently from February to May. No young were seen during the August–November dry season. The young usually were independent by 9 months. A captive was still living after 10 years (Jones 1982).

When the white-tailed mongoose dwells near a poultry raiser it may prove to be a pest. In captivity it is the shyest of mongooses, but it is said to become a pleasing pet if captured young.

CARNIVORA; HERPESTIDAE; **Genus ATILAX**
F. Cuvier, 1826

Marsh Mongoose, or Water Mongoose

The single species, *A. paludinosus*, is found from Senegal to Ethiopia, and south to South Africa (Baker 1992).

Head and body length is 440–620 mm, tail length is 250–430 mm, and weight is 2.5–4.1 kg (Baker 1992; Kingdon 1977). The pelage is long, coarse, and generally brown in color. A sprinkling of black guard hairs often gives a dark effect. In some individuals light rings on the hairs impart a grayish tinge. The head is usually lighter than the back, and the underparts are still paler.

Atilax is fairly heavily built. Although it is the most aquatic mongoose, it is the only one with toes that completely lack webbing. This feature may be associated with the habit of feeling for aquatic prey in mud or under stones. There are five digits on each limb, the soles are naked, and the claws are short and blunt. The anal area is large and naked, and there is a narrow, naked slit between the nose and the upper lip. Females usually have two pairs of mammae.

Atilax is found in a variety of general habitat types, but its basic requirement seems to be permanent water bordered by dense vegetation (Rosevear 1974). Favored haunts are marshes, reed-grown streambeds, and tidal estuaries. Grassy patches and floating masses of vegetation often serve as feeding places and as dry resting spots. Like other mongooses, *Atilax* does little or no climbing but does run up leaning tree trunks or other inclines easy of access. It is an excellent swimmer and diver. When hard pressed, it submerges, leaving only the tip of the nose exposed for breathing. Normally the head and part of the back are exposed when it is swimming. Food is sought in the water and in travels on regular pathways along the borders of streams and marshes. Activity is usually said to be nocturnal and crepuscular, but Rowe-Rowe (1978*b*) referred to *Atilax* as diurnal and stated that it does much of its hunting while walking in shallow water.

The diet consists of almost any form of animal life that can be caught and killed. Regular foods include insects, mussels, crabs, fish, frogs, snakes, eggs, small rodents, and fruit (Kingdon 1977; Rosevear 1974). *Atilax* may throw such creatures as snails and crabs against hard surfaces to break their shells. A captive was seen to take a piece of beef rib between its forefeet, rear onto its hind feet with its forefeet held high, and then forcefully throw the bone to the floor of the cage in an effort to break it.

Kingdon (1977) wrote that *Atilax* is usually seen alone and that individuals are widely spaced and undoubtedly highly territorial. The young are said to be born in burrows in stream banks or on masses of vegetation gathered into heaps among reed beds. There is no evidence of a particular breeding season in West Africa (Rosevear 1974). Data obtained in southern Africa indicate that young are born during the warm, wet months from about October to February (Skinner and Smithers 1990). The gestation period is 69–80 days (Baker 1992). The number of young per litter has been reported as one to three and usually seems to be two or three. The young weigh about 100 grams at birth, open their eyes after 9–14 days, are weaned at 30–46 days,

Marsh mongoose, or water mongoose *(Atilax paludinosus)*, photo by Ernest P. Walker.

and reach adult size at approximately 27 weeks (Baker 1992; Baker and Meester 1986). One water mongoose lived in captivity for just over 19 years (Marvin L. Jones, Zoological Society of San Diego, pers. comm., 1995).

Rosevear (1974) observed that within the last 50 years *Atilax* has probably declined substantially in the drier parts of its range because of human destruction of the available riverine habitat. *Atilax* is also widely hunted by people because it is reputed to be a poultry thief.

CARNIVORA; HERPESTIDAE; **Genus CYNICTIS**
Ogilby, 1833

Yellow Mongoose

The single species, *C. penicillata,* is found in extreme southern Angola, Namibia, Botswana, extreme western Zimbabwe, and South Africa (Taylor and Meester 1993).

Head and body length is 270–380 mm and tail length is 180–280 mm. Smithers (1971) listed weights of 440–797 grams, though Cavallini (1993) found some males weighing up to 1,000 grams. The hair is fairly long, especially on the tail, which is somewhat bushy. The general color is dark orange yellow to light yellow gray. The underfur is rich yellow, the chin is white, the underparts and limbs are lighter than the back, and the tail is tipped with white. The individual guard hairs are usually yellowish in the basal half, followed by a black band and a white tip. There is a seasonal color change in the coat: the summer pelage, typical in January and February, is reddish, short, and thin; the winter coat, typical from June to August, is yellowish, long, and thick; and the transitional coat, typical in November and December, is pale yellow. The forefeet have five digits and the hind feet have four. The first, or inner, digit on the forefoot is small and above the level of the other four, so it does not touch the ground. Females have three pairs of mammae (Ewer 1973).

The yellow mongoose frequents open country, preferably with loose soil, but may take refuge among rocks or in brush along the banks of streams when disturbed. It sometimes uses holes made by other animals, such as *Pedetes,* but usually excavates its own burrows. It is an energetic digger, constructing extensive underground systems that may cover 50 sq meters or more. These burrows may have 40 or more entrance holes, interconnecting tunnels on two or three levels to a depth of 1.5 meters, and enlarged nest chambers at intervals along the tunnels (Taylor and Meester 1993). Certain places within the burrow system are used for deposit of body wastes.

Cynictis is mainly diurnal but may be active at night when living near people. It basks in the sun and sits up on its haunches to obtain a better view of the surroundings. It is agile and capable of traveling at considerable speed. It seldom wanders more than 1 km from the burrow. Apparently, pairs and perhaps entire colonies seek new homes when food becomes scarce in the vicinity of a burrow. The diet consists mainly of insects and other invertebrates (Herzig-Straschil 1977; Smithers 1971). Other reported foods of *Cynictis* include lizards, snakes, birds, the eggs of birds and turtles, small rodents, and even mammals as large as itself.

Groups have been reported to include as many as 50 or more individuals, but Taylor and Meester (1993) indicated that such records result from confusion of *Cynictis* with *Suricata* and that colonies of the former usually have only 4–8 animals. There apparently is considerable variation in social structure. In a coastal area of limited resources, Cavallini (1993) found population density to be only about 1.2/100 ha. and observed little group organization. Adult males maintained separate home ranges of about 100 ha., about four times as large as the female ranges and encompassing several of the latter. Females in different dens maintained nonoverlapping ranges.

In South Africa, Earlé (1981) found the mean size of 5 colonies to be 8 individuals. The nucleus of each group consisted of an adult pair, their newest offspring, and 1 or 2 young or very old individuals. The other members had a loose, unclear association. The group hunting range of about 5–6 ha. corresponded to a territory, which was pa-

Yellow mongoose *(Cynictis penicillata)*, photo by Hans-Jürg Kuhn.

trolled each day by the adult male and marked with urine and secretion from the anal and cheek glands. Young animals from other groups were allowed to cross territorial boundaries if they showed submission by laying on their side and uttering high-pitched screams. Within the group, order of rank was: the adult male, the adult female, the youngest offspring, and then the other animals. The young showed reduced dominance by the age of 10 months. The mating season lasted from July to late September, and births occurred in October and November. Wenhold and Rasa (1994) found that most territorial defense and marking actually was carried out by younger, subordinate individuals. Subordinate males eventually dispersed to other colonies, and subordinate females temporarily left their natal territories to mate with males from other colonies. It was suggested that marking by these animals was partly a means of advertising for mates.

Long-term studies by Rasa et al. (1992) in South Africa and Namibia demonstrate that females commonly produce two litters annually within a period of 2–4 months. They are polyestrous and initiate a new cycle while still lactating. The first litter is born in October, the second from February to December. The gestation period is 60–62 days, mean litter size is 1.9 young (range 1–3), and lactation lasts 6–8 weeks. Smithers (1971) stated that there may be sporadic breeding throughout the year in Botswana, with a peak at some interval from October to April, and that the number of embryos per pregnant female averages 3.2 and ranges from 2 to 5. Taylor and Meester (1993) indicated that sexual maturity is attained after 1 year. One yellow mongoose lived in captivity for 15 years and 2 months (Jones 1982).

CARNIVORA; HERPESTIDAE; **Genus PARACYNICTIS**
Pocock, 1916

Gray Meerkat, or Selous's Mongoose

The single species, *P. selousi,* occurs in Angola, Zambia, Malawi, northern Namibia, Botswana, Zimbabwe, Mozambique, and eastern South Africa (Coetzee *in* Meester and Setzer 1977). This species originally was referred to *Cynictis.*

Head and body length is 390–470 mm and tail length is 280–400 mm. Smithers (1971) listed weights of 1.4–2.2 kg. The upper parts are dull buff gray, the belly is buffy, the feet are black, and the tail is white-tipped. There is no rufous in the coloring, nor are there any spots or stripes. Unlike *Cynictis, Paracynictis* has only four digits on each limb. The claws are long and slightly curved. These features are associated with a strong digging ability. *Paracynictis* can defend itself by expelling a strong-smelling secretion from its anal glands; its white-tipped tail, which makes the animal visible at night, may serve as a warning of this capability.

Paracynictis seems to prefer open scrub and woodland (Smithers 1971). It resides in labyrinthine burrows of its own construction. It is terrestrial and nocturnal but has been seen above ground by day. It has been described as shy and retiring. The diet consists of insects, other arthropods, frogs, lizards, and small rodents (Smithers 1971).

Apparently, each individual constructs its own burrow system, and there is less social activity than in *Cynictis.* On the basis of limited data, Smithers (1983) suggested that the births occur in the warm, wet months, probably from August to March, and that litter size is two to four young.

CARNIVORA; HERPESTIDAE; **Genus SURICATA**
Desmarest, 1804

Suricate, or Slender-tailed Meerkat

The single species, *S. suricatta,* occurs in southwestern Angola, Namibia, Botswana, and South Africa (Coetzee *in* Meester and Setzer 1977).

Head and body length is 250–350 mm and tail length is 175–250 mm. Smithers (1971) listed weights of 626–797 grams for males and 620–969 grams for females. The coloration is a light grizzled gray. The rear portion of the back is marked with black transverse bars, which result from the alternate light and black bands of individual hairs coinciding with similar markings of adjacent hairs. The head is almost white, the ears are black, and the tail is yellowish with a black tip. The coat is long and soft, and the underfur is dark rufous in color. The body is quite slender, though this feature is difficult to see because of the long fur. Scent glands peripheral to the anus open into a pouch that presumably stores the secretion (Ewer 1973). The forefeet have very long and powerful claws. Females have six mammae (Grzimek 1975).

The suricate inhabits dry, open country, commonly with hard or stony ground (Smithers 1971). It is an efficient digger. Colonies on the plains may excavate their own burrows or share the holes of African ground squirrels *(Xerus).* Self-excavated burrow systems average about 5 meters in di-

Selous's mongoose *(Paracynictis selousi)*, photo by Don Carter.

ameter, have approximately 15 entrance holes, and consist of two or three levels of tunnels extending to a depth of 1.5 meters and interconnected with chambers about 30 cm across. The home range of a colony may contain up to five such burrows (Van Staaden 1994). Colonies in stony areas live in crevices among the rocks. Outside activity is almost entirely diurnal. The suricate seems to enjoy basking in the sun, lying in various positions or sitting up on its haunches like a prairie dog *(Cynomys)*. Individuals generally forage near the burrow, turning over stones and rooting in crevices. However, a colony may travel up to 6 km during a day and use several burrows for sleeping (Skinner and Smithers 1990). The diet is primarily insectivorous (Ewer 1973) but also includes small vertebrates, eggs, and vegetable matter.

Home range of a colony is as large as 15 sq km (Van Staaden 1994). *Suricata* is highly social and occurs in troops of as many as 30 individuals. Usually, however, groups contain 2–3 family units and a total of 10–15 individuals. Each family contains a pair of adults and their young. The female may be larger than the male and may dominate him. Groups mark the vicinity of their burrows

Suricates *(Suricata suricatta)*, photo by Bernhard Grzimek.

with feces and secretions from the anal glands. Outside individuals are vigorously repulsed, and encounters between two groups are highly agonistic. Nonetheless, there is recruitment of new members, especially in the early summer, and both sexes occasionally change groups. At least 10 vocalizations have been identified, including a threatening growl and a shrill alarm bark. This call may be given by an individual that acts as a sentry and will cause the other members of the group to dive for their burrows or other cover (Ewer 1973; Skinner and Smithers 1990).

Breeding in captivity has been recorded throughout the year, with an interbirth interval of only about four months. Births in the wild occur during the warm, wet seasons, August–November and January–March, and there is no evidence of more than a single litter per year. The gestation period is 77 days, possibly less. In captivity litters number as many as seven young, but in the wild they number two to five, usually three or four (Van Staaden 1994). The young weigh 25–36 grams at birth, open their eyes after 10–14 days, and are weaned at 7–9 weeks (Ewer 1973). They attain sexual maturity by 1 year (Schliemann *in* Grzimek 1990). Two suricates living at the San Diego Zoo in December 1995 were 12 years and 8 months old (Marvin L. Jones, Zoological Society of San Diego, pers. comm., 1995).

Suricata tames readily, is affectionate, and enjoys the warmth of snuggling close to its master. It is sensitive to cold. It is often kept about homes in South Africa to kill mice and rats. It should not be confused with the yellow or thick-tailed meerkat *(Cynictis)*, with which it often associates. *Cynictis* is not as winning in its ways nor as pleasing as a pet.

CARNIVORA; HERPESTIDAE; **Genus CRYPTOPROCTA**
Bennett, 1833

Fossa

The single species, *C. ferox*, is found in Madagascar (Coetzee *in* Meester and Setzer 1977). *Cryptoprocta* sometimes has been placed in the cat family (Felidae), but recent debate has mostly been concerned with whether the genus should be assigned to the family Viverridae or to Herpestidae (see account thereof).

This is the largest carnivore of Madagascar. Head and body length is 610–800 mm, tail length is approximately the same, and shoulder height is about 370 mm. Weight is 7–12 kg (Albignac 1975). The fur is short, smooth, thick, soft, and usually reddish brown in color. Some black individuals have been captured. The mustache hairs are as long as the head. Like some other herpestids, *Cryptoprocta* has scent glands in the anal region that discharge a strong, disagreeable odor when the animal is irritated.

The general appearance is much like that of a large jaguarundi *(Felis yagouaroundi)* or small cougar *(Felis concolor)*. The curved claws are short, sharp, and retractile like those of a cat, but the head is relatively longer than in the Felidae. *Cryptoprocta* walks in a flat-footed manner on its soles, like bears, rather than on its toes, like cats.

The fossa dwells in forests and woodland savannahs, from coastal lowlands to mountainous areas at elevations of 2,000 meters (Coetzee *in* Meester and Setzer 1977). It is mainly nocturnal and crepuscular, is occasionally active by day, and often shelters in caves (Albignac 1972). A maternal den was located in an old termite mound and contained an unlined chamber about 70 cm deep, 100 cm wide, and 30 cm high (Albignac 1970*a*). The fossa is an excellent climber and pursues lemurs through the trees. Its diet consists mainly of small mammals and birds but also includes reptiles, frogs, and insects (Albignac 1972; Coetzee *in* Meester and Setzer 1977).

Cryptoprocta is solitary except during the reproductive

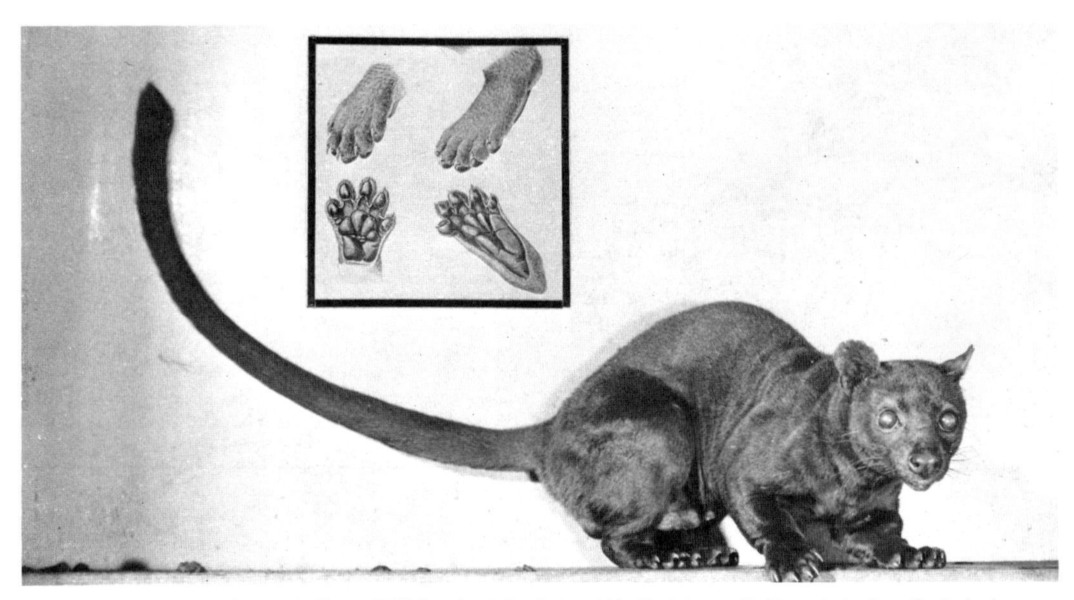

Fossa *(Cryptoprocta ferox)*, photo by Ernest P. Walker. Inset: forefoot and hind feet, top and bottom, photos from *Zoologie de Madagascar*, G. Grandidier and G. Petit.

season. Each individual has been estimated to require about 1 sq km. Mating occurs in September and October, and births take place in the austral summer. The gestation period lasts about 3 months. There are usually two young but sometimes three or four, each weighing about 100 grams at birth. The young leave the den after 4.5 months and are then weaned. Observations in captivity indicate that sexual maturity is not attained until 4 years. One specimen was still living after 20 years in captivity (Albignac 1972, 1975; Coetzee *in* Meester and Setzer 1977; Köhncke and Leonhardt 1986).

The fossa is a powerful predator but has an exaggerated reputation for savagery and destructiveness. It normally flees at the sight of a human, though it may be dangerous if wounded. It sometimes preys on poultry, and some accounts claim that it attacks wild hogs and even oxen. It is widely hunted by people and is now depleted in numbers. The IUCN now classifies it as vulnerable, noting that its habitat is declining because of human encroachment. It also is on appendix 2 of the CITES.

CARNIVORA; **Family HYAENIDAE**

Aardwolf and Hyenas

This family of four Recent genera and four species is found in Africa and southwestern and south-central Asia. The sequence of genera presented here follows that of Werdelin and Solounias (1991). Simpson (1945) and Wozencraft (*in* Wilson and Reeder 1993) recognized two subfamilies: the Protelinae, with the single genus *Proteles* (aardwolf); and the Hyaeninae, with *Parahyaena* (brown hyena), *Hyaena* (striped hyena), and *Crocuta* (spotted hyena). Some authorities, such as Coetzee (*in* Meester and Setzer 1977), Köhler and Richardson (1990), and Yalden, Largen, and Kock (1980), treat the Protelinae as a separate family.

Head and body length is 550–1,658 mm, tail length is 187–470 mm, and weight is 9–86 kg. The color pattern is striped in *Proteles* and *Hyaena*, spotted in *Crocuta*, and unmarked in *Parahyaena*, except for its barred feet and pale head. A well-developed mane is present in *Proteles*, *Parahyaena*, and *Hyaena*. In all genera the guard hairs are coarse and the tail is bushy.

The head and forequarters are large, but the hindquarters are rather weak. The forelimbs, which are slender in *Proteles* but powerfully built in the other genera, are longer than the hind limbs. *Proteles* has five digits on the forefoot and four on the hind foot; the other genera have four digits on each foot. The blunt claws are not retractile. Scent glands are present in the anal region. Males do not have a baculum, and females have one to three pairs of mammae.

All genera have strongly built jaws. In *Proteles* the cheek teeth are reduced and widely spaced and the carnassials are not developed, but the canines are sharp and fairly powerful. This genus may have as few as 24 teeth (Ewer 1973). In *Parahyaena*, *Hyaena*, and *Crocuta* the skull and jaws are massive and the teeth, including the carnassials, are powerfully developed for crushing bones. The dental formula of those three genera is: (i 3/3, c 1/1, pm 4/3, m 1/1) × 2 = 34. In all genera of the family the incisors are not specialized and the canines are elongate.

Hyenas and the aardwolf generally inhabit grassland or bush country, but they may also occur in open forest. They live in caves, dense vegetation, or the abandoned burrows of other animals. They are mainly nocturnal. They move about on their digits and seem to trot tirelessly. The aardwolf feeds primarily on insects, whereas hyenas are efficient scavengers and predators of large mammals. Individuals may occur alone, in pairs, or in groups. Females normally bear a single litter per year.

The phylogenetic origins of the family are uncertain; it may have evolved from a branch of the Viverridae or Herpestidae, or it could represent an entirely separate line of the aeluroid carnivores. The known geological range of the Hyaenidae is Miocene to Pleistocene in Europe, Miocene to Recent in Asia and Africa, Miocene to Pleistocene in Europe, and early Pleistocene in North America (Werdelin and Solounias 1991).

CARNIVORA; HYAENIDAE; **Genus PROTELES**
I. Geoffroy St.-Hilaire, 1824

Aardwolf

The single species, *P. cristatus*, is found from the southern border of Egypt to central Tanzania and from southern Angola and southern Zambia to the Cape of Good Hope (Coetzee *in* Meester and Setzer 1977; Corbet 1978; Kingdon 1977).

Head and body length is 550–800 mm, tail length is usually 200–300 mm, and shoulder height is usually 450–500 mm. Kingdon (1977) listed weight as 9–14 kg. The underfur is long, loose, soft, and wavy and is interspersed with larger, coarser guard hairs. The body is yellow gray with black stripes. The legs are banded with black, and the part below the knee and hock is entirely black. The tail is bushy and tipped with black, and the hair along the back is long and crestlike, as evidenced in the Boer vernacular name *manhaarjakkal*, or "maned jackal." The external ears and the auditory bullae are very large. The cheek teeth are vestigial and widely spaced, but the canine teeth are sharp and reasonably powerful. Köhler and Richardson (1990) noted that the large canines and relatively powerful jaws are associated more with intraspecific fighting and self-defense than with diet.

The aardwolf is most commonly found on open, sandy plains or in bush country. It dens in holes in the ground, usually abandoned burrows of the aardvark *(Orycteropus)*. It is primarily nocturnal and has been observed to cover just over 1 km/hr while foraging (Bothma and Nel 1980). Nightly movement is 8–12 km in summer and 3–8 km in winter (Köhler and Richardson 1990). The diet consists almost entirely of termites and insect larvae. The aardwolf lacks the powerful digging claws of other myrmecophagous mammals and seems to be specialized for feeding on nasute harvester termites *(Trinervitermes)*, which forage in dense concentrations on the surface of the soil and secrete a chemical that repels most other predators (Richardson 1987; Richardson and Levitan 1994). Although a captive juvenile killed a number of birds (Kingdon 1977), there are no substantive data on such behavior in the wild. *Proteles* has been accused of preying on chickens and lambs, but evidence to the contrary is overwhelming and it is known that the aardwolf can scarcely be induced to eat meat unless it is finely ground or cooked. When seen near carrion, *Proteles* is usually there to pick up carrion beetles, maggots, and other insects.

Skinner and Van Aarde (1988) reported a population density of about one adult per sq km and also that an adult female used a home range of 3.8 sq km over a period of 16 months. Köhler and Richardson (1990) stated that home range size is 1–4 sq km, varying inversely with termite density. The ranges are equivalent to defended territories

Aardwolf *(Proteles cristatus)*, photo by R. Pucholt.

and are commonly occupied year-round by a mated pair and their young, though there seems to be some variation in social structure. Kingdon (1977) noted that several females, together with their young, have been found in a single den. Bothma and Nel (1980) usually found a single animal per den but noted that dens could be less than 500 meters apart. Individuals from different territories sometimes form temporary aggregations, and there may even be mating between a female from one territory and an outside male, though generally adult males are hostile to one another and may fight viciously (Köhler and Richardson 1990). Both sexes mark their territories with secretions from anal scent glands. The hairs of the mane can be erected to make the aardwolf look twice its normal size for purposes of intraspecific display or defense against predators. Surprisingly loud growls and roars can be produced under stress (Smithers 1971). If attacked by dogs, the aardwolf ejects musky fluid from its anal glands and may fight effectively with its formidable canine teeth.

The following data on reproduction are available: estrus lasts 1–3 days and lactation up to 145 days (Hayssen, Van Tienhoven, and Van Tienhoven 1993); gestation is thought to last 90–110 days, an animal several months old was taken in Uganda in July, a birth in Kenya occurred in late May, the number of young per litter is usually two or three and ranges from one to five, and reports of larger litters may represent the utilization of one den by two mothers (Kingdon 1977); litters of two to four young are born in November and December in southern Africa (Ewer 1973); in Botswana lactating females have been taken in January and April, and pregnant females in July and October (Smithers 1971); and the young first emerge from the den at 1 month, begin to accompany the adults at about 4 months, and leave the natal territory at about 13 months (Köhler and Richardson 1990). One aardwolf lived at the Frankfurt Zoo for 18 years and 11 months (Marvin L. Jones, Zoological Society of San Diego, pers. comm., 1995).

Proteles is threatened in at least some areas by human hunting and habitat destruction. It was classified as rare in South Africa by Smithers (1986).

CARNIVORA; HYAENIDAE; **Genus PARAHYAENA**
Hendey, 1974

Brown Hyena

The single species, *P. brunnea,* occurs in Namibia, Botswana, western and southern Zimbabwe,southern Mozambique, and South Africa (Coetzee *in* Meester and Setzer 1977). *Parahyaena* often has been included in *Hyaena,* the name originally having been proposed as a subgenus, but the two were recognized as separate genera by Werdelin and Solounias (1990, 1991) and Wozencraft (*in* Wilson and Reeder 1993).

Head and body length is 1,100–1,356 mm, tail length is 187–265 mm, shoulder height is 640–880 mm, and weight is 37.0–47.5 kg (Skinner 1976; Smithers 1971). Males average slightly larger than females. The coat is long, coarse, and shaggy. The general coloration is dark brown. The head is gray, the neck and shoulders are tawny, and the lower legs and feet are gray with dark brown bars. There are erectile manes of long hairs up to 305 mm in length on the neck and back. Like *Hyaena, Parahyaena* has scent glands leading to an anal pouch that can be extruded for depositing secretions. The powerful jaws and teeth can crush the largest bones of cattle, and the digestive system can effectively utilize all parts of a carcass. Each front and hind foot has four digits.

Brown hyenas *(Parahyaena brunnea)*, photo by Bernhard Grzimek.

According to Werdelin and Solounias (1991), the following skeletal characters serve to distinguish *Parahyaena:* first upper molar tooth smaller, relative to fourth upper premolar, than in *Hyaena;* metastyle of fourth upper premolar longer than paracone (shorter than or equal to paracone in *Hyaena*); palate continues beyond last upper molar (ends at level of last upper molar in *Hyaena*); suture present between premaxillary and frontal on snout (absent in *Hyaena*); nasal wings of premaxilla diverge dorsad (vertically placed and parallel in *Hyaena*); sphenoid foramen and postpalatine located together in single depression (well separated in *Hyaena,* each in its own depression); processes for attachment of nuchal ligaments beneath terminus of sagittal crest much larger in *Parahyaena* than in *Hyaena;* supramastoid crest strongly developed (weak in *Hyaena*); axis and atlas have relatively long overlap (short overlap in *Hyaena*).

The brown hyena is found largely within arid habitat—open scrub, woodland savannah, grassland, and semidesert. It is mainly nocturnal and crepuscular, sheltering by day in a rocky lair, dense vegetation, or an underground burrow (Eaton 1976). It commonly uses an aardvark hole as a den but is capable of digging its own burrow. A favored maternal den site may be utilized for years (Owens and Owens 1979*a*). When foraging, *Parahyaena* travels at a walk of approximately 4 km/hr, but it is capable of running at 40–50 km/hr (M. G. L. Mills 1982). Foraging is done in a zigzag pattern along regular pathways. Nightly movements in the Kalahari average 31 km, range up to 54 km, and are usually longer in the dry season. Food is located mainly by scent, but hearing and night vision are also keen (Mills 1978, 1990; Owens and Owens 1978).

The brown hyena is primarily a scavenger of the remains of large mammals killed by other predators (Mills 1990; Mills and Mills 1978; Owens and Owens 1978; Skinner 1976). Other important foods include rodents, insects, eggs, and fruit. At certain times of the year vegetation may constitute up to half of the diet of some individuals. *Parahyaena* also is known to frequent shorelines and to feed on dead crabs, fish, and seals. It often stores excess food in shrubs or holes and usually recovers it within 24 hours. Individuals have been seen to raid ostrich nests and to cache the eggs at scattered locations.

Eaton (1976) wrote that maximum population density is about 1/130 sq km, but Skinner (1976) found at least six adults and two cubs within 20 sq km. Mills (1990) calculated a density of 1.8/100 sq km in the southern Kalahari. Social behavior evidently varies to some extent according to season (Mills 1978, 1982, 1990; Owens and Owens 1978, 1979*a*, 1979*b*). The animals in a given area are organized into a "clan" and are recognized by one another. They usually forage alone. In the dry season individual home range averages about 40 sq km. Individuals do not maintain territories but use common hunting paths and frequently meet and exchange greetings. At times, especially during the wet season, up to six clan members join to exploit carrion. Each clan has a central breeding den and defends a surrounding territory of approximately 300 sq km. Hyenas from neighboring clans behave aggressively toward one another.

A clan may include only one adult of each sex and associated young, but in some areas as many as four males and six females have been reported. In the latter situation, one of the males is thought to be dominant and the others subordinate. A clan is not a strictly closed system, and emigration is common, especially among the young. All males seem eventually to leave their natal clans, though sometimes not until after reaching adulthood, and they then may join another group. About a third of all adult males become nomadic, and these individuals frequently move into clans and mate with the females. Some females remain in their natal clans for life. There is a stable rank order maintained by ritualized displays of aggression that commonly involve biting at the back of the neck, where the skin is tough. Individuals regularly deposit scent from their anal glands as they move about. This activity seems to be mainly for communication of information to other clan members rather than demarcation of territory. Identified vocalizations include squeals, growls, yells, screams, and squeaks, all of which seem to be associated with conflict or

Top, brown hyena *(Parahyaena brunnea)* with a tumor near the left eye, photo from U.S. National Zoological Park. Bottom, striped hyena *(Hyaena hyaena)*, photo by Klaus Kussmann.

submissive behavior. Sounds are infrequent and mostly soft, there being no loud "whoop" as in *Crocuta.*

Females are apparently seasonally polyestrous. Mating is thought to occur mainly from May to August, with births from August to November. The gestation period is approximately 97 days. There are one to five, usually two or three, young per litter. They weigh about 1 kg at birth, open their eyes after 8 days, and emerge from the den after 3 months (Eaton 1976; M. G. L. Mills 1982, 1990; Skinner 1976). There evidently is some variation in rearing cubs. According to Owens and Owens (1979*a,* 1979*b*), the young of several litters are raised together in a communal den, each suckles from any lactating female, and all clan females participate in bringing food to the den. Mills (1990) reported that there usually is but one family per den, though the females of a clan tend to den within a few square kilometers of one another, and after about four months both the mother and other clan members begin to bring food to the cubs. There is no regurgitation. The cubs are not fully weaned and do not leave the vicinity of the den until they are about 14 months old. Females usually produce their first litter late in their second year, the interbirth interval is 12–41 months, natural mortality rates are relatively low until old age, and wild individuals are known to have lived at least 12 years (M. G. L. Mills 1982). A captive specimen was still living at about 29 years (Marvin L. Jones, Zoological Society of San Diego, pers. comm., 1995).

The brown hyena is classified as endangered by the USDI and is on appendix 1 of the CITES. Although protected and still widely distributed in Botswana, it has declined drastically in range and numbers in Namibia and South Africa (Eaton 1976). The main problem is killing by people, who consider it to be a predator of domestic animals. The damage that it does seems to have been greatly exaggerated; for example, on one large cattle ranch in the Transvaal where *Parahyaena* was studied no depredations were known to have occurred in 15 years (Skinner 1976).

CARNIVORA; HYAENIDAE; **Genus HYAENA**
Brünnich, 1771

Striped Hyena

The single species, *H. hyaena,* is found in open country from Morocco and Senegal to Egypt and Tanzania and from Asia Minor and the Arabian Peninsula to the Caucasus, southern Turkmenistan and Tajikistan, and eastern India (Coetzee *in* Meester and Setzer 1977; Corbet 1978; Heptner and Sludskii 1992). *Parahyaena* (see account thereof) often is included in *Hyaena.* Werdelin and Solounias (1991) considered the proper authority for the name *Hyaena* to be Zimmermann, 1777.

Except as noted, the information for the remainder of the account of this species was taken from the review papers by Rieger (1979, 1981). Head and body length is 1,036–1,190 mm, tail length is 265–470 mm, shoulder height is 600–942 mm, and weight is 25–55 kg. Males and females are about the same size. The general coloration is gray to pale brown, with dark brown to black stripes on the body and legs. The mane along the back is more distinct than in *Parahyaena.* The hairs of the mane are up to 200 mm long, while those of the rest of the body are about 70 mm long. Females have two or three pairs of mammae.

The striped hyena prefers open or rocky country and has an elevational range of up to 3,300 meters. It avoids true deserts and requires the presence of fresh water within 10 km. It is mainly crepuscular or nocturnal, resting by day in

Young striped hyena *(Hyaena hyaena),* photo by Reginald Bloom.

a temporary lair, usually under overhanging rocks (Kruuk 1976). Cubs are reared in natural caves, rocky crevices, or holes dug or enlarged by the parents. When searching for food the striped hyena moves in a zigzag pattern at 2–4 km/hr (Kruuk 1976). The diet varies by season and area but seems to consist mainly of mammalian carrion. The larger, more northerly subspecies prey on sheep, goats, donkeys, and horses. Other important foods are small vertebrates, insects, and fruit. Some food is stored in dense vegetation for later use. Food is often brought to the den, and eventually a large number of bones may accumulate (Skinner, Davis, and Ilani 1980).

The striped hyena may have a small defended territory around its breeding den, surrounded by a larger home range. In a radio-tracking study in the Serengeti, Kruuk (1976) determined home range to be about 44 sq km for a female and 72 sq km for a male. This species seems to be mainly solitary in East Africa, but there are indications of greater social activity farther north, where the animals are more likely to have a predatory life. In more than half of his sightings of *H. hyaena* in Israel, Macdonald (1978) saw more than one animal. An adult pair and their offspring may sometimes forage together, and a family unit may persist for several years. Adult females evidently are intolerant of one another and are dominant over males. Scent marking with anal glands is an important means of communication. Aggressive displays involve erection of the long hairs of the mane and tail. *H. hyaena* is much less vocal than *Crocuta* but growls, whines, and makes several other sounds.

Females are polyestrous and breed throughout the year. The estrous cycle may be 40–50 days long. Estrus lasts 1 day and may follow birth by 20–21 days. The gestation period is 88–92 days. The number of young per litter is 1–5, averaging 2.4. The young weigh about 700 grams at birth, open their eyes after 5–9 days, first take solid food at 30 days, and nurse for at least 4–5 months. Food is brought to the den for the young (Skinner and Ilani 1979). Mendelssohn (1992) found two wild females pregnant at an age of only about 15 months. However, sexual maturity usually has been reported at 2–3 years, and maximum longevity in captivity is 23–24 years.

The striped hyena has been known to attack and kill people, especially children. It can be tamed, however, and is said to become loyal and affectionate. Many of its parts are believed by some persons to have medicinal value. It is notorious in Israel for its destruction of melons, dates, grapes, apricots, peaches, and cucumbers (Kruuk 1976). It has been nearly or entirely eliminated in the Caucasus (Heptner and Sludskii 1992). The North African subspecies, *H. h. barbara,* is classified as endangered by the USDI. It has declined through habitat loss and persecution as a predator. Its range is now restricted to the highlands of Morocco, Algeria, and Tunisia.

CARNIVORA; HYAENIDAE; **Genus CROCUTA**
Kaup, 1828

Spotted Hyena

In historical time the single species, *C. crocuta,* was found throughout Africa south of the Sahara except in equatorial rainforests (Coetzee *in* Meester and Setzer 1977). Through the end of the Pleistocene the same species occurred in much of Europe and Asia (Kurten 1968).

Head and body length is 950–1,658 mm, tail length is 255–360 mm, shoulder height is 700–915 mm, and weight is 40–86 kg. On the average, females are 120 mm longer and 6.6 kg heavier than males (Kingdon 1977). The hair is coarse and woolly. The ground color is yellowish gray, and the round markings on the body are dark brown to black. There is no mane, or only a slight one. The jaws are probably the most powerful, in proportion to size, of any living mammal's. Each front and hind foot has four digits. From *Parahyaena* and *Hyaena, Crocuta* differs in its larger size, shorter and more rounded ears, paler and spotted coat, lack of a mane, larger and more swollen braincase, and greatly reduced upper molar tooth. In *Parahyaena* and *Hyaena* the diameter of the upper molar tooth is at least twice that of the first upper premolar, and there is a small metaconid on the first lower molar, whereas in *Crocuta* the upper molar is much smaller than the first premolar and there is no metaconid on the first lower molar.

The external genitalia of the female *Crocuta* so closely resemble those of the male that the two sexes are practically impossible to distinguish in the field (Kingdon 1977; Kruuk 1972): the clitoris looks like the penis, occupies the same position, and is capable of elongation and erection; in addition, the female has a pair of sacs, formed from fusion of the vaginal labia and filled with nonfunctional fibrous tissue, that looks very much like the scrotum and is located in the same place. The female has no external vagina and must urinate, mate, and deliver young through the urogenital canal traversing the clitoris (Glickman et al. 1992). Erection of the penis or "pseudopenis" apparently is a sign of submission displayed during greeting rituals between

Spotted hyena *(Crocuta crocuta),* photo by E. L. Button.

Spotted hyena *(Crocuta crocuta)*, photo by E. L. Button.

individuals (East, Hofer, and Wickler 1993). Both sexes have two anal scent glands that empty into the rectum. Females usually have a single pair of mammae. The sexual organs of female *Parahyaena* and *Hyaena*, unlike those of female *Crocuta*, do not closely resemble those of males.

The spotted hyena originally occupied nearly all the more open habitats of Africa south of the Sahara (Kingdon 1977). It is especially common in dry acacia bush, open plains, and rocky country. It is found at elevations of up to 4,000 meters. *Crocuta* is probably the most numerous of the large African predators because of its ability quickly to eat and digest entire carcasses, including skin and bones, and because the plasticity of its behavior allows it to function effectively either as a solitary scavenger and predator of small animals or as a group-living hunter of ungulates. Dens vary greatly in size, sometimes accommodating entire communities of hyenas. They are usually in abandoned aardvark holes but may also be in natural caves. Most activity is nocturnal or crepuscular. Up to 80 km may be covered in a night's foraging, though Mills (1990) found the average to be 27 km in the Kalahari. *Crocuta* has keen senses of sight, hearing, and smell.

Except as noted, the information for the remainder of this account was taken from Kruuk's (1972) report of his studies in the Serengeti National Park and the Ngorongoro Crater of northern Tanzania. In these areas the spotted hyena shelters in holes on the plains or in shady places with bushy vegetation on hillsides. Dens may have 12 or more entrances. Activity occurs mainly in the first half of the night, then declines, and then increases again toward dawn. The Ngorongoro population remains in one general area throughout the year, but the Serengeti hyenas may move in response to the seasonal migrations of their prey. During about half of the year Serengeti animals make "commuting trips" from their regular territories, lasting 3–10 days each and averaging about 40 km, to feed on the nearest migratory herds (Hofer and East 1993*a*, 1993*b*, 1993*c*).

The diet consists mainly of medium-sized ungulates, especially wildebeest *(Connochaetes)*, and mostly very young, very old, or otherwise inferior animals. One hyena commonly forces a herd of wildebeest to run, watches for a weak individual, and then begins a chase, which is soon joined by other hyenas. Zebras are hunted in a more organized manner by packs of 10–25 hyenas. Chases usually go for less than 2 km, with *Crocuta* averaging 40–50 km/hr. Maximum speed is about 60 km/hr. Roughly one-third of the hunts are successful. Far more animals are killed than are consumed as carrion. Although the spotted hyena is sometimes said to be a scavenger of the lion, most dead prey on which both hyenas and lions were seen feeding had been killed by the hyenas. In the Kalahari, Mills (1990) determined that at least 70 percent of the diet of *Crocuta* results from kills. The specialized teeth and digestive system of *Crocuta* allow it to crush and utilize much more bone than do other predators. *Crocuta* can consume 14.5 kg of food in one meal.

Population densities were estimated at 0.12/sq km in the Serengeti and 0.16–0.24/sq km in Ngorongoro. In the latter area *Crocuta* is clearly organized into large communities, or "clans," each with up to 80 individuals and each occupying a territory of about 30 sq km. Clans are usually divided into smaller hunting packs and individuals, but all members evidently are recognizable to one another. The members mark the borders of their territory with secretions from their anal scent glands and defend the area from other clans. Groups that move about together commonly contain 2–10 individuals, but larger groups form to hunt zebra. In the Serengeti the clan system is less well developed and average group size is smaller. In the Kalahari the average foraging group size is only 3 hyenas and average clan size is 8, but the average territory covers 1,095 sq km (Mills 1990). Within any group of *Crocuta,* females are dominant over males and adults are dominant over young. The animals eat together and may compete by shoving one another, but there generally is no fighting. The spotted hyena is extremely vocal. The well-known laughing sound is emitted by an animal that is being attacked or chased. A whoop or howl is usually given spontaneously by a lone individual with its head held close to the ground; it begins low and deep and increases in volume as it runs to higher pitches. Whoops are given by all members of a group, cubs mainly to request support, adult males generally for sexual advertisement, and adult females in agonistic interaction with other females or outside individuals (East and Hofer 1991*a,* 1991*b*).

Studies in Kenya (Frank 1986*a,* 1986*b;* Holekamp and Smale 1993; Smale, Frank, and Holekamp 1993) have provided further understanding of the internal structure of spotted hyena clans. Each of these groups is a stable community of related females, among which unrelated males reside for varying periods. Whereas females remain in their natal clan for life, males disperse around the time of puberty and temporarily join together in a nomadic group. They subsequently settle within a new clan, sometimes remaining for several years, but then may depart again. Within a clan there is a separate dominance hierarchy for each sex. The highest-ranking female and her descendants are dominant over all other animals. By the time they are 6–8 months old the juveniles of a group establish a hierarchy among themselves, and by 12–18 months they generally are able to dominate fully adult animals from matrilines of lower rank than their own and also males that have immigrated into the group. Although all resident males were observed to court females, only the highest-ranking male actually was seen to mate.

There apparently is no permanent pair bonding. Breeding may occur at any time of the year, perhaps with a peak in the wet season. Females are polyestrous and have an estrous cycle of 14 days. Gestation lasts about 110 days. There are usually two young per birth, occasionally one or three. Observations in captivity show that siblings begin to fight violently upon birth and that if both are of the same sex, one will usually be killed (Frank, Glickman, and Light 1991). At birth they weigh about 1.5 kg and their eyes are open. Each clan has a single central denning site where all females bear their cubs, but each mother suckles its own young. There is no cooperative rearing of young as sometimes found in *Parahyaena brunnea* (Mills 1985). Food is not carried to the den or regurgitated for the young. Weaning comes after 12–16 months, when the young are nearly full grown. Sexual maturity is attained at about 2 years by males and 3 years by females. A captive *Crocuta* lived 41 years and 1 month (Jones 1982).

Human relationships with the spotted hyena have varied (Kingdon 1977; Kruuk 1972). Some native peoples protected it as a valuable scavenger, but others regarded it with superstitious dread. Certain tribes put their dead out for hyenas to consume. Fatal attacks on living humans have occurred. In the twentieth century *Crocuta* has been considered to be a predator of domestic livestock and game and has been widely hunted, trapped, and poisoned. It has declined in numbers over much of its range and has been eliminated in parts of East and South Africa. It now is classified as conservation dependent by the IUCN. The well-studied and nominally protected population in Serengeti National Park may be jeopardized by human hunters who kill both the hyena and its prey when they move beyond the Park boundary (Hofer, East, and Campbell 1993). Kurten (1968) suggested that the disappearance of *Crocuta* in Eurasia in the early postglacial period was associated with the development of agriculture.

CARNIVORA; **Family FELIDAE**

Cats

This family of 4 Recent genera and 38 species has a natural distribution that includes all land areas of the world except the West Indies, Madagascar, Japan, most of the Philippines, Sulawesi and islands to the east, New Guinea, Australia, New Zealand, Antarctica, and most arctic and oceanic islands. Recognition here of 4 genera—*Felis, Neofelis, Panthera,* and *Acinonyx*—is based on Cabrera (1957), Chasen (1940), Corbet (1978), Corbet and Hill (1991), Ellerman and Morrison-Scott (1966), E. R. Hall (1981), Heptner and Sludskii (1992), Lekagul and McNeely (1977), Medway (1977, 1978), Meester et al. (1986), Skinner and Smithers (1990), Smithers (*in* Meester and Setzer 1977),

Leopard cat *(Felis bengalensis)* and young, photo from U.S. National Zoological Park.

Fishing cat *(Felis viverrina)*, photo from U.S. National Zoological Park.

Stains (1984), and Taylor (1934). There is, however, a great diversity of opinion on how the cats should be classified. Some authorities, such as Romer (1968), divide the family into only 2 genera, *Felis* and *Acinonyx*. Others, including some who recently have done extensive research on the subject, recognize many more genera. The main controversy involves the division of what is here considered to be the genus *Felis* (see account thereof) into genera and subgenera. All of the authorities cited here accept *Acinonyx* as a distinct genus, nearly all accept *Neofelis* (though with some disagreement about its content), and most accept *Panthera*.

The living Felidae are sometimes divided into subfamilies. Stains (1984) recognized the Acinonychinae, for *Acinonyx* (cheetah); the Felinae, for *Felis;* and the Pantherinae, for *Neofelis* and *Panthera*. Wozencraft (*in* Wilson and Reeder 1993) recognized the same three but included *Felis (Pardofelis) marmorata* in the Pantherinae. Leyhausen (1979) accepted the Acinonychinae, for *Acinonyx*, but put all other cats in the Felinae. The cheetah is usually placed last on lists of cats to express its aberrant characters. Actually, however, it seems in some respects to be little changed from the primitive stock that gave rise to all other cats. Locating it at the beginning of a systematic account, as was done by Leyhausen (1979) and Hemmer (1978), is thus fully appropriate. Moreover, placement of *Felis (Puma) concolor* next to *Acinonyx* by Hemmer (1978) and Salles (1992) may be fitting, as Adams (1979) has presented evidence that these two cats evolved from a common ancestor in North America. Analyses by Van Valkenburgh, Grady, and Kurten (1990) also suggest an evolutionary association of the two.

In the family Felidae head and body length is 337–2,800 mm, tail length is 51–1,100 mm, and weight is 1.5–306.0 kg. The color varies from gray to reddish and yellowish brown, and there are often stripes, spots, or rosettes. The pelage is soft and woolly. Its beautiful, glossy appearance is maintained by frequent cleaning with the tongue and paws. The tail is well haired but not bushy, and the whiskers are well developed.

Cats have a lithe, muscular, compact, and deep-chested body. The limbs range from short to long and sinewy. The forefoot has five digits and the hind foot has four. In most species the claws are retractile (to prevent them from becoming blunted), large, compressed, sharp, and strongly curved (to aid in holding living prey). In the cheetah *(Acinonyx)*, however, they are only semiretractile and relatively poorly developed. Except for the naked pads, the feet are well haired to assist in the silent stalking of prey. The baculum is vestigial or absent. Females have two to four pairs of mammae.

The head is rounded and shortened, the ears range from rounded to pointed, and the eyes have pupils that contract vertically. The tongue is suited for laceration and retaining food within the mouth, its surface being covered with sharp-pointed, recurved, horny papillae. The dental formula is: (i 3/3, c 1/1, pm 2–3/2, m 1/1) × 2 = 28 or 30. The incisors are small, unspecialized, and placed in a horizontal line. The canines are elongate, sharp, and slightly recurved. The carnassials, which cut the food, are large and well developed. The upper molar is small.

The dentition of the Felidae, with emphasis on the teeth used for seizing and cutting rather than on those used for grinding, reflects the highly predatory lifestyle of the family. Cats prey on almost any mammal or bird they can overpower and occasionally on fish and reptiles. They stalk their prey, or lie in wait, and then seize the quarry with a short rush. The cheetah *(Acinonyx)*, however, is adapted for a more lengthy pursuit at high speed. Cats walk or trot on their digits, often placing the hind foot in the track of the

A. Lion *(Panthera leo)*, photo by Leonard Lee Rue III. B. Clouded leopard *(Neofelis nebulosa)*, photo by Bernhard Grzimek. C. Leopard *(Panthera pardus)*, photo by Leonard Lee Rue III.

forefoot. They are agile climbers and good swimmers and have acute hearing and sight. Many are nocturnal, but some are active mainly during daylight. They shelter in trees, hollow logs, caves, crevices, abandoned burrows of other animals, or dense vegetation. They defend themselves with fang and claw or flee, sometimes seeking refuge in trees.

Cats usually are solitary but sometimes are found in pairs or larger groups. The females of most species are polyestrous and give birth once a year, some can have two litters annually, and those of the larger species sometimes breed only every 2–3 years. Gestation periods of 55–119 days have been reported, and litter size is usually 1–6 young. At birth the kittens are usually blind and helpless but are haired and often spotted. They remain with their mother until they can hunt for themselves. Potential longevity is probably at least 15 years for most species, and some individuals have lived more than 30 years.

Some of the larger species of cats have occasionally become a serious menace to human life in localized areas.

Various felids are considered by some people to be a threat to domestic animals. Certain species, especially those with spotted or striped skins, are of considerable value in the fur trade. The big cats, as well as some of the smaller ones, are sought as trophies by hunters. For all of these reasons the Felidae have been extensively hunted and killed by people. As a result, many species and subspecies have become rare or endangered in at least parts of their ranges. The entire family has been placed on appendix 2 of the CITES, except for those species on appendix 1 and the domestic cat *(Felis catus).*

The geological range of the Felidae is late Eocene to Recent in North America and Eurasia, early Eocene to Recent in Africa, and late Pliocene to Recent in South America (Stains 1984). There are several extinct groups, including the saber-toothed cats (subfamily Machairodontinae), which persisted through the end of the Pleistocene. *Smilodon fatalis,* which lived in North America until about 10,000 years ago, was a saber-toothed cat about the size of modern *Panthera leo.* It differed in its shorter and more powerful limbs and shorter tail. Rather than running down antelopes, it apparently preyed on young mammoths and other large, thick-skinned animals, leaping upon them from ambush, dragging them down, and sinking its huge fangs into their underparts (Heald and Shaw 1991).

CARNIVORA; FELIDAE; **Genus FELIS**
Linnaeus, 1758

Small Cats, Lynxes, and Cougar

There are 16 subgenera and 31 species (Cabrera 1957; Chasen 1940; Corbet 1978; Corbet and Hill 1992; Ellerman and Morrison-Scott 1966; Ewer 1973; Guggisberg 1975; E. R. Hall 1981; Hemmer 1978; Heptner and Sludskii 1992; Lekagul and McNeely 1977; Medway 1977, 1978; Roberts 1977; Smithers *in* Meester and Setzer 1977; Taylor 1934; Werdelin 1981):

subgenus *Felis* Linnaeus, 1758

F. silvestris (wild cat), France and Spain through Kazakhstan and the Arabian Peninsula to north-central China and central India, Great Britain, Balearic Islands, Sardinia, Corsica, Crete, woodland and savannah zones throughout Africa;
F. catus (domestic cat), worldwide in association with people;
F. bieti (Chinese desert cat), southern Mongolia, central China;
F. chaus (jungle cat), Volga River Delta and Egypt to Sinkiang and Indochina, Sri Lanka;
F. margarita (sand cat), desert zone from Morocco and northern Niger to southern Kazakhstan and Pakistan;
F. nigripes (black-footed cat), Namibia, Botswana, South Africa;

subgenus *Otocolobus* Brandt, 1841

F. manul (Pallas's cat), Caspian Sea and Iran to southeastern Siberia and Tibet;

subgenus *Lynx* Kerr, 1792

F. pardina (Spanish lynx), Spain, Portugal;
F. lynx (lynx), western mainland Europe to eastern Siberia and Tibet, possibly Sardinia, Sakhalin;
F. canadensis, Alaska, Canada, northern conterminous United States;
F. rufus (bobcat), southern Canada to Baja California and central Mexico;

subgenus *Caracal* Gray, 1843

F. caracal (caracal), Arabian Peninsula to Aral Sea and northwestern India, most of Africa;

subgenus *Leptailurus* Severtzov, 1858

F. serval (serval), Morocco, Algeria, most of Africa south of the Sahara;

subgenus *Pardofelis* Severtzov, 1858

F. marmorata (marbled cat), Nepal to Indochina and Malay Peninsula, Sumatra, Borneo;

subgenus *Catopuma* Severtzov, 1858

F. temmincki (Asian golden cat), Tibet and Nepal to southeastern China and Malay Peninsula, Sumatra;
F. badia (bay cat), Borneo;

subgenus *Profelis* Severtzov, 1858

F. aurata (African golden cat), Senegal to Kenya and northern Angola;

subgenus *Prionailurus* Severtzov, 1858

F. bengalensis (leopard cat), Ussuri region of southeastern Siberia, Manchuria, Korea, Quelpart and Tsushima islands (between Korea and Japan), most of China east of Tibet, Pakistan to Indochina and Malay Peninsula, Taiwan, Hainan, Sumatra, Java, Bali, Borneo, several islands in the western and central Philippines, Lombok;
F. rubiginosa (rusty-spotted cat), southern India, Sri Lanka;
F. viverrina (fishing cat), Pakistan to Indochina, Sri Lanka, Sumatra, Java;
F. planiceps (flat-headed cat), Malay Peninsula, Sumatra, Borneo;

subgenus *Mayailurus* Imaizumi, 1967

F. iriomotensis (Iriomote cat), Iriomote Island (southern Ryukyu Islands);

subgenus *Oreailurus* Cabrera, 1940

F. jacobita (mountain cat), Andes of southern Peru, southwestern Bolivia, northeastern Chile, and northwestern Argentina;

subgenus *Lynchailurus* Severtzov, 1858

F. colocolo (pampas cat), Ecuador and Mato Grosso region of Brazil to central Chile and Patagonia;

subgenus *Oncifelis* Severtzov, 1858

F. geoffroyi (Geoffroy's cat), Bolivia and extreme southern Brazil to Patagonia;
F. guigna (kodkod), central and southern Chile, southwestern Argentina;

A. Marbled cat *(Felis marmorata)*, photo from *Jour. Bombay Nat. Hist. Soc.* B. Serval *(F. serval)*, photo by Bernhard Grzimek. C. Golden cat *(F. temmincki)*, photo by Ernest P. Walker. D. Pallas's cat *(F. manul)*, photo by Howard E. Uible. E. Pumas *(F. concolor)*, photo by Ernest P. Walker.

Top, black-footed cat *(Felis nigripes),* photo by John Visser. B. Little spotted cat *(F. tigrina).* C. Jungle cat *(F. chaus),* photo by Ernest P. Walker. D. European wild cat *(F. silvestris),* photo by Bernhard Grzimek. E. Jaguarundi *(F. yagouaroundi),* photo by Ernest P. Walker.

A. Leopard cat *(Felis bengalensis)*, photo by Lim Boo Liat. B. Margay *(F. wiedii)*, photo by Ernest P. Walker. C. Pampas cat *(F. colocolo)*, photo from San Diego Zoological Garden. D. Fishing cat *(F. viverrina)*, photo from New York Zoological Society. E. Geoffroy's cat *(F. geoffroyi)*, photo by Ernest P. Walker. F. African wild cat *(F. silvestris libyca)*, photo by Bernhard Grzimek.

subgenus *Leopardus* Gray, 1842

F. pardalis (ocelot), Arizona and Texas to northern Argentina;
F. wiedii (margay), northern Mexico and possibly southern Texas to northern Argentina and Uruguay;
F. tigrina (little spotted cat), Costa Rica to northern Argentina;

subgenus *Herpailurus* Severtzov, 1858

F. yagouaroundi (jaguarundi), southern Arizona and southern Texas to northern Argentina;

subgenus *Puma* Jardine, 1834

F. concolor (cougar, puma, panther, or mountain lion), southern Yukon and Nova Scotia to southern Chile and Patagonia.

The sequence, systematic grouping, and individual acceptance of the species in this list is based on the authorities cited above and also gives some consideration to the detailed morphological analysis of Salles (1992). Most of the above authorities, as well as Corbet and Hill (1991), Jones et al. (1992), and Wozencraft (1989*a*), did not recognize the various subgenera in the list as being separate genera, except frequently *Lynx*. However, all of the above subgenera have been treated in the past as full genera, and there has been an increasing trend toward again accepting them as such, notably by authorities who have done primary research on the felids (Ewer 1973; Hemmer 1978; Leyhausen 1979; Salles 1992; Wayne et al. 1989; Wozencraft *in* Wilson and Reeder 1993). There seems to be little disagreement at the specific level, the main controversy involving the division of what is here considered to be the genus *Felis* into genera and subgenera. The production in captivity of hybrids between many of the species, including recent crossings of *F. pardalis* and *F. concolor* (Dubost and Royère 1993), has been one argument for not splitting *Felis*. Wozencraft (*in* Wilson and Reeder 1993) accepted all of the above subgenera as full genera with content as shown and also accepted all of the above species as valid, with the following exceptions: *F. catus* was included in *F. silvestris*, *Mayailurus* was not recognized and *F. iriomotensis* was included in *F. bengalensis*, and *Lynchailurus* was not recognized and *F. colocolo* was considered part of *Oncifelis*. In a recent response to Wozencraft, Leyhausen and Pfleiderer (1994) argued that *F. iriomotensis* is a highly distinctive and primitive species and that *Mayailurus* may warrant full generic status.

Corbet and Hill (1991), E. R. Hall (1981), and Jones et al. (1992) treated *Lynx* as a distinct genus. The preponderance of available information, however, suggests that there is no more (or less) justification for recognizing *Lynx* as a full genus than there is for elevating any other of the subgenera of *Felis* to generic level. The North American *F. canadensis* and the Iberian *F. pardina* often have been regarded as no more than subspecifically distinct from the Eurasian *F. lynx*, but studies of the evolution and taxonomy of the group by Werdelin (1981) and García-Perea (1992) show that all three, as well as *F. rufus*, are distinct species.

The African and most Asian populations of *F. silvestris* have been assigned to a separate species, *F. libyca*, by numerous authorities, including Smithers (*in* Meester and Setzer 1977). There is substantial evidence, however, that *F. silvestris* and *F. libyca* intergrade in the Middle East. A multistatistical analysis by Ragni and Randi (1986) showed broad morphological overlap of *silvestris*, *libyca*, and *catus*, indicating that all three should be regarded as conspecific. There also is much doubt that *F. bieti* of East Asia is distinct from *F. silvestris* (Corbet 1978). Leyhausen (1979) listed *F. silvestris* and *F. libyca* as separate species and considered the populations found from Iran to India to represent still another species, *F. ornata*. He united all three, however, in a single superspecies. Leyhausen (1979) also recognized *F. thinobia*, of Central Asia, and *F. tristis*, of Tibet, as species distinct from, respectively, *F. margarita* and *F. temmincki*. Heptner and Sludskii (1992) considered *F. euptilura*, found from southeastern Siberia to central China, a species distinct from *F. bengalensis*. García-Perea (1994) suggested that *F. (Lynchailurus) pajeros*, of the Andes and Patagonia, and *F. (L.) braccatus*, of central Brazil, Paraguay, and Uruguay, are species distinct from *F. (L.) colocolo*, which would then be restricted to northern and central Chile. Corbet and Hill (1992) treated *Neofelis* (see account thereof) as a synonym of *Pardofelis*.

For more than a century there have been unconfirmed reports of a large cat known as the "onza" from the highlands of Mexico. It has been variously associated with *Felis concolor*, *Panthera onca*, or an extinct relative of *Acinonyx jubatus*. In 1986 a specimen of an alleged onza was collected in the Sierra Madre. Published photographs suggested a slender, long-legged *F. concolor*, and an analysis of mitochondrial DNA (Dratch, Martenson, and O'Brien 1991) strongly supported referral to that species.

Except for *F. concolor* and some individuals of *F. lynx*, the members of the genus *Felis* are smaller than those of the three other genera of cats. Otherwise the characters of *Felis* are the same as those set forth for the family Felidae. From *Neofelis*, *Felis* is distinguished by its relatively shorter canine teeth and a smaller gap between the canines and the cheek teeth. From *Panthera* it is distinguished by a completely ossified hyoid apparatus without an elastic ligament. From *Acinonyx* it is distinguished by a greater gap between the canines and the cheek teeth and usually by shorter limbs and fully retractile claws. Additional information is provided separately for each species. Except as noted, the information for the following accounts was taken from Guggisberg (1975).

Felis silvestris (wild cat)

Head and body length is usually 500–750 mm, tail length is 210–350 mm, and weight is usually 3–8 kg. Males are generally larger than females. The fur is long and dense. In European populations the ground color is yellowish gray, the underparts are paler, and the throat is white. Four or five longitudinal stripes from the forehead to the nape merge into a dorsal line that ends near the base of the tail. The tail has several dark encircling marks and a blackish tip. The legs are transversely striped. In African and Asian populations the general coloration varies from pale sandy to gray brown and dark gray; there may be a pattern of distinct spots or stripes. European animals are generally about one-third larger than domestic cats and have longer legs, a broader head, and a relatively shorter, more bluntly ending tail. Females have eight mammae.

The wild cat occupies a variety of forested, open, and rocky country. It is mainly nocturnal and crepuscular, spending the day in a hollow tree, thicket, or rock crevice. It climbs with great agility and seems to enjoy sunning itself on a branch. It normally stays in one area, within which it has several dens and a system of hunting paths. It may hunt over a distance of 3–10 km each night (Heptner and Sludskii 1992; Nowell and Jackson 1996). It usually stalks its prey, attempting to approach to within a few bounds. The diet consists mainly of rodents and other small mammals and also includes birds, reptiles, and insects.

Average population density under optimal conditions is around 3–5/1,000 ha. (Stahl and Artois 1994). In contrast to *F. catus*, the wild cat is usually solitary, each individual having a well-defined home range. Males defend these areas but may wander outside of them during times of food shortage or to locate estrous females. Some variation in spacing has been reported (Nowell and Jackson 1996). In France the seasonal ranges of males averaged 5.7 sq km and overlapped the ranges of 3–5 females, which averaged 1.8 sq km. In Scotland, however, males and females had equivalent monthly home ranges with an average size of 1.75 sq km and little overlap.

Mating occurs from about January to March in Europe and Central Asia. Females are polyestrous, with heat lasting 2–8 days. Several males collect around a female in heat; there is considerable screeching and other vocalization and sometimes violent fighting. There is usually only a single litter per year, though occasionally a second is produced in the summer. Births in East Africa may occur at any time of year but seem to peak there and in southern Africa during the wet season (Kingdon 1977; Smithers 1971). The gestation period averages 66 days in Europe and about 1 week less in Africa. Litters usually contain two or three young in the wild. At the Berne Zoo, Meyer-Holzapfel (1968) observed that births occurred from March to August and that litter size averaged four and ranged from one to eight. The young weigh about 40 grams at birth, open their eyes after about 10 days, nurse for about 30 days, emerge from the den at 4–5 weeks, begin to hunt with the mother at 12 weeks, probably separate from her at 5 months, and attain sexual maturity at around 1 year. According to Kingdon (1977), captives have lived up to 15 years.

There is some uncertainty about the original range in historical times, partly because *F. silvestris* may be hard to distinguish from, and may even be conspecific with, *F. catus*. Some authorities consider the reported populations on Crete, Corsica, Sardinia, and the Balearic Islands to be distinctive, and now highly endangered, subspecies of *F. silvestris*, but others consider them to be feral domestic cats introduced by people centuries ago (Nowell and Jackson 1996). An interesting compromise position, suggested by Randi and Ragni (1991), is that these island populations do result from introduction but that the introduction took place in prehistoric times and involved the wild African subspecies *F. silvestris libyca*, which was the progenitor of *F. catus*.

The wild cat once occupied most of Europe but had withdrawn from Scandinavia and most of Russia by the Middle Ages because of climatic deterioration. In modern times, especially during the nineteenth century, the species was intensively hunted by persons who considered it to be a threat to game and domestic animals and so was eliminated from much of western and central Europe. Diversion of human activity during World Wars I and II apparently stimulated recovery in such places as Scotland and Germany (Smit and Van Wijngaarden 1981). *F. silvestris* has been utilized in the fur trade and is on appendix 2 of the CITES. Many thousands were taken during some years in the past, but at present there is thought to be little commerce in this species (Nowell and Jackson 1996).

By the 1980s the wild cat had been placed under complete legal protection in most countries of Europe and around the Mediterranean, populations generally seemed to be expanding, and reintroductions had been attempted in several places. Unfortunately, there now has been a reversal of this trend, with numbers and distribution again declining in many areas. Populations are mostly small, fragmented, and subject to illegal hunting and deteriorating habitat. There still are substantial numbers in eastern France and the Balkans but fewer than 2,000 in Germany and almost none in Poland. To the east, the wild cat survives only in the Caucasus and small parts of Ukraine and Moldova (Stahl and Artois 1994).

A major threat reported through most of the remaining range is hybridization with domestic *F. catus*. This problem is intensifying in the eastern Mediterranean region and even in sub-Saharan Africa but is of particular concern in Europe, where some authorities fear that very few pure *F. silvestris* remain (Nowell and Jackson 1996; Stahl and Artois 1994). The population in Scotland, the last in the British Isles and numbering a few thousand cats, long has been affected by interbreeding but apparently still includes many genetically distinct animals (Hubbard et al. 1992; Kitchener 1992). The subspecies in Scotland, *F. s. grampia*, is classified as vulnerable by the IUCN. A mysterious black, or "Kellas," cat, recently reported in Scotland, evidently has resulted from hybridization of the wild and domestic species (Kitchener and Easterbee 1992).

Felis catus (domestic cat)

According to the National Geographic Society (1981), there are more than 30 different breeds of domestic cat, and the average measurements of several popular breeds are: head and body length, 460 mm, and tail length, 300 mm. E. Jones (1977) found that feral males on Macquarie Island, south of Australia, averaged 522 mm in head and body length, 269 mm in tail length, and 4.5 kg in weight, while females there averaged 478 mm in head and body length, 252 mm in tail length, and 3.3 kg in weight. Ninety percent of the cats on Macquarie were orange or tabby, and the remainder were black or tortoiseshell. Female *F. catus* have four pairs of mammae.

The domestic cat evidently is descended primarily from the wild cat of Africa and extreme southwestern Asia, *F. silvestris libyca*. The latter may have been present in towns in Palestine as long ago as 7,000 years, and actual domestication occurred in Egypt about 4,000 years ago. Introduction to Europe began around 2,000 years ago, and some interbreeding occurred there with the wild subspecies *F. silvestris silvestris*. Domestication may originally have been associated with the cat's proclivity to prey on the rodents that threatened the stored grain upon which ancient civilizations depended but also seems to have a religious basis (Grzimek 1975; Kingdon 1977; Yurco 1990). The cat was the object of a passionate cult in ancient Egypt, where a city, Bubastis, was dedicated to its worship. The followers of Bastet, the goddess of pleasure, put bronze statues of cats in sanctuaries and carefully mummified the bodies of hundreds of thousands of the animals. The veneration of cats in Egypt intensified about 3,000 years ago and persisted at least into Roman times.

There have been relatively few detailed field studies of *F. catus*, but there is no reason to think that its behavior and ecology under noncaptive conditions differ greatly from what has been found for *F. silvestris*. On Macquarie Island, where the cat population has been feral since 1820, E. Jones (1977) obtained specimens in a variety of habitats by both day and night. The cats sheltered in rabbit burrows, thick vegetation, or piles of rocks. The diet consisted largely of rabbits (also introduced on the island) and also included rats, mice, birds, and carrion. Population density was estimated at 2–7/sq km.

In a rural area of southern Sweden, Liberg (1980) found

Domestic cat *(Felis catus)*, photo by William J. Allen.

a population density of 2.5–3.3/sq km. About 10 percent of the cats were feral, and the rest, including all of the females, were associated with human households. Adult females lived alone or in groups of up to 8 usually closely related individuals. Each member of a group had a home range of 30–40 ha. that overlapped extensively with the ranges of other members of the same group but not with the ranges of the cats in other groups. Most females spent their life in the area in which they were born, seldom wandering more than 600 meters away. Nonferal males remained in their area of birth, along with females, until they were 1.5–3.5 years old but then left and tried to settle somewhere else. Males living in the same group had separate home ranges. There were 6–8 feral males in the study area; their home ranges were 2–4 km across, partly overlapped one another, and sometimes included the areas used by several groups of females. According to Haspel and Calhoon (1989), home ranges of unrestrained urban cats are much smaller than those in rural areas. In Brooklyn, New York, range averaged 2.6 ha. for males and 1.7 ha. for females, the difference evidently being a function of body size. Population density in that area was up to about 5/ha. in habitat characterized by many abandoned buildings and voluminous, poorly contained refuse (Calhoon and Haspel 1989).

F. catus communicates through a variety of vocalizations. Purring differs from the other sounds in that it continues during inspiration as well as expiration and may occur simultaneously with voice production proper. The sound and vibration of purring evidently results primarily from laryngeal modulation of respiratory flow (Sissom, Rice, and Peters 1991).

According to Ewer (1973), the house cat is basically solitary, but individuals in a given area seem to have a social organization and hierarchy. A male newly introduced to an area normally must undergo a series of fights before its position is stabilized in relation to other males. Both males and females sometimes gather within a few meters of one another without evident hostility. A male and female may form a bond that extends beyond the mating process. Females are polyestrous and normally produce two litters annually. They may mate with more than one male per season, and if a litter is lost, they soon enter estrus again. The gestation period averages 65 days. The number of young per litter averages four and ranges from one to eight. Kittens weigh 85–110 grams at birth, open their eyes after 9–20 days, are weaned at 8 weeks, and attain independence at about 6 months. Hemmer (1976) listed age of sexual maturity in females as 7–12 months.

Although the cat sometimes has been venerated, it also has been associated with evil. Certain superstitions concerning it have persisted to modern times (Grzimek 1975). The species is now generally looked upon with favor by most cultures, but free-ranging cats often are considered to be among the greatest decimators of native wildlife, especially songbirds (Lowery 1974). On Kerguelen Island, a French possession of about 6,200 sq km in the southern Indian Ocean, a pair of cats introduced in 1956 has grown to a population of 10,000 that consumes an estimated 3 million birds annually (Chapuis, Boussès, and Barnaud 1994).

Among domestic animals *F. catus* is the species most commonly reported to be rabid. Although rabies in humans is now exceedingly rare in the United States, with an average of less than 1 indigenously acquired case annually for the past 30 years, efforts to control the disease cost more than $300 million each year. Moreover, 30,000–40,000 people thought to have been exposed to the virus receive the series of anti-rabies vaccinations each year in the United States. In some other countries, such as India, with an estimated 25,000 human deaths annually, rabies remains a serious health problem (Krebs, Wilson, and Child 1995; Squires 1995). In the last several years there has been an increase in human deaths from rabies in the United States, but mainly involving transmission from bats, not cats (notably *Eptesicus* and *Lasionycteris*).

Felis bieti (Chinese desert cat)

Head and body length is 685–840 mm and tail length is 290–350 mm. The build is stocky, the tail relatively short; one male weighed 9.0 kg, a female 6.5 kg (Nowell and Jackson 1996). The general coloration is yellowish gray, the back is somewhat darker, and there are practically no markings on the flanks. The tail is tipped with black and has three or four subterminal blackish rings.

According to Nowell and Jackson (1996), despite its com-

Jungle cat *(Felis chaus)*, photo from San Diego Zoological Society.

mon name, this cat is found mainly in alpine meadows and scrub, and sometimes in steppe and forest edge, at elevations of 2,800–4,100 meters. It is active mostly at night and in the early morning and rests and rears its young in burrows. The diet consists largely of rodents, some of which are dug out of subterranean tunnels. The sexes usually live separately, the mating season is January–March, young often are born in May, litter size is two to four, and age of independence is 7–8 months. *F. bieti* is poorly known but may not be rare, as its pelt is commonly found in local markets. It is on appendix 2 of the CITES.

Felis chaus (jungle cat)

Head and body length is 500–940 mm, tail length is 230–310 mm, and weight is 4–16 kg (Heptner and Sludskii 1992; Lekagul and McNeely 1977; Novikov 1962). The general coloration varies from sandy or yellowish gray to grayish brown and tawny red, usually with no distinct markings on the body. The tail has several dark rings and a black tip. Lekagul and McNeely (1977) noted that the legs are proportionately the longest of any felid's in Thailand and are thus helpful in running down prey.

The jungle cat is found in a variety of open and wooded habitats from sea level to elevations of 2,400 meters but generally is associated with dense vegetative cover and water. It swims well and may dive to catch fish (Nowell and Jackson 1996). It dens in thick vegetation or in the abandoned burrow of a badger, fox, or porcupine. It is active either by day or by night. The diet consists mainly of hares and other small mammals and also includes birds, frogs, and snakes. Mating occurs in Central Asia in February and March, but 3-week-old kittens have been found in Assam in January and February. The gestation period is 66 days, and the usual litter size is three to five young. The newborn weigh 83–161 grams and are weaned after about 3 months (Hayssen, Van Tienhoven, and Van Tienhoven 1993). Sexual maturity comes at 18 months. One specimen lived at the Copenhagen Zoo for 20 years (Marvin L. Jones, Zoological Society of San Diego, pers. comm., 1995).

The jungle cat adapts well to irrigated agriculture and often is found in the vicinity of human settlements (Nowell and Jackson 1996). It thus may be less threatened than most cats. In Central Asia and the Caucasus it is actively hunted because of alleged predation on game birds and poultry and also for its pelt (Heptner and Sludskii 1992). It is on appendix 2 of the CITES.

Felis margarita (sand cat)

Head and body length is 450–572 mm and tail length is 280–348 mm. Weight is 1.3–3.4 kg (Heptner and Sludskii 1992). The general coloration is pale sandy to gray straw ochre, the back is slightly darker, and the belly is white. A fulvous reddish streak runs across each cheek from the corner of the eye, and the tail has two or three subterminal rings and a black tip. The pelage is soft and dense. The soles of the feet are covered with dense hair. The limbs are short, and the ears are set low on the head.

The sand cat is adapted to extremely arid terrain, such as shifting dunes of sand. The padding on its soles facilitates progression over loose, sandy soil. Activity is mainly nocturnal and crepuscular, the day being spent in a shallow burrow. Prey consists of jerboas and other rodents and occasionally hares, birds, and reptiles. The sand cat apparently is able to subsist without drinking free water. A radiotracking study in Israel determined the presence of at least 22 individuals in an area of 100 sq km; the home range of one adult male was estimated at 16 sq km and overlapped those of neighboring males (Nowell and Jackson 1996).

Births in Central Asia occur in April (Heptner and Sludskii 1992). In Pakistan there may be two litters annually, as kittens have been found in both March–April and October (Roberts 1977). The estrous cycle is 46 days, estrus lasts 5 days, the gestation period is 59–67 days, litters usually contain two to four young, average weight at birth is 39 grams,

Sand cat *(Felis margarita)*, photo by Lothar Schlawe.

and the young open their eyes after 12–16 days (Hemmer 1976; Nowell and Jackson 1996). The young are thought to become independent when still quite small, at perhaps 6–8 months, and they may reach sexual maturity at 9 months; longevity is up to 13 years (Nowell and Jackson 1996).

The sand cat seems to be generally rare, but such a view may reflect a lack of knowledge. Thousands of skins have been taken for the fur trade during some years in Central Asia (Heptner and Sludskii 1992). The species is on appendix 2 of the CITES. The subspecies in Pakistan, *F. m. scheffeli*, was not discovered until 1966 and is now classified as endangered by the USDI and as near threatened by the IUCN. It reportedly declined drastically through uncontrolled exploitation by commercial animal dealers from 1967 to 1972, but recent observations suggest that populations have recovered (Nowell and Jackson 1996).

Felis nigripes (black-footed cat)

This is the smallest cat. Head and body length is 337–500 mm, tail length is 150–200 mm, and weight is 1.5–2.75 kg. The general coloration is dark ochre to pale ochre or sandy, being somewhat darker on the back and paler on the belly. A bold pattern of dark brown to black spots is arranged in rows on the flanks, throat, chest, and belly. There are two streaks across each cheek, two transverse bars on the forelegs, and as many as five transverse bars on the haunches. The tail has a black tip and two or three subterminal bands. The bottoms of the feet are black (since the animal walks on its toes, much of the black is usually visible).

The black-footed cat inhabits dry, open country. It shelters in old termite mounds and the abandoned burrows of other mammals. It is mainly nocturnal, but in captivity it has been found to be more active by day than most other small cats. The diet probably includes rodents, birds, and reptiles. Individuals have been observed to catch birds and hares and to cache them for later use (Nowell and Jackson 1996).

This felid seems to be highly unsocial. Even opposite sexes evidently come together only for 5–10 hours. However, a male's home range of 13 sq km reportedly overlapped a female's range of 12 sq km by about 50 percent (Nowell and Jackson 1996). Gestation lasts 59–68 days, and litters contain one to three young. Newborn weight is about 60–88 grams (Hayssen, Van Tienhoven, and Van Tienhoven 1993). The kittens leave the nest after 28–29 days and take their first solid food a few days later. Captive females have not initially entered heat until 15–21 months (Grzimek 1975).

F. nigripes is on appendix 1 of the CITES and is classified as rare in South Africa (Smithers 1986). It is not thought to be threatened with extinction but may be locally jeopardized by habitat degradation and by traps and poison put out to kill other predators (Nowell and Jackson 1996).

Felis manul (Pallas's cat)

Head and body length is 500–650 mm and tail length is 210–310 mm. Heptner and Sludskii (1992) stated that weight is 2.5–4.5 kg but that some animals may be heavier. The general coloration varies from light gray to yellowish buff and russet; the white tips of the hairs produce a frosted silvery appearance. There are two dark streaks across each side of the head and four rings on the dark-tipped tail. The coat is relatively longer and more dense than that of any other wild species of *Felis.* The fur is especially long near the end of the tail, and on the underparts of the body it is almost twice as long as on the back and sides. Such an

Black-footed cat *(Felis nigripes)*, photo by John Visser.

arrangement provides good insulation for an animal that spends much time lying on frozen ground and snow. The body is massive, the legs are short and stout, the head is short and broad, and the very short, bluntly rounded ears are set low and wide apart.

Pallas's cat inhabits steppes, deserts, and rocky country to elevations of more than 4,000 meters. It dens in a cave, crevice, or burrow dug by another animal. It usually is reported to be nocturnal or crepuscular but may be active by day since its main prey, pikas, are diurnal (Heptner and Sludskii 1992; Nowell and Jackson 1996). The diet also includes small rodents and ground-dwelling birds. According to Stroganov (1969), the young are born in Siberia in late April and May. Newborn weight has been reported as 89 grams (Hayssen, Van Tienhoven, and Van Tienhoven 1993). The estrous cycle lasts 46 days, the gestation period has been variously reported at 60–75 days, litter size is 1–6 young (usually 3 or 4), and females attain sexual maturity at 1 year (Heptner and Sludskii 1992; Nowell and Jackson 1996).

As many as 50,000 pelts of *F. manul* were being taken annually in Mongolia alone during the early twentieth century, but from the 1920s to the 1980s harvests in that country usually were less than 10,000 per year (Heptner and Sludskii 1992; Nowell and Jackson 1996). Broad, Luxmoore, and Jenkins (1988) reported that at least 20,000 skins of *F. manul* still were entering international trade each year and that such activity might be threatening the species. Concern about evident declining populations led to legal protection in Mongolia and China, and commerce in skins has largely ceased. The subspecies *F. m. ferrugineous*, found to the west, from Armenia and Iran to Uzbekistan, is designated as near threatened by the IUCN. The entire species also is on appendix 2 of the CITES.

Pallas's cat *(Felis manul)*, photo by Bernhard Grzimek.

Felis pardina (Spanish lynx)

Head and body length is 850–1,100 mm, tail length is 125–30 mm, and shoulder height is 600–700 mm. Average weight is 12.9 kg for males and 9.4 kg for females. The upper parts are yellowish red and the underparts are white. There are round black spots on the body, tail, and limbs. The ears have tufts, and the face has a prominent fringe of whiskers (Nowell and Jackson 1996; Smit and Van Wijngaarden 1981; Van Den Brink 1968). There originally seem to have been three different pelage patterns, one characterized by relatively large spots arranged roughly in lines, one with more randomly distributed large spots, and one with very small spots. The latter two patterns evidently have disappeared with the recent decline in the species (Beltrán and Delibes 1993).

According to Nowell and Jackson (1996), the Spanish lynx inhabits open woodland, thickets, and dense scrub. It is primarily nocturnal and travels about 7 km per night. It feeds primarily on rabbits and also takes small ungulates

Spanish lynx *(Felis pardina)*, photo by Lothar Schlawe.

and ducks. Reported population densities are about 4.5–16.0/100 sq km. Annual home range averages 18 sq km for males and 10 sq km for females. There is no overlap in the ranges of animals of the same sex but complete overlap between the ranges of opposite sexes. Mating occurs from January to July, peaking in January–February, and parturition peaks in March–April. Gestation lasts about 2 months, litter size is 2–3 young, and independence is attained at 7–10 months, but young remain in their natal territory until they are about 20 months old. Females are capable of breeding in their first winter but may have to wait until they can acquire a territory. Reproductive activity continues until 10 years of age and longevity is up to 13 years.

The Spanish lynx is classified as endangered by the IUCN and the USDI and is on appendix 1 of the CITES. It formerly occurred throughout the Iberian Peninsula, but the range has been reduced to less than 15,000 sq km in scattered mountainous areas and the Guadalquivir Delta. The total number of animals probably does not exceed 1,200, including about 350 breeding females (Nowell and Jackson 1996; Rodriguez and Delibes 1992). There was a major decline during the 1950s and 1960s, when the disease myxomatosis hit the rabbit populations. The decline is continuing as suitable rabbit and lynx habitat is replaced by cereal cultivation and forest plantations.

Felis lynx (Eurasian lynx)

Head and body length is 800–1,300 mm, tail length is 110–245 mm, shoulder height is 600–750 mm, and weight is 8–38 kg (Novikov 1962; Van Den Brink 1968). The average weight is 21.6 kg for males and 18.1 kg for females, about twice that of *F. canadensis* (Nowell and Jackson 1996). The fur is exceedingly dense, and coloration is more variable than in any other cat species (Heptner and Sludskii 1992). The upper parts may be reddish, brown, yellowish, gray, ashy blue, or almost white. Spots, variously darker than the rest of the coat, are almost always present. The underparts are usually white and the terminal part of the tail is black. There is a prominent black tuft on the ear. The summer coat is shorter than the winter coat, and the spots become more prominent. The legs are relatively long and the large feet are thickly haired in winter, producing a "snowshoe effect" for efficient travel through deep snow (Nowell and Jackson 1996).

According to Heptner and Sludskii (1992), the Eurasian lynx is typically associated with forests; it is most common where there is a mixture of spruce and deciduous trees but may sometimes penetrate the forest steppe and steppe zones. In the winter it follows its prey to lower elevations, and if game is scarce it may migrate up to 100 km. It is generally active at night and in the early morning, during which time it hunts over an average distance of about 10 km. Prey is followed for up to several days, then approached by stealth, and at last seized after a final rush. There is some controversy regarding diet. Heptner and Sludskii (1992) indicated that hares are by far the most important food for *F. lynx*, as they are for *F. canadensis*. Data compiled by Nowell and Jackson (1996), however, suggest that *F. lynx* preys primarily on small ungulates, particularly roe deer, chamois, and musk deer, and that smaller animals are eaten only when ungulates are not available. *F. lynx* is capable of killing animals three to four times its own size and in some areas preys mainly on large ungulates (mostly females and young), including red deer, reindeer, and argali *(Ovis ammon)*. The diet also includes rodents, pikas, and birds. Prey is usually dragged several hundred meters before being consumed, and a portion may be cached for later use.

Population densities of around 1–10/100 sq km have been reported in Russia, where numbers are said to fluctuate to some extent in association with the abundance of hares, as in North America. Home range size in Russia is 20–100 sq km, and such an area is traversed every 15–30 days (Heptner and Sludskii 1992). Density in Switzerland, where the species is scarce, is around 1/100 sq km and home ranges are very large. Average size is 264 sq km for males and 168 sq km for females. Female ranges have central core areas, averaging 72 sq km, which do not overlap one another and in which the residents spend most of their time. Male ranges overlap one another to some extent and also encompass most of a female's range, but a male tends to avoid the female's core area (Nowell and Jackson 1996). Heptner and Sludskii (1992) noted that sometimes a mated pair or a mother and young may hunt together, one animal chasing prey in the direction of the other. Mating occurs in Russia from January to March, with births from April to June. The gestation period is 67–74 days. There are 1–4 young, usually 2–3. They weigh about 250 grams at birth, open their eyes after 12 days, take some solid food at 50 days, and are weaned and begin to accompany their mother at about 3 months. They leave her just before the next mating season and attain sexual maturity in their second year. Studies by Kvam (1991) indicate that males in Norway normally reach sexual maturity by 31 months but that some do so at 21 months, and that about 50 percent of females are mature at 9 months, the rest by 21 months. Nowell and Jackson (1996) reported that females are reproductively active until 14 years, males until 16–17 years.

The Eurasian lynx has been intensively hunted and trapped for its valuable fur and because it is considered a threat to game and livestock. It is on appendix 2 of the CITES. The population in Russia has been estimated to number 40,000, and about 5,000–7,000 pelts are taken there annually (Heptner and Sludskii 1992; Nowell and Jackson 1996). The species also still is found in much of China and Central Asia, but it disappeared from most of Europe during the nineteenth century. To the west of Russia, major populations now occur only in Scandinavia and the Carpathian region. Small, isolated groups are present in northeastern Poland, the southern Balkans, and the French Pyrenees. Reintroductions have been carried out recently in parts of Germany, Austria, Switzerland, and Slovenia (Breitenmoser and Breitenmoser-Würsten 1994; Nowell and Jackson 1996; Smit and Van Wijngaarden 1981). The population reestablished in Switzerland now numbers more than 100 animals and is thought to be expanding into northwestern Italy (Guidali, Mingozzi, and Tosi 1990). However, Ragni, Possenti, and Mayr (1993) suggested that the current population in the Alps is different from the original and now extinct subspecies, which they designated *F. lynx alpina*. They also noted that the lynx evidently had disappeared from the Italian peninsula, south of the Alps, by about the beginning of the Bronze Age. The lynx may have occurred in Palestine before that area was largely deforested; in the Arabian region the species is now found only in northern Iraq, where it is rare (Harrison 1968).

Felis canadensis (Canada lynx)

Head and body length is about 800–1,000 mm, tail length is 51–138 mm, and weight is 5.1–17.2 kg (Banfield 1974; Burt and Grossenheider 1976). Average weight is 10.7 kg for males and 8.9 kg for females, about half that of the Eurasian *F. lynx* (Nowell and Jackson 1996). The coloration varies but is commonly yellowish brown; the upper parts may have a gray frosted appearance, the underparts are more buffy, and there is often a pattern of dark spots. The markedly short tail may have several dark rings and is tipped with black. The fur is long, lax, and thick. It is especially long on the lower cheeks in winter and gives the impression of a ruff around the neck. The triangular ears are tipped by tufts of black hairs about 40 mm long. The legs are relatively long. The paws are large and densely furred, an adaptation for moving over winter snow. Females have four mammae (Banfield 1974).

The lynx is generally found in tall forests with dense undergrowth but may also enter open forest, rocky areas, or tundra. For shelter it constructs a rough bed under a rock ledge, fallen tree, or shrub (Banfield 1974). It is mainly nocturnal; reports of average nightly movement range from 5 to 19 km (Ewer 1973). The lynx usually keeps to one area but may migrate under adverse conditions. The maximum known movement was by a female that was marked on 5 November 1974 in northern Minnesota and trapped on 20 January 1977 in Ontario, 483 km from the point of release (Mech 1977*c*). The lynx climbs well and is a good swimmer, sometimes crossing wide rivers. It hunts mainly by eye and also has well-developed hearing. It usually stalks its prey to within a few bounds, or it may wait in ambush for hours (Tumlison 1987). Reports from much of the range of the species indicate that leporids form a major part of the diet. The snowshoe rabbit *(Lepus americanus)* is of particular importance in North America. Deer and other ungulates are utilized heavily in certain areas, especially during the winter. Other foods include rodents, birds, and fish.

In Canada the numbers of lynxes seem to fluctuate, together with those of the snowshoe rabbit, in a regular cycle. Numerical peaks occur at an average interval of 9.6 years, though not at the same time all over Canada. Several authorities have suggested that this phenomenon is actually an artifact, perhaps associated with the intensity of human trapping, but Finerty (1979) upheld the traditional view. Studies in central Alberta (Brand and Keith 1979; Brand, Keith, and Fischer 1976; Nellis, Wetmore, and Keith 1972) indicate that the main direct cause of the cyclical drop in lynx numbers is postpartum mortality of kittens resulting from lack of food. There is also a reduced rate of pregnancy in females. Population density is this area varied from a low of 2.3/100 sq km in the winter of 1966–67 to a peak of 10/100 sq km in the winter of 1971–72.

The reported individual home range size is 4–70 sq km for males and 4–25 sq km for females (Nowell and Jackson 1996). The lynx is probably territorial, but the ranges of females may overlap one another to some extent, and the range of a male may include that of a female and her young. Adults tend to avoid one another except during the breeding season. At that time there is an increase in scent marking with urine and also in vocalizations—purring and meowing (Tumlison 1987). Females are monestrous and bear a single litter per year (Banfield 1974; Ewer 1973). Mating occurs mainly in February–March, parturition is usually in May–June, gestation lasts 9–10 weeks, and litters contain 1–8 young, averaging about 4–5 when prey is abundant and 2–3 when prey is scarce (Nowell and Jackson 1996). Banfield (1974) gave birth weight as 197–211 grams. Lactation lasts for 5 months, but some meat is eaten by

Canada lynx *(Felis canadensis)*, photo from San Diego Zoological Garden.

1-month-old kittens. The young usually remain with their mother until the winter mating season, and siblings may stay together for a while afterward. The age of sexual maturity is usually 22–23 months (Nowell and Jackson 1996). One wild female is known to have lived 14 years and 7 months (Chubbs and Phillips 1993), and a captive individual lived 26 years and 9 months (Rich 1983).

The range of *F. canadensis* once extended at least as far south as Oregon, southern Colorado, Nebraska, southern Indiana, and Pennsylvania (E. R. Hall 1981). The only substantial population now remaining in the conterminous United States is that of western Montana and nearby parts of northern Idaho and northeastern Washington. On occasion, especially during periods of cyclically high populations in Canada, the species is found in the states of the northern plains and upper Great Lakes regions. There are small numbers in northern New England and Utah and possibly in Oregon, Wyoming, and Colorado (Armstrong 1972; Cowan 1971; Deems and Pursley 1978; Gunderson 1978; Hunt 1974; Olterman and Verts 1972). From 1988 to 1990, 83 lynxes captured in the Yukon were released in the Adirondacks of New York in an attempt to restore the species (Brocke and Gustafson 1992). In 1994 the U.S. Fish and Wildlife Service rejected a petition to add the lynx in the conterminous United States to the List of Endangered and Threatened Wildlife. This action was taken at higher levels within the agency despite support for listing by the scientific community and by the agency's own field and regional offices with jurisdiction over Montana and other involved states. It may have been a response to political pressure from interests favoring logging, which would disturb lynx habitat and would have become subject to control on federal lands if listing had proceeded (Bourne 1995*b*). After losing a lawsuit on the issue, the Fish and Wildlife Service finally acknowledged that listing was warranted but claimed it could not proceed with such a measure because of higher-priority responsibilities (Knibb 1997).

The lynx has been exterminated in parts of southeastern Canada but still occurs regularly over most of that country. Its numbers there seem to be affected more by the cyclical availability of food than by pressure from human hunting. Populations evidently were relatively high until the early twentieth century, declined until mid-century, and then again increased (Cowan 1971). Subsequent highs and lows in the reported number of pelts taken in Canada have been as follows: 1949–50 trapping season, 3,734; 1954–55, 14,427; 1956–57, 8,748; 1962–63, 51,376; 1966–67, 13,038; 1971–72, 53,589; 1975–76, 13,162; and 1979–80, 34,366. The average price per pelt rose dramatically from $3.62 in 1953–54 to $30.52 in 1968–69 and to $336.36 in 1978–79 (Statistics Canada 1981). At autumn 1984 sales in Ontario the average value of lynx pelts surged to nearly $600.00 (Quinn and Parker 1987). Prices subsequently fell back to around $100 per skin in the early 1990s (data from U.S. Fish and Wildlife Service, 1994). *F. canadensis* is on appendix 2 of the CITES.

Felis rufus (bobcat)

Head and body length is 650–1,050 mm, tail length is 110–90 mm, and shoulder height is 450–580 mm. Banfield

Bobcat *(Felis rufus)*, photo by Lothar Schlawe.

(1974) gave the weight as 4.1–15.3 kg. There are various shades of buff and brown, spotted and lined with dark brown and black. The crown is streaked with black and the backs of the ears are heavily marked with black. The short tail has a black tip, but only on the upper side. *F. rufus* resembles *F. lynx* but is usually smaller and has slenderer legs, smaller feet, shorter fur, and ears that are tufted less conspicuously or not at all. As in *F. lynx*, a ruff of fur extends from the ears to the jowls, giving the impression of sideburns. Fully black individuals occasionally are found in Florida (Regan and Maehr 1990). Females have four mammae (Lowery 1974).

The bobcat is more ubiquitous than the lynx, occurring in forests, mountainous areas, semideserts, and brushland. Its den is usually concealed in a thicket, hollow tree, or rocky crevice. It is mainly nocturnal and terrestrial but climbs with ease. Nightly movements of about 3–11 km have been reported. Prey is usually stalked with great stealth and patience, then seized after a swift leap. The diet consists mainly of small mammals, especially rabbits, and birds. Larger prey, such as deer, is sometimes taken, especially in the winter. A study in Massachusetts found deer to be the most common winter food (McCord 1974).

Reported maximum population densities are 1/18.4 sq km in the western United States and 1/2.6 sq km in the Southeast (Jachowski 1981). Average minimum home range in Louisiana was found to be about 5 sq km for males and 1 sq km for females (Hall and Newsom 1978). In a study in southeastern Idaho, Bailey (1974) found average (and extreme) home range size to be 42.1 (6.5–107.9) sq km for males and 19.3 (9.1–45.3) sq km for females. Female ranges were almost exclusive of one another, but the ranges of males overlapped one another, as well as those of females. There was a land tenure system, seemingly based on prior right: no neighboring resident or transient permanently settled in an area already occupied by a resident. All observed changes in resident home ranges were attributed to the death of a resident. Territoriality was pronounced, especially in females, and scent marking was accomplished by use of feces, urine, scrapes, and anal gland secretions. Individuals were solitary, avoiding each other even in areas of range overlap, except during the mating season. The bobcat is usually silent but may emit loud screams, hisses, and other sounds during courtship.

Females apparently are seasonally polyestrous; they usually produce a single annual litter during the spring, but there is evidence of a second birth peak in late summer and early autumn, perhaps involving younger females or ones that lost their first litters (Banfield 1974). According to Fritts and Sealander (1978), mating may occur as early as November or as late as August. The estrous cycle is 44 days and estrus lasts 5–10 days (Nowell and Jackson 1996). The gestation period is 60–70 days (Hemmer 1976). The number of young per litter is one to six, commonly three. Weight at birth is 283–368 grams (Banfield 1974). The young open their eyes after 9–10 days, nurse for about 2 months, and begin to travel with the mother at 3–5 months. They separate from the mother in the winter (Jackson 1961), probably in association with the mating season. Females may reach sexual maturity by 1 year, but males do not mate until their second year of life. The oldest individuals captured in a Wyoming study were 12 years of age and still sexually active (Crowe 1975). A captive bobcat lived 32 years and 4 months (Jones 1982).

The bobcat occasionally preys on small domestic mammals and poultry and thus has been hunted and trapped by people. It has been exterminated in much of the Ohio Valley, the upper Mississippi Valley, and the southern Great Lakes region (Deems and Pursley 1978). The bobcat is uncommon in central Mexico (Leopold 1959); the subspecies there, *F. rufus escuinapae,* is listed as endangered by the USDI. It is intensively persecuted by sheepherders, and its habitat is being degraded (Nowell and Jackson 1996). The entire species is on appendix 2 of the CITES.

The value of bobcat fur has varied widely depending on fashion and economic conditions. The average price per pelt rose from about $10 in the 1970–71 season (Deems and Pursley 1978) to $145 in 1978–79 (Jachowski 1981). There was a corresponding increase in the number of bobcats being harvested. The total known annual kill in the United States was approximately 92,000 in the late 1970s. This rate continued or declined slightly in the 1980s. About two-thirds of these animals were being taken primarily for their fur, and most of the pelts were being exported. There was concern that bobcat populations were being seriously reduced in some areas and that little was being done to manage the resource. Jachowski (1981), however, reported substantial improvement in management since *F. rufus* was placed on appendix 2 of the CITES in 1977. Prior to that time few states had a closed season, and many had bounties. Presently there are no bounties, 10 states provide complete protection, and the rest allow a regulated harvest during a limited season. Jachowski estimated that there are between 725,000 and 1,020,000 bobcats in the United States and suggested that current known harvest levels are not jeopardizing overall populations in the country. Export of skins was halted briefly by legal action in 1981 but was subsequently restored. In the 1982–83 season about 75,000 skins were taken and sold for an average price of $142 (Rolley 1987). In 1991–92, 22,077 skins were taken and sold for an average of $63 (Linscombe 1994). Although they did not directly question the above population data, Nowell and Jackson (1996) indicated that there is still concern about whether commercial trapping is sustainable. They noted also that trade in bobcat fur is declining and that the European Community—the main market for the pelts—announced that after 1995 it would ban all importation of the skins of animals taken in leghold traps.

Felis caracal (caracal)

Head and body length is 600–915 mm, tail length is 230–310 mm, shoulder height is 380–500 mm, and weight is 6–19 kg (Kingdon 1977; Skinner and Smithers 1990). The pelage is dense but relatively short, and there are no side whiskers as in *F. lynx.* The general coloration is reddish brown. There is white on the chin, throat, and belly and a narrow black line from the eye to the nose. The ears are narrow, pointed, black on the outside, and adorned with black tufts up to 45 mm long. Smaller than *F. lynx, F. caracal* has a long and slender body and a tapering tail that is approximately one-third the length of the head and body.

The caracal is found mainly in dry country—woodland, savannah, and scrub—but avoids sandy deserts. Maternal dens are located in porcupine burrows, rocky crevices, or dense vegetation. This cat is largely nocturnal but sometimes is seen by day. It climbs and jumps well. It is mainly terrestrial and is apparently the fastest feline of its size. A radio-tracking study indicated that daily movement averaged 10.4 km for males and 6.6 km for females (Nowell and Jackson 1996). Prey is stalked and then captured after a quick dash or leap. The diet includes birds, rodents, and small antelopes.

In South Africa the home ranges of males were found to measure 5.1–48.0 sq km and to widely overlap one another and also the ranges of females, which measured 3.9–26.7 sq km and overlapped only slightly (Skinner and Smithers 1990). Much larger ranges have been reported in the Negev of Israel, the averages being 221 sq km for males and 57 sq km for females; in the latter area male ranges typically encompassed several female ranges (Nowell and Jackson

Caracals *(Felis caracal)*, photo by Bernhard Grzimek.

1996). The caracal has been reported to be territorial and to mark with urine. Vocalizations include miaows, growls, hisses, and coughing calls (Kingdon 1977). The species is usually seen alone, but Rowe-Rowe (1978*b*) reported a group of two adults and five young. Various observations suggest that the young may be born at any time of year, though there is a birth peak from October to February in South Africa. The estrous cycle is 14 days, estrus lasts 1–6 days, the gestation period has been reported at 69–81 days, and the number of young per litter is one to six, usually three, young. The kittens open their eyes after 10 days, are weaned at 10–25 weeks, and attain sexual maturity at 12–16 months. One female gave birth at 18 years (Kingdon 1977; Nowell and Jackson 1996; Skinner and Smithers 1990).

The caracal is easily tamed and has been used to assist human hunters in Iran and India. It sometimes raids poultry, however, and thus has been killed by people. It apparently has become scarce in North Africa, South Africa, and central and southwestern Asia. All Asian populations are on appendix 1 of the CITES, and African populations are on appendix 2.

Felis serval (serval)

Head and body length is 670–1,000 mm, tail length is 240–450 mm, shoulder height is 540–620 mm, and weight is 8.7–18.0 kg. Males are generally larger than females (Kingdon 1977). The general coloration of the upper parts ranges from off-white to dark gold, and the underparts are paler, often white (Smithers 1978). The entire pelage is marked either with small, dark spots or with large spots that tend to merge into longitudinal stripes on the head and back. The tail has several rings and a black tip. Melanistic individuals have been widely reported (Nowell and Jackson 1996). The build is light, the legs and neck are long, and the ears are large and rounded.

According to Smithers (1978), the serval is generally a species of the savanna zone and is found in the vicinity of streams with densely vegetated banks. It is primarily nocturnal and may move 3–4 km per night. It is mainly terrestrial and can run or bound swiftly for short distances. Prey is apparently located both by sight and by hearing. Birds up to 3 meters above the ground may be captured by remarkable leaps, but the diet seems to consist mostly of murid rodents. Van Aarde and Skinner (1986) reported home ranges of 2.1–2.7 sq km in South Africa. In Tanzania, however, a male was found to have a range of at least 11.6 sq km that overlapped the ranges of at least two females (Nowell and Jackson 1996).

The serval is basically solitary. It has a shrill cry and also growls and purrs. There is no definite mating season, but Smithers (1978) reported that births in Zimbabwe occurred mainly in the warm months from September to April. Kingdon (1977) suggested that there are two birth peaks in East Africa, in March–April and September–November. Observations at the Basel Zoo (Wackernagel 1968) show that females can give birth twice a year, with a minimum normal interval of 184 days. Estrus usually lasted only one day, and the average gestation period was determined to be 74 days. The number of young in 20 litters averaged 2.35 and ranged from 1 to 4. Five newborn weighed 230–60 grams each, and one opened its eyes at 9 days. One female at the Basel Zoo gave birth to her last litter at 14 years and died at about 19 years and 9 months.

The serval is on appendix 2 of the CITES. It is hunted for its skin in East Africa and now no longer occurs in areas heavily populated by people (Kingdon 1977). The species has been mercilessly hunted in farming areas of South Africa and is now considered to be rare in that country (Skinner, Fairall, and Bothma 1977). Because of its concentration in riparian vegetation, the serval is particularly vulnerable to both direct hunting and habitat disruption. Such problems may be especially severe in North Africa, where the species seems to survive only in the Atlas Mountains of Morocco (Nowell and Jackson 1996). The IUCN now classifies the subspecies there, *F. s. constantinus,* as endangered (using the generic name *Leptailurus*), noting that fewer than 250 mature individuals survive.

Felis marmorata (marbled cat)

Head and body length is 450–530 mm, tail length is 475–550 mm, and weight is 2–5 kg (Lekagul and McNeely 1977). The ground color is brownish gray to bright yellow or rufous brown. The sides of the body are marked with large, irregular, dark blotches, each margined with black. There are solid black dots on the limbs and underparts. The tail is spotted, tipped with black, long, and bushy. The pelage is thick and soft, and the ears are short and rounded. *F. marmorata* resembles *Neofelis nebulosa* in the appearance of its coat and in having relatively large canine teeth, and some authorities consider the two species to be closely related. Based on cranial characters and karyology, however, others suggest affinity with *Lynx* or *Panthera* (Nowell and Jackson 1996).

The marbled cat is a forest dweller, apparently nocturnal and partly arboreal in habit. It is thought to prey mostly on birds, to some extent on squirrels and rats, and possibly on lizards and frogs. According to Hayssen, Van Tienhoven, and Van Tienhoven (1993), reproduction is not seasonal, litters contain two young, and birth weight is 70–100 grams. Because of human disturbance and habitat destruction, *F. marmorata* evidently has declined and become very rare in much of its range (IUCN 1978). It is classified as endangered by the USDI and is on appendix 1 of the CITES.

Felis temmincki (Asian golden cat)

Head and body length is 730–1,050 mm and tail length is 430–560 mm. Weight is 8–15 kg (Lekagul and McNeely 1977; Nowell and Jackson 1996). The pelage is of moderate length, dense, and rather harsh. The general coloration varies from golden red to dark brown and gray. Some specimens, especially from the northern parts of the range, have a pattern of spots on the body. The face is marked with white and black streaks, and the underside of the terminal third of the tail is white. The ears are short and rounded.

According to Lekagul and McNeely (1977), the Asian golden cat occurs in deciduous forests, tropical rainforests, and occasionally more open habitats. It is usually terrestrial but is capable of climbing. The diet includes hares, small deer, birds, lizards, and domestic livestock. This cat often hunts in pairs, and the male is said to play an active role in rearing the young. There is no confirmed breeding season. Young weigh about 250 grams at birth. If a litter is lost, the female may produce another within 4 months. Nowell and Jackson (1996) recorded the following: estrous cycle, 39 days; estrus, 6 days; average gestation period, 80 days; litter size, range 1–3, mean, 1.11; age at sexual maturity, 18–24 months; longevity, up to 20 years.

Because of habitat destruction and inability to adjust to the presence of human activity, *F. temmincki* has declined in much of its range. It is classified as near threatened by the IUCN (under the genus *Catopuma*) and as endangered by the USDI and is on appendix 1 of the CITES.

Felis badia (bay cat)

Information provided by Nowell and Jackson (1996) and Sunquist et al. (1994) indicates that this species is known only by seven specimens, six of which were collected from 1855 to 1928. Head and body length is 533–670 mm in

Asian golden cat *(Felis temmincki)*, photo from San Diego Zoological Society.

three adults, tail length is 320–91 mm in four adults, and normal weight was estimated at 3–4 kg in a single adult. In most specimens the pelt is bright chestnut above and paler on the belly, with some obscure spots on the underparts and limbs. One specimen is grayish in color. The long and tapering tail has a whitish median streak down the middle of its lower surface and becomes pure white at the tip. *F. badia* closely resembles *F. temmincki* but is much smaller. All collections and precise sightings have been in highland areas, and at least three specimens were collected along rivers. The bay cat has been reported to inhabit dense primary forests, but there have been recent sightings at night in logged dipterocarp forest. Surveys in the 1980s failed to locate the species, which evidently is extremely rare, but an adult female was collected in November 1992. *F. badia* (using the generic name *Catopuma*) is classified as vulnerable by the IUCN, which estimates that fewer than 10,000 mature individuals survive, and is on appendix 2 of the CITES.

Felis aurata (African golden cat)

Head and body length is 616–1,016 mm, tail length is 160–460 mm, shoulder height is 380–510 mm, and weight is 5.3–16.0 kg. Males are generally larger than females (Kingdon 1977). The overall coloration varies from chestnut through fox red, fawn, gray brown, silver gray, and blue gray to dark slaty. The cheeks, chin, and underparts are white. Some specimens are marked all over with dark brown or dark gray dots; in others the spots are restricted to the belly and insides of the limbs. Black specimens have been recorded. The legs are long, the head is small, and the paws are large.

According to Kingdon (1977), the African golden cat is found mainly in forests, sometimes in mountainous areas. It is said to be active both by day and by night. It is mainly terrestrial but climbs well. Prey is taken by stalking and rushing. The diet includes birds, small ungulates, and domestic animals. *F. aurata* is normally solitary. A pregnant female was taken in Uganda in September. Two litters contained two young each, and two young weighed 195 and 235 grams (Hayssen, Van Tienhoven, and Van Tienhoven 1993). The species is on appendix 2 of the CITES.

Felis bengalensis (leopard cat)

Head and body length is 445–1,070 mm, tail length is 230–440 mm, and weight is 3–7 kg (Lekagul and McNeely 1977; Stroganov 1969). There is much variation in color,

Leopard cat *(Felis bengalensis)*, photo from New York Zoological Society.

but the upper parts are usually pale tawny and the underparts are white. The body and tail are covered with dark spots. There are usually four longitudinal black bands running from the forehead to behind the neck and breaking into short bands and rows of elongate spots on the shoulders. The tail is indistinctly ringed toward the tip. The head is small, the muzzle is short, and the ears are moderately long and rounded.

The leopard cat is found in many kinds of forested habitat at both high and low elevations. It dens in hollow trees or small caves or under overhangs or large roots. It is mainly nocturnal but often is seen by day. It is an excellent swimmer and has populated many offshore islands (Lekagul and McNeely 1977). It apparently hunts on the ground as well as in trees and feeds on hares, rodents, young deer, birds, reptiles, and fish. Four individuals radiotracked by Rabinowitz (1990) in Thailand moved about 500–1,000 meters per day. One female used a total area of 6.6 sq km over a 13-month period but had an average monthly home range of 1.8 sq km.

Births have been reported in May in both Siberia and India. According to Lekagul and McNeely (1977), however, breeding continues throughout the year in Southeast Asia. If one litter is lost, the female may mate and produce another within 4–5 months. The gestation period is 65–72 days. The number of young per litter is one to four, usually two or three. The father may participate in rearing the young. The latter open their eyes at about 10 days. Weight at birth is 75–120 grams (Hayssen, Van Tienhoven, and Van Tienhoven 1993). Sexual maturity may be attained at as early as 8 months and longevity is up to 15 years (Nowell and Jackson 1996).

The leopard cat is more adaptable to deforestation and other habitat alteration than most other Asian felids, and it often is found near villages. However, many island populations, notably that inhabiting Panay, Negros, and Cebu in the Philippines, which may be a distinctive subspecies, have declined drastically (Nowell and Jackson 1996). The species is common in most mainland regions and has been intensively exploited for the international fur trade. The annual take in China alone during 1985–88 is believed to have been around 400,000. The European Community, which had received most pelts from China, banned importation in 1988. Japan has since been the main consumer, importing 50,000 skins in 1989. About 1,000–2,000 pelts of the eastern Siberian form, *F. b. euptilura,* were being processed in the 1930s. In recent years the harvest of that subspecies has fallen to 100–300, apparently in association with a serious decline in the population (Heptner and Sludskii 1992). The subspecies *F. b. bengalensis,* of peninsular India and Southeast Asia, is listed as endangered by the USDI. The populations of that subspecies in India, Bangladesh, and Thailand are on appendix 1 of the CITES; all other populations of *F. bengalensis* are on appendix 2.

Felis rubiginosa (rusty-spotted cat)

This is one of the smallest cats. Head and body length is 350–480 mm and tail length is 150–250 mm (Grzimek 1975). Weight is 1.1–1.6 kg (Nowell and Jackson 1996). The upper parts are grizzled gray, with a rufous tinge of varying intensity, and marked with lines of brown, elongate blotches. The belly and insides of the limbs are white with large, dark spots. There are two dark streaks on the face, and four dark streaks run from the top of the head to the nape.

On the mainland of India the rusty-spotted cat seems to be found mostly in scrub, dry grassland, and open country. In Sri Lanka, however, it occurs in humid mountain forests. Nowell and Jackson (1996) suggested that this situation might be explained by the presence on the mainland of the closely related *F. bengalensis*, which is mainly a forest animal, and the presence in Sri Lanka of *F. chaus*, which prefers more open country. The smaller *F. rubiginosa* thus may be forced into whichever habitat is not dominated by the other species. It is nocturnal, frequently enters trees, and preys on birds and small mammals. Births occur in the spring in India (Grzimek 1975). Estrus lasts 5 days, the gestation period is 67 days, and average litter size is 1.5 young (Nowell and Jackson 1996). One specimen was still living after 16 years in captivity (Marvin L. Jones, Zoological Society of San Diego, pers. comm., 1995). The Indian population is on appendix 1 of the CITES and the Sri Lankan population is on appendix 2.

Felis viverrina (fishing cat)

Head and body length is 750–860 mm, tail length is 255–330 mm, shoulder height is 380–406 mm, and weight is 7.7–14.0 kg. The ground color is grizzled gray, sometimes tinged with brown, and there are elongate dark brown spots arranged in longitudinal rows. Six to eight dark lines run from the forehead, over the crown, and along the neck. The fur is short and rather coarse, the head is big and broad, and the tail is short and thick. The forefeet have moderately developed webbing between the digits. The claw sheaths are too small to allow the claws to retract completely.

The fishing cat is found in marshy thickets, mangrove swamps, and densely vegetated areas along creeks. It often wades in shallows and does not hesitate to swim in deep water. It catches prey fish by crouching on a rock or sandbank and using its paw as a scoop. The diet also includes crustaceans, mollusks, frogs, snakes, birds, and small mammals.

In northeastern India mating activity peaks in January–February, births in March–May (Nowell and Jackson 1996). At the Philadelphia Zoo births occurred in March and August (Ulmer 1968). Observations there indicate that the gestation period is 63 days and birth weight is about 170 grams. The young open their eyes by 16 days of age, eat their first meat at 53 days, and attain adult size at 264 days. Litter size averages 2.61 and ranges from 1 to 4, age at independence is 10 months, and average longevity is 12 years (Nowell and Jackson 1996).

There is some question about the original distribution of this species. Van Bree and Khan (1992) reported the first specimen known from mainland Malaysia but suggested that it may have been an escaped captive; they also indicated that presence of the species on Sumatra had not been confirmed and that the known population on Java could have resulted from ancient human introduction. Nowell and Jackson (1996) suggested that the discontinuous distribution may result from the strong association of the fishing cat with major river and coastal floodplains. These wetland systems are being rapidly disrupted by drainage for agriculture and settlement, pollution, and other human activity. *F. viverrina*, unlike *F. chaus*, does not adapt well to cultivated habitats and now is on the verge of extirpation in such areas as the Indus Basin and the southwestern coast of India. It is on appendix 2 of the CITES and is designated as near threatened by the IUCN (under the name *Prionailurus viverrinus*).

Felis planiceps (flat-headed cat)

Head and body length is 410–500 mm and tail length is 130–50 mm (Grzimek 1975). Two specimens from Malaysia weighed 1.6 and 2.1 kg (Muul and Lim 1970). The body is dark brown with a silvery tinge. The underparts are white, generally spotted and splashed with brown. The top of the head is reddish brown. The face is light reddish below the eyes, there are two narrow dark lines running across each cheek, and a yellow line runs from each eye to near the ear. The coat is thick, long, and soft. The pads of the feet are long and narrow, and the claws cannot be completely retracted. The skull is long, narrow, and flat; the nasals are short and narrow, and the orbits are placed well forward and close together (Lekagul and McNeely 1977).

The flat-headed cat is thought to be nocturnal and to hunt for frogs and fish along riverbanks. Observations of a captive kitten (Muul and Lim 1970) suggest that *F. planiceps* is a fishing cat. The kitten seemed to enjoy playing in water, took pieces of fish from the water, and captured live frogs but ignored live birds. Moreover, the long, narrow rostrum and the well-developed first upper premolar of the species would seem to be efficient for seizing slippery prey. Very little additional information is available, but Nowell and Jackson (1996) noted that the species also hunts rodents, that most records are from swamps and riverine forests, and that the gestation period is approximately 56 days. *F. planiceps* appears to be very rare and may be threatened by pollution and clearing of its wetland habitat. It is classified as vulnerable by the IUCN (using the generic name *Prionailurus*) and as endangered by the USDI and is on appendix 1 of the CITES.

Felis iriomotensis (Iriomote cat)

Head and body length is 600 mm and tail length is 200 mm. Average weight is 4.2 kg for males and 3.2 kg for females (Nowell and Jackson 1996). The ground color is dark dusky brown, and spots arranged in longitudinal rows tend to merge into bands. Five to seven lines run from the back of the neck to the shoulders. The body is relatively elongate, the legs and tail are short, and the ears are rounded. Various cranial and karyological characters have been interpreted by different authorities to mean that *F. iriomotensis* represents a monotypic genus or subgenus, a full species, or only a subspecies of *F. bengalensis*.

This cat dwells only on Iriomote, a Japanese-owned island of 292 sq km to the east of Taiwan. Its presence was not known to science until 1965. If it is indeed a full species, it has the smallest population and the most limited range of any species of cat in the world (Yasuma 1988). *F. iriomotensis* typically inhabits lowland, subtropical rainforest along the coast of Iriomote, rather than the interior mountains (Nowell and Jackson 1996). It is partially arboreal, swims well, and shelters in tree holes, on branches, or in rock crevices. Although primarily nocturnal, it sometimes hunts by day for the large skink, *Eumeces kishinouyei*. It also preys on small rodents, bats *(Pteropus)*, birds, amphibians, and crabs. The male's home range averages about 3 sq km and overlaps those of other males and females, and the animal tends to change its range after several months. Female ranges average about 1.75 sq km, seldom overlap one another, and are more stable. Mating apparently occurs in February–March and September–October, births have been observed only in late April and May, gestation lasts approximately 60–70 days, litter size is 1–4 young, and maximum known longevity is more than 10 years.

Although totally protected by Japanese law, the Iriomote cat is losing habitat to deforestation for agriculture and economic development of the island. Its numbers are estimated to be fewer than 100 but are thought to have remained stable since monitoring began in 1982 (Nowell and Jackson 1996). The species is classified as endangered by the IUCN (under the name *Prionailurus bengalensis iriomotensis*) and the USDI and is on appendix 2 of the CITES.

Felis jacobita (mountain cat)

Redford and Eisenberg (1992) reported that head and body length is 577–640 mm, tail length is 413–80 mm, and one

Flat-headed kitten *(Felis planiceps)*, photo from San Diego Zoological Society.

specimen weighed 4 kg. The coat is long, soft, and fine. The upper parts are silvery gray and marked by irregular brown or orange yellow spots and transverse stripes. The underparts are whitish and have blackish spots. The bushy tail is ringed with black or brown and has a light tip.

According to Nowell and Jackson (1996), this cat has been found only in the rocky arid and semiarid zones of the Andes above the timberline, generally at elevations exceeding 3,000–4,000 meters. Its range appears to coincide with the original distribution of viscachas and chinchillas, which probably are its main prey. The recent decline of these large rodents, especially the near extirpation of chinchillas by fur hunters, may have contributed to an evident rarity of *F. jacobita*. The species is classified as vulnerable by the IUCN (using the generic name *Oreailurus*), which estimates that fewer than 10,000 mature individuals survive, and as endangered by the USDI and is on appendix 1 of the CITES.

Felis colocolo (pampas cat)

Head and body length is 435–700 mm, tail length is 220–322 mm, and shoulder height is 300–350 mm. One individual weighed 3 kg (Redford and Eisenberg 1992). The coloration ranges from yellowish white and grayish yellow to brown, gray brown, silvery gray, and light gray. Transverse bands of yellow or brown run obliquely from the back to the flanks. Two bars run from the eyes across the cheeks and meet beneath the throat. The coat is long, the tail is bushy, the face is broad, and the ears are pointed.

The pampas cat inhabits open grassland in some areas but also enters humid forests and mountainous regions. According to Redford and Eisenberg (1992), it has a greater habitat range than any other South American cat, being found from low swamps and marshes to elevations above 5,000 meters. Grimwood (1969) reported it to occur throughout the Andes of Peru. It is mainly terrestrial but may climb a tree if pursued. It hunts at night, killing small mammals, especially guinea pigs *(Cavia)* and ground birds. Litters are said to contain one to three young. *F. colocolo* has been intensively hunted for its fur. It now is on appendix 2 of the CITES.

Felis geoffroyi (Geoffroy's cat)

Head and body length is 422–665 mm, tail length is 240–365 mm, and weight is 2–6 kg (Redford and Eisenberg 1992). The ground color varies widely, from brilliant ochre in the northern parts of the range to silvery gray in the south. The body and limbs are covered with small black

Geoffroy's cat *(Felis geoffroyi)*, photo by K. Rudloff through East Berlin Zoo.

spots. There may be several black streaks on the crown, two on each cheek, and one between the shoulders. The tail is ringed. Melanistic individuals are fairly common (Nowell and Jackson 1996).

Geoffroy's cat inhabits scrubby woodland, open bush country, and pampas grasslands. Ximenez (1975) reported the elevational range to be sea level to 3,300 meters. Data compiled by Nowell and Jackson (1996) indicate that the species is primarily nocturnal, readily enters water, and swims and climbs well. It seems usually to rest by day in the crook of a tree. The diet consists mainly of small mammals but also includes birds, reptiles, amphibians, and fish. Population density in southern Chile is approximately 1/10 sq km. A radio-tracking study in that region determined an average annual home range size of 9.2 sq km for males and 3.7 sq km for females. The ranges of females overlapped, but those of males did not (Johnson and Franklin 1991). The estrous cycle is 20 days and estrus averages 2.5 days (Nowell and Jackson 1996). Hemmer (1976) listed a gestation period of 74–76 days. The single annual litter contains two or three young and in Uruguay is produced from December to May (Ximenez 1975). The young reportedly weigh 65–123 grams at birth and are weaned 8–10 weeks later (Hayssen, Van Tienhoven, and Van Tienhoven 1993). A captive specimen lived to be nearly 21 years old (Marvin L. Jones, Zoological Society of San Diego, pers. comm., 1995).

F. geoffroyi has been heavily exploited for the international fur trade; more than 78,000 skins, mainly from Paraguay, were reported in commerce in 1983 (Broad 1987). The species subsequently was legally protected in most of the countries where it occurs and also placed on appendix 1 of the CITES. Although some exploitation continues, large-scale commercial traffic seems to have stopped (Nowell and Jackson 1996).

Felis guigna (kodkod)

Head and body length is 424–510 mm, tail length is 195–250 mm, and weight is 2.1–2.5 kg (Redford and Eisenberg 1992). The coat is buff or gray brown and is heavily marked with rounded, blackish spots on both the upper and lower parts. The tail has blackish rings. Melanistic individuals are not uncommon.

Nowell and Jackson (1996) indicated that the kodkod is strongly associated with moist forests of the southern Andes but also inhabits secondary forest and shrub country. Although it is usually considered nocturnal, there is evidence that it is also active by day. It is highly arboreal and shelters in trees when inactive. It apparently preys mainly on small rodents and birds and has been reported to raid henhouses. The gestation period is 72–78 days, litter size is one to four young, and longevity is up to 11 years. It may be threatened by deforestation, is classified as vulnerable by the IUCN (using the generic name *Oncifelis*), which estimates that fewer than 10,000 mature individuals survive, and is on appendix 2 of the CITES.

Felis pardalis (ocelot)

Head and body length is 550–1,000 mm and tail length is 300–450 mm (Grzimek 1975; Leopold 1959). Weight is 11.3–15.8 kg. The ground color ranges from whitish or tawny yellow to reddish gray and gray. Dark streaks and spots are arranged in small groups around areas that are darker than the ground color. There are two black stripes on each cheek and one or two transverse bars on the insides of the legs. The tail is either ringed or marked with dark bars on the upper surface.

The ocelot occurs in a great variety of habitats, from humid tropical forests to fairly dry scrub country. A consistent requirement of the species, however, is dense vegetative cover (Nowell and Jackson 1996). It is generally nocturnal, sleeping by day in a hollow tree, in thick vegetation, or on a branch. It is mainly terrestrial but climbs, jumps, and swims well. Mean daily travel distance is 1.8–7.6 km, with males moving up to twice as far as females (Nowell and Jackson 1996). The diet includes rodents, rabbits, young deer and peccaries, birds, snakes, and fish.

Reported population densities are 4 residents/5 sq km in lowland rainforest and 2/5 sq km in savannah, and reported home range size of adults in various areas has been about 1 to 14 sq km (Nowell and Jackson 1996). In a study in Peru, Emmons (1988) found adult females to occupy exclusive home ranges of about 2 sq km. Male ranges were several times greater and also were exclusive of one another but did overlap a number of female ranges. Individuals generally moved about alone but appeared to make contact frequently and probably maintained a network of social ties. The ocelot communicates by mewing and, during courtship, by yowls not unlike those of *F. catus.*

There is probably no seasonal breeding in the tropics (Grzimek 1975). In Mexico and Texas births are reported to occur in autumn and winter (Leopold 1959). According to Nowell and Jackson (1996), autumn breeding peaks have been noted in Paraguay and Argentina, the estrous cycle averages 25 days, estrus averages 4.6 days, gestation lasts 79–85 days, litter size averages 1.64 and ranges from 1 to 3 young, and age at sexual maturity is 18–22 months in females and 30 months in males. The young weigh 200–340 grams at birth and are weaned 6 weeks later (Hayssen, Van Tienhoven, and Van Tienhoven 1993). One individual lived at the Phoenix Zoo for 21 years and 5 months (Marvin L. Jones, Zoological Society of San Diego, pers. comm., 1995).

The ocelot is classified as endangered by the USDI and is on appendix 1 of the CITES. It was the spotted cat most heavily exploited by the international fur trade from the early 1960s to the mid-1970s, when as many as 200,000 were taken annually (Nowell and Jackson 1996). Prior to 1972, when importation into the United States was prohibited because of the animal's being listed as endangered, enormous numbers of skins were brought into the country—133,069 in 1969 alone. As late as 1975 Great Britain imported 76,838 skins. In the early 1980s ocelot fur coats sold for as high as U.S. $40,000 in West Germany. The ocelot also is reported to be in much demand for use as a pet, a live animal selling for $800 (Thornback and Jenkins 1982). Recent legal protection in most countries where the ocelot occurs and regulation of the international trade in its fur through CITES seem to have contributed to a substantial decline in market hunting and a partial recovery of some of the depleted populations. The species is adaptable and still occurs over most of its original range, though there is concern that its relatively low reproductive rate and its need for dense cover and abundant small prey could make it especially vulnerable to environmental disturbance. Based on a minimum estimated density of 1/5 sq km, there would be at least 800,000 ocelots in forested South America, with total numbers probably 1.5–3.0 million (Nowell and Jackson 1996).

The subspecies *F. p. albescens* is classified as endangered by the IUCN (using the generic name *Leopardus*). It probably once ranged over most of Texas and at least as far east as Arkansas and Louisiana (Lowery 1974). It declined rapidly during the nineteenth century because of hunting and loss of habitat to settlement and is currently restricted to the border region of extreme southern Texas and northeastern Mexico. Of the fewer than 250 mature individuals thought to survive, fewer than 100 are in Texas. The clearing of brush country for agricultural purposes is the main problem in this area.

Felis wiedii (margay)

Head and body length is 463–790 mm and tail length is 331–510 mm. Weight is 2.6–3.9 kg (Redford and Eisenberg 1992). The ground color is yellowish brown above and white below. There are longitudinal rows of dark brown spots, the centers of which are paler than the borders. The margay closely resembles the ocelot but is smaller, has a slenderer build, and has a relatively longer tail.

The margay is mainly, if not exclusively, a forest dweller. It is much more arboreal than the ocelot and is thought to forage in trees. The only specimen known from Texas was taken prior to 1852 and is thought to represent an individual that strayed far from the normal habitat (Leopold 1959). The arboreal acrobatics and effortless climbing of the margay are partly the result of limb structure (Grzimek 1975). The feet are broad and soft and have mobile metatarsals. The hind foot is much more flexible than that of other felids, being able to rotate 180°. The cat thus can hang vertically during descent like a squirrel. It preys on squirrels and other arboreal mammals and also eats birds, some terrestrial rodents, and fruit (Nowell and Jackson 1996; Redford and Eisenberg 1992).

According to Nowell and Jackson (1996), a radio-tracked adult male in Brazil maintained a home range of 16 sq km during a period of 18 months. The estrous cycle is 32–36 days, estrus lasts 4–10 days, the gestation period is 76–84 days, litter size is usually one and sometimes two young, and females first enter estrus at 6–10 months. The young weigh about 170 grams at birth and take solid food 52–57 days later (Hayssen, Van Tienhoven, and Van Tienhoven 1993). A captive margay lived to more than 21 years of age (Marvin L. Jones, Zoological Society of San Diego, pers. comm., 1995).

The margay is classified as endangered by the USDI and is on appendix 1 of the CITES. Leopold (1959) referred to the margay as "an exceedingly rare animal." Grimwood (1969) stated that exports of pelts from Peru were increasing. Paradiso (1972) indicated that at least 6,701 margays, including both live animals and skins, were imported into the United States in 1970. According to Broad, Luxmoore, and Jenkins (1988), the total number of skins known to be in international trade declined from at least 30,000 in 1977 to about 20,000 in 1980 and only 138 in 1985. Although commerce in the fur of the margay has been drastically reduced by legal protection over most of its range and CITES regulation, the species remains threatened by extensive illegal hunting and habitat destruction. Because of its arboreal nature, it probably suffers more than the ocelot from deforestation (Nowell and Jackson 1996; Thornback and Jenkins 1982).

Felis tigrina (little spotted cat)

Head and body length is 400–550 mm, tail length is 250–400 mm, and weight is 1.5–3.0 kg (Leyhausen *in* Grzimek 1990). The upper parts vary in color from light to rich ochre and have rows of large, dark spots. The underparts are paler and less spotted. The tail has 10 or 11 rings and a black tip. One-fifth of all specimens are melanistic.

The little spotted cat lives in forests and seems to prefer montane cloud forest. It generally occurs at higher elevations than the ocelot and the margay, having been found at elevations as high as 4,500 meters (Nowell and Jackson

1996). Its habits in the wild are not known. Captive females have an estrus of several days, a gestation period of 74–76 days, and litters of one or two young. The kittens develop slowly, opening their eyes at 17 days and starting to take solid food at 55 days. Longevity may exceed 17 years (Nowell and Jackson 1996).

Like the ocelot and the margay, the little spotted cat was subject to uncontrolled commerce until recently. At least 3,170 individuals, including both live animals and skins, were imported into the United States in 1970 (Paradiso 1972). The species subsequently was listed as endangered by the USDI, and importation was banned. However, in the early 1980s *F. tigrina* became the leading spotted cat in the international fur trade, with the number of skins in commerce peaking at nearly 84,500 in 1983 (Broad 1987). Eventual legal protection in most countries where *F. tigrina* occurs and addition of the entire species to appendix 1 of the CITES seem to have greatly reduced this market. The species may still be threatened by illegal hunting and loss of restricted forest habitat to agriculture and logging (Nowell and Jackson 1996; Thornback and Jenkins 1982). It is designated as near threatened by the IUCN (under the name *Leopardus tigrinus*).

Felis yagouaroundi (jaguarundi)

Head and body length is 550–770 mm, tail length is 330–600 mm, and weight is 4.5–9.0 kg. There are two color phases: blackish to brownish gray, and fox red to chestnut. The body is slender and elongate, the head is small and flattened, the ears are short and rounded, the legs are short, and the tail is very long. This cat is sometimes said to resemble a weasel or otter in external appearance.

The jaguarundi inhabits lowland forests and thickets. It hunts in the morning and evening and is much less nocturnal than most cats. Recent observations in Central America indicate that it is largely diurnal (McCarthy 1992). It forages mainly on the ground but is an agile climber. The diet includes birds, small mammals, and reptiles. Home range is surprisingly large. The ranges of two males in Belize were 88 and 100 sq km, with overlap less than 25 percent. A female there used a range of 13–20 sq km (Nowell and Jackson 1996). *F. yagouaroundi* has been reported to live in pairs in Paraguay but to be solitary in Mexico. According to Grzimek (1975), there is no definite reproductive season in the tropics, but in Mexico young are produced around March and August. It is not known whether one female gives birth in both seasons. The estrous cycle averages 54 days, estrus averages 3.2 days, gestation lasts 70–75 days, the average litter size is 1.8 and the range is 1–4 young, age at sexual maturity is 2–3 years, and longevity is up to 15 years (Nowell and Jackson 1996).

The pelt is of poor quality and little value (Leopold 1959). The jaguarundi is widespread and not subject to commercial exploitation (Paradiso 1972). Nonetheless, four subspecies—*cacomitli, tolteca, fossata,* and *panamensis*—which range from southern Texas and Arizona to Panama, are listed as endangered by the USDI. North American populations, which essentially comprise those four subspecies, are on appendix 1 of the CITES; South American populations are on appendix 2. The IUCN classifies *cacomitli*, of southern Texas and eastern Mexico, as endangered (using the generic name *Herpailurus*), noting that fewer than 250 mature individuals survive. The jaguarundi's status is not well understood, but the animal's numbers have declined in at least the northern portions of its range because of human persecution and habitat destruction (Thornback and Jenkins 1982). The species is not now known to be resident in Arizona, though one was sighted there in 1975 (Jay M. Sheppard, U.S. Fish and Wildlife Service, pers. comm., 1985), and only a few individuals survive in Texas. During the late Pleistocene the species occurred as far as Florida, and a small population may have become established there recently through introduction by human agency. There have been sporadic reports of a cat resembling the jaguarundi from other southeastern states.

Felis concolor (cougar, puma, panther, or mountain lion)

The names cougar, puma, panther, and mountain lion are used interchangeably for this species, and various other vernacular terms are applied in certain areas. This is by far the largest species in the genus *Felis,* averaging about the same size as the leopard *(Panthera pardus).* In males head and body length is 1,050–1,959 mm, tail length is 660–784 mm, and weight is 67–103 kg. In females head and body length is 966–1,517 mm, tail length is 534–815 mm, and weight is usually 36–60 kg. Shoulder height is 600–700 mm. Generally, the smallest animals are in the tropics and the largest are in the far northern and southern parts of the range. There are two variable color phases: one ranges from buff, cinnamon, and tawny to cinnamon rufous and ferrugineous; the other ranges from silvery gray to bluish and slaty gray. The body is elongate, the head is small, the face is short, and the neck and tail are long. The limbs are powerfully built; the hind legs are larger than the forelegs. The ears are small, short, and rounded. Females have three pairs of mammae (Banfield 1974).

The cougar has the greatest natural distribution of any mammal in the Western Hemisphere except *Homo sapiens.* It can thrive in montane coniferous forests, lowland tropical forests, swamps, grassland, dry brush country, or any other area with adequate cover and prey. The elevational range extends from sea level to at least 3,350 meters in California and 4,500 meters in Ecuador. There usually is no fixed den except when females are rearing young. Temporary shelter is taken in such places as dense vegetation, rocky crevices, and caves. The cougar is agile and has great jumping power: it may leap from the ground to a height of up to 5.5 meters in a tree. It swims well but commonly prefers not to enter water. Sight is the most acute sense and hearing is also good, but smell is thought to be poorly developed. Activity may be either nocturnal or diurnal. The cougar hunts over a large area, sometimes taking a week to complete a circuit of its home range (Leopold 1959). In Idaho, Seidensticker et al. (1973) found residents occupying fairly distinct but usually contiguous winter–spring and summer–autumn home areas. The latter area was generally larger and at a higher elevation, reflecting the summer movements of ungulate herds.

The cougar carefully stalks its prey and may leap upon the victim's back or seize it after a swift dash. Throughout the range of the cougar the most consistently important food is deer—*Odocoileus* in North America and *Blastoceros, Hippocamelus,* and *Mazama* in South America. Estimates of kill frequency vary from about one deer every 3 days for a female with large cubs to one every 16 days for a lone adult (Lindzey 1987). The diet also includes other ungulates, beavers, porcupines, and hares. The most common prey of the current population in Florida is wild hog (Maehr et al. 1990). The kill is usually dragged to a sheltered spot and then partly consumed. The remains are covered with leaves and debris, then visited for additional meals over the next several days.

Detailed studies in central Idaho (Hornocker 1969, 1970; Seidensticker et al. 1973) showed that the cougar population depends almost equally on mule deer *(Odocoileus hemionus)* and elk *(Cervus canadensis).* The study area contained 1 cougar for every 114 deer and 87 elk. About half of the animals killed were in poor condition. Deer and

Cougar *(Felis concolor)*, photo by Dom Deminick, Colorado Division of Wildlife.

elk populations increased during a four-year study period, evidently being affected more by food availability than by cougar predation; nonetheless, predation was thought to moderate prey oscillations and to remove less fit individuals. During the same period the cougar population remained stable, its numbers being regulated mainly by social factors rather than by food supply. Density was about 1 adult cougar per 35 sq km. The total area used by individuals varied from 31 to 243 sq km in winter–spring and from 106 to 293 sq km in summer–autumn. There was little overlap in the areas occupied by resident adult males, but the areas of resident females often overlapped one another completely and were overlapped by resident male areas. Young, transient individuals of both sexes moved through the areas used by residents. There was a land tenure system based on prior right; transients could not permanently settle in an occupied area unless the resident died. Dispersal and mortality of young individuals unable to establish an area for themselves seemed to limit the size of the cougar population.

According to Lindzey (1987), long-term population densities in other areas have been found to vary from about 1/21 sq km to 1/200 sq km, and reported annual home range has been as great as 1,826 sq km for one animal in Texas. Nowell and Jackson (1996) stated that resident male ranges typically cover several hundred sq km, usually do not overlap one another, and do overlap several resident female ranges, each of which is usually less than 100 sq km. Nowell and Jackson also cited a population density of 7/100 sq km in Patagonia, one of the highest ever reported. The cougar is essentially solitary, with individuals deliberately avoiding one another except during the brief period of courtship. Communication is mostly by visual and olfactory signals, and males regularly make scrapes in the soil or snow, sometimes depositing urine or feces therein. Vocalizations include growls, hisses, and birdlike whistles. A very loud scream has been reported on rare occasion, but its function is not known.

There is no specific breeding season, but most births in North America occur in late winter and early spring. In southern Chile all known births have occurred from February to June (Nowell and Jackson 1996). Females are seasonally polyestrous and usually give birth every other year (Banfield 1974). The estrous cycle averages about 23 days (Nowell and Jackson 1996). Estrus lasts about 9 days and the gestation period is 90–96 days. The number of young per litter is one to six, commonly three or four. The kittens weigh 226–453 grams at birth and are spotted until they are about 6 months old. They nurse for 3 months or more but begin to take some meat at 6 weeks. If born in the spring, they are able to accompany the mother by autumn and to make their own kills by the end of winter; nonetheless, they usually remain with their mother for several more months or even another year. Litter mates stay together for 2–3 months after leaving the mother. Females attain sexual maturity at about 2.5 years, but males may not mate until at least 3 years (Banfield 1974). Regular reproductive activity does not begin until a young animal establishes itself in a permanent home area (Seidensticker et al. 1973). Females may remain reproductively active until at least 12 years and males to at least 20 years; captives have lived more than 20 years (Currier 1983). One female killed in the wild was at least 18 years old (Nowell and Jackson 1996).

The cougar is generally considered harmless to people, but attacks have occurred and there is concern about increasing encounters between the two species (Braun 1991). Beier (1991) documented the deaths of nine persons in the United States and Canada from 1890 up to 1990. Two more people were killed in 1991 and 1992, and there have been numerous recent injuries and narrow escapes (Aune 1991; Olson 1991; Rollins and Spencer 1995). Two women were killed in separate attacks in California during 1994 (*New Orleans Times-Picayune,* 14 December 1994, A-10), and a 10-year-old boy died in Colorado in July 1997 (*Washington Post,* 19 July 1997).

The remainder of the account of this species is based largely on Nowak's (1976) report. Information compiled by Nowell and Jackson (1996) indicates that there has been little change in population and legal status in the 20 years since that report. The cougar has long been viewed as a threat to domestic animals, such as horses and sheep, and is also sometimes thought to reduce populations of game. The species has thus been intensively hunted since the arrival of European colonists in the Western Hemisphere. Most successful hunting is done by using dogs to pursue the cat until it seeks refuge in a tree, where it can easily be shot. By the early twentieth century the cougar apparently had been eliminated everywhere to the north of Mexico except in the mountainous parts of the West, in southern Texas, and in Florida. The species still occupies the same regions, and about 16,000 individuals may be present. California provides almost complete protection to the cougar, Texas still allows it to be killed at any time, and all other western states and provinces permit regulated hunting. About 2,100 cougars are killed annually for sport or predator control in the United States and Canada (Green 1991; Tully 1991). Public antipathy seems to have moderated in the last few decades, and the general pattern of decline may have been halted, but loss of habitat and conflict with agricultural interests are still problems.

The subspecies *F. c. coryi,* which formerly occurred from eastern Texas to Florida, and the subspecies *F. c. couguar,* of the northeastern United States and southeastern Canada, are classified as critically endangered by the IUCN (using the generic name *Puma*). These two subspecies, along with *F. c. costaricensis,* of Central America, are also listed as endangered by the USDI and are on appendix 1 of the CITES. All other subspecies are on appendix 2. Since the late 1940s, evidence has accumulated indicating the presence of small surviving cougar populations in south-central and southeastern Canada, the southern Appalachians, and the Ozark region and adjoining forests of Arkansas, southern Missouri, eastern Oklahoma, and northern Louisiana. The most consistent and best-documented reports have come from New Brunswick and Nova Scotia (Cumberland and Dempsey 1994; Stocek 1995).

In southern Florida numerous "panthers" have been killed or live-captured in the last two decades, and a federal-state program of study and conservation is under way (Belden 1986; Belden and Forrester 1980; Shapiro 1981). This program has involved the expenditure of many millions of U.S. dollars to purchase refuge lands, carry out research in the field and captivity, provide vaccinations and medical treatments to the animals, and construct tunnels to allow them to pass safely under a major highway through their range. Nonetheless, the isolated Florida population comprises only 30–50 individuals and is jeopardized by severe inbreeding depression, a reduced reproductive rate, disease and parasites, habitat disruption and fragmentation, and accidental killing on roads (Barone et al. 1994; Nowell and Jackson 1996; Roelke, Martenson, and O'Brien 1993). A controversial project to introduce western cougars to Florida was initiated in 1993; the project is intended to improve genetic viability but could modify the unique characters of the original population. There is some evidence that the population already has experienced genetic introgression as a result of the release of a few individuals of South American origin into the Everglades several decades ago (O'Brien et al. 1990).

CARNIVORA; FELIDAE; **Genus NEOFELIS**
Gray, 1867

Clouded Leopard

The range of the single species, *N. nebulosa,* extends from central Nepal northeastward to Shaanxi Province in central China and southeastward to Viet Nam and the Malay Peninsula and also includes Taiwan, Hainan, Sumatra, and Borneo. Most authorities, including Ellerman and Morrison-Scott (1966), Ewer (1973), Guggisberg (1975), Hemmer (1978), and Wozencraft (*in* Wilson and Reeder 1993) treat *Neofelis* as a distinct genus with one species. Simpson (1945) considered *Neofelis* to be a subgenus of *Panthera.* Leyhausen (1979) listed *Neofelis* as a full genus but included within it not only the clouded leopard but also the tiger (here referred to as *Panthera tigris*). Based on their shared unique and complex pelage, Corbet and Hill (1992) placed *nebulosa* together with *marmorata* (here referred to as *Felis marmorata*) in the genus *Pardofelis.*

Head and body length is about 616–1,066 mm, tail length is 550–912 mm, and weight is usually 16–23 kg. The coat is grayish or yellowish, with dark markings ("clouds") in such forms as circles, ovals, and rosettes. The markings on the shoulders and back are darker on their posterior margins than on their front margins, suggesting that stripes can be evolved from blotches or spots. The forehead, legs, and base of the tail are spotted, and the remainder of the tail is banded. Melanistic specimens have been reported (Medway 1977).

The tail is long, the legs are stout, the paws are broad, and the pads are hard. The skull is long, low, and narrow (Guggisberg 1975). The upper canine teeth are relatively longer than those of any other living cat, having a length about three times as great as the basal width at the socket. The first upper premolar is greatly reduced or absent, leaving a wide gap between the canine and the cheek teeth. Unlike that of *Panthera,* the hyoid of *Neofelis* is ossified (Guggisberg 1975).

The clouded leopard inhabits various kinds of forest, perhaps to elevations of up to 2,500 meters. It is usually said to be highly arboreal, to hunt in trees, and to spring on ground prey from overhanging branches. Nowell and Jackson (1996) noted that its arboreal talents rival those of *Felis wiedii* and that it even can run down tree trunks headfirst or move about on horizontal branches with its back to the ground. Information compiled by Guggisberg (1975), however, suggests that *Neofelis* is more terrestrial and diurnal than is generally assumed. Rabinowitz, Andau, and Chai (1987) indicated that it travels mainly on the ground and uses trees primarily as resting sites. A subadult male radio-tracked for eight days was fully terrestrial and even rested in dense patches of grass (Dinerstein and Mehta 1989). *Neofelis* has been reported to feed on birds, monkeys, pigs, cattle, young buffalo, goats, deer, and even porcupines.

Reproduction is known only through observations in captivity (Fellner 1965; Fontaine 1965; Guggisberg 1975; Murphy 1976; Yamada and Durrant 1989). Mating has occurred in all months except June and October, with a peak in December. Cubs have been produced in all months except December, with a peak in March. The estrous cycle averages 30 days, estrus 6 days, and the gestation period 93

Clouded leopard *(Neofelis nebulosa)*, photo from San Diego Zoological Garden.

days. The number of young per litter ranges from one to five but is commonly two. The young weigh about 140–70 grams at birth, open their eyes after 12 days, take some solid food at 10.5 weeks, and nurse for 5 months. The coat is initially black in the patterned areas, and full adult coloration is attained at about 6 months. Both males and females attain sexual maturity at about 26 months. A captive was still living at 19 years and 6 months (Irven 1993). Many captives are gentle and playful and like to be petted by their custodians. Three hundred individuals are known to be in captivity (Millard and Fletchall *in* Bowdoin et al. 1994).

The clouded leopard is classified as vulnerable by the IUCN (1978) and as endangered by the USDI and is on appendix 1 of the CITES. The main problem is loss of forest habitat to agriculture. The genus also has been excessively hunted in some areas for its beautiful pelt, which sells for up to U.S. $2,000 on the black market (Santiapillai 1989). It may already have disappeared from Hainan, and on Taiwan it is restricted to the wildest and most inaccessible parts of the central mountain range. After a survey on Taiwan, Rabinowitz (1988) reported that the most recent sighting was in 1983 but expressed hope that the genus does survive. *Neofelis* still is widespread in mainland China, but suitable habitat is fragmented into small patches and the remaining animals are illegally hunted for their pelts and for other parts that are used in traditional medicines and foods (Nowell and Jackson 1996). *Neofelis* had not been definitely recorded in Nepal since 1863, but a number of individuals were located in 1987–88 and observations suggested that the genus may be more common there than previously thought (Dinerstein and Mehta 1989).

CARNIVORA; FELIDAE; **Genus PANTHERA**
Oken, 1816

Big Cats

There are five subgenera and five species (Cabrera 1957; Corbet 1978; Ellerman and Morrison-Scott 1966; Guggisberg 1975; E. R. Hall 1981; Smithers *in* Meester and Setzer 1977):

subgenus *Uncia* Gray, 1854

P. uncia (snow leopard), mountainous areas from Afghanistan to Lake Baikal and eastern Tibet;

subgenus *Tigris* Frisch, 1775

P. tigris (tiger), eastern Turkey to southeastern Siberia and Malay Peninsula, Sumatra, Java, Bali;

subgenus *Panthera* Oken, 1816

P. pardus (leopard), western Turkey and Arabian Peninsula to southeastern Siberia and Malay Peninsula, Sri Lanka, Java, Kangean Island, most of Africa;

subgenus *Jaguarius* Severtzov, 1858

P. onca (jaguar), southern United States to Argentina;

subgenus *Leo* Brehm, 1829

P. leo (lion), found in historical time from the Balkan and Arabian peninsulas to central India and in almost all of Africa.

The use of the name *Panthera* for this genus is in keeping with Corbet (1978), Ellerman and Morrison-Scott (1966), Hemmer (1978), Leyhausen (1979), Mazak (1981), and most of the other authorities cited herein. A few authorities, such as Cabrera (1957) and Stains (1984), consider *Panthera* to be invalid for technical reasons of nomencla-

A. Lion *(Panthera leo)*, photo from P. D. Swanepoel. B. Lioness and cub *(P. leo)*, photo by Ernest P. Walker. C. Lion cubs *(P. leo)*, photo from San Diego Zoological Garden. D. Leopard *(P. pardus)*, photo from U.S. National Zoological Park.

Top, Chinese tiger *(Panthera tigris amoyensis).* Bottom, melanistic jaguar *(Panthera onca).* Photos from East Berlin Zoo.

ture and prefer to use the generic term *Leo.* Some authorities, including E. R. Hall (1981), do not consider *Panthera* to be generically distinct from *Felis. Uncia* was treated as a separate genus by Guggisberg (1975), Hemmer (1972, 1978), Heptner and Sludskii (1992), Leyhausen (1979), and Wozencraft (*in* Wilson and Reeder 1993) but not by Corbet (1978), Corbet and Hill (1991, 1992), Ellerman and Morrison-Scott (1966), and Ewer (1973). Leyhausen placed *P. tigris* in the genus *Neofelis.* Both Hemmer (1978) and Mazak (1981) divided *Panthera* into only two subgenera: *Tigris,* with *P. tigris;* and *Panthera,* with *P. pardus, P. onca,* and *P. leo.* Corbet and Hill (1992) noted that a single subfossil tooth suggests the former presence of *P. tigris* on Borneo, that other subfossil finds show that *P. leo* once occurred as far east as 87° E in Bengal, and that the presence of *P. pardus* on Kangean Island may have resulted from introduction.

The members of the genus *Panthera* have an incompletely ossified hyoid apparatus, with an elastic cartilaginous band replacing the bony structure found in other cats (Grzimek 1975). This elastic ligament usually has been considered to be the anatomical feature that allows roaring but limits purring to times of exhaling. At least some other cats can purr both when exhaling and when inhaling. The snow leopard *(P. uncia)* possesses an elastic ligament but does not roar. Detailed morphological studies (Hast 1989; Peters and Hast 1994) now demonstrate that the ability to roar depends on a complex of characters, especially a specialized larynx containing very long vocal folds with a thick pad of fibro-elastic tissue. Such characters are found in the leopard, the jaguar, the lion, and the tiger but not in the snow leopard or any other cat. According to Paul Leyhausen (Max-Planck Institut für Verhaltensphysiologie, pers. comm., c. 1980), certain small cats, such as *Felis nigripes,* produce a full roar but do not sound as formidable as the lion or tiger simply because of the size difference. *Panthera* is also distinguished from *Felis* by having hair that extends to the front edge of the nose (Grzimek 1975). Additional information is provided separately for each species.

Panthera uncia (snow leopard)

Except as noted, the information for the account of this species was taken from Guggisberg (1975), Hemmer (1972), and Schaller (1977). Head and body length is 1,000–1,300 mm, tail length is 800–1,000 mm, shoulder height is about 600 mm, and weight is 25–75 kg. Males usually weigh 45–55 kg, females 35–40 kg (Nowell and Jackson 1996). The ground color varies from pale gray to creamy smoke gray, and the underparts are whitish. On the head, neck, and lower limbs there are solid spots, and on the back, sides, and tail are large rings or rosettes, many enclosing some small spots. The coat is long and thick and the head is relatively small.

The snow leopard is found in the high mountains of Central Asia. In summer it occurs commonly in alpine meadows and rocky areas at elevations of 2,700–6,000 meters. In the winter it may follow its prey down into the forests below 1,800 meters. It sometimes dens in a rocky cavern or crevice. It is often active by day, especially in the early morning and late afternoon. It is graceful and agile and has been reported to leap as far as 15 meters. It tends to remain in a relatively small area for 7–10 days and then shift activity to another part of its home range (Nowell and Jackson 1996). Prey is either stalked or ambushed. The diet includes mountain goats and sheep, deer, boars, marmots, pikas, and domestic livestock.

Population density estimates across the range of *P. uncia* vary from about 0.5 to 10.0 individuals per 100 sq km (Nowell and Jackson 1996). In a study in Nepal, Jackson and Ahlborn (1988) found that an area of 100 sq km supported 5–10 snow leopards. The home ranges of five individuals in this region measured about 12–39 sq km; these ranges overlapped almost entirely both between and within sexes, but the animals kept well apart. Socially the snow leopard is thought to be like the tiger, essentially solitary but not unsociable. The snow leopard does not roar but has several vocalizations, including a loud moaning associated with attraction of a mate.

Births usually occur from April to June both in the wild and in captivity after a gestation period of 90–103 days. The estrous cycle is 15–39 days, estrus 2–12 days (Nowell and Jackson 1996). The young are born in a rocky shelter lined with the mother's fur. The number of young per litter is one to five, usually two or three. The cubs weigh about 450 grams at birth, open their eyes after 7 days, eat their first solid food at 2 months, and follow their mother at 3 months. They hunt with the mother at least through their first winter and attain sexual maturity at about 2 years. Reproductive activity continues until 15 years, and longevity is up to 21 years (Nowell and Jackson 1996).

The snow leopard is classified as endangered by the IUCN, the USDI, and Russia and is on appendix 1 of the CITES. It has declined in numbers through hunting by people because it is considered to be a predator of domestic stock, it is valued as a trophy, and its fur is in demand by commerce. Like those of *P. tigris* (see account thereof), the bones of *P. uncia* are used in certain traditional Oriental medicines (Nowell and Jackson 1996). Although it is protected in China, it continues to be hunted there, and its skin is sold on the open market (Schaller et al. 1988). Contributing to the problem in northwestern India and Nepal, each with perhaps 400 individuals, are a reduction in natural prey and increased use of alpine pastures for livestock,

Snow leopard *(Panthera uncia)*, photo by Ernest P. Walker.

Tiger *(Panthera tigris)*, photo from East Berlin Zoo.

leading to predation on the latter and retaliatory measures by people (Fox et al. 1991; Jackson 1979). It recently has been estimated that a total of about 4,500–7,300 snow leopards remain in the wild and that they occupy 1.9 million sq km of habitat (Nowell and Jackson 1996). There are another 580 in captivity (Wharton and Blomqvist *in* Bowdoin et al. 1994).

Panthera tigris (tiger)

Head and body length is 1,400–2,800 mm, tail length is 600–1,100 mm, and shoulder height is 800–1,100 mm (Leyhausen *in* Grzimek 1990). The subspecies found in southeastern Siberia and Manchuria, *P. t. altaica,* is the largest living cat. The other mainland subspecies are also large, but those of the East Indies are much smaller. In *P. t. altaica* males weigh 180–306 kg, females 100–167 kg; in *P. t. tigris* of India and adjoining countries males weigh 180–258 kg, females 100–160 kg; in *P. t. sumatrae* of Sumatra males weigh 100–140 kg, females 75–110 kg; and in *P. t. balica* of Bali males weigh 90–100 kg, females 65–80 kg (Mazak 1981). The ground color of the upper parts ranges from reddish orange to reddish ochre, and the underparts are creamy or white. The head, body, tail, and limbs have a series of narrow black, gray, or brown stripes. On the flanks the stripes generally run in a vertical direction; and in some specimens the stripes are much reduced on the shoulders, forelegs, and anterior flanks (Guggisberg 1975). There also is a rare but much publicized variant with a chalky white coat, dark stripes, and icy blue eyes; 103 are now in captivity (Roychoudhury 1987). There have been no records of white tigers in the wild since 1958 (Patnaik and Acharjyo 1990).

Except as noted, the information for the remainder of this account was taken from Mazak (1981). The tiger is tolerant of a wide range of environmental conditions, its only requirements being adequate cover, water, and prey. It is found in such habitats as tropical rainforests, evergreen forests, mangrove swamps, grasslands, savannahs, and rocky country. An individual may have one or more favored dens within its territory in such places as caves, hollow trees, and dense vegetation. The tiger usually does not climb trees but is capable of doing so. It has been reported to cover up to 10 meters in a horizontal leap. It seems to like water and can swim well, easily crossing rivers 6–8 km

Top, Siberian tiger *(Panthera tigris altaica).* Bottom, Sumatran tiger *(P. t. sumatrae).* Photos from East Berlin Zoo.

wide and sometimes swimming up to 29 km. It is mainly nocturnal but may be active in daylight, especially in winter in the northern part of its range. Siberian animals have moved up to 60 km per day. In Nepal, Sunquist (1981) determined the usual daily movement to be 10–20 km.

To hunt, the tiger depends more on sight and hearing than on smell. It usually carefully stalks its prey, approaching from the side or rear and attempting to get as close as possible. It then leaps upon the quarry and tries simultaneously to throw it down and grab its throat. Killing is by strangulation or a bite to the back of the neck. The carcass is often dragged to an area within cover or near water. One individual dragged an adult gaur *(Bos gaurus)* 12 meters, and later 13 men tried to pull the carcass but could not move it (Grzimek 1975). After eating its fill, the tiger may cover the remains with grass or debris and then return for additional meals over the next several days. The diet consists mainly of large mammals, such as pigs, deer, antelopes, buffalo, and gaur. Smaller mammals and birds occasionally are taken. A tiger can consume up to 40 kg of meat at one time, but individuals in zoos are given 5–6 kg per day. Although the tiger is an excellent hunter, it fails in at least 90 percent of its attempts to capture animals. The tiger thus cannot eliminate entire prey populations. Indeed, Sunquist (1981) found that prey numbers, with the exception of one species, were not even being limited by tiger predation.

In Kanha National Park, in central India, Schaller (1967) determined that 10–15 adult tigers were regularly resident in an area of about 320 sq km. In Royal Chitawan National Park, in Nepal, Sunquist (1981) found an overall population density of 1 adult per 36 sq km. Observations in these and other areas indicate much variation in home range size and social behavior, evidently depending on habitat conditions and prey availability. In India individual home range seems usually to be 50–1,000 sq km. In Manchuria and southeastern Siberia the usual size is 500–4,000 sq km and the maximum reported is 10,500 sq km. In Nepal, Smith, McDougal, and Sunquist (1987) found home range size to be 19–151 sq km for males and 10–51 sq km for females. These ranges essentially corresponded to defended territories in that there was no overlap between those of adults of the same sex. A male range, however, overlapped the ranges of several females. Mothers evidently allowed their daughters to establish adjacent territories, but eventually the two generations may become agonistic. Schaller's studies indicate that the same kind of situation may exist in central India but also that females are not territorial there, adults of the same sex sometimes share a home range, and there are transient animals that lack an established range. All of these authorities suggested the presence of a land tenure system based on prior right, by which a resident animal is never replaced on its range until its death. Territorial boundaries are not patrolled, but individuals do visit all parts of their ranges over a period of days or weeks. These areas are marked with urine and feces. Avoidance, rather than fighting, seems to be the rule for tigers; nonetheless, one individual transplanted from its normal range to a different area evidently was killed soon thereafter by another tiger (Seidensticker et al. 1976).

Except for courting pairs and females with young, the tiger is essentially solitary. Even individuals that share a range usually keep 2–5 km apart (Sunquist 1981). The tiger is not unsociable, however, and the animals in a given area (probably close relatives) may know one another and have a generally amicable relationship (Schaller 1967). Several adults may come together briefly, especially to share a kill. Limited evidence suggests that a tiger roars to announce to its associates that it has made a kill. An additional function of roaring seems to be the attraction of the opposite sex. There are a number of other vocalizations, such as purrs and grunts, and the tiger also communicates by marking with urine, feces, and scratches.

Mating may occur at any time but is most frequent from November to April. Females usually give birth every 2–2.5 years and occasionally wait 3–4 years; if all the newborn are lost, however, another litter can be produced within 5 months (Schaller 1967). Females enter estrus at intervals of 3–9 weeks, and receptivity lasts 3–6 days. The gestation period is usually 104–6 days but ranges from 93 to 111 days. Births occur in caves, rocky crevices, or dense vegetation. The number of young per litter is usually two or three and ranges from one to six. The cubs weigh 780–1,600 grams at birth, open their eyes after 6–14 days, nurse for 3–6 months, and begin to travel with the mother at 5–6 months. They are taught how to hunt prey and apparently are capable killers at 11 months (Schaller 1967). They usually separate from the mother at 2 years but may wait another year. Sexual maturity is attained at 3–4 years by females and at 4–5 years by males. About half of all cubs do not survive more than 2 years, but maximum known longevity is about 26 years both in the wild and in captivity.

The tiger probably has been responsible for more human deaths through direct attack than any other wild mammal. In this regard perhaps the most dangerous place in modern history was Singapore and nearby islands, where following extensive settlement during the 1840s more than 1,000 persons were being killed annually (McDougal 1987). About 1,000 more people were reportedly killed each year in India during the early twentieth century. Guggisberg (1975) questioned the accuracy of these statistics, but there seems to be little doubt that some tigers have preyed extensively, or almost exclusively, on people. One individual is said to have killed 430 persons in India. Although such man-eaters have declined with the general reduction in tiger numbers in the twentieth century, the problem persists. In 1972, for example, India's production of honey and beeswax dropped by 50 percent when at least 29 persons who gathered these materials were devoured (Mainstone 1974).

Tigers currently seem to be especially dangerous in the Sundarbans mangrove forest, at the mouth of the Ganges River. Hendrichs (1975) reported that 129 persons were killed in this area from 1969 to 1971 but noted that only 1 percent of the tigers there actually seem to seek out human prey. P. Jackson (1985) wrote that 429 persons had been killed in the same area during the previous 10 years, but Chakrabarti (1984) noted that unofficial estimates put the average annual toll at 100. A table compiled by Khan (1987) indicates a unique mammalian situation: in some recent years the number of people killed by tigers in the Sundarbans considerably exceeded the number of tigers killed by people.

Because it is considered to be a threat to human life and domestic livestock and also is valued as a big-game trophy, the tiger has been relentlessly hunted, trapped, and poisoned. Some European hunters and Indian maharajahs killed hundreds of tigers each. After World War II, hunting became even more widespread than previously (Guggisberg 1975). The commercial trade in tiger skins intensified in the 1960s, and by 1977 a pelt brought as much as U.S. $4,250 in Great Britain (IUCN 1978). Listing of the tiger as an endangered species by the United States in 1972, comparable protective measures by other countries, and international regulation through CITES contributed to a major decline in the fur market.

However, it subsequently became apparent that an equally devastating problem is posed by the utilization of the tiger's body parts for medicinal preparations and consequent widespread poaching and illegal commerce. Of particular importance are the bones, which are reputed to give strength, relieve pain, and cure numerous diseases, such as rheumatism. They are crushed and processed into various pills, tablets, liquid medicines, and wines. Uses exist for many other parts as well, including even the penis, which is made into a soup and sold for up to U.S. $300 per serving as a supposed aphrodisiac (Norchi and Bolze 1995). Research by Mulliken and Haywood (1994) suggests that there is a demand for such products in Oriental communities throughout the world and that CITES has not been effective in controlling the associated trade. From 1990 to 1992 a minimum of 27 million items containing tiger derivatives were recorded in trade, most of them exported from China and imported by Japan. South Korea and Taiwan also are heavily involved in the trade.

Although tiger products have been in use for centuries, the market long seems to have been supplied mainly by *P. tigris amoyensis*, the native subspecies of China (see below). The large-scale elimination of that subspecies between the 1950s and 1970s forced the market to go farther afield and also to attract much more attention. Tigers now are being killed illegally throughout the range of the species, and their parts smuggled into China and other places for conversion into consumer items. The consequent alarming decline in the species led to new international efforts at conservation, including U.S. trade sanctions against Taiwan (Bender 1994). That country, South Korea, and China all now legally prohibit commerce in tiger products. However, Peter Jackson (1995), chairman of the IUCN Species Survival Commission Cat Specialist Group, warned that illegal activity is continuing in response to a massive demand from persons who believe in the effectiveness of tiger-based medicines and that responsible countermeasures have been lacking. He suggested that if present trends continue, viable tiger populations could be eliminated by the end of the century.

Even if the tiger had no commercial value, its survival would be threatened by the destruction of its habitat. With the expansion of human populations, the logging of forests, the elimination of natural prey, and the spread of agriculture, there is continuous conflict between people and the tiger, and the latter species is almost always the loser. It is estimated that in 1920 there were about 100,000 tigers in the world. Estimates for wild populations ranged as low as 4,000 in the 1970s (Fisher 1978; Jackson 1978) but subsequently increased to about 6,000–8,000 (Foose 1987; P. Jackson 1985; Luoma 1987). The apparent increase was thought to have resulted largely from an intensive effort to protect the species and establish reserves in India (Karanth 1987; Panwar 1987), though there were doubts that viable populations could be maintained there for long in the face of growing human encroachment (Ward 1987). The reserves in India are becoming increasingly isolated from one another, and the tiger populations therein are subject to poaching and loss of reproductive viability through inbreeding depression (Norchi and Bolze 1995; Smith and McDougal 1991). More recent analysis has indicated that the higher estimates of the 1980s were overly optimistic and that protective mechanisms have not been as effective as believed (Jackson 1993). This factor, together with the spread of the market for tiger parts, has led to a new estimate of 5,000–7,400 (P. Jackson 1994). The tiger is classified as endangered by the IUCN (except for the three critically endangered subspecies indicated below) and the USDI and is on appendix 1 of the CITES.

Except as noted, the information for the following summaries of the status of the eight subspecies of *P. tigris* was taken from the IUCN (1978), P. Jackson (1993, 1994), Luoma (1987), and Mazak (1981):

P. t. virgata (Caspian tiger), occurred in modern times from eastern Turkey and the Caucasus to the mountains of Kazakhstan and Sinkiang, in the Middle Ages may have reached Ukraine (Heptner and Sludskii 1992), also one report from northern Iraq in 1892 (Kock 1990), a few individuals still present in Turkey in the 1970s, now probably extinct;

P. t. tigris (Bengal tiger), originally found from Pakistan to western Burma, exterminated in Pakistan by 1906 (Roberts 1977), 2,750–3,750 individuals now estimated to survive in India, 300–460 in Bangladesh, 300 in Burma (Seal, Jackson, and Tilson 1987), 150–250 in Nepal, and 50–240 in Bhutan;

P. t. corbetti (Indochinese tiger), still found from eastern Burma to Viet Nam and the Malay Peninsula, 1,050–1,750 individuals estimated to remain but recent studies in Thailand suggest that the lower figure is the more reliable (Rabinowitz 1993);

P. t. amoyensis (Chinese tiger), classified as critically endangered by the IUCN, formerly occurred throughout eastern China, an estimated 4,000 individuals still survived in 1949 (Lu 1987), now confined largely to the Yangtze (Changjiang) Valley and apparently near extinction, with only 30–80 animals left in the wild, another 40 in captivity in China (Tan 1987; Xiang, Tan, and Jia 1987);

P. t. altaica (Siberian tiger), classified as critically endangered by the IUCN, formerly found from Lake Baikal to the Pacific coast and Korea, also sporadically on Sakhalin (Heptner and Sludskii 1992), apparently now very rare or absent in Manchuria and Korea, protection in the Ussuri region of Siberia seemed to result in an increase in numbers and distribution after the 1940s (Heptner and Sludskii 1992; Prynn 1980), only 20–30 survived in 1947 (Prynn 1993) but approximately 500 individuals present in the wild by 1990 (Quigley and Hornocker 1994), subsequent drastic decline as regulatory controls faltered and poaching for Chinese market intensified, now 150–200 in wild, also 632 in captivity (Olney, Ellis, and Fisken 1994);

P. t. sumatrae (Sumatran tiger), classified as critically endangered by the IUCN, population declined rapidly from an estimated 1,000 in the 1970s to about 400 now in the wild, another 194 in captivity (Olney, Ellis, and Fisken 1994);

P. t. sondaica (Javan tiger), almost all suitable habitat destroyed, only 4 or 5 individuals survived in the 1970s, now probably extinct;

P. t. balica (Bali tiger), probably extinct, last known specimen taken in 1937, though Van Den Brink (1980) held out some hope that it survives.

Panthera pardus (leopard)

Head and body length is 910–1,910 mm, tail length is 580–1,100 mm, and shoulder height is 450–780 mm. Males weigh 37–90 kg, and females, 28–60 kg. There is much variation in color and pattern. The ground color ranges from pale straw and gray buff to bright fulvous, deep ochre, and chestnut. The underparts are white. The shoulders, upper arms, back, flanks, and haunches have dark spots arranged in rosettes, which usually enclose an area darker than the ground color. The head, throat, and chest are marked with small black spots, and the belly has large black blotches. Melanistic leopards (black panthers) are common, especially in moist, dense forests (Guggisberg 1975; Kingdon 1977).

The leopard can adapt to almost any habitat that provides it with sufficient food and cover. It occupies lowland forests, mountains, grasslands, brush country, and deserts. A specimen was found at an elevation of 5,638 meters on Kilimanjaro. *P. pardus* is partly sympatric with *P. uncia*, the snow leopard, on the slopes of the Himalayas. The leopard is usually nocturnal, resting by day on the branch of a tree, in dense vegetation, or among rocks. It may move 25 km in a night, or up to 75 km if disturbed. However, typical daily movement for radio-tracked individuals in both South Africa and Thailand was only about 1 or 2 km (Bailey 1993; Rabinowitz 1989). The leopard generally progresses by a slow, silent walk but can briefly run at speeds of more than 60 km/hr. It has been reported to leap more than 6 meters horizontally and more than 3 meters vertically. It climbs with great agility and can descend headfirst. It is a strong swimmer but is not as fond of water as the tiger. Vision and hearing are acute, and the sense of smell seems to be better developed than in the tiger (Guggisberg 1975).

Hunting is accomplished mainly by stalking and stealthily approaching as close as possible to the quarry. Larger animals are seized by the throat and killed by strangulation. Smaller prey may be dispatched by a bite to the back of the neck. The diet is varied but seems to consist mainly of whatever small or medium-sized ungulates are available, such as gazelles, impalas, wildebeests, deer, wild goats and pigs, and domestic livestock. Monkeys and baboons also are commonly taken. If necessary, the leopard can switch to such prey as rodents, rabbits, birds, and even arthropods. Food is frequently stored in trees for later use. Such is the strength of the leopard that it can ascend a tree carrying a carcass larger than itself (Guggisberg 1975; Kingdon 1977). In a study in South Africa, Bailey (1993) found each adult leopard to kill one large prey animal, usually an impala, about every seven days. Leopards themselves are frequently pursued by lions and sometimes by hyenas and wild dogs *(Lycaon)*. Large trees or other refuge sites are thus important, especially to a female raising cubs, both for immediate escape and as a place to keep food from the reach of rival species.

Population density is usually about 1/20–30 sq km, but under exceptionally favorable conditions it reportedly has been as high as 1/sq km. Most documented home range sizes are 8–63 sq km. Individuals apparently keep to a re-

Leopard *(Panthera pardus)*, photo from Zoological Society of London.

Jaguar *(Panthera onca)*, photo from Zoological Society of London.

stricted area, which they usually defend against others of the same sex. Territories are marked with urine, and severe intraspecific fighting has been observed. The range of a male may include the range of one or more females. Several males sometimes follow and fight over a female. Apparently, the leopard is normally a solitary species, but there have been reports of males remaining with females after mating and even helping to rear the young. There are a variety of vocalizations, the most common being a coughing grunt and a rasping sound, which seem to function in communication (Grzimek 1975; Guggisberg 1975; Kingdon 1977; Muckenhirn and Eisenberg 1973; Myers 1976; Schaller 1972; R. M. Smith 1977).

In an extended radio-tracking study in Kruger National Park, South Africa, Bailey (1993) determined population density to average 1/28.7 sq km for the whole park and to reach 1/3.3 sq km in an area of especially abundant prey. Social structure was found to be based on a three-tiered land tenure system. First, adult females maintained permanent home ranges of 5.6–29.9 sq km, sometimes exclusive but averaging about 18 percent overlap with one another. These areas were essentially limited territories that assured each resident access to adequate prey, cover, and den sites for raising young. Second, adult males occupied ranges of 16.5–96.1 sq km that overlapped little with ranges of other adult males and seemed to serve mainly to assure access to females. Each adult male range overlapped part or all of up to six female ranges. Third, superimposed on the mosaic of adult ranges were the more fluctuating ranges of young independent animals of both sexes, which essentially were waiting for vacancies in the permanent mosaic. There was extensive overlap of adult and subadult ranges but no contact between the occupants. Indeed, other than mothers with cubs or courting couples (which stayed together for up to five days), leopards were almost always found at least 1 km apart. Spacing was facilitated by vocalizations and by scent marking with ground scrapes and urine. Fighting was almost never recorded.

Breeding occurs throughout the year in most of Africa and India, though there may be peaks in some areas. In the northern portion of the Eurasian range mating apparently takes place mostly from December to February, parturition from March to May (Heptner and Sludskii 1992). In South Africa, Bailey (1993) observed most male-female association in the late dry season, from July to October. A female may give birth every 1–2 years. The estrous cycle averages about 46 days, and heat lasts 6–7 days; the gestation period is 90–105 days. Births occur in caves, crevices, hollow trees, or thickets. The number of young per litter is one to six, usually two or three. The cubs weigh 500–600 grams each at birth, open their eyes after 10 days, are weaned at 3 months, and usually separate from the mother at 18–24 months. Full size and sexual maturity are attained at around 3 years. Maximum longevity in captivity is more than 23 years (Grzimek 1975; Guggisberg 1975; Kingdon 1977).

The leopard seems to be more adaptable to the presence and activities of people than is the tiger and still occurs over a greater portion of its original range. Nonetheless, it is confronted by the same problems—persecution as a predator, value as a trophy, commercial demand for its beautiful fur, and loss of habitat and prey base. Man-eating leopards, though representing only a tiny percentage of the species, have undeniably been a menace in some areas. One in India, for example, is said to have killed more than 200 people during the 1850s (Guggisberg 1975). And from 1982 to 1989, reportedly 170 people were killed by leopards in India (Nowell and Jackson 1996). Intensification of agriculture, along with elimination of natural prey, sets up conflicts between herders and the leopard, usually to the detriment of the latter. The pelt of the leopard has been sought since ancient times, but in the 1960s there was a substantial increase in the worldwide market for the furs of spotted cats. Illegal hunting became rampant, and some leopard populations were decimated. As many as 50,000 leopard skins were marketed annually, of which nearly 10,000 were imported into the United States in some years (Myers 1976; Paradiso 1972). National laws and international agreements subsequently seem to have reduced this traffic. More recently, as in the case of *Panthera tigris* (see account thereof), Asian leopard populations have been subject to intensified poaching to obtain their body parts for use in traditional medicinal preparations (Nowell and Jackson 1996).

The leopard is said to be still relatively common in parts of East and Central Africa. Martin and de Meulenaer (1988) estimated its total numbers in sub-Saharan Africa at about 700,000, including 226,000 in Zaire alone, but these and other such high figures have been questioned (see, e.g., Bailey 1993; Meadows 1991; and Norton 1990). It has been pointed out that Martin and de Meulenaer improperly assumed that all suitable habitat was still occupied, that the leopard always occurred at the maximum density the habitat could support, that relaxation of human killing would result in rapid recovery, and that leopard density was higher in areas with higher rainfall. Stuart and Wilson (1988) estimated numbers at fewer than 15,000 for that part of Africa south of the Zambezi River, a region for which Martin and de Meulenaer had listed figures totaling about 75,000. Norton (1986) reported that the species has been eliminated from most of South Africa. Bailey (1993) determined that there were only about 700 leopards in Kruger National Park, the largest area of high-quality habitat in South Africa, and doubted that there could be even as many as 3,000 leopards in the whole country, whereas Martin and de Meulenaer had set the figure at over 24,000. Bailey anticipated further declines throughout Africa in response to human and livestock encroachment on habitat, with populations in the eastern and southern parts of the continent becoming restricted to isolated protected areas. Myers (1987) indicated that the leopard populations of Ethiopia, Kenya, Namibia, and Zimbabwe were reduced by 90 percent during the 1970s largely because of poisoning by livestock interests.

Data compiled by Nowell and Jackson (1996) indicate that the leopard also has been greatly reduced in West Africa and completely eliminated in the Sahel zone south of the Sahara. In the entire vast region from North Africa to Central and southwest Asia the species is now rare or absent. Although small, fragmented populations still occur over most of the region, the only country known to have a substantial population is Turkmenistan, with about 150 animals. Numbers also have been greatly reduced in Pakistan and Bangladesh, but there are about 14,000 in India. The leopard still is found in scattered parts of China and has even been taken recently within 50 km of Beijing. Farther to the northeast the species has all but vanished from Manchuria and Korea, and only 25–30 individuals survive in southeastern Siberia. The leopard's status is not well understood in Southeast Asia, but the population on Java is estimated to number 350–700. Santiapillai, Chambers, and Ishwaran (1982) noted that only 400–600 individuals survived in Sri Lanka and that the future of the leopard there looked bleak.

The IUCN no longer has a general designation for *P. pardus* but does classify the following subspecies as indicated: *P. p. orientalis* (southeastern Siberia, Manchuria, Korea), critically endangered; *P. p. japonensis* (northern China), endangered; *P. p. saxicolor* (Turkmenistan, Afghanistan, Iran,

Iraq), endangered; *P. p. melas* (Java), endangered; *P. p. kotiya* (Sri Lanka), endangered; *P. p. tulliana* (Turkey, Caucasus, Syria, Jordan), critically endangered; *P. p. nimr* (Arabian Peninsula), critically endangered; and *P. p. panthera* (northwestern Africa), critically endangered. All of these subspecies have declined drastically, with fewer than 50 *orientalis* and fewer than 250 *tulliana, nimr,* and *panthera* thought to survive. Another subspecies sometimes recognized, *P. p. jarvisi,* of the Sinai Peninsula and Negev, apparently now is represented by only about 17 individuals in southern Israel (Nowell and Jackson 1996). In captivity there are 40 *japonensis,* 19 *kotiya,* 120 *orientalis,* and 126 *saxicolor* (Olney, Ellis, and Fisken 1994). The USDI lists the entire species *P. pardus* as endangered except in that part of Africa south of a line corresponding with the southern borders of Equatorial Guinea, Cameroon, Central African Republic, Sudan, Ethiopia, and Somalia. In that region the leopard is listed as threatened, and regulations allow the importation into the United States, under the provisions of the CITES, of trophy leopards taken in this region for purposes of sport hunting; importation for commercial purposes is prohibited. *P. pardus* is on appendix 1 of the CITES.

Panthera onca (jaguar)

Head and body length is 1,120–1,850 mm and tail length is 450–750 mm. Reported weights range from 36 to 158 kg. In Venezuela males usually weigh 90–120 kg, and females, 60–90 kg. The ground color varies from pale yellow through reddish yellow to reddish brown and pales to white or light buff on the underparts. There are black spots on the head, neck, and limbs and large black blotches on the underparts. The shoulders, back, and flanks have spots forming large rosettes that enclose one or more dots in a field darker than the ground color. Along the midline of the back is a row of elongate black spots that may merge into a solid line. Melanistic individuals are common, but the spots on such animals can still be seen in oblique light. *P. onca* averages larger than *P. pardus* and has a relatively shorter tail, a more compact and more powerfully built body, and a larger and broader head (Grzimek 1975; Guggisberg 1975; Mondolfi and Hoogesteijn 1986).

The jaguar is commonly found in forests and savannahs but at the northern extremity of its range may enter scrub country and even deserts. It seems usually to require the presence of much fresh water and is an excellent swimmer. It may den in a cave, canyon, or ruin of a human building. Although it sometimes is said to be nocturnal, radiotelemetry has shown that it is often active in the daytime, with activity peaks around dawn and dusk (Nowell and Jackson 1996). It is not known to migrate, but individuals sometimes shift their range because of seasonal habitat changes, and males have been known to wander for hundreds of kilometers. Daily movement in Belize was found generally to be 2–5 km (Seymour 1989). The jaguar climbs well and is almost as arboreal as the leopard, but most hunting is done on the ground. Prey is stalked or ambushed, and carcasses may be dragged some distance to a sheltered spot. The most important foods are peccaries and capybaras. Tapirs, crocodilians, and fish are also taken (Grzimek 1975; Guggisberg 1975).

In a radio-tracking study in southwestern Brazil, Schaller (1980*a*, 1980*b*) found a population density of about 1/25 sq km. There was a land tenure system much like that of the cougar and tiger. Females had home ranges of 25–38 sq km, which overlapped one another. Resident males used areas twice as large, which overlapped the ranges of several females. Studies in Belize indicate that 25–30 jaguars were present in about 250 sq km; male home ranges there averaged 33.4 sq km and overlapped extensively, while two females had nonoverlapping ranges of 10–11 sq km. Estimates of home range in other areas have been as high as 390 sq km (Seymour 1989). The jaguar seems to be basically solitary and territorial. It marks its territory with urine and tree scrapes and has a variety of vocalizations, including roars, grunts, and mews (Guggisberg 1975). A female in estrus may wander far from her regular home range and may sometimes be accompanied briefly by several males. Estrus lasts about 6–17 days (Mondolfi and Hoogesteijn 1986).

Births occur throughout the year in captivity and perhaps also in the wild, especially in tropical areas, but there is evidence for a breeding season in the more northerly and southerly parts of the range. Most births in Paraguay take place in November–December; in Brazil, in December–May; in Argentina, in March–July; in Mexico, in July–September; and in Belize, in June–August, the rainy season (Seymour 1989). The estrous cycle averages 37 days and estrus lasts 6–17 days (Nowell and Jackson 1996). The gestation period commonly is 93–105 days, and litters contain one to four, often two, young. The offspring weigh 700–900 grams each at birth, open their eyes after 3–13 days, suckle for 5–6 months, stay with the mother about 2 years, and attain full size and sexual maturity at 2–4 years (Grzimek 1975; Guggisberg 1975; Mondolfi and Hoogesteijn 1986). A captive specimen at the Wuppertal Zoo in Germany reportedly lived to an age of 24 years (Marvin L. Jones, Zoological Society of San Diego, pers. comm., 1995).

Until the end of the Pleistocene *P. onca* occurred throughout the southern United States, and it seems to have been especially common in Florida. There is some evidence that the species still inhabited the southeastern United States in historical time (Daggett and Henning 1974; Nowak 1975*b*). Breeding may have continued in Arizona until about 1950 (Brown 1983). Otherwise resident populations had disappeared from the United States by the early twentieth century, though for many years wanderers continued to enter the country from Mexico, whereupon they usually were quickly shot. The most recent confirmed records are from southeastern Arizona and involve an animal killed in December 1986 (Nowak 1994) and jaguars photographed in March and September 1996 (Rabinowitz 1997). The jaguar now has been exterminated in most of Mexico, much of Central America, and, at the other end of its range, in most of eastern Brazil, Uruguay, and all but the northernmost parts of Argentina (Roig 1991; Swank and Teer 1989). Its numbers also have been greatly reduced in the vast Patanal wetland of Brazil, Bolivia, and Paraguay but reportedly are increasing farther south, in the Paraguayan Gran Chaco. Surprisingly, one of the most viable remaining populations, some 600–1,000 cats, is in Belize, the second smallest mainland country in the New World (Nowell and Jackson 1996). There may be another 500 in Guatemala but no more than 500 in all of Mexico (Emmons 1991).

The jaguar declined because of the same factors that affected other large cats—persecution as a predator, habitat loss, and commercial fur hunting. The jaguar is thought to have killed people on rare occasion, but no individuals are known to have become systematic man-eaters. The spe-cies does have a reputation as a serious menace to domestic cattle; indeed, it may actually have increased in colonial times, when livestock was introduced to the savannahs of South America, a continent lacking vast natural herds of ungulates (Guggisberg 1975). As land was cleared and opened to ranching and other human activity, inevitable conflicts arose between the jaguar and people. As in the case of the leopard and other spotted cats, there was a great increase in the commercial demand for jaguar skins in the 1960s. An estimated 15,000 jaguars were then being killed annually in the Amazonian region of Brazil alone. The recorded number of pelts entering the United States reached 13,516 in 1968. Subsequent national and international conservation measures seem to have reduced the kill, but the problem continues (Thornback and Jenkins 1982). *P. onca* is classified as endangered by the USDI and as near threatened by the IUCN and is on appendix 1 of the CITES.

Panthera leo (lion)

Except as noted, the information for the account of this species was taken from Grzimek (1975), Guggisberg (1975), Kingdon (1977), and Schaller (1972). In males head and body length is 1,700–2,500 mm, tail length is 900–1,050 mm, shoulder height is about 1,230 mm, and weight is 150–250 kg. In females head and body length is 1,400–1,750 mm, tail length is 700–1,000 mm, shoulder height is about 1,070 mm, and weight is 120–82 kg. The coloration varies widely, from light buff and silvery gray to yellowish red and dark ochraceous brown. The underparts and insides of the limbs are paler, and the tuft at the end of the tail is black. The male's mane, which apparently serves to protect the neck in intraspecific fighting, is usually yellow, brown, or reddish brown in younger animals but tends to darken with age and may be entirely black.

The preferred habitats of the lion are grassy plains, savannahs, open woodlands, and scrub country. It sometimes enters semideserts and forests and has been recorded in mountains at elevations of up to 5,000 meters. It normally walks at about 4 km/hr and can run for a short distance at 50–60 km/hr. Leaps of up to 12 meters have been reported. The lion readily enters trees by jumping but is not an adept climber. Its senses of sight, hearing, and smell are all thought to be excellent. Activity may occur at any hour but is mainly nocturnal and crepuscular; in places where the lion is protected from human harassment it is commonly seen by day. The average period of inactivity is about 20–21 hours per day. Nightly movements in Nairobi National Park were found to cover 0.5–11.2 km (Rudnai 1973). In the Serengeti most lions remain in a single area throughout the year, but about one-fifth of the animals are nomadic, following the migrations of ungulate herds.

The lion usually hunts by a slow stalk, alternately creeping and freezing, utilizing every available bit of cover; it then makes a final rush and leaps upon the objective. If the intended victim cannot be caught in a chase of 50–100 meters, the lion usually tires and gives up, but pursuits of up to 500 meters have been observed. Small prey may be dispatched by a swipe of the paw; large animals it seizes by the throat and strangles, or it suffocates them by clamping its jaws over the mouth and nostrils. Two lions sometimes approach prey from opposite directions; if one misses, the other tries to capture the victim as it flees by. An entire pride may fan out and then close in on a quarry from all sides. Groups are about twice as likely as lone individuals to capture prey. Most hunts fail: of 61 stalks observed by Rudnai (1973) only 10 were successful. The lion eats anything it can catch and kill, but it depends mostly on animals weighing 50–300 kg. Important prey species are wildebeests, impalas, other antelopes, giraffes, buffalo, wild hogs, and zebras. Carrion is readily taken. An adult male can consume as much as 40 kg of meat at one meal. After making a kill, a lion may rest in the vicinity of the carcass for several days. About 10–20 large animals per lion are killed each year.

Population density in the whole Serengeti ecosystem of East Africa was determined to be 1/10.0–12.7 sq km. Reported densities in most other areas vary from 1/2.6 sq km to 1/50.0 sq km. Nonmigratory lions in the Serengeti live in prides, each with a home range of 20–400 sq km. All or part of the range is a territory, which is vigorously defended against other lions. Nomadic lions have ranges of up to 4,000 sq km; there is much overlap in these areas, and individuals behave amicably. Nomads are commonly found in groups of two to four animals, and membership changes freely.

The basis of a resident pride is a group of related females and their young. These associations may persist for many years, being generally closed to strange females. Most daughters of group members are recruited into the pride, but young males depart as they approach maturity. Several related males frequently leave at about the same time and maintain a nomadic "coalition" until adulthood; sometimes unrelated males join to form a larger and potentially more powerful unit (Packer et al. 1991). Such a group, or occasionally a single male, eventually is able to gain residence with a pride of females and young for an indefinite period. The males cooperatively defend the pride against the approach of outside males. Some males associate with and defend several prides. Males of a coalition, especially if related, show remarkable tolerance in allowing one another to mate (Bertram 1991). Eventually, usually within three years, the pride males are driven off by another group of

African lions *(Panthera leo)*, photo by Bernhard Grzimek.

African lions *(Panthera leo)*, photo by Bernhard Grzimek.

males (Bertram 1975). Studies by Hanby and Bygott (1987) indicate that the factor stimulating departure of subadult males from a pride is the arrival of a new group of adult males and that some subadult females also may depart at such time if they are not yet ready to mate. The newly arrived males generally kill those young that are unable to escape. As a result, females of a pride tend to mate at the same time, to give birth synchronously, and to rear their young communally (Packer, Scheel, and Pusey 1990).

In the Serengeti the average number of lions in a pride was found to be 15, the range was 4–37, and the number of adult males present was 2–4. Prides often were divided into widely scattered smaller groups of about 5 individuals each. Studies by Packer, Scheel, and Pusey (1990) in the Serengeti suggest that such units frequently comprise females that are staying together to forage in order to defend their young against attack by outside males; cooperation to improve hunting success thus would not be the primary function of sociality in lions. In Nairobi National Park, Rudnai (1973) observed a single adult male to be associated with four prides of females. There was a rank order among the females, and a female led each group, even when the male was present, but the male was dominant with respect to access to food. Males living within a pride allow the females to do almost all of the hunting, but they arrive subsequent to a kill and sometimes drive the others away. Lions appear to behave asocially at a kill, there being much quarreling and snapping and little tolerance shown to subordinates and cubs.

The lion has at least nine distinct vocalizations, including a series of grunts that apparently serve to maintain contact as a pride moves about. The roar, which can be heard by people up to 9 km away, is usually given shortly after sundown for about an hour and then again following a kill and after eating. It apparently has a territorial function. The lion also proclaims its territory by scent marking through urination, defecation, and rubbing its head against a bush.

Breeding occurs throughout the year in India and in Africa south of the Sahara. In any one pride, however, females tend to give birth at about the same time (Bertram 1975). Females are polyestrous, and heat lasts about 4 days. A female normally gives birth every 18–26 months, but if an entire litter is lost, she may mate again within a few days. The gestation period is 100–119 days, and litters contain one to six young, usually three or four. The newborn weigh about 1,300 grams; their eyes may be open at birth, or they may take up to 2 weeks to open. Cubs follow their mother after 3 months, suckle from any lactating female in the pride, and usually are weaned by 6–7 months. They begin to participate in kills at about 11 months, are fully dependent on the adults for food until 16 months, and probably are not capable of surviving on their own until at least 30 months. Sexual maturity is attained at around 3–4 years, but growth continues until about 6 years. The average longevity in zoos is about 13 years, but some captives have lived nearly 30 years.

With the exception of people, their domestic animals, and their commensals, the lion attained the greatest geographical distribution of any terrestrial mammal. Various populations, known from fossils by such names as *Panthera atrox* and *P. spelaea*, are now regarded as conspecific with *P. leo* (Hemmer 1974; Kurten 1985). About 10,000 years ago the lion apparently occurred in most of Africa, in all of Eurasia except probably the southeastern forests, throughout North America, and at least in northern South America. The lion is thought to have disappeared from most of Europe because of the development there of dense forests (Guggisberg 1975). It probably vanished from the Western Hemisphere when many of the large mammals on which it preyed were exterminated through the spread of advanced human hunters at the close of the Pleistocene. It was eliminated on the Balkan Peninsula, its last major stronghold in Europe, about 2,000 years ago and in Palestine at the time of the Crusades.

The continued decline of *P. leo* in modern times has resulted primarily from the expansion of human activity and domestic livestock and the consequent persecution of the lion as a predator. A few lions became regular man-eaters—for example, a pair killed 124 people in Uganda in 1925—and thus gave a sinister reputation to the entire species. Hunting for sport was also a major factor in some areas; for example, one person killed more than 300 lions in India in the mid–nineteenth century. At that time *P. leo* was still common from Asia Minor to central India and in North Africa. The last wild individual of the North African subspecies, *P. leo leo*, was shot in Morocco in 1920, though subsequently animals of the same type were found to exist in captivity (Leyhausen *in* Grzimek 1990). By about 1940 the lion also had been eliminated in Asia except in the Gir Forest, Gujarat State, western India. Vigorous conservation efforts resulted in the number of lions there stabilizing at around 200 for some years and recently increasing to more than 250. Unfortunately, the Gir lions have been under continuous pressure from livestock interests, and conflicts recently have intensified. There have been increasing attacks on people, 20 being killed from 1988 to 1991 (Saberwal et al. 1994). To help relieve these problems, as well as to limit the loss of genetic viability, reintroduction has been proposed at several sites within the former range of the lion in India (Chellam and Johnsingh 1993). There are 119 lions descended from the Gir population in captivity (Olney, Ellis, and Fisken 1994). The subspecies involved, *P. leo persica*, is classified as endangered by the IUCN and the USDI and is on appendix 1 of the CITES. All other lion populations are classified as vulnerable by the IUCN and are on appendix 2.

To the south of the Sahara the lion has become rare in West Africa (Rosevear 1974) and has been exterminated in most of South Africa and much of East Africa. It is still present over a large region but is widely hunted and poisoned by persons owning livestock. Myers (1975*b*) wrote that since 1950 the number of lions in Africa may have been reduced by half, to 200,000 or fewer; later Myers (1984) set the estimate at only 50,000. He cautioned that the species was rapidly losing ground to agriculture and that by the end of the century it might number only a few thousand individuals and survive only in major parks and reserves. Nowell and Jackson (1996) agreed that numbers are almost certainly below 100,000, perhaps as low as 50,000, and that the species is becoming increasingly rare outside of protected areas. Stuart and Wilson (1988) suggested that in southern Africa there was little hope for the lion outside of these conservation areas; they estimated numbers south of the Zambezi River to be 6,100–9,100, of which about 4,200 already were in the protected reserves. Myers (1987) warned, however, that such restriction and fragmentation of the lion population might result in inbreeding and loss of genetic viability. A severe new problem in East Africa is the loss of many lions to an epidemic of canine distemper, evidently spread by *Canis familiaris* (Woodford 1994).

CARNIVORA; FELIDAE; **Genus ACINONYX**
Brookes, 1828

Cheetah

The single species, *A. jubatus*, originally occurred from Palestine and the Arabian Peninsula to Tajikistan and central India, as well as throughout Africa except in the tropical forest zone and the central Sahara (Ellerman and Morrison-Scott 1966; Guggisberg 1975; Kingdon 1977; Smithers *in* Meester and Setzer 1977).

Except as noted, the information for the remainder of this account was taken from Caro (1994), Eaton (1974), Grzimek (1975), Guggisberg (1975), and Kingdon (1977). Head and body length is 1,120–1,500 mm, tail length is 600–800 mm, shoulder height is 670–940 mm, and weight is 21–72 kg; on the average, males are larger than females. The ground color of the upper parts is tawny to pale buff or grayish white, and the underparts are paler, often white. The pelage is generally marked by round, black spots set closely together and not arranged in rosettes. A black stripe extends from the anterior corner of the eye to the mouth. The last third of the tail has a series of black rings. The coat is coarse. The hair is somewhat longer on the nape than elsewhere, forming a short mane; in young cubs the mane is much more pronounced and extends over the head, neck, and back. *Acinonyx* has a slim body, very long legs, a rounded head, and short ears. The pupil of the eye is round. The paws are very narrow compared with those of other cats and look something like those of dogs. The claws are blunt, only slightly curved, and only partly retractile.

An additional species, *A. rex* (king cheetah), was described in 1927. It was based on specimens that differed from other cheetahs in having longer and softer hair and partial replacement of the normal spots by dark bars. Only 14 skins have been recorded from the wild, most from Zimbabwe and adjacent areas but 1 from Burkina Faso. It is now generally accepted that these specimens represent merely a variety of *A. jubatus* (Frame 1992; Hills and Smithers 1980). Indeed, individuals with the king cheetah markings now have been bred in captivity and recorded within otherwise normal litters (Brand 1983; Skinner and Smithers 1990; Van Aarde and Van Dyk 1986).

The habitat of the cheetah varies widely, from semidesert through open grassland to thick bush. Activity is mostly diurnal, and shelter is sought in dense vegetation. Recorded daily movements are about 3.7 km for a female with cubs and 7.1 km for adult males. The cheetah is capable of climbing and often plays about in trees. It is the fastest terrestrial mammal. Reported estimates of maximum speed range from 80 to 112 km/hr; such velocities, however, cannot be maintained for more than a few hun-

Cheetah *(Acinonyx jubatus)*, photo by Bernhard Grzimek.

dred meters. Unlike most cats, the cheetah does not usually ambush its prey or approach to within springing distance. It stalks an animal and then charges from about 70–100 meters away. It is seldom successful if it attacks from a point more than 200 meters distant, and it can only continue a chase for about 500 meters. Most hunts fail. If an animal is overtaken, it is usually knocked down by the force of the cheetah's charge and then seized by the throat and strangled. The diet consists mainly of gazelles, especially *Gazella thomsoni,* and also includes impalas, other small and medium-sized ungulates, and the calves of large ungulates. A female with cubs may kill such an animal every day, whereas lone adults hunt every two to five days. Hares, other small mammals, and birds are sometimes taken. The cheetah seems to work harder for its living than do the other big cats of Africa and thus may be more vulnerable to environmental changes brought about by human disturbance.

Population density in good habitat varies from about 1/20 sq km to 1/100 sq km. Seasonal concentrations may reach 1/2.5 sq km in some areas. In marginal habitat, however, density may be only 1/250 sq km or less (Myers 1975*a*). Some home ranges have been reported to measure about 50–130 sq km, but there appears to be considerable spacing variation. The cheetah occurs alone or in small groups. Solitary males and females are usually seminomadic and may occupy large home ranges of 700–1,500 sq km that overlap with one another and with the territories of resident animals. The groups seem usually to be either a female with cubs or two to four related adult males (litter mates that have remained together). The male "coalitions" commonly defend a territory against other males, perhaps thus facilitating access to prey and mates (Caro and Collins 1987). An unrelated male sometimes is accepted into the group. Coalitions persist through the lifetime of the members and seem to function primarily to secure mates. A single male may sometimes acquire a territory if there are no coalitions in the vicinity (Caro 1991). Such a territory is located in an area of high prey density, generally where females concentrate to hunt, and usually measures only about 40 sq km. In contrast to those of most cats, male territories of cheetahs, whether of an individual or a coalition, are usually much smaller than female ranges. Young females normally establish a range overlapping that of their mother, but young males leave their natal range.

Groups avoid one another and mark the area they are using at a given time. Marking is accomplished by regular urination on prominent objects. Such activity also serves to communicate sexual information. The cheetah is normally amicable toward others of its kind, but several males sometimes gather near an estrous female and fight over her. Members of male coalitions, however, usually show little overt competition in such situations. They join forces to control the movements of females, fight together against other males, rest in close proximity, groom one another, and search for their companions if they become separated. There does not seem to be a strong dominance hierarchy. There are a number of antagonistic vocalizations, purrs of contentment, a chirping sound made by a female to her cubs, and an explosive yelp that can be heard by people 2 km away.

Breeding occurs throughout the year, though a birth peak has been reported during the rainy season in East Africa (Nowell and Jackson 1996; Skinner and Smithers 1990). Females are polyestrous, with an average estrous cycle of approximately 12 days and an estrus of 1–3 days. Wild females normally give birth at intervals of 17–20 months; however, if all young are lost, the mother may soon mate and bear another litter. The gestation period is 90–95 days. The number of young per litter is one to eight, usually three to five. The cubs are born in thick vegetation or some other shelter, weigh 150–400 grams at birth, open their eyes after 4–11 days, and are weaned at 3–6 months. Cheetah cubs have an unusually high mortality rate for a large carnivore, especially because of lion predation: only 5 percent may reach independence. They begin to follow their mother after about 6 weeks, and the family may then shift its place of shelter on an almost daily basis. The mother teaches her cubs to hunt. They separate from her at 13–20 months and usually remain together for several months (as noted above, the males sometimes continue to associate for life). Both sexes may be physiologically capable of reproduction at about 18 months, but they seem not to attain full sexual maturity until they are 2–3 years old. A 15-year-old female was still reproductively active. Captives have lived up to 19 years.

People have tamed the cheetah and used it to run down game for at least 4,300 years. It was employed in ancient Egypt, Sumeria, and Assyria and has been used more recently by the royalty of Europe and India. It is usually hooded, like a falcon, when taken out for the chase, then freed when the game is in sight. If the hunt is successful, the cheetah is rewarded with a portion of the kill. If the cheetah should attempt to escape, it soon tires and can be easily caught by persons on horseback. Tame individuals are usually playful and affectionate.

The removal of live cheetahs from the wild has contributed to a decline in the species. Other factors are excessive hunting of both the cheetah and its prey, the spread of people and their livestock, and the fur market. In Namibia and other parts of southern Africa the cheetah is considered a serious predator of domestic sheep, goats, and young cattle, and it is persecuted accordingly (Nowell and Jackson 1996). Some studies also have indicated that *Acinonyx* has an unusually low degree of genetic variation, perhaps reflecting a severe population contraction during evolutionary history, leaving the species especially vulnerable to environmental disruption (Cohn 1986; S. J. O'Brien 1983, 1994; O'Brien et al. 1985; Wayne, Modi, and O'Brien 1986). Analysis of mitochondrial DNA and associated back calculation (Menotti-Raymond and O'Brien 1993) suggest that this hypothetical genetic bottleneck occurred near the end of the Pleistocene, about 10,000 years ago, a period characterized by widespread extinction of large mammals. Nonetheless, Merola (1994) has presented a reasonable argument that genetic constitution in the cheetah is not significantly less variable than in some other carnivores and that the real threat to the species is human habitat disruption. In addition, Caro and Laurenson (1994) reviewed recent studies of the cheetah and found no substantial evidence that genetic problems are contributing to poor reproduction, susceptibility to disease, or mortality of young.

The cheetah seems to be much less adaptable than the leopard to the presence of people. It evidently has disappeared in Asia except in Iran and possibly adjacent parts of Pakistan, Afghanistan, and Turkmenistan. In the mid-1970s the population in Iran was estimated to include more than 250 individuals and was considered to be well protected (IUCN 1976); this population still exists but has declined because of both direct hunting and destruction of its prey (Karami 1992; Nowell and Jackson 1996). In Africa the species remains widely distributed but has become very rare in the northern and western parts of the continent. There still is a population of perhaps 500 animals in the mountains of Mali, Niger, Chad, and southern Algeria (Nowell and Jackson 1996). The species has been extirpated in most of South Africa and much of East Africa, though Hamilton (1986) thought it was holding its own in Kenya and had proved to be more resilient than anticipated. Myers (1975a) suggested that a century earlier the sub-Saharan population comprised about 100,000 individuals. Myers (1987) estimated the total number remaining at just 10,000–15,000, and Stuart and Wilson (1988) estimated 6,200–8,500 for that part of the continent south of the Zambezi River. According to Grisham and Marker-Kraus (*in* Bowdoin et al. 1994), there are 9,000–12,000 in the wild, including 2,500 in Namibia, which has the largest single national population. Olney, Ellis, and Fisken (1994) reported that there also are 699 in captivity, 555 of which were born in that state. The IUCN classifies the cheetah generally as vulnerable, the northwest African subspecies *(A. j. hecki)* as endangered, and the Asiatic subspecies *(A. j. venaticus)* as critically endangered. The entire species is listed as endangered by the USDI and is on appendix 1 of the CITES.

World Distribution of Carnivores

For maximum usefulness, it has been necessary to devise the simplest practicable outline of the approximate distribution of the genera in the sequence used in the text. The tabulation should be regarded as an index guide to the carnivores or to geographic regions. At the same time it gives a good overall picture of the general distribution of Order Carnivora.

The major geographic distribution of the genera of Recent carnivores that appears in the tabulation is designed to show their natural distribution at present or within comparatively recent times. Most of the animals occupy only a portion of the geographic region that appears at the head of the column. Some are limited to the tropical regions, others to temperate zones, and still others to the colder areas. Also, many restricted ranges cannot be designated either by letters to show the general area or by footnotes because of limited space on the tabulation. *It therefore should not be assumed that a mark indicating that an animal occurs within a geographic region implies that it inhabits that entire area.* For more detailed outlines of the ranges of the respective genera, it is necessary to consult the generic texts.

Explanation of Geographic Column Headings

Europe and Asia constitute a single land mass, but this land mass comprises widely different types of zoogeographic areas created by high mountain ranges, plateaus, latitudes, and prevailing winds. The general distribution of Recent carnivores can be shown much more accurately by two columns, headed "Europe" and "Asia," than by a single column headed "Eurasia."

Most islands are included with the major land masses nearby unless otherwise specified, although in many instances some of the carnivores indicated for the continental mass do not occur on the islands.

With Europe are included the British Isles and other adjacent islands, including those in the Arctic.

With Asia are included the Japanese Islands, Taiwan, Hainan, Sri Lanka, and other adjacent islands, including those in the Arctic.

With North America are included Mexico and Central America south to Panama, adjacent islands, the Aleutian chain, the islands in the arctic region, and Greenland but not the West Indies.

With South America are included Trinidad, the Netherlands Antilles, and other small adjacent islands but not the Falkland and Galapagos Islands unless named in footnotes.

With Africa are included only Zanzibar Island and small islands close to the continent but not the Cape Verde or Canary Islands.

The island groups treated separately are:

Southeastern Asian islands, in which are included the Andamans, the Nicobars, the Mentawais, Sumatra, Java, the Lesser Sundas, Borneo, Sulawesi, the Moluccas, and the many other adjacent small islands;

New Guinea and small adjacent islands;

the Australian region, which includes Australia, Tasmania, and adjacent small islands;

the Philippine Islands and small adjacent islands;

the West Indies;

Madagascar and small adjacent islands;

other islands that have only one or a few forms of carnivores and are named in footnotes.

Footnotes indicate the major easily definable deviations from the distribution indicated in the tables.

Symbols

†	The carnivores are extinct.
■	The carnivores occur on or adjacent to the land or in the water area.
N	Northern portion
S	Southern portion
E	Eastern portion
W	Western portion
Ne	Northeastern portion
Se	Southeastern portion
Sw	Southwestern portion
Nw	Northwestern portion
C	Central portion

Examples: "N,C" = northern and central; "Nc" = north-central. Numerals refer to footnotes indicating clearly defined limited ranges within the general area.

Genera of Recent Mammals	page	North America	West Indies	South America	Madagascar	Africa	Europe	Asia	Southeast Asia Islands	Philippine Islands	New Guinea	Australian Region	Antarctic Region	Arctic Region	Atlantic Ocean	Indian Ocean	Pacific Ocean
CARNIVORA CANIDAE																	
Vulpes	72	■				■	■	■									
Fennecus	79					■N		■Sw									
Alopex	80	■N					■N	■N						■			
Urocyon	83	■		■N													
Lycalopex	84			■C													
Pseudalopex	84			■													
Dusicyon	86														■1		
Cerdocyon	87			■													
Nyctereutes	88							■E									
Atelocynus	89			■N													
Speothos	90	■2		■													
Canis	91	■				■	■	■									
Chrysocyon	108			■													
Otocyon	109					■E,S											
Cuon	110							■	■3								
Lycaon	112					■											
CARNIVORA URSIDAE																	
Tremarctos	116			■W													
Ursus	117	■				■Nw	■	■	■					■			
Ailuropoda	129							■Ec									
CARNIVORA PROCYONIDAE																	
Ailurus	131							■Ec									
Bassariscus	132	■W															
Procyon	134	■	■4	■													
Nasua	136	■S		■													
Nasuella	137			■Nw													
Potos	138	■S		■N													
Bassaricyon	139	■S		■N													
CARNIVORA MUSTELIDAE																	
Mustela	141	■		■N		■N	■	■	■								
Vormela	150						■Se	■W,C									
Martes	152	■					■	■	■								
Eira	156	■S		■													
Galictis	157	■S		■													
Lyncodon	158			■S													
Ictonyx	159					■											
Poecilictis	160					■N											
Poecilogale	160					■C,S											
Gulo	162	■					■N	■N									
Mellivora	163					■		■S									
Meles	164						■	■									
Arctonyx	166							■C	■5								
Mydaus	166								■	■W							
Taxidea	167	■W,C															
Melogale	168							■Se	■6								
Spilogale	169	■															
Mephitis	171	■															
Conepatus	172	■		■													
Lutra	173					■	■	■	■								

1. Falkland Islands. 2. Panama only. 3. Sumatra and Java only. 4. Lesser Antilles and New Providence Island only. 5. Sumatra only. 6. Java, Bali, and Borneo only.

Genera of Recent Mammals	page	North America	West Indies	South America	Madagascar	Africa	Europe	Asia	Southeast Asia Islands	Philippine Islands	New Guinea	Australian Region	Antarctic Region	Arctic Region	Atlantic Ocean	Indian Ocean	Pacific Ocean
CARNIVORA MUSTELIDAE Continued																	
Lutrogale	175							■S	■								
Lontra	176	■		■													
Pteronura	178			■													
Aonyx	179					■		■S	■	■W							
Enhydra	181	■W						■Ne									■N
CARNIVORA VIVERRIDAE																	
Viverra	185							■S	■	■							
Civettictis	186					■											
Viverricula	187							■S	■								
Genetta	187					■	■Sw	■Sw									
Osbornictis	190					■1											
Poiana	191					■W,C											
Prionodon	192							■Se	■								
Nandinia	192					■											
Arctogalidia	193							■Se	■								
Paradoxurus	194							■S	■	■							
Paguma	195							■S	■								
Macrogalidia	196								■2								
Arctictis	197							■Se	■	■W							
Hemigalus	198							■Se	■3								
Diplogale	199								■4								
Chrotogale	199							■Se									
Cynogale	200							■Se	■3								
Fossa	201				■												
Eupleres	202				■												
CARNIVORA HERPESTIDAE																	
Galidia	204				■												
Galidictis	205				■												
Mungotictis	205				■W												
Salanoia	207				■Ne												
Herpestes	207					■	■Sw	■S	■	■W							
Galerella	209					■											
Mungos	210					■											
Crossarchus	211					■W,C											
Liberiictis	212					■5											
Helogale	213					■E,S											
Dologale	214					■C											
Bdeogale	215					■		■6									
Rhynchogale	216					■E,S											
Ichneumia	216					■		■Sw									
Atilax	217					■											
Cynictis	218					■S											
Paracynictis	219					■S											
Suricata	219					■S											
Cryptoprocta	221				■												
CARNIVORA HYAENIDAE																	
Proteles	222					■E,S											
Parahyaena	223					■S											
Hyaena	226					■		■S									
Crocuta	227					■											

1. Zaire only. 2. Sulawesi only. 3. Sumatra and Borneo only. 4. Borneo only. 5. Liberia and probably Ivory Coast only.
6. Yemen only.

Genera of Recent Mammals	page	North America	West Indies	South America	Madagascar	Africa	Europe	Asia	Southeast Asia Islands	Philippine Islands	New Guinea	Australian Region	Antarctic Region	Arctic Region	Atlantic Ocean	Indian Ocean	Pacific Ocean
CARNIVORA FELIDAE																	
Felis	232	■		■		■	■	■	■	■							
Neofelis	256							■Se	■1								
Panthera	257	■S		■		■	■2	■	■								
Acinonyx	270					■		■S									

1. Sumatra and Borneo only. 2. No longer present.

Appendix

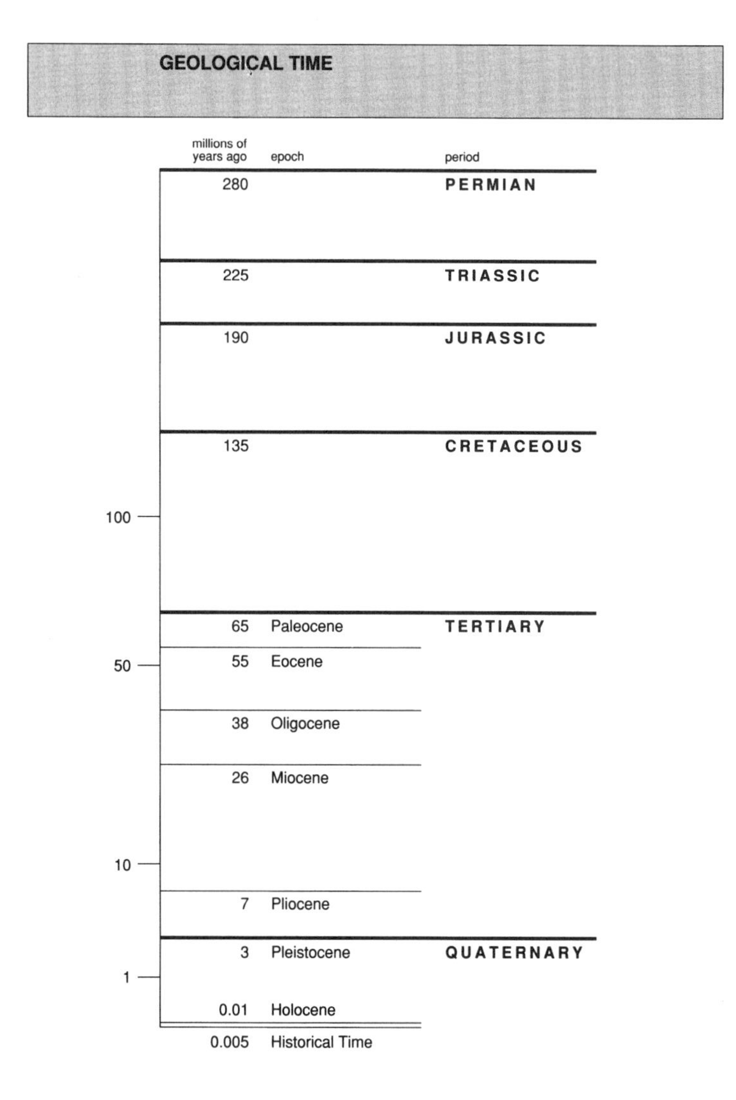

GEOLOGICAL TIME
millions of years ago
epoch
period
280
PERMIAN
225
TRIASSIC
190
JURASSIC
135
CRETACEOUS
100
65
Paleocene
TERTIARY
50
55
Eocene
38
Oligocene
26
Miocene
10
7
Pliocene
3
Pleistocene
QUATERNARY
1
0.01
Holocene
0.005
Historical Time

LENGTH

scales for comparison of metric and U.S. units of measurement

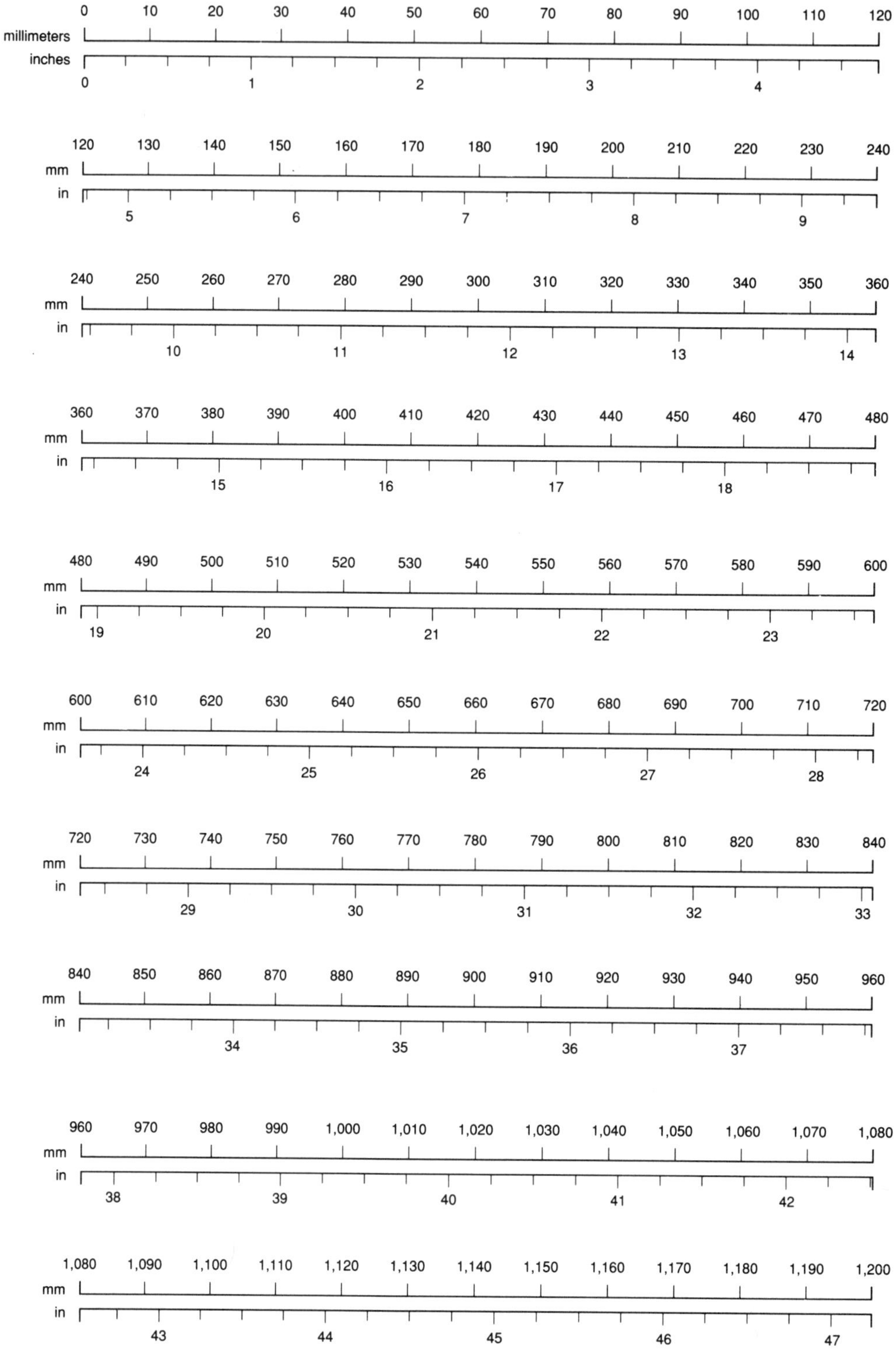

0 10 20 30 40 50 60 70 80 90 100 110 120
millimeters
inches
0 1 2 3 4

120 130 140 150 160 170 180 190 200 210 220 230 240
mm
in
5 6 7 8 9

240 250 260 270 280 290 300 310 320 330 340 350 360
mm
in
10 11 12 13 14

360 370 380 390 400 410 420 430 440 450 460 470 480
mm
in
15 16 17 18

480 490 500 510 520 530 540 550 560 570 580 590 600
mm
in
19 20 21 22 23

600 610 620 630 640 650 660 670 680 690 700 710 720
mm
in
24 25 26 27 28

720 730 740 750 760 770 780 790 800 810 820 830 840
mm
in
29 30 31 32 33

840 850 860 870 880 890 900 910 920 930 940 950 960
mm
in
34 35 36 37

960 970 980 990 1,000 1,010 1,020 1,030 1,040 1,050 1,060 1,070 1,080
mm
in
38 39 40 41 42

1,080 1,090 1,100 1,110 1,120 1,130 1,140 1,150 1,160 1,170 1,180 1,190 1,200
mm
in
43 44 45 46 47

LENGTH

scales for comparison of metric and U.S. units of measurement

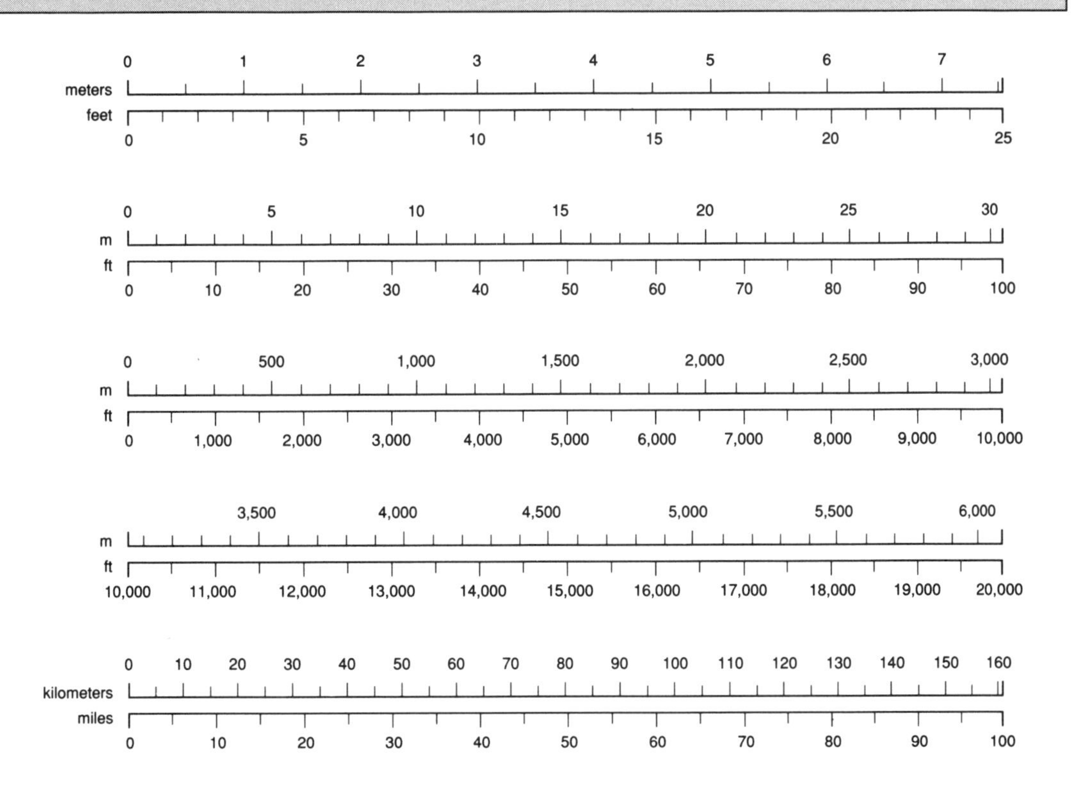

CONVERSION TABLES

	U.S. to Metric: *to convert*	*multiply by*	Metric to U.S.: *to convert*	*multiply by*
LENGTH	in. to mm.	25.4	mm. to in.	0.039
	in. to cm.	2.54	cm. to in.	0.394
	ft. to m.	0.305	m. to ft.	3.281
	yd. to m.	0.914	m. to yd.	1.094
	mi. to km.	1.609	km. to mi.	0.621
AREA	sq. in. to sq. cm.	6.452	sq. cm. to sq. in.	0.155
	sq. ft. to sq. m.	0.093	sq. m. to sq. ft.	10.764
	sq. yd. to sq. m.	0.836	sq. m. to sq. yd.	1.196
	sq. mi. to ha.	258.999	ha. to sq. mi.	0.004
VOLUME	cu. in. to cc.	16.387	cc. to cu. in.	0.061
	cu. ft. to cu. m.	0.028	cu. m. to cu. ft.	35.315
	cu. yd. to cu. m.	0.765	cu. m. to cu. yd.	1.308
CAPACITY (liquid)	fl. oz. to liter	0.03	liter to fl. oz.	33.815
	qt. to liter	0.946	liter to qt.	1.057
	gal. to liter	3.785	liter to gal.	0.264
MASS (weight)	oz. avdp. to g.	28.35	g. to oz. avdp.	0.035
	lb. avdp. to kg.	0.454	kg. to lb. avdp.	2.205
	ton to t.	0.907	t. to ton	1.102
	l. t. to t.	1.016	t. to l. t.	0.984

Abbreviations

avdp.	avoirdupois
cc.	cubic centimeter(s)
cm.	centimeter(s)
cu.	cubic
ft.	foot, feet
g.	gram(s)
gal.	gallon(s)
ha.	hectare(s)
in.	inch(es)
kg.	kilogram(s)
lb.	pound(s)
l. t.	long ton(s)
m.	meter(s)
mi.	mile(s)
mm.	millimeter(s)
oz.	ounce(s)
qt.	quart(s)
sq.	square
t.	metric ton(s)
yd.	yard(s)

WEIGHT

scales for comparison of metric and U.S. units of measurement

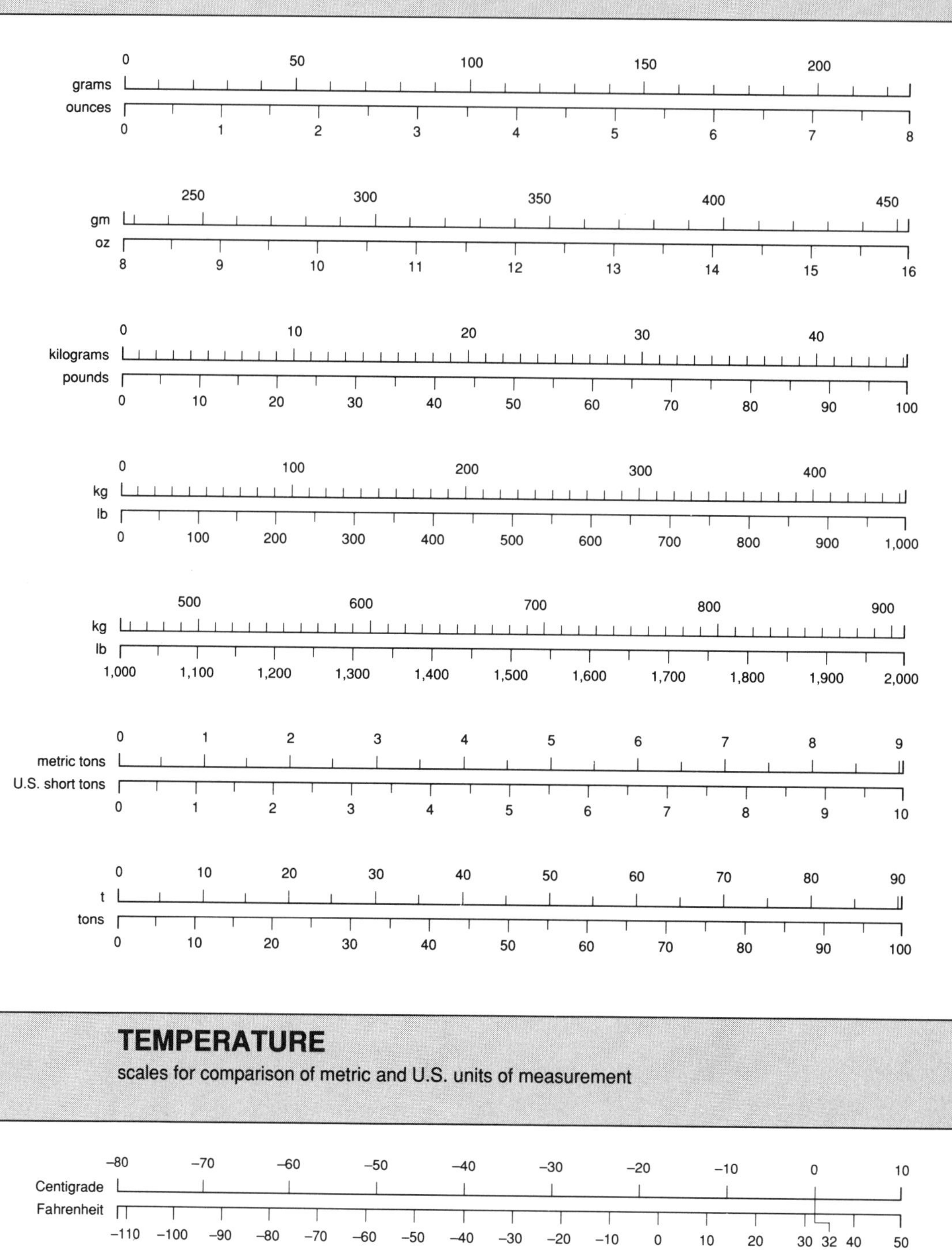

TEMPERATURE

scales for comparison of metric and U.S. units of measurement

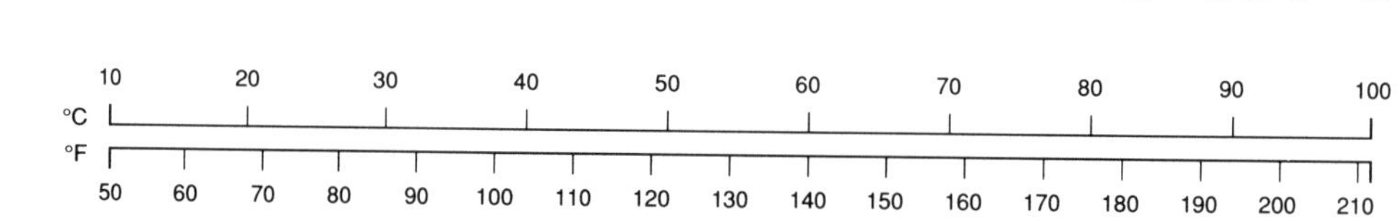

AREA

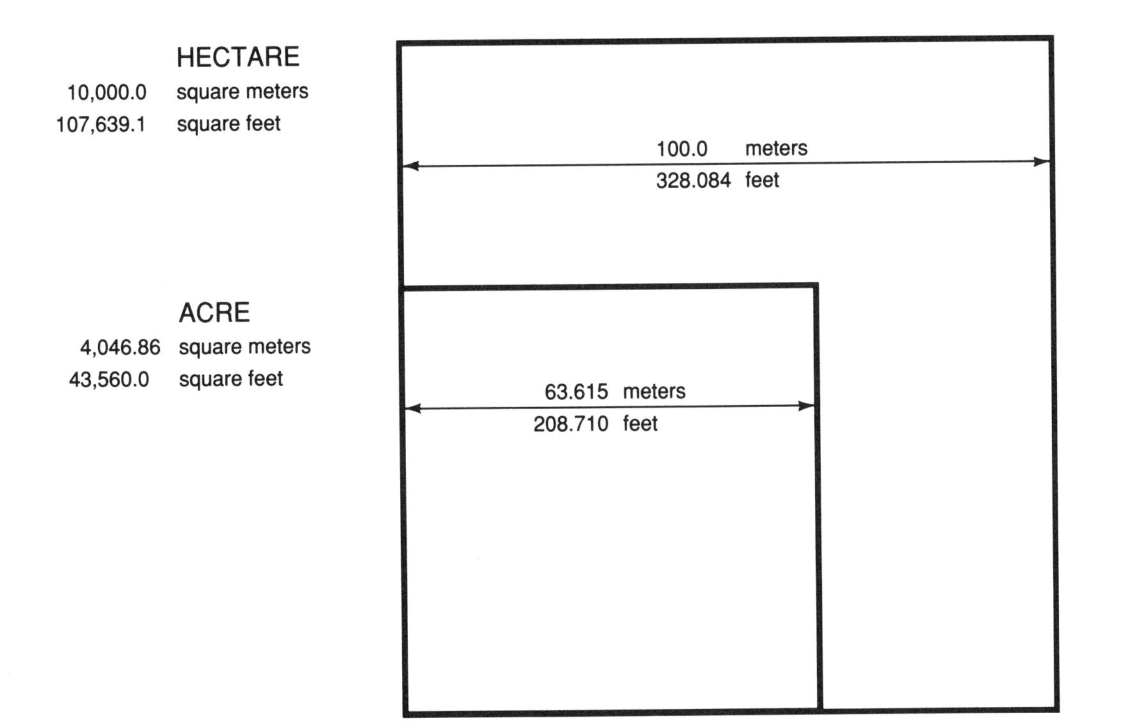
HECTARE
10,000.0 square meters
107,639.1 square feet
100.0 meters
328.084 feet
ACRE
4,046.86 square meters
43,560.0 square feet
63.615 meters
208.710 feet

Literature Cited

A

Ables, E. D. 1975. Ecology of the red fox in North America. *In* Fox (1975), 216–36.

Adams, D. B. 1979. The cheetah: native American. Science 205:1155–58.

Ahnlund, H. 1980. Sexual maturity and breeding season of the badger, *Meles meles,* in Sweden. J. Zool. 190:77–95.

Alberico, M. S. 1994. New locality record for the Columbian weasel *(Mustela felipei).* Small Carnivore Conserv. 10:16–17.

Albignac, R. 1969. Notes ethologiques sur quelques carnivores Malgaches: le *Galidia elegans* I. Geoffroy. Terre Vie 23:202–15.

———. 1970*a*. Notes ethologiques sur quelques carnivores Malgaches: le *Cryptoprocta ferox* (Bennett). Terre Vie 24:395–402.

———. 1970*b*. Notes ethologiques sur quelques carnivores Malgaches: le *Fossa fossa* (Schreber). Terre Vie 24:383–94.

———. 1971. Notes ethologiques sur quelques carnivores Malgaches: le *Mungotictis lineata* Pocock. Terre Vie 25:328–43.

———. 1972. The Carnivora of Madagascar. *In* Battistini and Richard-Vindard (1972), 667–82.

———. 1974. Observations éco-ethologiques sur le genre *Eupleres,* viverridide de Madagascar. Terre Vie 28:321–51.

———. 1975. Breeding the fossa *Cryptoprocta ferox* at Montpelier Zoo. Internatl. Zoo Yearbook 15:147–50.

———. 1976. L'écologie de *Mungotictis decemlineata* dans les forêts décidues de l'ouest de Madagascar. Terre Vie 30:347–76.

Allardyce, D. A. 1995. Twelve-month finding for a petition to list the swift fox as endangered. Federal Register 60:31663–66.

Allen, G. M. 1942. Extinct and vanishing mammals of the Western Hemisphere with the marine species of all the oceans. Spec. Publ. Amer. Comm. Internatl. Wildl. Protection, no. 11, xv + 620 pp.

Amstrup, S. C., and J. Beecham. 1976. Activity patterns of radio-collared black bears in Idaho. J. Wildl. Mgmt. 40:340–48.

Amstrup, S. C., G. W. Garner, and G. M. Durner. 1995. Polar bears in Alaska. *In* LaRoe et al. (1995), 351–53.

Amstrup, S. C., I. Stirling, and J. W. Lentfer. 1986. Past and present status of polar bears in Alaska. Wildl. Soc. Bull. 14:241–54.

Andelt, W. F. 1985. Behavioral ecology of coyotes in south Texas. Wildl. Monogr., no. 94, 45 pp.

Anderson, E. 1970. Quaternary evolution of the genus *Martes* (Carnivora, Mustelidae). Acta Zool. Fennica 130:1–132.

———. 1973. Ferret from the Pleistocene of central Alaska. J. Mamm. 54:778–79.

———. 1989. The phylogeny of mustelids and the systematics of ferrets. *In* Seal, U. S., E. T. Thorne, M. A. Bogan, and S. H. Anderson, eds., Conservation biology and the black-footed ferret, Yale Univ. Press, New Haven, 10–20.

———. 1994. Evolution, prehistoric distribution, and systematics of *Martes. In* Buskirk et al. (1994), 13–25.

Andrews, R. D., and E. K. Boggess. 1978. Ecology of coyotes in Iowa. *In* Bekoff (1978), 249–65.

Ansell, W. F. H. 1978. The mammals of Zambia. Natl. Parks and Wildl. Serv., Chilanga, Zambia, ii + 126 pp.

Arden-Clarke, C. H. G. 1986. Population density, home range size, and spatial organization of the Cape clawless otter, *Aonyx capensis,* in a marine habitat. J. Zool. 209: 201–11.

Armstrong, D. M. 1972. Distribution of mammals in Colorado. Monogr. Mus. Nat. Hist. Univ. Kansas, no. 3, x + 415 pp.

Armstrong, D. M., J. K. Jones, Jr., and E. C. Birney. 1972. Mammals from the Mexican state of Sinaloa. III. Carnivora and Artiodactyla. J. Mamm. 53:48–61.

Armstrong, J. J. 1979. The California sea otter: emerging conflicts in resource management. San Diego Law Rev. 16:249–85.

Artois, M., and M.-J. Duchêne. 1982. Première identification du chien viverrin (*Nyctereutes procyonoides* Gray, 1834) en France. Mammalia 46:265–67.

Asa, C. S., and M. P. Wallace. 1990. Diet and activity pattern of the Sechuran desert fox *(Dusicyon sechurae).* J. Mamm. 71:69–72.

Asdell, S. A. 1964. Patterns of mammalian reproduction. Cornell Univ. Press, Ithaca, viii + 670 pp.

Ashraf, N. V. K., A. Kumar, and A. J. T. Johnsingh. 1993. Two endemic viverrids of the Western Ghats, India. Oryx 27:109–14.

Aune, K. E. 1991. Increasing mountain lion populations and human-mountain lion interactions in Montana. *In* Braun (1991), 86–94.

B

Bailey, E. 1993. Introduction of foxes to Alaskan islands—history, effects on avifauna, and eradication. U.S. Fish and Wildl. Serv. Resource Publ., no. 193, iv + 53 pp.

Bailey, T. N. 1974. Social organization in a bobcat population. J. Wildl. Mgmt. 38:435–46.

———. 1993. The African leopard: ecology and behavior of a solitary felid. Columbia Univ. Press, New York, xviii + 429 pp.

Bailey, V. 1930. Animal life of Yellowstone National Park. Charles C. Thomas, Springfield, Illinois, 241 pp.

Baker, C. M. 1992. *Atilax paludinosus.* Mammalian Species, no. 408, 6 pp.

Baker, C. M., and J. Meester. 1986. Postnatal physical development of the water mongoose *(Atilax paludinosus).* Z. Saugetierk. 51:236–43.

Bakeyev, N. N., and A. A. Sinitsyn. 1994. Status and conservation of sables in the Commonwealth of Independent States. *In* Buskirk et al. (1994), 246–54.

Balharry, D. 1993. Social organization in martens: an inflexible system? Symp. Zool. Soc. London 65:321–45.

Ballard, W. B., R. Farnell, and R. O. Stephenson. 1983. Long distance movement by gray wolves, *Canis lupus.* Can. Field-Nat. 97:333.

Banci, V., and A. Harestad. 1988. Reproduction and natality of wolverine *(Gulo gulo)* in Yukon. Ann. Zool. Fennici 25:265–70.

———. 1990. Home range and habitat use of wolverines *Gulo gulo* in Yukon, Canada. Holarctic Ecol. 13:195–200.

Banfield, A. W. F. 1974. The mammals of Canada. Univ. Toronto Press, xxv + 438 pp.

Bangs, E. E., and S. H. Fritts. 1996. Reintroducing the gray wolf to central Idaho and Yellowstone National Park. Wildl. Soc. Bull. 24:402–13.

Banholzer, U. 1976. Water balance, metabolism, and heart rate in the fennec. Naturwissenschaften 63:202.

Banks, E. 1978. Mammals from Borneo. Brunei Mus. J. 4:165–227.

Baranauskas, K., E. Mickevicius, S. M. Macdonald, and C. F. Mason. 1994. Otter distribution in Lithuania. Oryx 28:128–30.

Barone, M. A., M. E. Roelke, J. Howard, J. L. Brown, A. E. Anderson, and D. E. Wildt. 1994. Reproductive characteristics of male Florida panthers: comparative studies from Florida, Texas, Colorado. Latin America, and North American zoos. J. Mamm. 75:150–62.

Barrat, J., and M. F. A. Aubert. 1993. Current status of fox rabies in Europe. Onderstepoort J. Veterinary Res. 60:357–64.

Batten, P., and A. Batten. 1966. Notes on breeding the small-toothed palm civet *Arctogalidia trivirgata* at Santa Cruz Zoo. Internatl. Zoo Yearbook 6:172–73.

Bayha, K., and J. Kormendy, technical coordinators. 1990. Sea Otter Symposium: Proceedings of a symposium to evaluate the response effort on behalf of sea otters after the T/V *Exxon Valdez* oil spill into Prince William Sound, Anchorage, Alaska, 17–19 April 1990. U.S. Fish and Wildl. Serv. Biol. Rept., no. 90(12), x + 485 pp.

Beattie, M. H. 1995. Endangered and threatened wildlife and plants; 90-day finding for a petition to list as endangered or threatened the contiguous United States population of the North American wolverine. Federal Register 60:19567–68.

Beck, A. M. 1973. The ecology of stray dogs: a study of free-ranging urban animals. York Press, Baltimore, xiv + 98 pp.

———. 1975. The ecology of "feral" and free-roving dogs in Baltimore. *In* Fox (1975), 380–90.

Beebe, B. F. 1978. Two new Pleistocene mammal species from Beringia. Amer. Quaternary Assoc. Abstr., 5th Bien. Mtg., 159.

Beier, P. 1991. Cougar attacks on humans in the United States and Canada. Wildl. Soc. Bull. 19:403–12.

Bekoff, M. 1975. Social behavior and ecology of the African Canidae: a review. *In* Fox (1975), 120–42.

Bekoff, M., and M. C. Wells. 1980. The social ecology of coyotes. Sci. Amer. 242(4):130–48.

———. 1982. Behavioral ecology of coyotes: social organization, rearing patterns, space use, and resource defense. Z. Tierpsychol. 60:281–305.

Belden, R. C. 1986. Florida panther recovery plan implementation—a 1983 progress report. *In* Miller and Everett (1986), 159–72.

Belden, R. C., and D. J. Forrester. 1980. A specimen of *Felis concolor coryi* from Florida. J. Mamm. 61:160–61.

Beltrán, J. F., and M. Delibes. 1993. Physical characteristics of Iberian lynxes *(Lynx pardinus)* from Doñana, southwestern Spain. J. Mamm. 74:852–62.

Ben-David, M., S. Hellwing, and H. Mendelssohn. 1988*a*. Home-range and spacing pattern of the marbled polecat *(Vormela peregusna syriaca)* in Israel. Deutsche Gesselschaft Saugetierk. 62:7.

———. 1988*b*. Reproduction and growth of the marbled polecat *(Vormela peregusna syriaca)* in Israel. J. Reprod. Fert. Abstract Ser., no. 1, 20.

Bender, M. 1988. Regional news. Endangered Species Tech. Bull. 13(6–7):2–8.

———. 1994. United States imposes limited trade sanctions on Taiwan for continued trade in endangered species. Endangered Species Tech. Bull. 19(3):1, 10.

Ben-Yaacov, R., and Y. Yom-Tov. 1983. On the biology of the Egyptian mongoose, *Herpestes ichneumon,* in Israel. Z. Saugetierk. 48:34–45.

Berg, W. E., and R. A. Chesness. 1978. Ecology of coyotes in northern Minnesota. *In* Bekoff (1978), 229–47.

Berman, M., and I. Dunbar. 1983. The social behaviour of free-ranging suburban dogs. Appl. Anim. Ethol. 10:5–17.

Berta, A. 1982. *Cerdocyon thous.* Mammalian Species, no. 186, 4 pp.

———. 1984. The Pleistocene bush dog *Speothos pacivorus* (Canidae) from the Lagoa Santa Caves, Brazil. J. Mamm. 65:549–59.

———. 1986. *Atelocynus microtis.* Mammalian Species, no. 256, 3 pp.

———. 1987. Origin, diversity, and zoogeography of the South American Canidae. Fieldiana Zool., n.s., 39:455–71.

Bertram, B. 1975. Social factors influencing reproduction in wild lions. J. Zool. 177:463–82.

———. 1991. Kin selection in lions. *In* Seidensticker and Lumpkin (1991), 90–93.

Bevanger, K. 1992. Report on the Norwegian wolverine (*Gulo gulo* L.). Small Carnivore Conserv. 6:8–10.

Bibikov, D. I. 1980. Wolves in the USSR. Nat. Hist. 89(6):58–63.

Bibikov, D. I., A. N. Filimonov, and A. N. Kudaktin. 1983. Territoriality and migration of the wolf in the USSR. Acta Zool. Fennica 174:267–68.

Bibikov, D. I., N. G. Ovsyannikov, and A. N. Filimonov. 1983. The status and management of the wolf population in the USSR. Acta Zool. Fennica 174:269–71.

Birney, E. C. 1974. Twentieth century records of wolverine in Minnesota. Loon 46:78–81.

Bjorge, R. R., J. R. Gunson, and W. M. Samuel. 1981. Population characteristics and movements of striped skunks *(Mephitis mephitis)* in central Alberta. Can. Field-Nat. 95:149–55.

Blacher, C. 1994. Strategic reproduction of *Lutra longicaudis.* IUCN (World Conservation Union) Otter Specialist Group Bull. 9:6.

Blanchard, B. M., and R. R. Knight. 1991. Movements of Yellowstone grizzly bears. Biol. Conserv. 58:41–67.

Blanchard, H. M. 1974. The marten makes a comeback. Maine Fish and Game 16(4):21.

Blanco, J. C., S. Reig, and L. de la Cuesta. 1992. Distribution, status, and conservation problems of the wolf *Canis lupus* in Spain. Biol. Conserv. 60:73–80.

Blandford, P. R. S. 1987. Biology of the polecat *Mustela putorius:* a literature review. Mamm. Rev. 17:155–98.

Bloxam, Q. 1977. Breeding the spectacled bear *Tremarctos ornatus* at Jersey Zoo. Internatl. Zoo Yearbook 17:158–61.

Blus, L. J., L. Fitzner, and R. E. Fitzner. 1993. Wolverine specimen from south-central Washington state. Northwestern Nat. 74:22.

Boggess, E. K., F. R. Henderson, and J. R. Choate. 1980. A black-footed ferret from Kansas. J. Mamm. 61:571.

Boitani, L. 1992. Wolf research and conservation in Italy. Biol. Conserv. 61:125–32.

Boppel, P. J., and C. A. Long. 1994. Status of the spotted skunk *(Spilogale putorius)* in its northeastern range, north-central United States. Small Carnivore Conserv. 11:11–12.

Bothma, J. D. P. 1966. Food of the silver fox *Vulpes chama.* Zool. Afr. 2:205–10.

———. 1971*a.* Control and ecology of the black-backed jackal *Canis mesomelas* in the Transvaal. Zool. Afr. 6:187–93.

———. 1971*b.* Food of *Canis mesomelas* in South Africa. Zool. Afr. 6:195–203.

Bothma, J. D. P., and J. A. J. Nel. 1980. Winter food and foraging behaviour of the aardwolf *Proteles cristatus* in the Namib-Naukluft Park. Madoqua 12:141–49.

Bourne, J. 1995*a.* Hyde County's wolf war. Defenders 70(2):10–17.

———. 1995*b.* Loggers versus the lynx. Defenders 70(3):42–51.

Bowdoin, J., R. J. Wiese, K. Willis, and M. Hutchins, eds. 1994. AZA Annual Report on Conservation and Science, 1993–94. American Association of Zoological Parks and Aquariums, Bethesda, Maryland, 375 pp.

Bowen, W. D. 1982. Home range and spatial organization of coyotes in Jasper National Park, Alberta. J. Wildl. Mgmt. 46:201–16.

Boyd, D. K., and M. D. Jimenez. 1994. Successful rearing of young by wild wolves without mates. J. Mamm. 75:14–17.

Brady, C. A. 1978. Reproduction, growth, and parental care in crab-eating foxes *Cerdocyon thous.* Internatl. Zoo Yearbook 18:130–34.

———. 1979. Observations on the behavior and ecology of the crab-eating fox *(Cerdocyon thous). In* Eisenberg (1979), 161–71.

———. 1981. The vocal repertoires of the bush dog *(Speothos venaticus),* crab-eating fox *(Cerdocyon thous),* and maned wolf *(Chrysocyon brachyurus).* Anim. Behav. 29:649–69.

Brady, C. A., and M. K. Ditton. 1979. Management and breeding of maned wolves *Chrysocyon brachyurus* at the National Zoological Park, Washington. Internatl. Zoo Yearbook 19:171–76.

Brady, J. R., and H. W. Campbell. 1983. Distribution of coyotes in Florida. Florida Field Nat. 11:40–41.

Brady, J. R., and D. S. Maehr. 1985. Distribution of black bears in Florida. Florida Field Nat. 13:1–7.

Brand, C. J., and L. B. Keith. 1979. Lynx demography during a snowshoe hare decline in Alberta. J. Wildl. Mgmt. 43:827–49.

Brand, C. J., L. B. Keith, and C. A. Fischer. 1976. Lynx responses to changing snowshoe hare densities in central Alberta. J. Wildl. Mgmt. 40:416–28.

Brand, D. J. 1983. A "king cheetah" born at the cheetah breeding and research centre of the National Zoological Gardens of South Africa, Pretoria. Zool. Garten 53:366–68.

Braun, A.-J. 1990. The European mink in France: past and present. Mustelid & Viverrid Conserv. 3:5–8.

Braun, C. E., ed. 1991. Mountain lion–human interaction: symposium and workshop. Colorado Division of Wildlife, vi + 114 pp.

Breitenmoser, U., and C. Breitenmoser-Würsten. 1994. SCALP—status and conservation of the Alpine lynx population. Council of Europe Environ. Encounters, no. 17, 111–12.

Brisbin, I. L., Jr., R. P. Coppinger, M. H. Feinstein, S. N. Austad, and J. J. Mayer. 1994. The New Guinea singing dog: taxonomy, captive studies, and conservation priorities. Sci. New Guinea 20:27–38.

Broad, S. 1987. International trade in skins of Latin American spotted cats. Traffic Bull. 9:56–63.

Broad, S., R. Luxmoore, and M. Jenkins. 1988. Significant trade in wildlife: a review of selected species in CITES appendix II. Volume 1: Mammals. IUCN (World Conservation Union), Gland, Switzerland, xix + 183 pp.

Brocke, R. H., and K. A. Gustafson. 1992. Lynx in New York state. Cat News 17:14.

Brooks, D. M. 1992. Notes on group size, density, and habitat association of the pampas fox *(Dusicyon gymnocercus)* in the Paraguayan Chaco. Mammalia 56:314–16.

Brooks, T. M., and G. C. L. Dutson. 1994. A sighting of a masked palm civet *(Paguma larvata).* Small Carnivore Conserv. 11:19.

Brown, D. E. 1983. On the status of the jaguar in the southwest. Southwestern Nat. 28: 459–60.

Brown, J. H., and R. C. Lasiewski. 1972. Metabolism of weasels: the cost of being long and thin. Ecology 53:939–43.

Brownell, R. L., Jr., and G. B. Rathbun. 1988. California sea otter translocation: a status report. Endangered Species Tech. Bull. 13(4): 1, 6.

Bryant, H. N., A. P. Russell, and W. D. Fitch. 1993. Phylogenetic relationships within the extant Mustelidae (Carnivora): appraisal of the cladistic status of the Simpsonian subfamilies. Zool. J. Linnean Soc. 108:301–34.

Brzezinski, M., W. Jedrzejewski, and B. Jedrzejewska. 1992. Winter home ranges and movements of polecats *Mustela putorius* in Bialowieza Primeval Forest, Poland. Acta Theriol. 37:181–91.

Buden, D. W. 1986. Distribution of mammals of the Bahamas. Florida Field Nat. 14:53–84.

Bueler, L. E. 1973. Wild dogs of the world. Stein & Day, New York, 274 pp.

Buk, K. 1994. Zambia wild dog project. Canid News 2:26–27.

Bulir, L. 1972. Breeding binturongs *Arctictis binturong* at Liberec Zoo. Internatl. Zoo Yearbook 12:11–12.

Burrows, R., H. Hofer, and M. L. East. 1994. Demography, extinction, and intervention in a small population: the case of the Serengeti wild dogs. Proc. Roy. Soc. London, ser. B, 256:281–92.

Burt, W. H., and R. P. Grossenheider. 1976. A field guide to the mammals. Houghton Mifflin, Boston, xxv + 289 pp.

Buskirk, S. W. 1994. Introduction to the genus *Martes. In* Buskirk et al. (1994), 1–10.

C

Cabrera, A. 1957, 1961. Cat logo de los mamiferos de América del Sur. Rev. Mus. Argentino Cien. Nat. "Bernardo Rivadavia," 4:1–732.

Calhoon, R. E., and C. Haspell. 1989. Urban cat populations compared by season, subhabitat, and supplemental feeding. J. Anim. Ecol. 58:321–28.

California Department of Fish and Game. 1978. At the crossroads. Sacramento, 103 pp.

Camenzind, F. J. 1978. Behavioral ecology of coyotes on the National Elk Refuge, Jackson, Wyoming. *In* Bekoff (1978), 267–94.

Campbell, R. R. 1988. Status of the sea mink, *Mustela macrodon,* in Canada. Can. Field-Nat. 102:304–6.

Canivenc, R., and M. Bonnin. 1979. Delayed implantation is under environmental control in the badger (*Meles meles* L.). Nature 278:849–50.

Canivenc, R., C. Mauget, M. Bonnin, and R. J. Aitken. 1981. Delayed implantation in the beech marten *(Martes foiana).* J. Zool. 193:325–32.

Carbyn, L. N., ed. 1983. Wolves in Canada and Alaska: their status, biology, and management. Can. Wildl. Serv. Rept. Ser., no. 45, 135 pp.

———. 1987. Gray wolf and red wolf. *In* Novak, Baker, et al. (1987), 358–77.

———. 1989. Coyote attacks on children in western North America. Wildl. Soc. Bull. 17:444–46.

Carbyn, L. N., H. J. Armbruster, and C. Mamo. 1994. The swift fox reintroduction program in Canada from 1983 to 1992. *In* Bowles, M., and C. Whelan, eds., Restoration of endangered species, Cambridge Univ. Press, 247–71.

Carbyn, L. N., and P. C. Paquet. 1986. Long distance movement of a coyote from Riding Mountain National Park. J. Wildl. Mgmt. 50:89.

Carey, J. 1987. The sea otter's uncertain future. Natl. Wildl. 25(1):16–20.

Carley, C. J. 1979. Status summary: the red wolf *(Canis rufus)*. U.S. Fish and Wildl. Serv., Albuquerque, Endangered Species Rept., no. 7, iv + 36 pp.

Carnio, J. 1989. Liberian mongoose specimen found. Amer. Assoc. Zool. Parks Aquar. Newsl. 30(12):22.

Carnio, J., and M. Taylor. 1988. In search of the Liberian mongoose *(Liberiictis kuhni)*. Unpubl. ms. distributed at Conference on Breeding Endangered Species in Captivity, Cincinnati, October 1988.

Caro, T. M. 1991. Cheetahs. *In* Seidensticker and Lumpkin (1991), 138–47.

———. 1994. Cheetahs of the Serengeti Plains: group living in an asocial species. Univ. Chicago Press, xxi + 478 pp.

Caro, T. M., and D. A. Collins. 1987. Male cheetah social organization and territoriality. Ethology 74:52–64.

Caro, T. M., and M. K. Laurenson. 1994. Ecological and genetic factors in conservation: a cautionary tale. Science 263:485–86.

Carpaneto, G. M. 1990. The Indian grey mongoose *(Herpestes edwardsi)* in the Circeo National Park: a case of incidental introduction. Mustelid & Viverrid Conserv. 2:10.

Carr, A., III. 1986. Introduction. *In* The black-footed ferret, Great Basin Nat. Mem. 8:1–7.

Cavallini, P. 1992. *Herpestes pulverulentus*. Mammalian Species, no. 409, 4 pp.

———. 1993. Spatial organization of the yellow mongoose *Cynictis penicillata* in a coastal area. Ethol. Ecol. Evol. 5:501–9.

Cavallini, P., and J. A. J. Nel. 1990. Ranging behaviour of the Cape grey mongoose *Galerella pulverulenta* in a coastal area. J. Zool. 222:353–62.

Ceballos, G., and D. Navarro L. 1991. Diversity and conservation of Mexican mammals. *In* Mares, M. A., and D. J. Schmidly, eds., Latin American mammalogy: history, biodiversity, and conservation, Univ. Oklahoma Press, Norman, 167–98.

Chadwick, D. H. 1995. Montana's wolf recovery. Defenders 70(1):18–27.

Chafee, J. H. 1995. A turning tide? Defenders 70(4):36–37.

Chakrabarti, K. 1984. The Sundarbans tiger. J. Bombay Nat. Hist. Soc. 81:459–60.

———. 1993. An ecological review of otter in the mangrove ecosystem of the Sundarbans (India). Tigerpaper 20(1):19–21.

Chanin, P. R. F., and D. J. Jefferies. 1978. The decline of the otter *Lutra lutra* L. in Britain: an analysis of hunting records and discussion of causes. Biol. J. Linnean Soc. 10:305–28.

Channing, A., and D. T. Rowe-Rowe. 1977. Vocalizations of South African mustelines. Z. Tierpsychol. 44:283–93.

Chapuis, J. L., P. Boussèes, and G. Barnaud. 1994. Alien mammals, impact and management in the French Subantarctic islands. Biol. Conserv. 67:97–104.

Charles-Dominique, P. 1978. Écologie et vie sociale de *Nandinia binotata* (Carnivores, Viverrides): comparaison avec les prosimiens sympatriques du Gabon. Terre Vie 32:477–528.

Chase, A. 1986. The grizzly and the juggernaut. Outside 11(1):28–34, 55–65.

Chasen, F. N. 1940. A handlist of Malaysian mammals. Bull. Raffles Mus., Singapore, no. 15, xx + 209 pp.

Chellam, R., and A. J. T. Johnsingh. 1993. Management of Asiatic lions in the Gir Forest, India. Symp. Zool. Soc. London 65:409– 24.

Chesemore, D. L. 1972. History and economic importance of the white fox, *Alopex*, fur trade in northern Alaska, 1798–1963. Can. Field-Nat. 86:259–67.

———. 1975. Ecology of the Arctic fox *(Alopex lagopus)* in North America—a review. *In* Fox (1975), 143–63.

Chestin, I. E., Y. P. Gubar, V. E. Sokolov, and V. S. Lobachev. 1992. The brown bear (*Ursus arctos* L.) in the USSR: numbers, hunting, and systematics. Ann. Zool. Fennici 29:57–68.

Childes, S. L. 1988. The past history, present status, and distribution of the hunting dog *Lycaon pictus* in Zimbabwe. Biol. Conserv. 44:301–16.

Cholley, B. 1982. Une martre (*Martes martes* L.) en Corse. Mammalia 46:267.

Chorn, J., and R. S. Hoffmann. 1978. *Ailuropoda melanoleuca*. Mammalian Species, no. 110, 6 pp.

Christopherson, D., and S. C. Torbit. 1993. Proposed establishment of a nonessential experimental population of black-footed ferrets in north-central Montana. Federal Register 58:19220–31.

Chubbs, T. E., and F. R. Phillips. 1993. An apparent longevity record for Canada lynx, *Lynx canadensis*. Can. Field-Nat. 107:367–68.

———. 1996. Apparent longevity records for red foxes, *Vulpes vulpes*, in Labrador. Can. Field-Nat. 110:348–49.

Cicnjak, L., and R. L. Ruff. 1990. Human-bear conflicts in Yugoslavia. Trans. 19th IUGB Congress, Trondheim, 573–80.

Clark, T. W. 1976. The black-footed ferret. Oryx 13:275–80.

———. 1978. Current status of the black-footed ferret in Wyoming. J. Wildl. Mgmt. 42:128–34.

———. 1987*a*. Black-footed ferret recovery: a progress report. Conserv. Biol. 1:8–11.

———. 1987*b*. Restoring balance between the endangered black-footed ferret *(Mustela nigripes)* and human use of the Great Plains and intermountain West. J. Washington Acad. Sci. 77:168–73.

Clark, T. W., E. Anderson, C. Douglas, and M. Strickland. 1987. *Martes americana*. Mammalian Species, no. 289, 8 pp.

Clark, T. W., J. Grensten, M. Gorges, R. Crete, and J. Gill. 1987. Analysis of black-footed ferret translocation sites in Montana. Prairie Nat. 19:43–56.

Clements, E. D., E. G. Neal, and D. W. Yalden. 1988. The national badger sett survey. Mamm. Rev. 18:1–9.

Clevenger, A. P. 1993*a*. The European pine marten *Martes martes* in the Balearic Islands, Spain. Mamm. Rev. 23:65–72.

———. 1993*b*. Status of martens and genets in the Balearic and Pityusic islands, Spain. Small Carnivore Conserv. 9:18–19.

Clutton-Brock, J., G. B. Corbet, and M. Hills. 1976. A review of the family Canidae, with a classification by numerical methods. Bull. British Mus. (Nat. Hist.), Zool. 29:117–99.

Clutton-Brock, J., A. C. Kitchener, and J. M. Lynch. 1994. Changes in the skull morphology of the arctic wolf, *Canis lupus arctos*, during the twentieth century. J. Zool. 233:19–36.

Cohen, J. A. 1977. A review of the biology of the dhole or Asiatic wild dog (*Cuon alpinus* Pallas). Anim. Reg. Studies 1:141–58.

———. 1978. *Cuon alpinus*. Mammalian Species, no. 100, 3 pp.

Cohen, J. A., M. W. Fox, A. J. T. Johnsingh, and B. D. Barnett. 1978. Food habits of the dhole in south India. J. Wildl. Mgmt. 42:933–36.

Cohn, J. P. 1986. Surprising cheetah genetics. Bioscience 36:358–62.

Cole, G. F. 1976. Management involving grizzly and black bears in Yellowstone National Park, 1970–75. U.S. Natl. Park Serv. Nat. Res. Rept., no. 9, 26 pp.

Collins, L. R. 1989. Black-footed ferrets born at the National Zoo's Conservation and Research Center. Amer. Assoc. Zool. Parks Aquar. Newsl. 30(12):23.

Collins, P. W. 1991*a*. Interaction between island foxes *(Urocyon littoralis)* and Indians on islands off the coast of southern California: I. Morphologic and archaeological evidence of human assisted dispersal. J. Ethnobiol. 11:51–81.

———. 1991*b*. Interaction between island foxes *(Urocyon littoralis)* and Native Americans on islands off the coast of southern California: II. Ethnographic, archaeological, and historical evidence. J. Ethnobiol. 11:205–29.

———. 1992. Taxonomic and biogeographic relationships of the island fox *(Urocyon littoralis)* and gray fox *(U. cinereoargenteus)* from western North America. *In* Hochberg, F. G., ed., Recent advances in California Islands research: Proceedings of the Third California Islands Symposium, Santa Barbara Museum of Natural History, 351–90.

Colyn, M., N. Dethier, P. Ngegueu, O. Perpete, and H. Van Rompaey. 1995. First observations of *Crossarchus platycephalus* (Goldman, 1984) in the Zaire/Congo system (Dja River, southeastern Cameroon). Small Carnivore Conserv. 12:10–11.

Colyn, M., and H. Van Rompaey. 1990. *Crossarchus ansorgei nigricolor,* a new subspecies of Ansorge's cusimanse (Carnivora, Viverridae) from south-central Zaire. Z. Saugetierk. 55:94–98.

Connolly, G. 1992. Sheep and goat losses to predators in the United States. Proc. East. Wildl. Damage Control Conf. 5:75–82.

Cook, R. S., ed. 1993. Ecological issues on reintroducing wolves into Yellowstone National Park. U.S. Natl. Park Serv. Sci. Monogr., NPS/NRYELL/NRSM-93/22, viii + 328 pp.

Corbet, G. B. 1978. The mammals of the Palaearctic Region: a taxonomic review. British Mus. (Nat. Hist.), London, 314 pp.

———. 1984. The mammals of the Palaearctic Region: a taxonomic review. Supplement. British Mus. (Nat. Hist.) Publ., no. 944, vi + 45 pp.

Corbet, G. B., and J. E. Hill. 1991. A world list of mammalian species. Natural History Museum Publ., London, and Oxford Univ. Press, viii + 243 pp.

———. 1992. The mammals of the Indomalayan region: a systematic review. Oxford Univ. Press, viii + 488 pp.

Corbett, L. K. 1988. Social dynamics of a captive dingo pack: population regulation by dominant female infanticide. Ethology 78:177–98.

———. 1995. The dingo in Australia and Asia. Comstock/Cornell Univ. Press, Ithaca, viii + 200 pp.

Corbett, L. K., and A. Newsome. 1975. Dingo society and its maintenance: a preliminary analysis. *In* Fox (1975), 369–79.

Cottrell, W. 1978. The fisher *(Martes pennanti)* in Maryland. J. Mamm. 59:886.

Coulter, M. W. 1974. Maine's "black cat." Maine Fish and Game 16(3):23–26.

Cowan, I. M. 1971. Summary of the symposium on the native cats of North America. *In* Jorgenson, S. E., and L. D. Mech, eds., Proceedings of a symposium on the native cats of North America, U.S. Bur. Sport Fish. and Wildl., 1–8.

———. 1972. The status and conservation of bears (Ursidae) of the world—1970. *In* Herrero (1972), 343–67.

Craighead, F. C., Jr. 1976. Grizzly bear ranges and movement as determined by radiotracking. *In* Pelton, Lentfer, and Folk (1976), 97–109.

Craighead, F. C., Jr., and J. J. Craighead. 1972. Grizzly bear prehibernation and denning activities as determined by radiotracking. Wildl. Monogr., no. 32, 35 pp.

Craighead, J. J., F. C. Craighead, Jr., and J. Sumner. 1976. Reproductive cycles and rates in the grizzly bear, *Ursus arctos horribilis,* of the Yellowstone ecosystem. *In* Pelton, Lentfer, and Folk (1976), 337–56.

Craighead, J. J., J. R. Varney, and F. C. Craighead, Jr. 1974. A population analysis of the Yellowstone grizzly bears. Montana Cooperative Wildl. Res. Unit, 20 pp.

Crawford-Cabral, J. 1981. A new classification of genets. Afr. Small Mamm. Newsl., no. 6, pp. 8–10.

———. 1989. The prior scientific name of the larger red mongoose (Carnivora: Viverridae: Herpestinae). Garcia de Orta, Ser. Zool., Lisbon, 14(2):1–2.

Crawford-Cabral, J., and A. P. Pacheco. 1989. A craniometrical study on some water rats of the genus *Dasymys* (Mammalia, Rodentia, Muridae). Sep. Garcia de Orta, Ser. Zool., Lisbon, 15(1):11–24.

Crespo, J. A. 1975. Ecology of the pampas gray fox and the large fox (culpeo). *In* Fox (1975), 179–91.

Cresswell, P., S. Harris, R. G. H. Bunce, and D. J. Jefferies. 1989. The badger *(Meles meles)* in Britain: present status and future population changes. Biol. J. Linnean Soc. 38:91–101.

Cronin, M. A. 1993. Mitochondrial DNA in wildlife taxonomy and conservation biology: cautionary notes. Wildl. Soc. Bull. 21:339–48.

Crooks, K. 1994. Demography and status of the island fox and the island spotted skunk on Santa Cruz Island, California. Southwestern Nat. 39:257–62.

Crowe, D. M. 1975. Aspects of ageing, growth, and reproduction of bobcats from Wyoming. J. Mamm. 56: 177–98.

Cumberland, R. E., and J. A. Dempsey. 1994. Recent confirmation of a cougar, *Felis concolor,* in New Brunswick. Can. Field-Nat. 108:224–26.

Currier, M. J. P. 1983. *Felis concolor.* Mammalian Species, no. 200, 7 pp.

D

Daggett, P. M., and D. R. Henning. 1974. The jaguar in North America. Amer. Antiquity 39:465–69.

Dalquest, W. W. 1953. Mammals of the Mexican state of San Luis Potosi. Louisiana State Univ. Studies, Biol. Ser., no. 1, 229 pp.

Dang, N. X., P. T. Anh, and D. H. Huynh. 1992. The biology and status of Owston's palm civet in Vietnam. Small Carnivore Conserv. 6:5–6.

Daniels, T. J. 1983*a*. The social organization of free-ranging urban dogs. I. Non-estrous social behavior. Appl. Anim. Ethol. 10:341–63.

———. 1983*b*. The social organization of free-ranging urban dogs. II. Estrous groups and the mating system. Appl. Anim. Ethol. 10:365–73.

Daniels, T. J., and M. Bekoff. 1989. Population and social biology of free-ranging dogs, *Canis familiaris.* J. Mamm. 70:754–62.

Daniloff, R. 1986. Russian sables: "soft gold." Internatl. Wildl. 16(4):20–23.

Da Silveira, E. K. P. 1968. Notes on the care and breeding of the maned wolf *Chrysocyon brachyurus* at Brasilia Zoo. Internatl. Zoo Yearbook 8:21–23.

Dathe, H. 1970. A second generation birth of captive sun bears *Helarctos malayanus* at East Berlin Zoo. Internatl. Zoo Yearbook 10:79.

Davidar, E. R. C. 1975. Ecology and behavior of the dhole or Indian wild dog *Cuon alpinus* (Pallas). *In* Fox (1975), 109–19.

Davis, D. D. 1962. Mammals of the lowland rain-forest of north Borneo. Bull. Singapore Natl. Mus., no. 31, 129 pp.

Davis, W. B. 1966. The mammals of Texas. Texas Parks and Wildl. Dept. Bull., no. 41, 267 pp.

Dawes, P. R., M. Elander, and M. Ericson. 1986. The wolf *(Canis lupus)* in Greenland: a historical review and present status. Arctic 39:119–32.

Dean, F. C. 1976. Aspects of grizzly bear population ecology in Mount McKinley National Park. *In* Pelton, Lentfer, and Folk (1976), 111–19.

Decker, D. M. 1991. Systematics of the coatis, genus *Nasua* (Mammalia: Procyonidae). Proc. Biol. Soc. Washington 104:370–86.

Decker, D. M., and W. C. Wozencraft. 1991. Phylogenetic analysis of Recent procyonid genera. J. Mamm. 72:42–55.

Deems, E. F., Jr., and D. Pursley, eds. 1978. North American furbearers. Internatl. Assoc. Fish and Wildl. Agencies, Univ. Maryland Press, College Park, x + 165 pp.

DeMaster, D. P., and I. Stirling. 1981. *Ursus maritimus.* Mammalian Species, no. 145, 7 pp.

Derocher, A. E., and I. Stirling. 1990*a*. Distribution of polar bears *(Ursus maritimus)* during the ice-free period in western Hudson Bay. Can. J. Zool. 68:1395–1403.

———. 1990*b*. Observations of aggregating behaviour in adult male polar bears *(Ursus maritimus)*. Can. J. Zool. 68:1390–94.

Dhungel, S. K., and W. D. Edge. 1985. Notes on the natural history of *Paradoxurus hermaphroditus*. Mammalia 49:302–3.

Dietz, J. M. 1984. Ecology and social organization of the maned wolf *(Chrysocyon brachyurus)*. Smithson. Contrib. Zool., no. 392, iv + 51 pp.

———. 1985. *Chrysocyon brachyurus*. Mammalian Species, no. 234, 4 pp.

Dinerstein, E., and J. N. Mehta. 1989. The clouded leopard in Nepal. Oryx 23:199–201.

Donnelly, B. G., and J. H. Grobler. 1976. Notes on food and anvil using behaviour by the Cape clawless otter, *Aonyx capensis*, in the Rhodes Matopos National Park, Rhodesia. Arnoldia 7(37):1–7.

Dorst, J., and P. Dandelot. 1969. A field guide to the larger mammals of Africa. Houghton Mifflin, Boston, 287 pp.

Douglas, C. W., and M. A. Strickland. 1987. Fisher. *In* Novak, Baker, et al. (1987), 510–29.

Dowling, T. E., B. D. DeMarais, W. L. Minckley, M. E. Douglas, and P. C. Marsh. 1992. Use of genetic characters in conservation biology. Conserv. Biol. 6:7–8.

Dowling, T. E., W. L. Minckley, M. E. Douglas, P. C. Marsh, and B. D. DeMarais. 1992. Response to Wayne, Nowak, and Phillips and Henry: use of molecular characters in conservation biology. Conserv. Biol. 6:600–603.

Dragoo, J. W., R. D. Bradley, R. L. Honeycutt, and J. W. Templeton. 1993. Phylogenetic relationships among the skunks: a molecular perspective. J. Mamm. Evol. 1:255–67.

Dragoo, J. W., J. R. Choate, T. L. Yates, and T. P. O'Farrell. 1990. Evolutionary and taxonomic relationships among North American arid-land foxes. J. Mamm. 71:318–32.

Dratch, P. A., J. S. Martenson, and S. J. O'Brien. 1991. The Mexican onza: what cat is that? Animal Genetics 22(suppl. 1):66.

Drew, L. 1989. Are we loving the panda to death? Natl. Wildl. 27(1):14–17.

Dubost, H., and J.-Y. Royère. 1993. Hybridization between ocelot *(Felis pardalis)* and puma *(Felis concolor)*. Zoo Biol. 12:277–83.

Duckworth, W. 1992. Mammals found in south-west Ecuador. *In* Best, B. J., ed., The threatened forests of south-west Ecuador, Biosphere Publ., Leeds, U.K., 121–35.

Duplaix, N. 1980. Observations on the ecology and behavior of the giant river otter *Pteronura brasiliensis* in Suriname. Rev. Ecol. (Terre Vie) 34:496–620.

Duplaix-Hall, N. 1975. River otters in captivity: a review. *In* Martin (1975*b*), 315–27.

Durran, J. C., P. E. Cattan, and J. L. Yáñez. 1985. The grey fox *Canis griseus* (Gray) in Chilean Patagonia (southern Chile). Biol. Conserv. 34:141–48.

E

Earlé, R. A. 1981. Aspects of the social and feeding behaviour of the yellow mongoose *Cynictis penicillata* (G. Cuvier). Mammalia 45:143–55.

East, M. L., and H. Hofer. 1991*a*. Loud calling in a female-dominated mammalian society: I. Structure and composition of whooping bouts of spotted hyaenas, *Crocuta crocuta*. Anim. Behav. 42:637–49.

———. 1991*b*. Loud calling in a female-dominated mammalian society: II. Behavioural contexts and functions of whooping of spotted hyaenas, *Crocuta crocuta*. Anim. Behav. 42:651–69.

East, M. L., H. Hofer, and W. Wickler. 1993. The erect "penis" is a flag of submission in a female-dominated society: greetings in Serengeti spotted hyenas. Behav. Ecol. Sociobiol. 33:355–70.

Eaton, R. L. 1974. The cheetah. Van Nostrand Reinhold, New York, xii + 178 pp.

———. 1976. The brown hyena: a review of biology, status, and conservation. Mammalia 40:377–99.

Eberhardt, L. E., and W. C. Hanson. 1978. Long-distance movements of arctic foxes tagged in northern Alaska. Can. Field-Nat. 92:386–89.

Egbert, A. L., and A. W. Stokes. 1976. The social behavior of brown bears on an Alaskan salmon stream. *In* Pelton, Lentfer, and Folk (1976), 41–56.

Egoscue, H. J. 1979. *Vulpes velox*. Mammalian Species, no. 122, 5 pp.

Egoscue, H. J., J. G. Bittmenn, and J. A. Petrovich. 1970. Some fecundity and longevity records for captive small mammals. J. Mamm. 51:622–23.

Eisenberg, J. F. 1989. Mammals of the neotropics: the northern neotropics. Univ. Chicago Press, x + 449 pp.

Elgmork, K. 1978. Human impact on a brown bear population (*Ursus arctos* L.). Biol. Conserv. 13:81–103.

Ellerman, J. R., and T. C. S. Morrison-Scott. 1966. Checklist of Palaearctic and Indian mammals. British Mus. (Nat. Hist.), London, 810 pp.

Emmons, L. H. 1988. A field study of ocelots *(Felis paradalis)* in Peru. Rev. Ecol. (Terre Vie) 43:133–57.

______. 1991. Jaguars. *In* Seidensticker and Lumpkin (1991), 116–23.

Erdbrink, D. P. 1953. A review of fossil and Recent bears of the Old World with remarks on their phylogeny based upon their dentition. Deventer, Netherlands, 597 pp.

Erlinge, S. 1967. Home range of the otter *Lutra lutra* L. in southern Sweden. Oikos 18:186–209.

———. 1968. Territoriality of the otter *Lutra lutra* L. Oikos 19:81–98.

———. 1977. Spacing strategy in stoat *Mustela erminea*. Oikos 28:32–42.

Estes, J. A. 1980. *Enhydra lutris*. Mammalian Species, no. 133, 8 pp.

———. 1986. Marine otters and their environment. Ambio 15:181–83.

———. 1990. Growth and equilibrium in sea otter populations. J. Anim. Ecol. 59:385–401.

Estes, J. A., and G. R. VanBlaricom. 1985. Sea otters and shellfisheries. *In* Beddington, Beverton, and Levigne (1985), 187–235.

Evans, G. D., and E. W. Pearson. 1980. Federal coyote control methods used in the western United States, 1971–77. Wildl. Soc. Bull. 8:34–39.

Ewer, R. F. 1973. The carnivores. Cornell Univ. Press, Ithaca, xv + 494 pp.

Ewer, R. F., and C. Wemmer. 1974. The behaviour in captivity of the African civet, *Civettictis civetta* (Schreber). Z. Tierpsychol. 34:359–94.

F

Fanshawe, J. H., L. H. Frame, and J. R. Ginsberg. 1991. The wild dog—Africa's vanishing carnivore. Oryx 25:137–46.

Faust, R., and C. Scherpner. 1967. A note on the breeding of the maned wolf. Internatl. Zoo Yearbook 7:119.

Fellner, K. 1965. Natural rearing of clouded leopards *Neofelis nebulosa* at Frankfurt Zoo. Internatl. Zoo Yearbook 5:111–13.

Ferguson, J. W. H., J. A. J. Nel, and M. J. de Wet. 1983. Social organization and movement patterns of black-backed jackals *Canis mesomelas* in South Africa. J. Zool. 199:487–502.

Ferguson, W. W. 1981. The systematic position of *Canis aureus lupaster* (Carnivora: Canidae) and the occurrence of *Canis lupus* in North Africa, Egypt, and Sinai. Mammalia 45:459–65.

Ferrari, S. F., and M. A. Lopes. 1992. A note on the behaviour of the weasel *Mustela* cf. *africana* (Carnivora, Mustelidae), from Amazonas, Brazil. Mammalia 56:482–83.

Field, R. J., and G. Feltner. 1974. Wolverine. Colorado Outdoors 23(2):1–6.

Finerty, J. P. 1979. Cycles in Canadian lynx. Amer. Nat. 114:453–55.

Fisher, J. 1978. Tiger! Tiger! Internatl. Wildl. 8(3):4–12.

Flannery, T. F. 1990. Mammals of New Guinea. Robert Brown & Assoc. Carina, Queensland, iii + 439 pp.

Floyd, B. L., and M. R. Stromberg. 1981. New records of the swift fox *(Vulpes velox)* in Wyoming. J. Mamm. 62:650–51.

Folk, G. E., Jr., A. Larson, and M. A. Folk. 1976. Physiology of hibernating bears. *In* Pelton, Lentfer, and Folk (1976), 373–80.

Fontaine, P. A. 1965. Breeding clouded leopards *Neofelis nebulosa* at Dallas Zoo. Internatl. Zoo Yearbook 5:113–14.

Foose, T. 1987. Species survival plans and overall management strategies. *In* Tilson and Seal (1987), 304–16.

Ford, L. S., and R. S. Hoffmann. 1988. *Potos flavus.* Mammalian Species, no. 321, 9 pp.

Foresman, K. R., and R. A. Mead. 1973. Duration of post-implantation in a western subspecies of the spotted skunk *(Spilogale putorius).* J. Mamm. 54:521–23.

Forrest, S. C., D. E. Biggins, L. Richardson, T. W. Clark, T. M. Campbell, III, K. A. Fagerstone, and E. T. Thorne. 1988. Population attributes for the black-footed ferret *(Mustela nigripes)* at Meeteetse, Wyoming, 1981–1985. J. Mamm. 69:261–73.

Foster-Turley, P., S. Macdonald, and C. Mason, eds. 1990. Otters: an action plan for their conservation. IUCN (World Conservation Union), Gland, Switzerland, iv + 126 pp.

Fox, J. L., S. P. Sinha, R. S. Chundawat, and P. K. Das. 1991. Status of the snow leopard *Panthera uncia* in northwest India. Biol. Conserv. 55:283–98.

Fox, M. W., ed. 1975. The wild canids. Van Nostrand Reinhold, New York, xvi + 508 pp.

———. 1978. The dog: its domestication and behavior. Garland, New York, vii + 296 pp.

———. 1984. The whistling hunters: field studies of the Asiatic wild dog *(Cuon alpinus).* State Univ. New York Press, Albany, viii + 150 pp.

Fox, M. W., A. M. Beck, and E. Blackman. 1975. Behavior and ecology of a small group of urban dogs *(Canis familiaris).* Appl. Anim. Ethol. 1:119–37.

Frafjord, K., D. Becker, and A. Angerbjörn. 1989. Interactions between arctic and red foxes in Scandinavia—predation and aggression. Arctic 42:354–56.

Frame, G. 1992. First record of king cheetah outside southern Africa. Cat News 16:2–3.

Frame, L. H., and G. W. Frame. 1976. Female African wild dogs emigrate. Nature 263:227–29.

Frame, L. H., J. R. Malcolm, G. W. Frame, and H. Van Lawick. 1979. Social organization of African wild dogs *(Lycaon pictus)* on the Serengeti Plains, Tanzania (1967–1978). Z. Tierpsychol. 50:225–49.

Frampton, G. T., Jr. 1995*a*. Importation of polar bear trophies from Canada: proposed rule on legal and scientific findings to implement section 104(c)(5)(A) of the 1994 amendments to the Marine Mammal Protection Act. Federal Register 60:36382–400.

———. 1995*b*. Proposed policy on giant panda permits. Federal Register, 60:16487–98.

Frank, L. G. 1986*a*. Social organization of the spotted hyaena *(Crocuta crocuta).* I. Demography. Anim. Behav. 34:1500–1509.

———. 1986*b*. Social organization of the spotted hyaena *Crocuta crocuta.* II. Dominance and reproduction. Anim. Behav. 34: 1510–27.

Frank, L. G., S. E. Glickman, and P. Light. 1991. Fatal sibling aggression, precocial development, and androgens in neonatal spotted hyenas. Science 252:702–4.

Fritts, S. H. 1982. Wolf depredations on livestock in Minnesota. U.S. Fish and Wildl. Serv. Resource Publ., no. 145, 11 pp.

———. 1983. Record dispersal by a wolf from Minnesota. J. Mamm. 64:166–67.

Fritts, S. H., and L. D. Mech. 1981. Dynamics, movements, and feeding ecology of a newly protected wolf population in northwestern Minnesota. Wildl. Monogr., no. 80, 79 pp.

Fritts, S. H., W. J. Paul, L. D. Mech, and D. P. Scott. 1992. Trends and management of wolf-livestock conflicts in Minnesota. U.S. Fish and Wildl. Serv. Resource Publ., no. 181, iv + 27 pp.

Fritts, S. H., and J. A. Sealander. 1978. Reproductive biology and population characteristics of bobcats *(Lynx rufus)* in Arkansas. J. Mamm. 59:347–53.

Fritzell, E. K. 1978. Aspects of raccoon *(Procyon lotor)* social organization. Can. J. Zool. 56:260–71.

Fritzell, E. K., and K. J. Haroldson. 1982. *Urocyon cinereoargenteus.* Mammalian Species, no. 189, 8 pp.

Fuller, T. K. 1978. Variable home-range sizes of female gray foxes. J. Mamm. 59:446–49.

Fuller, T. K., W. E. Berg, G. L. Radde, M. S. Lenarz, and G. B. Joselyn. 1992. A history and current estimate of wolf distribution and numbers in Minnesota. Wildl. Soc. Bull. 20:42–55.

Fuller, T. K., A. R. Biknevicius, and P. W. Kat. 1990. Movements and behavior of large spotted genets (*Genetta maculata* Gray 1830) near Elmenteita, Kenya (Mammalia Viverridae). Trop. Zool. 3:13–19.

Fuller, T. K., A. R. Biknevicius, P. W. Kat, B. Van Valkenburgh, and R. K. Wayne. 1989. The ecology of three sympatric jackal species in the Rift Valley of Kenya. Afr. J. Ecol. 27: 313–23.

Fuller, T. K., W. E. Johnson, W. L. Franklin, and K. A. Johnson. 1987. Notes on the Patagonian hog-nosed skunk *(Conepatus humboldti)* in southern Chile. J. Mamm. 68: 864–67.

Fuller, T. K., P. W. Kat, J. B. Bulger, A. H. Maddock, J. R. Ginsberg, J. W. McNutt, and M. G. L. Mills. 1992. Population dynamics of African wild dogs. *In* McCullough, D. R., and R. H. Barrett, eds., Wildlife 2001: populations, Elsevier, New York, 1125–39.

Fuller, T. K., and L. B. Keith. 1980. Wolf population dynamics and prey relationships in northwestern Alberta. J. Wildl. Mgmt. 44: 583–601.

G

Gallagher, M. 1992. The white-tailed mongoose in Oman, eastern Arabia. Small Carnivore Conserv. 7:8–9.

Gangloff, B. 1975. Beitrag zur Ethologie der Schleichkatzen (Banderlinsang, *Prionodon linsang* [Hardw.] und Banderpalmenroller, *Hemigalus derbyanus* [Gray]). Zool. Garten 45:329–76.

Garcelon, D. K., R. K. Wayne, and B. J. Gonzales. 1992. A serologic survey of the island fox *(Urocyon littoralis)* on the Channel Islands, California. J. Wildl. Diseases 28:223–29.

García-Perea, R. 1992. New data on the systematics of lynxes. Cat News 16:15–16.

———. 1994. Pampas cats: how many species? Cat News 20:21–24.

Gardner, A. L. 1976. The distributional status of some Peruvian mammals. Occas. Pap. Mus. Zool. Louisiana State Univ., no. 48, 18 pp.

Gardner, C. L., W. B. Ballard, and H. Jessup. 1986. Long distance movement by an adult wolverine. J. Mamm. 67:603.

Garrott, R. A., L. E. Eberhardt, and W. C. Hanson. 1983. Arctic fox den identification and characteristics in northern Alaska. Can. J. Zool. 61:423–26.

Garshelis, D. L. 1995. Formosan black bear workshop. Internatl. Bear News 4(1):7.

Garshelis, D. L., and J. A. Garshelis. 1984. Movements and management of sea otters in Alaska. J. Wildl. Mgmt. 48:665–78.

Garshelis, D. L., A. M. Johnson, and J. A. Garshelis. 1984. Social organization of sea otters in Prince William Sound, Alaska. Can. J. Zool. 62:2648–58.

Gasperetti, J., D. L. Harrison, and W. Büttiker. 1985. The Carnivora of Arabia. Fauna Saudi Arabia 7:397–461.

Gee, C. K., R. Magleby, W. R. Bailey, R. L.

Gum, and L. M. Arthur. 1977. Sheep and lamb losses to predators and other causes in the western United States. U.S. Dept. Agric., Agric. Econ. Rept., no. 369, 41 pp.

Geffen, E. 1994. *Vulpes cana.* Mammalian Species, no. 462, 4 pp.

Geffen, E., A. A. Degen, M. Kam, R. Hefner, and K. A. Nagy. 1992. Daily energy expenditure and water flux of free-living Blanford's foxes *(Vulpes cana),* a small desert carnivore. J. Animal Ecol. 61:611–17.

Geffen, E., R. Hefner, D. W. Macdonald, and M. Ucko. 1992*a.* Diet and foraging behavior of Blanford's foxes, *Vulpes cana,* in Israel. J. Mamm. 73:395–402.

———. 1992*b.* Morphological adaptations and seasonal weight changes in Blanford's fox, *Vulpes cana.* J. Arid Environ. 23:287–92.

———. 1993. Biotope and distribution of Blanford's fox. Oryx 27:104–8.

Geffen, E., and D. W. Macdonald. 1992. Small size and monogamy: spatial organization of Blanford's foxes, *Vulpes cana.* Anim. Behav. 44:1123–30.

———. 1993. Activity and movement patterns of Blanford's foxes. J. Mamm. 74: 455–63.

Geffen, E., A. Mercure, D. J. Girman, D. W. MacDonald, and R. K. Wayne. 1992. Phylogenetic relationships of the fox-like canids: mitochondrial DNA restriction fragment, site and cytochrome *b* sequence analyses. J. Zool. 228:27–39.

Geist, V. 1993. The cave bear. *In* Stirling (1993), 18.

Gensch, W. 1963. Successful rearing of the binturong *Arctictis binturong.* Internatl. Zoo Yearbook 4:79–80.

———. 1965. Birth and rearing of a spectacled bear *Tremarctos ornatus* at Dresden Zoo. Internatl. Zoo Yearbook 5:11.

George, S. B., and R. K. Wayne. 1991. Island foxes: a model for conservation genetics. Terra 30(1):18–23.

Gerhardt, G. 1995. *Canis lupus* comes home. Defenders 70(1):8–14.

Gese, E. M., and L. D. Mech. 1991. Dispersal of wolves *(Canis lupus)* in northeastern Minnesota, 1969–1989. Can. J. Zool. 69: 2946–55.

Ghose, R. K. 1965. A new species of mongoose (Mammalia: Carnivora: Viverridae) from West Bengal, India. Proc. Zool. Soc. Calcutta 18:173–78.

Ghose, R. K., and U. Chaturvedi. 1972. Extension of range of the mongoose *Herpestes palustris* Ghose (Mammalia: Carnivora: Viverridae), with a note on its endoparasitic nematode. J. Bombay Nat. Hist. Soc. 69:412–13.

Gibilisco, C. J. 1994. Distributional dynamics of modern *Martes* in North America. *In* Buskirk et al. (1994), 59–71.

Gier, H. T. 1975. Ecology and behavior of the coyote *(Canis latrans). In* Fox (1975), 247–62.

Gilbert, B. K., and L. D. Roy. 1977. Prevention of black bear damage to beeyards using aversive conditioning. *In* Phillips and Jonkel (1977), 93–102.

Ginsberg, J. R. 1993. Mapping wild dogs. Canid News 1:2–6.

Ginsberg, J. R., and D. W. Macdonald. 1990. Foxes, wolves, jackals, and dogs: an action plan for the conservation of canids. IUCN (World Conservation Union), Gland, Switzerland, iv + 116 pp.

Gipson, P. S. 1985. Ecology of wolverines in an arctic ecosystem. Natl. Geogr. Soc. Res. Rept. 20:253–63.

Gipson, P. S., and J. A. Sealander. 1977. Ecological relationships of white-tailed deer and dogs in Arkansas. *In* Phillips and Jonkel (1977), 3–16.

Gjertz, I., and E. Persen. 1987. Confrontations between humans and polar bears in Svalbard. Polar Res., n.s., 5:253–56.

Glatston, A. R. 1994. The red panda, olingos, coatis, raccoons, and their relatives: status survey and conservation action plan for procyonids and ailurids. IUCN (World Conservation Union), Gland, Switzerland, viii + 103 pp.

Glenn, L. P., J. W. Lentfer, J. B. Faro, and L. H. Miller. 1976. Reproductive biology of female brown bears *(Ursus arctos),* McNeil River, Alaska. *In* Pelton, Lentfer, and Folk (1976), 381–90.

Glickman, S. E., L. G. Frank, P. Licht, T. Yalcinkaya, P. K. Siiteri, and J. Davidson. 1992. Sexual differentiation of the female spotted hyena: one of nature's experiments. Ann. New York Acad. Sci. 662:135–59.

Goldman, C. A. 1984. Systematic revision of the African mongoose genus *Crossarchus* (Mammalia: Viverridae). Can. J. Zool. 62: 1618–30.

———. 1987. *Crossarchus obscurus.* Mammalian Species, no. 290, 5 pp.

Goldman, C. A., and M. E. Taylor. 1990. *Liberiictis kuhni.* Mammalian Species, no. 348, 3 pp.

Goldstein, I. 1991. Are spectacled bear's tree nests feeding platforms or resting places? Mammalia 55:433–34.

Gollan, K. 1984. The Australian dingo: in the shadow of man. *In* Archer and Clayton (1984), 921–27.

Gompper, M. E. 1995. *Nasua narica.* Mammalian Species, no. 487, 10 pp.

Gompper, M. E., and J. S. Krinsley. 1992. Variation in social behavior of adult male coatis *(Nasua narica)* in Panama. Biotrópica 24: 216–19.

Goodwin, G. G., and A. M. Greenhall. 1961. A review of the bats of Trinidad and Tobago. Bull. Amer. Mus. Nat. Hist. 122:187–302.

Gorman, M. L. 1976. Seasonal changes in the reproductive pattern of feral *Herpestes auropunctatus* (Carnivora: Viverridae) in the Fijian Islands. J. Zool. 178:237–46.

Gottelli, D., and C. Sillero-Zubiri. 1992. The Ethiopian wolf—an endangered endemic canid. Oryx 26:205–14.

———. 1994*a.* Highland gods, but for how long? Wildl. Conserv. 97(4):44–53.

———. 1994*b.* Hybridization: an emergency for the Ethiopian wolf. Canid News 2:33–35.

Gottelli, D., C. Sillero-Zubiri, G. D. Appelbaum, M. S. Roy, D. J. Girman, J. Marcia-Moreno, E. A. Ostrander, and R. K. Wayne. 1994. Molecular genetics of the most endangered canid: the Ethiopian wolf *Canis simensis.* Molecular Ecol. 3:301–12.

Graham, R. W., and M. A. Graham. 1994. Late Quaternary distribution of *Martes* in North America. *In* Buskirk et al. (1994), 26–58.

Grayson, D. K. 1984. Time of extinction and nature of adaptation of the noble marten, *Martes nobilis.* Carnegie Mus. Nat. Hist. Spec. Publ. 8:233–40.

Green, K. A. 1991. Summary: mountain lion–human interaction questionnaire, 1991. *In* Braun (1991), 4–9.

Griffiths, H. I., and D. H. Thomas. 1993. The status of the badger *Meles meles* (L., 1758) (Carnivora, Mustelidae) in Europe. Mamm. Rev. 23:17–58.

Grimwood, I. R. 1969. Notes on the distribution and status of some Peruvian mammals. Spec. Publ. Amer. Comm. Internatl. Wildl. Protection, no. 21, v + 86 pp.

———. 1976. The Palawan stink badger. Oryx 13:297.

Groves, C. P. 1976. The origin of the mammalian fauna of Sulawesi (Celebes). Z. Saugetierk. 41:201–16.

Grzimek, B., ed. 1975. Grzimek's animal life encyclopedia: mammals, I–IV. Van Nostrand Reinhold, New York, vols. 10–13.

———, ed. 1990. Grzimek's encyclopedia of mammals. McGraw-Hill, New York, 5 vols.

Guggisberg, C. A. W. 1975. Wild cats of the world. Taplinger, New York, 328 pp.

Guidali, F., T. Mingozzi, and G. Tosi. 1990. Historical and recent distributions of lynx (*Lynx lynx* L.) in northwestern Italy, during the 19th and 20th centuries. Mammalia 54:587–96.

Gum, R. L., L. M. Arthur, and R. S. Magleby. 1978. Coyote control: a simulation evaluation of alternative strategies. U.S. Dept. Agric., Agric. Econ. Rept., no. 408, 49 pp.

Gunderson, H. L. 1978. A mid-continent irruption of Canada lynx, 1962–1963. Prairie Nat. 10:71–80.

Gunson, J. R. 1983. Status and management of wolves in Alberta. *In* Carbyn (1983), 25–29.

Gunson, J. R., and R. R. Bjorge. 1979. Winter denning of the striped skunk in Alberta. Can. Field-Nat. 93:252–58.

Guo Yinfeng. 1995. Conservation and medicinal use of bears in China. *In* Rose and Gaski (1995), 120–30.

H

Hagenbeck, C., and K. Wunnemann. 1992. Breeding the giant otter *Pteronura brasiliensis* at Carl Hagenbeck's tierpark. Internatl. Zoo Yearbook 31:240–45.

Hall, E. R. 1946. Mammals of Nevada. Univ. California Press, Berkeley, xi + 710 pp.

———. 1951. American weasels. Univ. Kansas Publ. Mus. Nat. Hist. 4:1–466.

———. 1955. Handbook of mammals of Kansas. Univ. Kansas Mus. Nat. Hist. Misc. Publ., no. 7, 303 pp.

———. 1981. The mammals of North America. John Wiley & Sons, New York, 2 vols.

———. 1984. Geographic variation among brown and grizzly bears *(Ursus arctos)* in North America. Univ. Kansas Mus. Nat. Hist. Spec. Publ., no. 13, 16 pp.

Hall, E. R., and W. W. Dalquest. 1963. The mammals of Veracruz. Univ. Kansas Publ. Mus. Nat. Hist. 14:165–362.

Hall, H. T., and J. D. Newsom. 1978. Summer home ranges and movements of bobcats in bottomland hardwoods of southern Louisiana. Proc. Ann. Conf. Southeast. Assoc. Fish and Wildl. Agencies 30:427–36.

Haltenorth, T., and H. H. Roth. 1968. Short review of the biology and ecology of the red fox. Saugetierk. Mitt. 16:339–52.

Hamilton, P. H. 1986. Status of the cheetah in Kenya, with reference to sub-Saharan Africa. *In* Miller and Everett (1986), 65–76.

Hanby, J. P., and J. D. Bygott. 1987. Emigration of subadult lions. Anim. Behav. 35:161–69.

Hancox, M. 1992*a*. Badgers and bovine tuberculosis: an irreconcilable farming/conservation dilemma? Small Carnivore Conserv. 6:18–19.

———. 1992*b*. Some aspects of the distribution and breeding biology of honey badgers. Small Carnivore Conserv. 6:19.

Handley, C. O., Jr. 1976. Mammals of the Smithsonian Venezuelan Project. Brigham Young Univ. Sci. Bull., Biol. Ser., 20(5):1–89.

———. 1980. Mammals. *In* Linzey, D. W., ed., Endangered and threatened plants and animals of Virginia, Virginia Polytechnic Inst., Blacksburg, 483–621.

Harbo, S. J., Jr., and F. C. Dean. 1983. Historical and current perspectives on wolf management in Alaska. *In* Carbyn (1983), 51–64.

Harden, R. H. 1985. The ecology of the dingo in north-eastern New South Wales. I. Movements and home range. Austral. Wildl. Res. 12:25–37.

Harding, L. E. 1976. Den-site characteristics of arctic coastal grizzly bears (*Ursus arctos* L.) on Richards Island, Northwest Territories, Canada. Can. J. Zool. 54:1357–63.

Harington, C. R. 1968. Denning habits of the polar bear (*Ursus maritimus* Phipps). Can. Wildl. Serv. Rept. Ser., no. 5, 30 pp.

Harper, F. 1945. Extinct and vanishing mammals of the Old World. Spec. Publ. Amer. Comm. Internatl. Wildl. Protection, no. 12, xv + 850 pp.

Harrington, F. H., and L. D. Mech. 1979. Wolf howling and its role in territory maintenance. Behaviour 68:207–49.

Harris, S., and G. C. Smith. 1987. Demography of two urban fox *(Vulpes vulpes)* populations. J. Appl. Ecol. 24:75–86.

Harrison, D. L. 1968. The mammals of Arabia. Ernest Benn, London, 3 vols.

Harrison, D. L., and P. J. Bates. 1991. The mammals of Arabia. Harrison Zoological Museum, Sevenoaks, Kent, England, xvi + 354 pp.

Hart, J. A., and R. M. Timm. 1978. Observations on the aquatic genet in Zaire. Carnivore 1:130–31.

Haspel, C., and R. E. Calhoon. 1989. Home ranges of free-ranging cats *(Felis catus)* in Brooklyn, New York. Can. J. Zool. 67:178–81.

Hast, M. H. 1989. The larynx of roaring and non-roaring cats. J. Anat. 163:117–21.

Hatcher, R. T. 1982. Distribution and status of red foxes (Canidae) in Oklahoma. Southwestern Nat. 27:183–86.

Hayashi, T. 1984. The people attacked by the Japanese black bear—instances in Fukui Prefecture from 1970 to 1983. J. Mamm. Soc. Japan 10:55–62.

Hayssen, V., A. Van Tienhoven, and A. Van Tienhoven. 1993. Asdell's patterns of mammalian reproduction: a compendium of species-specific data. Comstock/Cornell Univ. Press, Ithaca, viii + 1023 pp.

Heald, F., and C. Shaw. 1991. Sabertooth cats. *In* Seidensticker and Lumpkin (1991), 26–27.

Heard, S., and H. Van Rompaey. 1990. Rediscovery of the crested genet. Mustelid & Viverrid Conserv. 3:1–4.

Heinrich, R. E., and K. D. Rose. 1995. Partial skeleton of the primitive carnivoran *Miacis petilus* from the early Eocene of Wyoming. J. Mamm. 76:148–62.

Helle, E., and K. Kauhala. 1991. Distribution history and present status of the raccoon dog in Finland. Holarctic Ecol. 14:278–86.

———. 1993. Age structure, mortality, and sex ratios of the raccoon dog in Finland. J. Mamm. 74:936–42.

———. 1995. Reproduction in the raccoon dog. J. Mamm. 76:1036–46.

Hellgren, E. C. 1993. Status, distribution, and summer food habits of black bears in Big Bend National Park. Southwestern Nat. 38:77–80.

Hemmer, H. 1972. *Uncia uncia*. Mammalian Species, no. 20, 5 pp.

———. 1974. Untersuchungen zur Stammesgeschichte der Pantherkatzen (Pantherinae). III. Zur Artgeschichte des Löwen *Panthera (Panthera) leo* (Linnaeus 1758). Veroff. Zool. Staatssaml. München 17:167–280.

———. 1976. Gestation period and postnatal development in felids. *In* Eaton, R. L., ed., Proc. 3rd Internatl. Symp. World's Cats, vol. 3, Carnivore Res. Inst., Univ. Washington, 143–65.

———. 1978. The evolutionary systematics of living Felidae: present status and current problems. Carnivore 1:71–79.

Henderson, F. R., P. F. Springer, and R. Adrian. 1969. The black-footed ferret in South Dakota. South Dakota Dept. Game, Fish, and Parks Tech. Bull., no. 4, 37 pp.

Hendey, Q. H. 1977. Fossil bear from South Africa. S. Afr. J. Sci. 73:112–16.

Hendrichs, H. 1975. The status of the tiger *Panthera tigris* (Linne, 1758) in the Sundarbans Mangrove Forest (Bay of Bengal). Saugetierk. Mitt. 23:161–99.

Henry, V. G. 1993. Proposed revision of the special rule for nonessential experimental populations of red wolves in North Carolina and Tennessee. Federal Register 58:62086–91.

———. 1995. Red wolves in North Carolina and Tennessee; revision of the special rule for nonessential experimental populations; final rule. Federal Register 60:18940–48.

Heptner, V. G., and A. A. Sludskii. 1992. Mammals of the Soviet Union. Volume II, Part 2. Carnivora (hyaenas and cats). Smithsonian Inst. Libraries, Washington, D.C., xxiv + 784 pp.

Herman, T., and K. Fuller. 1974. Observations of the marten, *Martes americana,* in the

Mackenzie District, Northwest Territories. Can. Field-Nat. 88:501–3.

Herrero, S. 1970. Human injury inflicted by grizzly bears. Science 170:593–98.

———, ed. 1972. Bears—their biology and management. Internatl. Union Conserv. Nat. Publ., n.s., no. 23, 371 pp.

———. 1976. Conflicts between man and grizzly bears in the national parks of North America. *In* Pelton, Lentfer, and Folk (1976), 121–45.

Herrero, S., C. Schroeder, and M. Scott-Brown. 1986. Are Canadian foxes swift enough? Biol. Conserv. 36:159–67.

Herrmann, M. 1994. Habitat use and spatial organization by the stone marten. *In* Buskirk et al. (1994), 122–36.

Hersteinsson, P. 1989. Population genetics and ecology of different colour morphs of arctic foxes *Alopex lagopus* in Iceland. Finnish Game Res. 46:64–78.

Hersteinsson, P., A. Angerbjörn, K. Frafjord, and A. Kaikusalo. 1989. The arctic fox in Fennoscandia and Iceland: management problems. Biol. Conserv. 49:67–81.

Hersteinsson, P., and D. W. Macdonald. 1982. Some comparisons between red and arctic foxes, *Vulpes vulpes* and *Alopex lagopus*, as revealed by radio tracking. *In* Cheeseman and Mitson (1982), 259–89.

———. 1992. Interspecific competition and the geographical distribution of red and arctic foxes *Vulpes vulpes* and *Alopex lagopus*. Oikos 64:505–15.

Herzig-Straschil, B. 1977. Notes on the feeding habits of the yellow mongoose *Cynictis penicillata*. Zool. Afr. 12:225–26.

Higgins, A. K. 1990. Breeding of the polar wolf in Greenland. Polar Record 26:55–56.

Highley, K., and S. C. Highley. 1995. China's bear farms and the trade in parts. *In* Rose and Gaski (1995), 131–49.

Hill, E. P., P. W. Sumner, and J. B. Wooding. 1987. Human influences on range expansion of coyotes in the Southeast. Wildl. Soc. Bull. 15:521–24.

Hillman, C. N., and J. W. Carpenter. 1980. Masked mustelid. Nature Conservancy News 30(2):20–23.

Hillman, C. N., and T. W. Clark. 1980. *Mustela nigripes*. Mammalian Species, no. 126, 3 pp.

Hillman, C. N., R. L. Linder, and R. B. Dahlgren. 1979. Prairie dog distribution in areas inhabited by black-footed ferrets. Amer. Midl. Nat. 102:185–87.

Hills, D. M., and R. H. N. Smithers. 1980. The "king cheetah": a historical review. Arnoldia 9(1):1–22.

Hilton, H. 1978. Systematics and ecology of the eastern coyote. *In* Bekoff (1978), 209–28.

Hines, T. D., and R. M. Case. 1991. Diet, home range, movements, and activity periods of swift fox in Nebraska. Prairie Nat. 23:131–38.

Hoagland, D. B., G. R. Horst, and C. W. Kilpatrick. 1989. Biogeography and population biology of the mongoose in the West Indies. *In* Woods (1989*a*), 611–34.

Hofer, H., and M. L. East. 1993*a*. The commuting system of Serengeti spotted hyaenas: how a predator copes with migratory prey. I. Social organization. Anim. Behav. 46:547–57.

———. 1993*b*. The commuting system of Serengeti spotted hyaenas. II. Intrusion pressure and commuters' space use. Anim. Behav. 46:559–74.

———. 1993*c*. The commuting system of Serengeti spotted hyaenas. III. Attendance and maternal care. Anim. Behav. 46:575–89.

Hofer, H., M. L. East, and K. L. I. Campbell. 1993. Snares, commuting hyaenas, and migratory herbivores: humans as predators in the Serengeti. Symp. Zool. Soc. London 65:347–66.

Hoi-Leitner, M., and E. Kraus. 1989. Der Goldschakal, *Canis aureus* (Linnaeus 1758), in Österreich. Bonner Zool. Beitr. 40:197–204.

Holekamp, K. E., and L. Smale. 1993. Ontogeny of dominance in free-living spotted hyaenas: juvenile rank relations with other immature individuals. Anim. Behav. 46:451–66.

Holzman, S., M. J. Conroy, and J. Pickering. 1992. Home range, movements, and habitat use of coyotes in southcentral Georgia. J. Wildl. Mgmt. 56:139–46.

Hoppe-Dominik, B. 1990. On the occurrence of the honey-badger *(Mellivora capensis)* and the viverrids in the Ivory Coast. Mustelid & Viverrid Conserv. 3:9–13.

Hornocker, M. G. 1969. Winter territoriality in mountain lions. J. Wildl. Mgmt. 33:457–64.

———. 1970. An analysis of mountain lion predation upon mule deer and elk in the Idaho Primitive Area. Wildl. Monogr., no. 21, 39 pp.

———. 1983. Tracking the truth about wolverines. Natl. Wildl. 21(5):32–39.

Hornocker, M. G., and H. S. Hash. 1981. Ecology of the wolverine in northwestern Montana. Can. J. Zool. 59:1286–1301.

Houseknecht, C. R., and J. R. Tester. 1978. Denning habits of striped skunks *(Mephitis mephitis)*. Amer. Midl. Nat. 100:424–30.

Hsu Lunghui and Wu Jiayan. 1981. A new subspecies of *M. flavigula* from Hainan Island. Acta Theriol. Sinica 1:145–48.

Hubbard, A. L., S. McOrist, T. W. Jones, R. Boid, R. Scott, and N. Easterbee. 1992. Is survival of European wildcats *Felis silvestris* in Britain threatened by interbreeding with domestic cats? Biol. Conserv. 61:203–8.

Hubbard, W. P., and S. Harris. 1960. Notorious grizzly bears. Sage Books, Denver, 205 pp.

Humphrey, S. R., ed. 1992. Rare and endangered biota of Florida. Volume I. Mammals. Univ. Press of Florida, Gainesville, xxviii + 392 pp.

Humphrey, S. R., and H. W. Setzer. 1989. Geographic variation and taxonomic revision of mink *(Mustela vison)* in Florida. J. Mamm. 70:241–52.

Hunt, J. H. 1974. The little-known lynx. Maine Fish and Game 16(2):28.

Hunt, R. M., Jr., and R. H. Tedford. 1993. Phylogenetic relationships within the aeluroid Carnivora and implications of their temporal and geographic distribution. *In* Szalay, Novacek, and McKenna (1993*b*), 53–73.

Husson, A. M. 1978. The mammals of Suriname. E. J. Brill, Leiden, xxxiv + 569 pp.

I

Ikeda, H. 1986. Old dogs, new treks. Nat. Hist. 95(8):38–45.

Imaizumi, Y., and M. Yoshiyuki. 1989. Taxonomic status of the Japanese otter (Carnivora, Mustelidae), with a description of a new species. Bull. Natl. Sci. Mus. (Tokyo), ser. A, 15:177–88.

Ingle, M. 1994. Wolves of the islands. Defenders 69(2):24–31.

Insley, H. 1977. An estimate of the population density of the red fox *(Vulpes vulpes)* in the New Forest, Hampshire. J. Zool. 183: 549–53.

Irven, P. 1993. The clouded leopard *(Neofelis nebulosa)*: a short review of the species' desperate situation at present. Ratel 20:93–98.

IUCN (World Conservation Union). 1972–78. Red data book. I. Mammalia. Morges, Switzerland.

Izor, R. J., and L. de la Torre. 1978. A new species of weasel *(Mustela)* from the highlands of Colombia, with comments on the evolution and distribution of South American weasels. J. Mamm. 59:92–102.

Izor, R. J., and N. E. Peterson. 1985. Notes on South American weasels. J. Mamm. 66:788–90.

J

Jachowski, R. L. 1981. Proposal to remove the bobcat from Appendix II of the Convention on International Trade in Endangered Species of Wild Fauna and Flora. Federal Register 46:45652–56.

Jackson, H. H. T. 1961. Mammals of Wisconsin. Univ. Wisconsin Press, Madison, xii + 504 pp.

Jackson, P. 1985. Man-eaters. Internatl. Wildl. 15(6):4–11.

———. 1993. The status of the tiger in 1993 and threats to its future. Cat News 19:5–11.

———. 1994. Status of tiger *Panthera tigris* (Linnaeus 1758), 1994. Cat News 21:13.

———. 1995. No respite for the tiger. Cat News 22:1.

Jackson, P. F. R. 1978. Scientists hunt the Bengal tiger—but only in order to trace and save it. Smithsonian 9(5):28–37.

Jackson, R. 1979. Snow leopards in Nepal. Oryx 15:191–95.

Jackson, R., and G. G. Ahlborn. 1988. Observations on the ecology of snow leopard in west Nepal. Proc. 5th Internatl. Snow Leopard Symp., 65–87.

Jameson, R. J. 1989. Movements, home range, and territories of male sea otters off central California. Mar. Mamm. Sci. 5:159–72.

———. 1993. Survey of a translocated sea otter population. Internatl. Union Conserv. Nat. Otter Specialist Group Bull. 8:2–4.

Jameson, R. J., K. W. Kenyon, A. M. Johnson, and H. M. Wight. 1982. History and status of translocated sea otter populations in North America. Wildl. Soc. Bull. 10:100–107.

Janzen, D. H., and W. Hallwachs. 1982. The hooded skunk, *Mephitis macroura*, in lowland northwestern Costa Rica. Brenesia 19–20:549–52.

Jefferies, D. J. 1989. The changing otter population of Britain, 1700–1989. Biol. J. Linnean Soc. 38:61–69.

Jenks, S. M., and R. K. Wayne. 1992. Problems and policy for species threatened by hybridization: the red wolf as a case study. *In* McCullough, D. R., and R. H. Barrett, eds., Wildlife 2001: populations, Elsevier, London, 237–51.

Jenness, S. E. 1985. Arctic wolf attacks scientist—a unique Canadian incident. Arctic 38:129–32.

Jensen, W. F., T. K. Fuller, and W. L. Robinson. 1986. Wolf, *Canis lupus*, distribution on the Ontario-Michigan border near Sault Ste. Marie. Can. Field-Nat. 100:363–66.

Johnsingh, A. J. T. 1982. Reproductive and social behaviour of the dhole, *Cuon alpinus* (Canidae). J. Zool. 198:443–63.

Johnson, A. S. 1970. Biology of the raccoon (*Procyon lotor varius* Nelson and Goldman) in Alabama. Auburn Univ. Agric. Exp. Sta. Bull., no. 402, vi + 148 pp.

Johnson, K. G., G. B. Schaller, and Hu Jinchu. 1988. Comparative behaviour of red and giant pandas in the Wolong Reserve, China. J. Mamm. 69:552–64.

Johnson, R. E. 1977. An historical analysis of wolverine abundance and distribution in Washington. Murrelet 58:13–16.

Johnson, W. E. 1992. Patagonia's little foxes. Nat. Hist. 92(4):27–31.

Johnson, W. E., and W. L. Franklin. 1991. Feeding and spatial ecology of *Felis geoffroyi* in southern Patagonia. J. Mamm. 72:815–20.

———. 1994. Role of body size in the diets of sympatric gray and culpeo foxes. J. Mamm. 75:163–74.

Jones, E. 1977. Ecology of the feral cat, *Felis catus* (L.) (Carnivora: Felidae), on Macquarie Island. Austral. Wildl. Res. 4:249–62.

———. 1990. Physical characteristics and taxonomic status of wild canids, *Canis familiaris*, from the eastern highlands of Victoria. Austral. Wildl. Res. 17:69–81.

Jones, J. K., Jr., R. S. Hoffmann, D. W. Rice, C. Jones, R. J. Baker, and M. D. Engstrom. 1992. Revised checklist of North American mammals north of Mexico, 1991. Occas. Pap. Mus. Texas Tech Univ., no. 146, 23 pp.

Jones, M. J., and J. B. Theberge. 1982. Summer home range and habitat utilisation of the red fox *(Vulpes vulpes)* in a tundra habitat, northwest British Columbia. Can. J. Zool. 60:807–12.

Jones, M. L. 1982. Longevity of captive mammals. Zool. Garten 52:113–28.

Jonkel, C. J. 1978. Black, brown (grizzly), and polar bears. *In* Schmidt and Gilbert (1978), 227–48.

———. 1987. Brown bear. *In* Novak, Baker, et al. (1987), 456–73.

Jonkel, C. J., and I. M. Cowan. 1971. The black bear in the spruce-fir forest. Wildl. Monogr., no. 27, 57 pp.

Jonkel, C. J., and F. L. Miller. 1970. Recent records of black bears *(Ursus americanus)* on the barren grounds of Canada. J. Mamm. 51:826–28.

Jonkel, C. J., P. Smith, I. Stirling, and G. B. Kolenosky. 1976. The present status of the polar bear in the James Bay and Belcher Islands area. Can. Wildl. Serv. Occas. Pap., no. 26, 42 pp.

Julien-Laferriere, D. 1993. Radio-tracking observations on ranging and foraging patterns by kinkajous *(Potos flavus)* in French Guiana. J. Trop. Ecol. 9:19–32.

K

Karami, M. 1992. Cheetah distribution in Khorasan Province, Iran. Cat News 16:4.

Karanth, K. U. 1987. Tigers in India: a critical review of field censuses. *In* Tilson and Seal (1987), 118–32.

Kaufmann, J. H. 1962. Ecology and social behavior of the coati, *Nasua narica*, on Barro Colorado Island, Panama. Univ. California Publ. Zool. 60:95–222.

———. 1987. Ringtail and coati. *In* Novak, Baker, et al. (1987), 500–509.

Kaufmann, J. H., D. V. Lanning, and S. E. Poole. 1976. Current status and distribution of the coati in the United States. J. Mamm. 57:621–37.

Kauhala, K. 1994. The raccoon dog: a successful canid. Canid News 2:37–40.

Kauhala, K., E. Helle, and K. Taskinen. 1993. Home range of the raccoon dog *(Nyctereutes procyonoides)* in southern Finland. J. Zool. 231:95–106.

Kellnhauser, J. T. 1983. The acceptance of *Lontra* Gray for the New World river otters. Can. J. Zool. 61:278–79.

Kemp, G. A. 1976. The dynamics and regulation of black bear *Ursus americanus* populations in northern Alberta. *In* Pelton, Lentfer, and Folk (1976), 191–97.

Kennelly, J. J. 1978. Coyote reproduction. *In* Bekoff (1978), 73–93.

Kenyon, K. W. 1969. The sea otter in the eastern Pacific Ocean. N. Amer. Fauna, no. 68, ix + 352 pp.

Khan, M. A. R. 1987. The problem tiger of Bangladesh. *In* Tilson and Seal (1987), 92–96.

Kilgore, D. L., Jr. 1969. An ecological study of the swift fox *(Vulpes velox)* in the Oklahoma Panhandle. Amer. Midl. Nat. 81:512–34.

King, C. 1975. The home range of the weasel *(Mustela nivalis)* in an English woodland. J. Anim. Ecol. 44:639–69.

King, C. M. 1983. *Mustela erminea*. Mammalian Species, no. 195, 8 pp.

Kingdon, J. 1977. East African mammals: an atlas of evolution in Africa. III(A). Carnivores. Academic Press, London, viii + 475 pp.

Kipp, H. 1965. Beitrag zur Kenntnis der Gattung *Conepatus* Molina, 1782. Z. Saugetierk. 30:193–232.

Kistchinski, A. A. 1972. Life history of the brown bear (*Ursus arctos* L.) in north-east Siberia. *In* Herrero (1972), 67–73.

Kitchener, A. 1992. The Scottish wildcat *Felis silvestris:* decline and recovery. *In* Mansard, P., ed., Cats, Ridgeway Trust for Endangered Cats, Hastings, U.K., 21–41.

Kitchener, C., and N. Easterbee. 1992. The taxonomic status of black wild felids in Scotland. J. Zool. 227:342–46.

Kleiman, D. G. 1972. Social behavior of the maned wolf *(Chrysocyon brachyurus)* and bush dog *(Speothos venaticus):* a study in contrast. J. Mamm. 53:791–806.

———. 1983. Ethology and reproduction of captive giant pandas *(Ailuropoda melanoleuca).* Z. Tierpsychol. 62:1–46.

Knap, J. J. 1975. Martens on the move. Internatl. Wildl. 5(5):32–35.

Knibb, D. 1997. Listing the lynx. Wildl. Conserv. 100(6):16.

Knight, R. R., B. M. Blanchard, and K. C. Kendall. 1981. Yellowstone grizzly bear investigations: report of the Interagency Study Team. U.S. Natl. Park Serv., v + 55 pp.

Knowlton, F. F. 1972. Preliminary interpretations of coyote population mechanics with some management implications. J. Wildl. Mgmt. 36:369–82.

Knudsen, B. 1978. Time budgets of polar bears *(Ursus maritimus)* on North Twin Island, James Bay, during summer. Can. J. Zool. 56:1627–28.

Kock, D. 1990. Historical record of a tiger, *Panthera tigris* (Linnaeus, 1758), in Iraq. Zoology in the Middle East, Mammalia, 4:11–15.

Koehler, G. M., M. G. Hornocker, and H. S. Hash. 1980. Wolverine marking behavior. Can. Field-Nat. 94:339–41.

Koenig, L. 1970. Zur Fortpflanzung und Jugendentwicklung des Wüstenfuchses *(Fennecus zerda* Zimm. 1780). Z. Tierpsychol. 27:205–46.

Köhler, C. E., and P. R. K. Richardson. 1990. *Proteles cristatus.* Mammalian Species, no. 363, 6 pp.

Köhncke, M., and K. Leonhardt. 1986. *Cryptoprocta ferox.* Mammalian Species, no. 254, 5 pp.

Kolenosky, G. B. 1987. Polar bear. *In* Novak, Baker, et al. (1987), 474–87.

Kortlucke, S. M. 1973. Morphological variation in the kinkajou, *Potos flavus* (Mammalia: Procyonidae), in Middle America. Occas. Pap. Mus. Nat. Hist. Univ. Kansas, no. 17, 36 pp.

Krebs, J. W., T. W. Strine, and J. E. Childs. 1993. Rabies surveillance in the United States during 1992. J. Amer. Veterinary Med. Assoc. 203:1718–31.

Krebs, J. W., M. L. Wilson, and J. E. Childs. 1995. Rabies—epidemiology, prevention, and future research. J. Mamm. 76:681–94.

Krott, P. 1992. The black colour phase of the red fox *(Vulpes vulpes)* in Europe. Saugetierk. Mitt. 34:23–30.

Kruuk, H. 1972. The spotted hyena: a study of predation and social behavior. Univ. Chicago Press, xvi + 335 pp.

———. 1976. Feeding and social behaviour of the striped hyaena *(Hyaena vulgaris* Desmarest). E. Afr. Wildl. J. 14:91–111.

———. 1978*a*. Foraging and spatial organization of the European badger, *Meles meles* L. Behav. Ecol. Sociobiol. 4:75–89.

———. 1978*b*. Spatial organization and territorial behaviour of the European badger, *Meles meles.* J. Zool. 184:1–19.

Kruuk, H., and A. Morehouse. 1991. The spatial organization of otters *(Lutra lutra)* in Shetland. J. Zool. 224:41–57.

Krystufek, B., and S. Petkovski. 1990. New record of the jackal *Canis aureus* Linnaeus, 1758 in Macedonia (Mammalia, Carnivora). Fragmenta Balcanica, Mus. Macedon. Sci. Nat. 14:131–38.

Krystufek, B., and N. Tvrtkovic. 1990. Range expansion by Dalmatian jackal population in the twentieth century *(Canis aureus* Linnaeus, 1758). Folia Zool. 39:291–96.

———. 1992. New information on the introduction into Europe of the small Indian mongoose, *Herpestes auropunctatus.* Small Carnivore Conserv. 7:16.

Kubiak, H. 1965. The appearance of a raccoon dog, *Nyctereutes procyonides* (Gray, 1834) in Cracow District (Poland). Prezelgl. Zool. 9:417–22.

Kucherenko, S. P. 1976. The common otter *(Lutra lutra)* in the Amur-Ussury district. Zool. Zhur. 55:904–11.

Kucherenko, S. P., and V. G. Yudin. 1973. Distribution, population density, and economical importance of the raccoon-dog *(Nyctereutes procyonides)* in the Amur-Ussury district. Zool. Zhur. 52:1039–45.

Kuntzsch, V., and J. A. J. Nel. 1992. Diet of bat-eared foxes *Otocyon megalotis* in the Karoo. Koedoe 35:37–48.

Kurten, B. 1968. Pleistocene mammals of Europe. Aldine, Chicago, viii + 317 pp.

———. 1973. Transberingean relationships of *Ursus arctos* Linne (brown and grizzly bears). Commentat. Biol. 65:1–10.

———. 1985. The Pleistocene lion of Beringia. Ann. Zool. Fennici 22:117–21.

Kurten, B., and E. Anderson. 1980. Pleistocene mammals of North America. Columbia Univ. Press, New York, 442 pp.

Kuschinski, L. 1974. Breeding binturongs *Arctictis binturong* at Glasgow Zoo. Internatl. Zoo Yearbook 14:124–26.

Kuyt, E. 1972. Food habits and ecology of wolves on barren-ground caribou range in the Northwest Territories. Can. Wildl. Serv. Rept. Ser., no. 21, 36 pp.

Kvam, T. 1991. Reproduction in the European lynx, *Lynx lynx.* Z. Saugetierk. 56: 146–58.

L

LaBarge, T., A. Baker, and D. Moore. 1990. Fisher *(Martes pennanti):* birth, growth, and development in captivity. Mustelid & Viverrid Conserv. 2:1–3.

Laing, R. I., and G. L. Holroyd. 1989. The status of the black-footed ferret in Canada. Blue Jay 47:121–25.

Lamotte, M., and M. Tranier. 1983. Un spécimen de *Genetta (Paragenetta)* johnstoni collecté dans la région du Nimba (Côte d'Ivoire). Mammalia 47:430–32.

Lamprecht, J. 1979. Field observations on the behaviour and social system of the bat-eared fox *Otocyon megalotis* Desmarest. Z. Tierpsychol. 49:260–84.

Langguth, A. 1975. Ecology and evolution in the South American canids. *In* Fox (1975), 192–206.

Lanning, D. V. 1976. Density and movements of the coati in Arizona. J. Mamm. 57:609–11.

Larivière, S., and M. Crête. 1993. The size of eastern coyotes *(Canis latrans):* a comment. J. Mamm. 74:1072–74.

Larkin, P., and M. Roberts. 1979. Reproduction in the ring-tailed mongoose *Galidia elegans* at the National Zoological Park, Washington. Internatl. Zoo Yearbook 19:189–93.

Larsen, T. 1975. Polar bear den surveys in Svalbard in 1973. Norsk Polarinst. Arbok, 1973, 101–12.

Latinen, K. 1987. Longevity and fertility of the polar bear, *Ursus maritimus* Phipps, in captivity. Zool. Garten 57:197–99.

Laundré, J. W., and B. L. Keller. 1984. Home-range size of coyotes: a critical review. J. Wildl. Mgmt. 48:127–39.

Laurie, A., and J. Seidensticker. 1977. Behavioural ecology of the sloth bear *(Melursus ursinus).* J. Zool. 182:187–204.

Laurie, E. M. O., and J. E. Hill. 1954. List of land mammals of New Guinea, Celebes, and adjacent islands, 1758–1952. British Mus. (Nat. Hist.), London, 175 pp.

Lawrence, B., and W. H. Bossert. 1967. Multiple character analysis of *Canis lupus, latrans,* and *familiaris,* with a discussion of the relationships of *Canis niger.* Amer. Zool. 7:223–32.

———. 1975. Relationships of North American *Canis* shown by a multiple character analysis of selected populations. *In* Fox (1975), 73–86.

Lay, D. M. 1967. A study of the mammals of Iran. Fieldiana Zool. 54:1–282.

Layne, J. N., ed. 1978. Rare and endangered biota of Florida. I. Mammals. Univ. Presses of Florida, Gainesville, xx + 52 pp.

Leatherwood, S., L. J. Harrington-Coulombe,

and C. L. Hubbs. 1978. Relict survival of the sea otter in central California and evidence of its recent dispersal south of Point Conception. Bull. S. California Acad. Sci. 77:109–15.

Lehman, N., P. Clarkson, L. D. Mech, T. J. Meier, and R. K. Wayne. 1992. A study of the genetic relationships within and among wolf packs using DNA fingerprinting and mitochondrial DNA. Behav. Ecol. Sociobiol. 30:83–94.

Lehman, N., A. Eisenhawer, K. Hansen, L. D. Mech, R. O. Peterson, P. J. P. Gogan, and R. K. Wayne. 1991. Introgression of coyote mitochondrial DNA into sympatric North American gray wolf populations. Evolution 45: 104–19.

Lehner, P. N. 1978. Coyote communication. *In* Bekoff (1978), 128–62.

Lekagul, B., and J. A. McNeely. 1977. Mammals of Thailand. Sahakarnbhat, Bangkok, li + 758 pp.

Lensing, J. E., and E. Joubert. 1977. Intensity distribution patterns for five species of problem animals in South West Africa. Madoqua 10:131–41.

Leonard, R. D. 1986. Aspects of reproduction of the fisher, *Martes pennanti*, in Manitoba. Can. Field-Nat. 100:32–44.

Leopold, A. S. 1959. Wildlife of Mexico. Univ. California Press, Berkeley, xiii + 568.

Leslie, G. 1971. Further observations on the oriental short-clawed otter *Amblonyx cinerea* at Aberdeen Zoo. Internatl. Zoo Yearbook 11:112–13.

Lever, C. 1985. Naturalized mammals of the world. Longman, London, xvii + 487 pp.

Leyhausen, P. 1979. Cat behavior. Garland, New York, 340 pp.

Leyhausen, P., and M. Pfleiderer. 1994. The taxonomic status of the Iriomote cat. Cat News 21:18–20.

Liberg, O. 1980. Spacing patterns in a population of rural free roaming domestic cats. Oikos 35:336–49.

Licht, D. S., and S. H. Fritts. 1994. Gray wolf *(Canis lupus)* occurrences in the Dakotas. Amer. Midl. Nat. 132:74–81.

Liers, E. E. 1966. Notes on breeding the Canadian otter *Lutra canadensis* in captivity and longevity records of beavers *Castor canadensis*. Internatl. Zoo Yearbook 6:171–72.

Lindsay, I. M., and D. W. Macdonald. 1986. Behaviour and ecology of the Ruppell's fox, *Vulpes ruppelli*, in Oman. Mammalia 50: 461–74.

Lindzey, F. 1987. Mountain lion. *In* Novak, Baker, et al. (1987), 656–69.

Lindzey, F. G., and E. C. Meslow. 1976. Winter dormancy in black bears in southwestern Washington. J. Wildl. Mgmt. 40:408–15.

———. 1977*a*. Home range and habitat use by black bears in southwestern Washington. J. Wildl. Mgmt. 41:413–25.

———. 1977*b*. Population characteristics of black bears on an island in Washington. J. Wildl. Mgmt. 41:408–12.

Linscombe, G. 1994. U.S. fur harvest (1970–1992) and fur value (1974–1992) statistics by state and region. Louisiana Department of Wildlife and Fisheries, Baton Rouge, 29 pp.

Lloyd, H. G. 1975. The red fox in Britain. *In* Fox (1975), 207–15.

Long, C. A. 1972. Taxonomic revision of the North American badger, *Taxidea taxus*. J. Mamm. 53:725–59.

———. 1973. *Taxidea taxus*. Mammalian Species, no. 26, 4 pp.

———. 1978. A listing of Recent badgers of the world, with remarks on taxonomic problems in *Mydaus* and *Melogale*. Rept. Fauna and Flora Wisconsin, Univ. Wisconsin Mus. Nat. Hist., 14:1–6.

———. 1995. Stone marten *(Martes foiana)* in southeast Wisconsin, U.S.A. Small Carnivore Conserv. 13:14.

Long, C. A., and C. A. Killingley. 1983. The badgers of the world. Charles C. Thomas, Springfield, Ill., xxiv + 404 pp.

Lotze, J. 1979. The raccoon *(Procyon lotor)* on St. Catherines Island, Georgia. 4. Comparisons of home ranges determined by livetrapping and radiotracking. Amer. Mus. Novit., no. 2664, 25 pp.

Lotze, J., and S. Anderson. 1979. *Procyon lotor*. Mammalian Species, no. 119, 8 pp.

Loughlin, T. R. 1980. Home range and territoriality of sea otters near Monterey, California. J. Wildl. Mgmt. 44:576–82.

Lowery, G. H., Jr. 1974. The mammals of Louisiana and its adjacent waters. Louisiana State Univ. Press, xxiii + 565 pp.

Lu Houji. 1987. Habitat availability and prospects for tigers in China. *In* Tilson and Seal (1987), 71–74.

Luoma, J. R. 1987. The state of the tiger. Audubon 89(4):61–63.

Lynch, J. M. 1995. Conservation implications of hybridisation between mustelids and their domesticated counterparts: the example of polecats and feral ferrets in Britain. Small Carnivore Conserv. 13:17–18.

M

Ma Yi-ching. 1983. The status of bears in China. Acta Zool. Fennica 174:165–66.

Ma Yiqing and Li Xu. 1994. Distribution and conservation of sables in China. *In* Buskirk et al. (1994), 255–61.

Maas, B. 1994. Bat-eared foxes in the Serengeti. Canid News 2:3–7.

Macdonald, D. W. 1978. Observations on the behaviour and ecology of the striped hyaena, *Hyaena hyaena*, in Israel. Israel J. Zool. 27:189–98.

MacDonald, S. M., and C. F. Mason. 1994. Status and needs of the otter *(Lutra lutra)* in the western Palaearctic. Council of Europe Nature and Environment Ser., no. 67, 54 pp.

Macintosh, N. W. G. 1975. The origin of the dingo: an enigma. *In* Fox (1975), 87–106.

Macpherson, A. H. 1965. The barren-ground grizzly bear and its survival in northern Canada. Can. Audubon 27(1):2–8.

———. 1969. The dynamics of Canadian arctic fox populations. Can. Wildl. Serv. Rept. Ser., no. 8, 52 pp.

Madhusudan, M. D. 1995. Sighting of the Nilgiri marten *(Martes gwatkinsi)* at Eravikulam National Park, Kerala, India. Small Carnivore Conserv. 13:6–7.

Maehr, D. S., R. C. Belden, E. D. Land, and L. Wilkins. 1990. Food habits of panthers in southwest Florida. J. Wildl. Mgmt. 54:420–23.

Mainstone, B. J. 1974. Tigers reduce honey production. Malayan Nat. J. 28:36.

Malcolm, J. R. 1980*a*. African wild dogs play every game by their own rules. Smithsonian 11(8):62–71.

———. 1980*b*. Food caching by African wild dogs *(Lycaon pictus)*. J. Mamm. 61:743–44.

———. 1986. Socio-ecology of bat-eared foxes *(Otocyon megalotis)*. J. Zool. 208:457–67.

Malcolm, J. R., and K. Marten. 1982. Natural selection and the communal rearing of pups in African wild dogs *(Lycaon pictus)*. Behav. Ecol. Sociobiol. 10:1–13.

Malcolm, J. R., and H. Van Lawick. 1975. Notes on wild dogs *(Lycaon pictus)* hunting zebras. Mammalia 39:231–40.

Mallinson, J. J. C. 1973. The reproduction of the African civet *Viverra civetta* at Jersey Zoo. Internatl. Zoo Yearbook 13:147–50.

———. 1974. Establishing mammal gestation at the Jersey Zoological Park. Internatl. Zoo Yearbook 14:184–87.

Manning, R. W., J. K. Jones, Jr., and R. R. Hollander. 1986. Northern limits of distribution of the hog-nosed skunk, *Conepatus mesoleucus*, in Texas. Texas J. Sci. 38:289–91.

Manville, R. H. 1966. The extinct sea mink, with taxonomic notes. Proc. U.S. Natl. Mus. 122:1–12.

Maran, T. 1992. The European mink, *Mustela lutreola*, in protected areas in the former Soviet Union. Small Carnivore Conserv. 7:10–12.

———. 1994. On the status and management of the European mink *Mustela lutreola*. Council of Europe Environ. Encounters, no. 17, 84–90.

Martin, L. D. 1989. Fossil history of the terrestrial Carnivora. *In* Gittleman (1989), 536–68.

Martin, R. B., and T. de Meulenaer. 1988. Survey of the status of the leopard *(Panthera pardus)* in sub-Saharan Africa. Secretariat of the Convention on International Trade in Endangered Species of Wild Fauna and Flora, Lausanne, Switzerland, xx + 106 pp.

Martinka, C. J. 1974. Population characteristics of grizzly bears in Glacier National Park, Montana. J. Mamm. 55:21–29.

———. 1976. Ecological role and management of grizzly bears in Glacier National Park, Montana. *In* Pelton, Lentfer, and Folk (1976), 147–56.

Mason, C. F., and S. M. Macdonald. 1986. Otters: ecology and conservation. Cambridge Univ. Press, vii + 236 pp.

Mattson, D. J., B. M. Blanchard, and R. R. Knight. 1992. Yellowstone grizzly bear mortality, human habituation, and whitebark pine seed crops. J. Wildl. Mgmt. 56:432–42.

Mattson, D. J., and M. M. Reid. 1991. Conservation of the Yellowstone grizzly bear. Conserv. Biol. 5:364–72.

Mattson, D. J., R. G. Wright, K. C. Kendall, and C. J. Martinka. 1995. Grizzly bears. *In* LaRoe et al. (1995), 103–5.

May, R. M. 1986. The cautionary tale of the black-footed ferret. Nature 320:13–14.

Mazak, V. 1981. *Panthera tigris*. Mammalian Species, no. 152, 8 pp.

McBride, R. T. 1980. The Mexican wolf *(Canis lupus baileyi)*: a historical review and observations on its status and distribution. U.S. Fish and Wildl. Serv., Albuquerque, Endangered Species Rept., no. 8, vi + 38 pp.

McCarley, H., and C. J. Carley. 1979. Recent changes in distribution and status of wild red wolves *(Canis rufus)*. U.S. Fish and Wildl. Serv., Albuquerque, Endangered Species Rept., no. 4, v + 38 pp.

McCarthy, T. J. 1992. Notes concerning the jagua-rundi cat *(Herpailurus yagouaroundi)* in the Caribbean lowlands of Belize and Guatemala. Mammalia 56:302–6.

McCord, C. 1974. Selection of winter habitat by bobcats *(Lynx rufus)* on the Quabbin Reservoir, Massachusetts. J. Mamm. 55: 428–37.

McCracken, C., D. A. Rose, and K. A. Johnson. 1995. Status, management, and commercialization of the American black bear *(Ursus americanus)*. Traffic USA, Washington, D.C., x + 132 pp.

McCusker, J. S. 1974. Breeding Malayan sun bears *Helarctos malayanus* at Fort Worth Zoo. Internatl. Zoo Yearbook 14:118–19.

McDougal, C. W. 1987. The man-eating tiger in geographical and historical perspective. *In* Tilson and Seal (1987), 435–48.

McGrew, J. C. 1979. *Vulpes macrotis*. Mammalian Species, no. 123, 6 pp.

McShane, L. J., J. A. Estes, M. L. Riedman, and M. M. Staedler. 1995. Repertoire, structure, and individual variation of vocalizations in the sea otter. J. Mamm. 76:414–27.

Mead, R. A. 1968*a*. Reproduction in eastern forms of the spotted skunk (genus *Spilogale*). J. Zool. 156:119–36.

———. 1968*b*. Reproduction in western forms of the spotted skunk (genus *Spilogale*). J. Mamm. 49:373–90.

———. 1994. Reproduction in *Martes*. *In* Buskirk et al. (1994), 404–22.

Meadows, R. 1991. How many leopards are there in Africa? *In* Seidensticker and Lumpkin (1991), 114.

Mech, L. D. 1961. The marten: symbol of wilderness. Anim. Kingdom 44(5):133–37.

———. 1970. The wolf: the ecology and behavior of an endangered species. Natural History Press, Garden City, N.Y., xx + 384 pp.

———. 1974*a*. *Canis lupus*. Mammalian Species, no. 37, 6 pp.

———. 1974*b*. A new profile for the wolf. Nat. Hist. 83(4):26–31.

———. 1977*a*. Population trend and winter deer consumption in a Minnesota wolf pack. *In* Phillips and Jonkel (1977), 55–83.

———. 1977*b*. Productivity, mortality, and population trends of wolves in northeastern Minnesota. J. Mamm. 58:559–74.

———. 1977*c*. Record movement of a Canadian lynx. J. Mamm. 58:676–77.

———. 1977*d*. A recovery plan for the eastern timber wolf. Natl. Parks and Conserv. Mag. 50(1):17–21.

———. 1977*e*. Wolf-pack buffer zones as prey reservoirs. Science 198:320–21.

———. 1979. Why some deer are safe from wolves. Nat. Hist. 88(1):70–77.

———. 1989. Wolf longevity in the wild. Endangered Species Tech. Bull. 14(5):8.

———. 1990. Who's afraid of the big bad wolf? Audubon 92(2):82–85.

———. 1995. The wolf's world brightens. Defenders 70(1):28–29.

Mech, L. D., L. D. Frenzel, Jr., R. R. Ream, and J. W. Winship. 1971. Movements, behavior, and ecology of timber wolves in northeastern Minnesota. *In* Mech, L. D., and L. D. Frenzel, Jr., eds., Ecological studies of the timber wolf in northeastern Minnesota, U.S. Forest Serv. Res. Pap., no. NC-52, 1–35.

Mech, L. D., and S. H. Fritts. 1987. Parovirus and heartworm found in Minnesota wolves. Endangered Species Tech. Bull. 12(5–6):5–6.

Mech, L. D., and P. D. Karns. 1977. Role of the wolf in a deer decline in the Superior National Forest. U.S. Forest Serv. Res. Pap., no. NC-148, 23 pp.

Mech, L. D., and M. Korb. 1978. An unusually long pursuit of a deer by a wolf. J. Mamm. 59:860–61.

———. 1989. Polygyny in a wild wolf pack. J. Mamm. 70:675–76.

———. 1990. Non-family wolf, *Canis lupus*, packs. Can. Field-Nat. 104:484–83.

Mech, L. D., and R. M. Nowak. 1981. Return of the gray wolf to Wisconsin. Amer. Midl. Nat. 105:408–9.

Mech, L. D., D. H. Pletscher, and C. J. Martinka. 1995. Gray wolves. *In* LaRoe et al. (1995), 98–100.

Mech, L. D., and L. L. Rogers. 1977. Status, distribution, and movements of martens in northeastern Minnesota. U.S. Forest Serv. Res. Pap., no. NC-143, 7 pp.

Medel, R. G., J. J. Jiménez, F. M. Jaksic, J. L. Yáñez, and J. J. Armesto. 1990. Discovery of a continental population of the rare Darwin's fox, *Dusicyon fulvipes* (Martin, 1837) in Chile. Biol. Conserv. 51:71–77.

Medjo, D. C., and L. D. Mech. 1976. Reproductive activity in nine- and ten-month-old wolves. J. Mamm. 57:406–8.

Medway, Lord. 1977. Mammals of Borneo. Monogr. Malaysian Branch Roy. Asiatic Soc., no. 7, xii + 172 pp.

———. 1978. The wild mammals of Malaya (peninsular Malaysia) and Singapore. Oxford Univ. Press, Kuala Lumpur, xxii + 128 pp.

Meester, J. 1976. South African red data book—small mammals. S. Afr. Natl. Sci. Programmes Rept., no. 11, vi + 59 pp.

Meester, J., I. L. Rautenbach, N. J. Dippenaar, and C. M. Baker. 1986. Classification of southern African mammals. Transvaal Mus. Monogr., no. 5, x + 359 pp.

Meester, J., and H. W. Setzer. 1977. The mammals of Africa: an identification manual. Smithson. Inst. Press, Washington, D.C.

Mehrer, C. F. 1976. Gestation period in the wolverine, *Gulo gulo*. J. Mamm. 57:570.

Melisch, R., P. B. Asmoro, and L. Kusumawardhani. 1994. Major steps taken towards otter conservation in Indonesia. IUCN (World Conservation Union) Otter Specialist Group Bull. 10:21–24.

Melquist, W. E., and A. E. Dronkert. 1987. River otter. *In* Novak, Baker, et al. (1987), 626–41.

Melquist, W. E., and M. G. Hornocker. 1983. Ecology of river otters in west central Idaho. Wildl. Monogr., no. 83, 60 pp.

Mendelssohn, H. 1992. The present situation of the hyaena *(Hyaena hyaena suriaca)* in Israel. Internatl. Union. Conserv. Nat. Hyaena Specialist Group Newsl. 5:24–25.

Mendelssohn, H., Y. Yom-Tov, G. Ilany, and D. Meninger. 1987. On the occurrence of Blanford's fox, *Vulpes cana* Blanford, 1877, in Israel and Sinai. Mammalia 51:459–67.

Menotti-Raymond, M., and S. J. O'Brien. 1993. Dating the genetic bottleneck of the African cheetah. Proc. Natl. Acad. Sci. 90: 3172–76.

Mercure, A., K. Ralls, K. P. Koepfli, and R. K. Wayne. 1993. Genetic subdivisions among small canids: mitochondrial DNA differentiation of swift, kit, and arctic foxes. Evolution 47:1313–28.

Merola, M. 1994. A reassessment of homozygosity and the case for inbreeding depression in the cheetah, *Acinonyx jubatus:* implications for conservation. Conserv. Biol. 8:961–71.

Messick, J. P. 1987. North American badger. *In* Novak, Baker, et al. (1987), 586–97.

Messick, J. P., and M. G. Hornocker. 1981. Ecology of the badger in southwestern Idaho. Wildl. Monogr., no. 76, 53 pp.

Meyer-Holzapfel, M. 1968. Breeding the European wild cat *Felis s. silvestris* at Berne Zoo. Internatl. Zoo Yearbook 8:31–38.

Mikkola, H. 1974. The raccoon dog spreads to western Europe. Wildlife 16:344–45.

Mikuriya, M. 1976. Notes on the Japanese otter, *Lutra lutra whitleyi* (Gray). J. Mamm. Soc. Japan 6:214–17.

Miller, B. J., S. H. Anderson, M. W. DonCarlos, and E. T. Thorne. 1988. Biology of the endangered black-footed ferret and the role of captive propagation in its conservation. Can. J. Zool. 66:765–73.

Miller, S. D., J. Rottmann, K. J. Raedeke, and R. D. Taber. 1983. Endangered mammals of Chile: status and conservation. Biol. Conserv. 25:335–52.

Mills, J. A. 1993. Bears as pets, food and medicine. *In* Stirling (1993), 176–81.

———. 1994. The sum of their parts: Asia's drug demand threatens North America's bears. Pacific Discovery 47(1):16–19.

———. 1995. Asian dedication to the use of bear bile as medicine. *In* Rose and Gaski (1995), 4–16.

Mills, M. G. L. 1978. Foraging behaviour of the brown hyaena (*Hyaena brunnea* Thunberg, 1820) in the southern Kalahari. Z. Tierpsychol. 48:113–41.

———. 1982. *Hyaena brunnea.* Mammalian Species, no. 194, 5 pp.

———. 1985. Related spotted hyaenas forage together but do not cooperate in rearing young. Nature 316:61–62.

———. 1990. Kalahari hyaenas: comparative behavioural ecology of two species. Unwin Hyman, London, xvi + 304 pp.

Mills, M. G. L., and M. E. J. Mills. 1978. The diet of the brown hyaena *Hyaena brunnea* in the southern Kalahari. Koedoe 21:125–49.

Minta, S. C., K. A. Minta, and D. F. Lott. 1992. Hunting associations between badgers *(Taxidea taxus)* and coyotes *(Canis latrans).* J. Mamm. 73:814–20.

Mitchell, J. G. 1987. The photographer who got too close. Audubon 89(2):28–34.

Mitchell, R. M. 1977. Accounts of Nepalese mammals and analysis of the host-ectoparasite data by computer techniques. Ph.D. diss., Iowa State Univ., 557 pp.

Mittermeier, R. A., H. de Macedo-Ruiz, B. A. Luscombe, and J. Cassidy. 1977. Rediscovery and conservation of the Peruvian yellow-tailed woolly monkey *(Lagothrix flavicauda).* *In* Rainier III and Bourne (1977), 95–115.

Moehlman, P. D. 1978. Jackals of the Serengeti. Wildl. News, African Wildl. Leadership Foundation, 13(3):2–6.

———. 1983. Socioecology of silverbacked and golden jackals *(Canis mesomelas* and *Canis aureus).* Amer. Soc. Mamm. Spec. Publ., no. 7, pp. 423–53.

———. 1987. Social organization in jackals. Amer. Sci. 75:366–75.

Mohler, L. L. 1974. Threatened wildlife of Idaho. Idaho Wildl. Rev. 26(5):3–5.

Mondolfi, E. 1989. Notes on the distribution, habitat, food habits, status, and conservation of the spectacled bear (*Tremarctos ornatus* Cuvier) in Venezuela. Mammalia 53:525–44.

Mondolfi, E., and R. Hoogesteijn. 1986. Notes on the biology and status of the jaguar in Venezuela. *In* Miller and Everett (1986), 85–123.

Mones, A., and J. Olazarri. 1990. Confirmacion de la existencia de *Chrysocyon brachyurus* (Illiger) en el Uruguay. Comun. Zool. Mus. Hist. Nat. Montevideo, no. 174, 6 pp.

Monge-Nájera, J., and B. M. Brenes. 1987. Why is the coyote *(Canis latrans)* expending its range? A critique of the deforestation hypothesis. Rev. Biol. Trop. 35:169–71.

Moore, C. M., and P. W. Collins. 1995. *Urocyon littoralis.* Mammalian Species, no. 489, 7 pp.

Moore, G. C., and J. S. Millar. 1984. A comparative study of colonizing and longer established eastern coyote populations. J. Wildl. Mgmt. 48:691–99.

Moore, R. E., and N. S. Martin. 1980. A recent record of the swift fox *(Vulpes velox)* in Montana. J. Mamm. 61:161.

Morgan, G. S., and C. A. Woods. 1986. Extinction and the zoogeography of West Indian mammals. Biol. J. Linnean Soc. 28:167–203.

Morrell, S. 1972. Life history of the San Joaquin kit fox. California Fish and Game 53:162–74.

Morris, P. A., and J. R. Malcolm. 1977. The Simien fox in the Bale Mountains. Oryx 14:151–60.

Muckenhirn, N. A., and J. F. Eisenberg. 1973. Home ranges and predation of the Ceylon leopard. *In* Eaton, R. L., ed., The world's cats, vol. 1, World Wildlife Safari, Winston, Oregon, 142–75.

Mugaas, J. N., J. Seidensticker, and K. P. Mahlke-Johnson. 1993. Metabolic adaptation to climate and distribution of the raccoon *Procyon lotor* and other Procyonidae. Smithson. Contrib. Zool., no. 542, iv + 34 pp.

Mulliken, T., and M. Haywood. 1994. Recent data on trade in rhino and tiger products, 1988–1992. Traffic Bull. 14:99–106.

Munthe, K., and J. H. Hutchison. 1978. A wolf-human encounter on Ellesmere Island, Canada. J. Mamm. 59:876–78.

Murie, A. 1981. The grizzlies of Mount McKinley. U.S. Natl. Park Serv. Sci. Monogr., no. 14, xvi + 251 pp.

Murphy, E. T. 1976. Breeding the clouded leopard *Neofelis nebulosa* at Dublin Zoo. Internatl. Zoo Yearbook 16:122–24.

Mussehl, T. W., and F. W. Howell. 1971. Game management in Montana. Montana Fish and Game Dept., Helena, xi + 238 pp.

Muul, I., and Lim Boo Liat. 1970. Ecological and morphological observations of *Felis planiceps.* J. Mamm. 51:806–8.

Myers, N. 1975*a*. The cheetah *Acinonyx jubatus* in Africa. Internatl. Union Conserv. Nat. Monogr., no. 4, 88 pp.

———. 1975*b*. The silent savannahs. Internatl. Wildl. 5(5):411.

———. 1976. The leopard *Panthera pardus* in Africa. Internatl. Union Conserv. Nat. Monogr., no. 5, 79 pp.

———. 1984. Cats in crisis. Internatl. Wildl. 14(6):42–48.

———. 1987. Africa. *In* Kingdom of cats, Natl. Wildl. Fed., Washington, D.C., 124–77.

N

Nader, I. A. 1979. The present status of the viverrids of the Arabian Peninsula (Mammalia: Carnivora: Viverridae). Senckenberg. Biol. 59:311–16.

Nader, I. A., and M. Al-Safadi. 1991. The bushy-tailed mongoose, *Bdeogale crassicauda* Peters, 1850, a new record for the Arabian Peninsula (Mammalia: Carnivora: Herpestidae). Zool. Anz. 226:202–4.

National Geographic Society. 1981. Book of mammals. Spec. Publ. Div., Natl. Geogr. Soc., Washington, D.C., 2 vols.

Navarro L., D., and M. Suarez. 1989. A survey of the pygmy raccoon *(Procyon pygmaeus)*. Mammalia 53:458–61.

Nead, D. M., J. C. Halfpenny, and S. Bissell. 1985. The status of wolverines in Colorado. Northwest Sci. 8:286–89.

Neal, E. 1970. The banded mongoose, *Mungos mungo* Gmelin. E. Afr. Wildl. J. 8:63–71.

Neal, W. A. 1993. Proposed designation of critical habitat for the Louisiana black bear. Federal Register 58:63560–69.

Nel, J. A. J. 1978. Notes on the food and foraging behavior of the bat-eared fox, *Otocyon megalotis.* Bull. Carnegie Mus. Nat. Hist., no. 6, 132–37.

Nel, J. A. J., and M. H. Bester. 1983. Communication in the southern bat-eared fox *Otocyon m. megalotis* (Desmarest, 1822). Z. Saugetierk. 48:277–90.

Nellis, C. H., and L. B. Keith. 1976. Population dynamics of coyotes in central Alberta, 1964–1968. J. Wildl. Mgmt. 40:389–99.

Nellis, C. H., S. P. Wetmore, and L. B. Keith. 1972. Lynx-prey interactions in central Alberta. J. Wildl. Mgmt. 36:320–29.

Nellis, D. W. 1989. *Herpestes auropunctatus.* Mammalian Species, no. 342, 6 pp.

Nellis, D. W., N. F. Eichholz, T. W. Regan, and C. Feinstein. 1978. Mongoose in Florida. Wildl. Soc. Bull. 6:249–50.

Nesbitt, W. H. 1975. Ecology of a feral dog pack on a wildlife refuge. *In* Fox (1975), 391–95.

Newsome, A. E., and L. K. Corbett. 1982. The identity of the dingo. II. Hybridization with domestic dogs in captivity and in the wild. Austral. J. Zool. 30:365–74.

———. 1985. The identity of the dingo. III. The incidence of dingoes, dogs, and hybrids and their coat colours in remote and settled regions of Australia. Austral. J. Zool. 33: 363–75.

Newsome, A. E., L. K. Corbett, and S. M. Carpenter. 1980. The identity of the dingo. I. Morphological discriminants of dingo and dog skulls. Austral. J. Zool. 28:615–25.

Niebauer, T. J., and O. J. Rongstad. 1977. Coyote food habits in northwestern Wisconsin. *In* Phillips and Jonkel (1977), 237–51.

Nobbe, G., and D. Garshelis. 1994. The shaggy bear. Wildl. Conserv. 97(6):32–39.

Norchi, D., and D. Bolze. 1995. Saving the tiger: a conservation strategy. Wildl. Conserv. Soc. (New York) Policy Rept., no. 3, 24 pp.

Nordstrom, L. H. 1995. Ninety-day finding for a petition to list as endangered or threatened the contiguous United States population of the North American wolverine. Federal Register 60:19567–68.

Norton, P. M. 1986. Historical changes in the distribution of leopards in the Cape Province, South Africa. Bontebok 5:1–9.

———. 1990. How many leopards? A criticism of Martin and de Meulenaer's population estimates from Africa. S. Afr. J. Sci. 86:218–20.

Novacek, M. J. 1992. Mammalian phylogeny: shaking the tree. Nature 356:121–25.

Novak, M., M. E. Obbard, J. G. Jones, R. Newman, A. Booth, A. J. Satterthwaite, and G. Linscombe. 1987. Furbearer harvests in North America, 1600–1984. Ontario Ministry Nat. Res., Toronto, xvi + 270 pp.

Novaro, A. 1993. Culpeo foxes in Patagonia. Canid News 1:15–17.

Novikov, G. A. 1962. Carnivorous mammals of the fauna of the U.S.S.R. Israel Progr. Sci. Transl., Jerusalem, 284 pp.

Nowak, E. 1984. Verbreitungs- und Bestandsentwicklung des Marderhundes, *Nyctereutes procyonides* (Gray, 1834) in Europa. Z. Jagdwiss 30:137–54.

Nowak, R. M. 1972. The mysterious wolf of the south. Nat. Hist. 81(1):50–53, 74–77.

———. 1973. Return of the wolverine. Natl. Parks and Conserv. Mag. 47(2):20–23.

———. 1974. Red wolf: our most endangered mammal. Natl. Parks and Conserv. Mag. 48(8):9–12.

———. 1975*a*. The cosmopolitan wolf. Natl. Rifle Assoc. Conserv. Yearbook, 76–82.

———. 1975*b*. Retreat of the jaguar. Natl. Parks and Conserv. Mag. 49(12):10–13.

———. 1976. The cougar in the United States and Canada. Rept. to U.S. Fish and Wildl. Serv., mimeographed, 190 pp.

———. 1978. Evolution and taxonomy of coyotes and related *Canis. In* Bekoff (1978), 3–16.

———. 1979. North American Quaternary *Canis.* Monogr. Mus. Nat. Hist. Univ. Kansas, no. 6, 154 pp.

———. 1992. The red wolf is not a hybrid. Conserv. Biol. 6:593–95.

———. 1994. Jaguars in the United States. Endangered Species Tech. Bull. 19(5):6.

Nowell, K., and P. Jackson. 1996. Wild cats: status survey and conservation action plan. IUCN (World Conservation Union), Gland, Switzerland, xxiv + 382 pp.

O

O'Brien, S. J. 1983. The cheetah is depauperate in genetic variation. Science 221:459–62.

———. 1993. The molecular evolution of bears. *In* Stirling (1993), 26–35.

———. 1994. The cheetah's conservation controversy. Conserv. Biol. 8:1153–55.

O'Brien, S. J., M. E. Roelke, L. Marker, A. Newman, C. A. Winkler, D. Meltzer, L. Colly, J. F. Evermann, M. Bush, and D. E. Wildt. 1985. Genetic basis for species vulnerability in the cheetah. Science 227:1428–34.

O'Brien, S. J., M. E. Roelke, N. Yuhki, K. W. Richards, W. E. Johnson, W. L. Franklin, A. E. Anderson, O. L. Bass, Jr., R. C. Belden, and J. S. Martenson. 1990. Genetic introgression within the Florida panther *Felis concolor coryi.* Natl. Geogr. Res. Explor. 6:485–94.

O'Farrell, T. P. 1987. Kit fox. *In* Novak, Baker, et al. (1987), 422–31.

Ognev, S. I. 1962–64. Mammals of eastern Europe and northern Asia. Israel Progr. Sci. Transl., Jerusalem, 8 vols.

Okarna, H. 1993. Status and management of the wolf in Poland. Biol. Conserv. 66:153–58.

Olney, P. J. S., P. Ellis, and F. A. Fisken. 1994. Census of rare animals in captivity. Internatl. Zoo Yearbook 33:408–53.

Olsen, S. J. 1985. Origins of the domestic dog: the fossil record. Univ. Arizona Press, Tucson, xiv + 118 pp.

Olsen, S. J., and J. W. Olsen. 1977. The Chinese wolf, ancestor of New World dogs. Science 197:533–35.

Olson, P. D. 1991. Welcome and overview. *In* Braun (1991), 2.

Olson, S. L., and G. K. Pregill. 1982. Introduction to the paleontology of Bahaman vertebrates. *In* Olson, S. L., ed., Fossil vertebrates from the Bahamas, Smithson. Contrib. Paleobiol., no. 48, 1–7.

Olterman, J. H., and B. J. Verts. 1972. Endangered plants and animals of Oregon. IV. Mammals. Oregon State Univ. Agric. Exp. Sta. Spec. Rept., no. 364, 47 pp.

Osgood, W. H. 1943. The mammals of Chile.

Field Mus. Nat. Hist. Publ., Zool. Ser., 30:1–268.

Owens, D. D., and M. J. Owens. 1979*a*. Communal denning and clan associations in brown hyenas (*Hyaena brunnea*, Thunberg) of the central Kalahari Desert. Afr. J. Ecol. 17:35–44.

———. 1979*b*. Notes on social organization and behavior in brown hyenas (*Hyaena brunnea*). J. Mamm. 60:405–8.

Owens, M. J., and D. D. Owens. 1978. Feeding ecology and its influence on social organization in brown hyenas (*Hyaena brunnea*, Thunberg) of the central Kalahari Desert. E. Afr. Wildl. J. 16:113–35.

P

Pacheco, V., H. de Macedo, E. Vivar, C. Ascorra, R. Arana-Cardó, and S. Solari. 1995. Lista anotada de los mamíferos Peruanos. Conservation International Occas. Pap., no. 2, 35 pp.

Packer, C., D. A. Gilbert, A. E. Pusey, and S. J. O'Brien. 1991. A molecular genetic analysis of kinship and cooperation in African lions. Nature 351:562–65.

Packer, C., D. Scheel, and A. E. Pusey. 1990. Why lions form groups: food is not enough. Amer. Nat. 136:1–19.

Palomares, F., and M. Delibes. 1988. Time and space use by two common genets (*Genetta genetta*) in the Doñana National Park, Spain. J. Mamm. 69:635–37.

———. 1993. Social organization in the Egyptian mongoose: group size, spatial behaviour, and inter-individual contacts in adults. Anim. Behav. 45:917–25.

———. 1994. Spatio-temporal ecology and behavior of European genets in southwestern Spain. J. Mamm. 75:714–24.

Panwar, H. S. 1987. Project tiger: the reserves, the tigers, and their future. *In* Tilson and Seal (1987), 110–17.

Paradiso, J. L. 1972. Status report on cats (Felidae) of the world, 1971. U.S. Bur. Sport Fish. and Wildl. Spec. Sci. Rept.—Wildl., no. 157, iv + 43 pp.

Paradiso, J. L., and R. M. Nowak. 1972. *Canis rufus*. Mammalian Species, no. 22, 4 pp.

Parker, C. 1979. Birth, care, and development of Chinese hog badgers *Arctonyx collaris albogularis* at Metro Toronto Zoo. Internatl. Zoo Yearbook 19:182–85.

Parker, G. R. 1973. Distribution and densities of wolves within barren-ground caribou range in northern mainland Canada. J. Mamm. 54:341–48.

Parsons, D. R. 1996. Endangered and threatened wildlife and plants: proposed establishment of a nonessential experimental population of the Mexican gray wolf in Arizona and New Mexico. Federal Register 61:19237–48.

Pasitschniak-Arts, M. 1993. *Ursus arctos*. Mammalian Species, no. 439, 10 pp.

Pasitschniak-Arts, M., and S. Larivière. 1995. *Gulo gulo*. Mammalian Species, no. 499, 10 pp.

Patnaik, S. K., and L. N. Acharjyo. 1990. White tiger in India—its past and present. Tigerpaper 17(1):8–10.

Paulraj, S., N. Sundarajan, A. Manimozhi, and S. Walker. 1992. Reproduction of the Indian wild dog (*Cuon alpinus*) in captivity. Zoo Biol. 235–41.

Payne, J., C. M. Francis, and K. Phillipps. 1985. A field guide to the mammals of Borneo. Sabah Society with World Wildlife Fund Malaysia, 332 pp.

Pearson, A. M. 1975. The northern interior grizzly bear *Ursus arctos* L. Can. Wildl. Serv. Rept. Ser., no. 34, 86 pp.

———. 1976. Population characteristics of the arctic mountain grizzly bear. *In* Pelton, Lentfer, and Folk (1976), 247–60.

Pearson, E. W. 1978. A 1974 coyote harvest estimate for seventeen western states. Wildl. Soc. Bull. 6:25–32.

Pelton, M. R., C. D. Scott, and G. M. Burghardt. 1976. Attitudes and opinions of persons experiencing property damage and/or injury by black bears in the Great Smoky Mountains National Park. *In* Pelton, Lentfer, and Folk (1976), 157–67.

Penrod, B. 1976. Fisher in New York. Conservationist 31(2):20.

Peres, C. A. 1991. Observations on hunting by small-eared (*Atelocynus microtis*) and bush dogs (*Speothos venaticus*) in central-western Amazonia. Mammalia 55:635–39.

Peters, G. 1982. A note on the vocal behaviour of the giant panda, *Ailuropoda melanoleuca* (David, 1869). Z. Saugetierk. 47:236–46.

Peters, G., and M. H. Hast. 1994. Hyoid structure, laryngeal anatomy, and vocalization in felids (Mammalia: Carnivora: Felidae). Z. Saugetierk. 59:87–104.

Peters, R. P., and L. D. Mech. 1975. Scent-marking in wolves. Amer. Sci. 63:628–37.

Petersen, L. R., M. A. Martin, and C. M. Pils. 1977. Status of fishers in Wisconsin, 1975. Wisconsin Dept. Nat. Resources Res. Rept., no. 92, 9 pp.

Peterson, R. O. 1977. Wolf ecology and prey relationships on Isle Royale. U.S. Natl. Park Serv. Sci. Monogr., no. 11, xx + 210 pp.

Peterson, R. O., and R. E. Page. 1988. The rise and fall of Isle Royale wolves, 1975–1986. J. Mamm. 69:89–99.

Peterson, R. O., and J. M. Thurber. 1993. The size of eastern coyotes (*Canis latrans*): a rebuttal. J. Mamm. 64:1075–76.

Peyton, B. 1980. Ecology, distribution, and food habits of spectacled bears, *Tremarctos ornatus*, in Peru. J. Mamm. 61:639–52.

———. 1986. Spectacled bear news. Species 6:15–16.

Phillips, M. K. 1994. Reestablishment of red wolves in North Carolina, 14 September 1987 to 31 December 1993. U.S. Fish and Wildlife Serv., Atlanta, 10 pp.

Phillips, M. K., and V. G. Henry. 1992. Comments on red wolf taxonomy. Conserv. Biol. 6:596–97.

Piekielek, W., and T. S. Burton. 1975. A black bear population study in northern California. California Fish and Game 61:4–25.

Pienaar, U. D. V. 1970. A note on the occurrence of bat-eared fox *Otocyon megalotis megalotis* (Desmarest) in the Kruger National Park. Koedoe 13:23–27.

Pilgrim, W. 1980. Fisher, *Martes pennanti* (Carnivora: Mustelidae) in Labrador. Can. Field-Nat. 94:468.

Pils, C. M., and M. A. Martin. 1978. Population dynamics, predator-prey relationships, and management of the red fox in Wisconsin. Wisconsin Dept. Nat. Res. Tech. Bull., no. 105, 56 pp.

Pimlott, D. H., ed. 1975. Wolves. Internatl. Union Conserv. Nat. Suppl. Pap., n. s., no. 43, 144 pp.

Pimlott, D. H., J. A. Shannon, and G. B. Kolenosky. 1969. The ecology of the timber wolf in Algonquin Provincial Park. Ontario Dept. Lands and Forests, Toronto, 92 pp.

Pine, R. H., S. D. Miller, and M. L. Schamberger. 1979. Contributions to the mammalogy of Chile. Mammalia 43:339–76.

Poché, R. M., S. J. Evans, P. Sultana, M. E. Hague, R. Sterner, and M. A. Siddique. 1987. Notes on the golden jackal (*Canis aureus*) in Bangladesh. Mammalia 51:259–70.

Poelker, R. J., and H. D. Hartwell. 1973. Black bear of Washington. Washington State Game Dept. Biol. Bull., no. 14, viii + 180 pp.

Poglayen-Neuwall, I. 1966. Notes on care, display, and breeding of olingos *Bassaricyon*. Internatl. Zoo Yearbook 6:169–71.

———. 1973. Preliminary notes on maintenance and behaviour of the Central American cacomistle *Bassariscus sumichrasti*. Internatl. Zoo Yearbook 13:207–11.

———. 1975. Copulatory behavior, gestation, and parturition of the tayra (*Eira barbara* L., 1758). Z. Saugetierk. 40:176–89.

———. 1989. Notes on reproduction, aging,

and longevity of *Bassaricyon* sp. (Procyonidae). Zool. Garten 59:122–28.

———. 1991. Notes on reproduction of captive *Bassariscus sumichrasti* (Procyonidae). Z. Saugetierk. 56:193–99.

———. 1992. Notes on the estrous cycles of captive tayras (Mustelidae: *Eira barbara*). Zool. Garten 62:60–62.

———. 1993. Additional observations on reproduction of the Central American cacomixtle, *Bassariscus sumichrasti* (Procyonidae). Zool. Garten 63:388–98.

Poglayen-Neuwall, I., B. S. Durrant, M. L. Swansen, R. C. Williams, and R. A. Barnes. 1989. Estrous cycle of the tayra, *Eira barbara*. Zoo Biol. 8:171–77.

Poglayen-Neuwall, I., and I. Poglayen-Neuwall. 1976. Postnatal development of tayras (Carnivora: *Eira barbara* L., 1758). Zool. Beitr. 22:345–405.

———. 1980. Gestation period and parturition of the ringtail *Bassariscus astutus* (Liechtenstein, 1830). Z. Saugetierk. 45:73–81.

———. 1993. Behavior, reproduction, and postnatal development of *Bassariscus astutus* (Carnivora; Procyonidae) in captivity. Zoo. Garten 63:73–125.

———. 1995. Observations on the ethology and biology of the Central American cacomixtle, *Bassariscus sumichrasti* (de Saussure, 1860), in captivity, with notes on its ecology. Zool. Garten 65:11–49.

Porton, I. J., D. G. Kleiman, and M. Rodden. 1987. Aseasonality of bush dog reproduction and the influence of social factors on the estrous cycle. J. Mamm. 68:867–71.

Powell, R. A. 1981. *Martes pennanti*. Mammalian Species, no. 156, 6 pp.

———. 1982. The fisher: life history, ecology, and behavior. Univ. Minnesota Press, Minneapolis, xvi + 217 pp.

———. 1994. Structure and spacing of *Martes* populations. *In* Buskirk et al. (1994), 101–21.

Preston, E. M. 1975. Home range defense in the red fox *Vulpes vulpes* L. J. Mamm. 56: 645–52.

Prestrud, P. 1991. Adaptations by the arctic fox *(Alopex lagopus)* to the polar winter. Arctic 44:132–38.

Prynn, D. 1980. Tigers and leopards in Russia's Far East. Oryx 15:496–503.

———. 1993. Tigers in the Russian Far East face renewed dangers. Oryx 27:130–31.

Pulliainen, E. 1965. On the distribution and migrations of the arctic fox (*Alopex lagopus* L.) in Finland. Aquilo, Ser. Zool., 2:25–40.

———. 1968. Breeding biology of the wolverine (*Gulo gulo* L.) in Finland. Ann. Zool. Fennici 5:338–44.

———. 1975. Wolf ecology in northern Europe. *In* Fox (1975), 292–99.

———. 1980. The status, structure, and behaviour of populations of the wolf (*Canis l. lupus* L.) along the Fenno-Soviet border. Ann. Zool. Fennici 17:107–12.

Pulliainen, E., and P. Ovaskainen. 1975. Territory marking by a wolverine *(Gulo gulo)* in northeastern Lapland. Ann. Zool. Fennici 12:268–70.

Q

Quigley, H., and M. Hornocker. 1994. The Siberian tiger project: saving endangered species through international cooperation. Endangered Species Tech. Bull. 19(3):4–11.

Quinn, N. W. S., and G. Parker. 1987. Lynx. *In* Novak, Baker, et al. (1987), 682–94.

R

Rabinowitz, A. 1988. The clouded leopard in Taiwan. Oryx 22:46–47.

———. 1989. The density and behavior of large cats in a dry tropical forest mosaic in Huai Kha Khaeng Wildlife Sanctuary, Thailand. Nat. Hist. Bull. Siam Soc. 37:235–51.

———. 1990. Notes on the behavior and movements of leopard cats, *Felis bengalensis*, in a dry tropical forest mosaic in Thailand. Biotrópica 22:397–403.

———. 1991. Behaviour and movements of sympatric civet species in Huai Kha Khaeng Wildlife Sanctuary, Thailand. J. Zool. 223: 281–98.

———. 1993. Estimating the Indochinese tiger *Panthera tigris corbetti* population in Thailand. Biol. Conserv. 65:213–17.

———. 1997. Home on the range. Wildl. Conserv. 100(5):24–27.

Rabinowitz, A., P. Andau, and P. P. K. Chai. 1987. The clouded leopard in Malaysian Borneo. Oryx 21:107–11.

Ragni, B., M. Possenti, and S. Mayr. 1993. The lynx in the Italian Alps. Cat News 19:21–25.

Ragni, B., and E. Randi. 1986. Multivariate analysis of craniometric characters in European wild cat, domestic cat, and African wild cat (genus *Felis*). Z. Saugetierk. 51:243–51.

Ralls, K., J. Ballou, and R. L. Brownell, Jr. 1983. Diversity in California sea otters: theoretical considerations and management implications. Biol. Conserv. 25:209–32.

Ramsay, M. A., and I. Stirling. 1988. Reproductive biology and ecology of female polar bears *(Ursus maritimus)*. J. Zool. 214:601–34.

Randi, E., and B. Ragni. 1991. Genetic variability and biochemical systematics of domestic and wild cat populations (*Felis silvestris:* Felidae). J. Mamm. 72:79–88.

Rasa, O. A. E. 1972. Aspects of social organization in captive dwarf mongooses. J. Mamm. 53:181–85.

———. 1973. Marking behaviour and its social significance in the African dwarf mongoose, *Helogale undulata rufula*. Z. Tierpsychol. 32:293–318.

———. 1975. Mongoose sociology and behaviour as related to zoo exhibition. Internatl. Zoo Yearbook 15:65–73.

———. 1976. Invalid care in the dwarf mongoose *(Helogale undulata rufula)*. Z. Tierpsychol. 42:337–42.

———. 1977. The ethology and sociology of the dwarf mongoose *(Helogale undulata rufula)*. Z. Tierpsychol. 43:337–406.

———. 1983. A case of invalid care in wild dwarf mongooses. Z. Tierpsychol. 62:235–40.

———. 1986. Ecological factors and their relationship to group size, mortality, and behaviour in the dwarf mongoose *Helogale undulata* (Peters, 1852). Cimbebasia, ser. A, 8:15–21.

Rasa, O. A. E., B. A. Wenhold, P. Howard, A. Marais, and J. Pallett. 1992. Reproduction in the yellow mongoose revisited. S. Afr. J. Zool. 27:192–96.

Rasmussen, J. L., and R. L. Tilson. 1984. Food provisioning by adult maned wolves *(Chrysocyon brachyurus)*. Z. Tierpsychol. 65:346–52.

Rausch, R. A., and A. M. Pearson. 1972. Notes on the wolverine in Alaska and the Yukon Territory. J. Wildl. Mgmt. 36:249–68.

Rausch, R. L. 1953. On the status of some arctic mammals. Arctic 6:91–148.

———. 1963. Geographic variation in size in North American brown bears, *Ursus arctos* L., as indicated by condylobasal length. Can. J. Zool. 41:33–45.

Ray, J. C. 1995. *Civettictis civetta*. Mammalian Species, no. 488, 7 pp.

Raybourne, J. W. 1987. Black bear: home in the highlands. *In* Kallman (1987), 105–17.

Reading, R. P., T. W. Clark, A. Vargas, L. R. Hanebury, B. J. Miller, and D. Biggins. 1996. Recent directions in black-footed ferret recovery. Endangered Species Update 13(10–11):1–7.

Ream, R. R. 1980. Wolf ecology project: annual report. Wilderness Inst., Univ. Montana, 58 pp.

Redford, K. H., and J. F. Eisenberg, eds. 1992. Mammals of the neotropics: the southern cone. Univ. Chicago Press, ix + 430 pp.

Rees, M. D. 1989. Red wolf recovery effort

intensifies. Endangered Species Tech. Bull. 14(1–2):3.

Regan, T. W., and D. S. Maehr. 1990. Melanistic bobcats in Florida. Florida Field Nat. 18:84–87.

Reid, D., M. Jiang, Q. Teng, Z. Qin, and J. Hu. 1991. Ecology of the Asiatic black bear (*Ursus* thibetanus) in Sichuan, China. Mammalia 55:221–37.

Reid, D., Hu Jinchu, and Huang Yan. 1991. Ecology of the red panda *Ailurus fulgens* in the Wolong Reserve, China. J. Zool. 225: 347–64.

Ricciuti, E. R. 1978. Dogs of war. Internatl. Wildl. 8(5):36–40.

Rice, D. W. 1977. A list of the marine mammals of the world (third edition). U.S. Natl. Mar. Fish. Serv., NOAA Tech. Rept. NMFS SSRF-711, iii + 15 pp.

Rich, M. S. 1983. The longevity record for *Lynx canadensis* Kerr, 1792. Zool. Garten 53:365.

Richardson, L., T. W. Clark, S. C. Forrest, and T. M. Campbell, III. 1987. Winter ecology of black-footed ferrets *(Mustela nigripes)* at Meeteetse, Wyoming. Amer. Midl. Nat. 117: 225–39.

Richardson, P. R. K. 1987. Aardwolf: the most specialized myrmecophagous mammal? S. Afr. J. Sci. 83:643–46.

Richardson, P. R. K., and C. D. Levitan. 1994. Tolerance of aardwolves to defense secretions of *Trinervitermes trinervoides*. J. Mamm. 75:84–91.

Ride, W. D. L. 1970. A guide to the native mammals of Australia. Oxford Univ. Press, Melbourne, xiv + 249 pp.

Riedman, M. L., and J. A. Estes. 1990. The sea otter *(Enhydra lutris):* behavior, ecology, and natural history. U.S. Fish and Wildl. Serv. Biol. Rept., no. 90(14), iii + 126 pp.

Riedman, M. L., J. A. Estes, M. M. Staedler, A. A. Giles, and D. R. Carlson. 1994. Breeding patterns and reproductive success of California sea otters. J. Wildl. Mgmt. 58:391–99.

Rieger, I. 1979. A review of the biology of striped hyaenas, *Hyaena hyaena* (Linne, 1758). Saugetierk. Mitt. 27:81–95.

———. 1981. *Hyaena hyaena*. Mammalian Species, no. 150, 5 pp.

Riffel, M. 1991. An update on the Javan ferret-badger *Melogale orientalis* (Horsfield 1821). Mustelid & Viverrid Conserv. 5:2–3.

Riley, G. A., and R. T. McBride. 1975. A survey of the red wolf *(Canis rufus)*. *In* Fox (1975), 263–77.

Roben, P. 1974. Zum vorkommen des Otters, *Lutra lutra* (Linne, 1758) in der Bündesrepublik Deutschland. Saugetierk. Mitt. 22: 29–36.

———. 1975. Zur ausbreitung des Waschbaren, *Procyon lotor* (Linne, 1758), und der Marderhundes, *Nyctereutes procyonides* (Gray, 1834), in der Bündesrepublik Deutschland. Saugetierk. Mitt. 23:93–101.

Roberts, M. S. 1975. Growth and development of mother-reared red pandas *Ailurus fulgens*. Internatl. Zoo Yearbook 15:57–63.

———. 1980. Breeding the red panda *(Ailurus fulgens)* at the National Zoological Park. Zool. Garten 50:253–63.

Roberts, M. S., and J. L. Gittleman. 1984. *Ailurus fulgens*. Mammalian Species, no. 222, 8 pp.

Roberts, T. J. 1977. The mammals of Pakistan. Ernest Benn, London, xxvi + 361 pp.

Robertshaw, J. D., and R. H. Harden. 1985. The ecology of the dingo in north-eastern New South Wales. II. Diet. Austral. Wildl. Res. 12:39–50.

Robinson, W. L., and G. J. Smith. 1977. Observations on recently killed wolves in upper Michigan. Wildl. Soc. Bull. 5:25–26.

Rodriguez, A., and M. Delibes. 1992. Current range and status of the Iberian lynx *Felis pardina* Temminck, 1824 in Spain. Biol. Conserv. 61:189–96.

Roelke, M. E., J. S. Martenson, and S. J. O'Brien. 1993. The consequences of demographic reduction and genetic depletion in the endangered Florida panther. Current Biol. 3:340–50.

Rogers, L. L. 1987. Effects of food supply and kinship on social behavior, movements, and population growth of black bears in northeastern Minnesota. Wildl. Monogr., no. 97, 72 pp.

Rogers, L. L., L. D. Mech, D. K. Dawson, J. M. Peek, and M. Korb. 1980. Deer distribution in relation to wolf pack territory edges. J. Wildl. Mgmt. 44:253–58.

Rohwer, S. A., and D. L. Kilgore, Jr. 1973. Interbreeding in the arid-land foxes, *Vulpes velox* and *V. macrotis*. Syst. Zool. 22:157–65.

Roig, V. G. 1991. Desertification and distribution of mammals in the southern cone of South America. *In* Mares, M. A., and D. J. Schmidly, eds., Latin American mammalogy: history, biodiversity, and conservation, Univ. Oklahoma Press, Norman, 239–79.

Rolley, R. E. 1987. Bobcat. *In* Novak, Baker, et al. (1987), 670–81.

Rollins, C. E., and D. E. Spencer. 1995. A fatality and the American mountain lion: bite mark analysis and profile of the offending lion. J. Forensic Sci. 40:486–89.

Romanowski, J. 1990. Minks in Poland. Mustelid & Viverrid Conserv. 2:13.

Romer, A. S. 1968. Notes and comments on vertebrate paleontology. Univ. Chicago Press, viii + 304 pp.

Rood, J. P. 1974. Banded mongoose males guard young. Nature 248:176.

———. 1975. Population dynamics and food habits of the banded mongoose. E. Afr. Wildl. J. 13:89–111.

———. 1978. Dwarf mongoose helpers at the den. Z. Tierpsychol. 48:277–87.

———. 1980. Mating relationships and breeding suppression in the dwarf mongoose. Anim. Behav. 28:143–50.

Rosatte, R. C. 1987. Striped, spotted, hooded, and hog-nosed skunk. *In* Novak, Baker, et al. (1987), 598–613.

———. 1988. Rabies in Canada: history, epidemiology, and control. Can. Vet. J. 29: 362–65.

Rose, D. A., and A. L. Gaski. 1995. Proceedings of the International Symposium on the Trade of Bear Parts for Medicinal Use. Traffic USA, World Wildlife Fund, Washington, D.C., iv + 167 pp.

Rosenberg, H. 1971. Breeding the bat-eared fox *Otocyon megalotis* at Utica Zoo. Internatl. Zoo Yearbook 11:101–2.

Rosevear, D. R. 1974. The carnivores of West Africa. British Mus. (Nat. Hist.), London, xii + 548 pp.

Rothman, R. J., and L. D. Mech. 1979. Scent-marking in lone wolves and newly formed pairs. Anim. Behav. 27:750–60.

Rotterman, L. M., and T. Simon-Jackson. 1988. Sea otter: *Enhydra lutris*. *In* Lentfer, J. W., ed., Selected marine mammals of Alaska, Mar. Mamm. Comm., Washington, D.C., 239–75.

Rounds, R. C. 1987. Distribution and analysis of colourmorphs of the black bear *(Ursus americanus)*. J. Biogeogr. 14:521–38.

Rovner, S. 1992. When a dog is not always man's best friend. Washington Post Health, 31 March, 7.

Rowe-Rowe, D. T. 1977. Prey capture and feeding behaviour of South African otters. Lammergeyer, no. 23, 13–21.

———. 1978*a*. Reproduction and post-natal development of South African mustelines (Carnivora: Mustelidae). Zool. Afr. 13:103–14.

———. 1978*b*. The small carnivores of Natal. Lammergeyer, no. 25, 48 pp.

———. 1982. Home range and movements of black-backed jackals in an African montane region. S. Afr. J. Wildl. Res. 12:79–84.

———. 1990. The African weasel: a red data book species in South Africa. Mustelid & Viverrid Conserv. 2:6–7.

Roy, M. S., E. Geffen, D. Smith, E. A. Ostrander, and R. K. Wayne. 1994. Patterns of differentiation and hybridization in North

American wolflike canids, revealed by analysis of microsatellite loci. Molecular Biol. Evol. 11:553–70.

Roy, M. S., D. J. Girman, and R. K. Wayne. 1994. The use of museum specimens to reconstruct the genetic variability and relationships of extinct populations. Experientia 50:1–7.

Roychoudhury, A. K. 1987. White tigers and their conservation. *In* Tilson and Seal (1987), 380–88.

Rozhnov, V. V. 1994. Notes on the behaviour and ecology of the binturong *(Arctictis binturong)* in Vietnam. Small Carnivore Conserv. 10:4–5.

Rozhnov, V. V., G. V. Kuznetzov, and P. T. Anh. 1992. New distributional information on Owston's palm civet. Small Carnivore Conserv. 6:7.

Rubin, H. D., and A. M. Beck. 1982. Ecological behavior of free-ranging urban dogs. Appl. Anim. Ethol. 8:161–68.

Rudnai, J. A. 1973. The social life of the lion. Washington Square East, Publishers, Wallingford, Pennsylvania, 122 pp.

Ruiz-Olmo, J., and S. Palazon. 1991. New information on European and American minks in the Iberian Peninsula. Mustelid & Viverrid Conserv. 5:13.

Russell, J. K. 1981. Exclusion of adult male coatis from social groups: protection from predation. J. Mamm. 62:206–8.

Russell, R. H. 1975. The food habits of polar bears of James Bay and southwest Hudson Bay in summer and autumn. Arctic 28:117–29.

Russell, W. C., E. T. Thorne, R. Oakleaf, and J. D. Ballou. 1994. The genetic basis of black-footed ferret reintroduction. Conserv. Biol. 8:263–66.

S

Sabean, B. 1989. The eastern coyote in Nova Scotia. NS Conservation 13(1):5–8.

Saberwal, V. K., J. P. Gibbs, R. Chellam, and A. J. T. Johnsingh. 1994. Lion-human conflict in the Gir Forest, India. Conserv. Biol. 8:501–7.

Salles, L. O. 1992. Felid phylogenetics: extant taxa and skull morphology (Felidae, Aeluroidea). Amer. Mus. Novit., no. 3047, 67 pp.

Sanborn, C. C. 1952. Philippine Zoological Expedition 1946–1947: mammals. Fieldiana Zool. 33:89–158.

Sanderson, G. C., and A. V. Nalbandov. 1973. The reproductive cycle of the raccoon in Illinois. Illinois Nat. Hist. Surv. Bull. 31:29–85.

Santiapillai, C. 1989. The status and conservation of the clouded leopard *(Neofelis nebulosa diardi)* in Sumatra. Tigerpaper 16(1): 1–7.

Santiapillai, C., M. R. Chambers, and N. Ishwaran. 1982. The leopard *Panthera pardus fusca* (Meyer 1794) in the Ruhuna National Park, Sri Lanka, and observations relevant to its conservation. Biol. Conserv. 23:5–14.

Sargeant, A. B., and D. W. Warner. 1972. Movements and denning habits of a badger. J. Mamm. 53:207–10.

Schaller, G. B. 1967. The deer and the tiger. Univ. Chicago Press, 370 pp.

———. 1972. The Serengeti lion: a study of predator-prey relations. Univ. Chicago Press, xiii + 480 pp.

———. 1977. Mountain monarchs: wild sheep and goats of the Himalaya. Univ. Chicago Press, xviii + 425 pp.

———. 1980*a*. Epitaph for a jaguar. Anim. Kingdom 83(2):4–11.

———. 1980*b*. Movement patterns of jaguar. Biotrópica 12:161–68.

———. 1981. Pandas in the wild. Natl. Geogr. 160:735–49.

———. 1993. The last panda. Univ. Chicago Press, xx + 291 pp.

Schaller, G. B., Hu Jinchu, Pan Wenshi, and Zhu Jing. 1985. The giant pandas of Wolong. Univ. Chicago Press, xix + 298 pp.

Schaller, G. B., Li Hong, Talipu, Ren Junrang, and Qiu Mingjiang. 1988. The snow leopard in Xinjiang, China. Oryx 22:197–204.

Schaller, G. B., R. Tulgat, and B. Navantsatsvalt. 1993. Observations on the Gobi brown bear in Mongolia. Proc. 6th Conf. Specialists Studying Bears, Central Forest State Reserve, Ministry of Environmental Protection, Russian Federation, Moscow, 110–24.

Schempf, P. F., and M. White. 1977. Status of six furbearer populations in the mountains of northern California. U.S. Forest Serv., Washington, D.C., iv + 51 pp.

Schlawe, L. 1980. Zur geographischen Verbreitung der Ginsterkatzen, Gattung *Genetta* Cuvier, 1816. Faun. Abhandl. (Dresden) 7:147–61.

———. 1981. Material, Fundorte, Text- und Bildquellan als Grundlagen für eine Artenliste zur Revision der Gattung *Genetta* G. Cuvier, 1816 (Mammalia, Carnivora, Viverridae). Zool. Abhandl. (Dresden) 37:85–182.

Schlitter, D. A. 1974. Notes on the Liberian mongoose, *Liberiictis kuhni* Hayman, 1958. J. Mamm. 55:438–42.

Schmidly, D. J. 1983. Texas mammals east of the Balcones Fault Zone. Texas A&M Univ. Press, College Station, xviii + 400 pp.

Schneider, D. G., L. D. Mech, and J. R. Tester. 1971. Movements of female raccoons and their young as determined by radio-tracking. Anim. Behav. Monogr. 4:1–43.

Schoen, J. W., S. D. Miller, and H. V. Reynolds, III. 1987. Last stronghold of the grizzly. Nat. Hist. 96(1):50–61.

Schreiber, A., R. Wirth, M. Riffel, and H. Van Rompaey. 1989. Weasels, civets, mongooses, and their relatives: an action plan for the conservation of mustelids and viverrids. Internatl. Union Conserv. Nat., Gland, Switzerland, iv + 99 pp.

Scott, M. D., and K. Causey. 1973. Ecology of feral dogs in Alabama. J. Wildl. Mgmt. 37: 253–65.

Scott, P. A., C. V. Bentley, and J. J. Warren. 1985. Aggressive behavior by wolves toward humans. J. Mamm. 66:807–9.

Seal, U. S., P. Jackson, and R. L. Tilson. 1987. A global tiger conservation plan. *In* Tilson and Seal (1987), 487–98.

Seal, U. S., E. T. Thorne, M. A. Bogan, and S. H. Anderson. 1989. Conservation biology and the black-footed ferret. Yale Univ. Press, New Haven, 302 pp.

Sealander, J. A. 1979. A guide to Arkansas mammals. River Road Press, Conway, Arkansas, x + 313 pp.

Searls, D., and S. C. Torbit. 1993. Proposed establishment of a nonessential experimental population of black-footed ferrets in southwestern South Dakota. Federal Register 58: 29176–86.

Seidensticker, J., R. K. Lahiri, K. C. Das, and A. Wright. 1976. Problem tiger in the Sundarbans. Oryx 13:267–73.

Seidensticker, J., and S. Lumpkin, eds. 1991. Great cats: majestic creatures of the wild. Rodale Press, Emmaus, Pennsylvania, 240 pp.

Seidensticker, J. C., IV, M. G. Hornocker, W. V. Wiles, and J. P. Messick. 1973. Mountain lion social organization in the Idaho Primitive Area. Wildl. Monogr., no. 35, 60 pp.

Serez, M., and M. Eroglu. 1994. A new threatened wolf species, *Cuon alpinus hesperius* Afanasiev and Zolatarev, 1935, in Turkey. Council of Europe Environ. Encounters, no. 17, 103–6.

Servheen, C. 1989. The status and conservation of the bears of the world. Internatl. Conf. Bear Res. Mgmt. Monogr. Ser., no. 2, 32 pp.

Seymour, K. 1989. *Panthera onca*. Mammalian Species, no. 340, 9 pp.

Shahi, S. P. 1982. Status of the grey wolf (*Canis lupus pallipes* Sykes) in India—a preliminary survey. J. Bombay Nat. Hist. Soc. 79:493–502.

Shapiro, A. E. 1981. Florida panther population studies. Endangered Species Tech. Bull. 6(7):1, 3.

Shaw, J. H., and P. A. Jordan. 1977. The wolf that lost its genes. Nat. Hist. 86(10):80–88.

Sheffield, S. R., and C. M. King. 1994. *Mustela nivalis*. Mammalian Species, no. 454, 10 pp.

Sillero-Zubiri, C., and D. Gottelli. 1994. *Canis simensis*. Mammalian Species, no. 485, 6 pp.

Simpson, C. D. 1964. Notes on the banded mongoose, *Mungos mungo* (Gmelin). Arnoldia 1(19):1–8.

Simpson, G. G. 1945. The principles of classification and a classification of the mammals. Bull. Amer. Mus. Nat. Hist. 85:i–xvi + 1–350.

Sissom, D. E. F., D. A. Rice, and G. Peters. 1991. How cats purr. J. Zool. 223:67–78.

Sitek, H. 1995. Breeding of the Libyan striped weasel *Poecilictis libyca* at Poznan Zoo, Poland. Small Carnivore Conserv. 13:8–9.

Skinner, J. D. 1976. Ecology of the brown hyaena *Hyaena brunnea* in the Transvaal with a distribution map for southern Africa. S. Afr. J. Sci. 72:262–69.

Skinner, J. D., S. Davis, and G. Ilani. 1980. Bone collecting by striped hyaenas, *Hyaena hyaena*, in Israel. Paleontol. Afr. 23:99–104.

Skinner, J. D., N. Fairall, and J. D. P. Bothma. 1977. South African red data book—large mammals. S. Afr. Natl. Sci. Programmes Rept., no. 18, v + 29 pp.

Skinner, J. D., and G. Ilani. 1979. The striped hyaena *Hyaena hyaena* of the Judean and Negev deserts and a comparison with the brown hyaena *H. brunnea*. Israel J. Zool. 28:229–32.

Skinner, J. D., and R. H. N. Smithers. 1990. The mammals of the southern African subregion. Univ. Pretoria, xxxii + 771 pp.

Skinner, J. D., and R. J. Van Aarde. 1988. The use of space by the aardwolf *Proteles cristatus*. J. Zool. 214:299–301.

Skoog, P. 1970. The food of the Swedish badger, *Meles meles* L. Viltrevy 7:1–120.

Slobodyan, A. A. 1976. The European brown bear in the Carpathians. *In* Pelton, Lentfer, and Folk (1976), 313–19.

Slough, B. G. 1994. Translocations of American martens: an evaluation of factors in success. *In* Buskirk et al. (1994), 165–78.

Smale, L., L. G. Frank, and K. E. Holecamp. 1993. Ontogeny of dominance in free-living spotted hyaenas: juvenile rank relations with adult females and immigrant males. Anim. Behav. 46:467–77.

Smeeton, C. 1993. Mee yah chah, the swift fox. Canid News 1:7–9.

Smielowski, J. M. 1985. Longevity of carnivores from South America. Zool. Garten 55:177.

———. 1986. Longevity of large Indian civet, *Viverra zibetha* L., 1758. Zool. Garten 56:436–37.

Smit, C. J., and A. Van Wijngaarden. 1981. Threatened mammals in Europe. Akademische Verlagsgesellschaft, Wiesbaden, 259 pp.

Smith, A. T., and D. M. Cary. 1982. Distribution of Everglades mink. Florida Sci. 45:106–12.

Smith, J. L. D., and C. W. McDougal. 1991. The contribution of variance in lifetime reproduction to effective population size in tigers. Conserv. Biol. 5:484–90.

Smith, J. L. D., C. W. McDougal, and M. E. Sunquist. 1987. Female land tenure system in tigers. *In* Tilson and Seal (1987), 97–109.

Smith, K. G., and J. D. Clark. 1994. Black bears in Arkansas: characteristics of a successful translocation. J. Mamm. 75:309–20.

Smith, K. G., J. D. Clark, and S. C. Shull. 1993. The reintroduction of black bears in Arkansas. *In* Stirling (1993), 222–23.

Smith, P. A., and C. J. Jonkel. 1975. Résumé of the trade in polar bear hides in Canada, 1973–74. Can. Wildl. Serv. Progress Notes, no. 48, 5 pp.

Smith, R. M. 1977. Movement patterns and feeding behaviour of leopard in the Rhodes Matopos National Park, Rhodesia. Arnoldia 8(13):1–16.

Smith, T. G. 1976. Predation of ringed seal pups *(Phoca hispida)* by the arctic fox *(Alopex lagopus)*. Can. J. Zool. 54:1610–16.

Smithers, R. H. N. 1971. The mammals of Botswana. Trustees Natl. Museums and Monuments Rhodesia Mus. Mem., no. 4, 340 pp.

———. 1978. The serval, *Felis serval* Schreber, 1776. S. Afr. J. Wildl. Res. 8:29–37.

———. 1983. The mammals of the southern African subregion. Univ. Pretoria, xxii + 736 pp.

———. 1986. South African red data book—terrestrial mammals. S. Afr. Natl. Sci. Programmes Rept., no. 125, ix + 216 pp.

Smythe, N. 1970. The adaptive value of the social organization of the coati *(Nasua narica)*. J. Mamm. 51:818–20.

Solt, V. 1981. Black-footed ferret discovered on Wyoming ranch. Fish and Wildl. News, Oct.–Nov., 3.

Sosnovskii, I. P. 1967. Breeding the red dog *Cuon alpinus* at Moscow Zoo. Internatl. Zoo Yearbook 7:120–22.

Soutiere, E. C. 1979. Effects of timber harvesting on marten in Maine. J. Wildl. Mgmt. 43:850–60.

Spassov. N. 1989. The position of jackals in the *Canis* genus and life-history of the golden jackal (*Canis aureus* L.) in Bulgaria and on the Balkans. Hist. Nat. Bulgarica 1:44–56.

Spiess, A. 1976. Labrador grizzly (*Ursus arctos* L.): first skeletal evidence. J. Mamm. 57:787–90.

Springer, J. T. 1980. Fishing behavior of coyotes on the Columbia River, southcentral Washington. J. Mamm. 61:373–74.

Squires, S. 1995. Warnings reissued in wake of rabies deaths. Washington Post Health, 7 November, 8–11.

Stahl, P., and M. Artois. 1994. Status and conservation of the wildcat *(Felis silvestris)* in Europe and around the Mediterranean rim. Council of Europe Nature and Environment Ser., no. 69, 76 pp.

Stains, H. J. 1967. Carnivores and pinnipeds. *In* Anderson and Jones (1967), 325–54.

———. 1984. Carnivores. *In* Anderson and Jones (1984), 491–522.

Stanley, W. C. 1963. Habits of the red fox in northeastern Kansas. Univ. Kansas Mus. Nat. Hist. Misc. Publ., no. 34, 31 pp.

Stardom, R. R. P. 1983. Status and management of wolves in Manitoba. *In* Carbyn (1983), 30–34.

Statistics Canada. 1981. Fur production. Can. Min. Supply and Services, Ottawa.

Steck, F., and A. Wandeler. 1980. The epidemiology of fox rabies in Europe. Epidemiol. Rev. 2:71–95.

Sterner, R. T., and S. A. Shumake. 1978. Coyote damage-control research: a review and analysis. *In* Bekoff (1978), 297–325.

Stewart, P. 1993. Mapping the dhole. Canid News 1:18–21.

———. 1994. Mapping the dhole update. Canid News 2:35–36.

Stirling, I. 1974. Midsummer observations on the behavior of wild polar bears *(Ursus maritimus)*. Can. J. Zool. 52:1191–98.

———, ed. 1993. Bears: majestic creatures of the wild. Rodale Press, Emmaus, Pennsylvania, 240 pp.

Stirling, I., W. Calvert, and D. Andriashek. 1980. Population ecology studies of the polar bear in the area of southeastern Baffin Island. Can. Wildl. Serv. Occas. Pap., no. 44, 33 pp.

Stirling, I., and A. E. Derocher. 1993. Possible impacts of climatic warming on polar bears. Arctic 46:240–45.

Stirling, I., C. Jonkel, P. Smith, R. Robertson, and D. Cross. 1977. The ecology of the polar bear *(Ursus maritimus)* along the western coast of Hudson Bay. Can. Wildl. Serv. Occas. Pap., no. 33, 64 pp.

Stirling, I., and H. P. L. Kiliaan. 1980. Population ecology studies of the polar bear in northern Labrador. Can. Wildl. Serv. Occas. Pap., no. 42, 21 pp.

Stocek, R. F. 1995. The cougar, *Felis concolor,* in the Maritime provinces. Can. Field-Nat. 109:19–22.

Storer, T. I., and L. P. Tevis, Jr. 1955. California grizzly. Univ. California Press, Berkeley, vii + 335 pp.

Storm, G. L., R. D. Andrews, R. L. Phillips, R. A. Bishop, D. B. Siniff, and J. R. Tester. 1976. Morphology, reproduction, dispersal, and mortality of midwestern red fox populations. Wildl. Monogr., no. 49, 81 pp.

Storm, G. L., and G. G. Montgomery. 1975. Dispersal and social contact among red foxes: results from telemetry and computer simulation. *In* Fox (1975), 237–46.

Strahl, S. D., J. L. Silva, and I. R. Goldstein. 1992. The bush dog *(Speothos venaticus)* in Venezuela. Mammalia 56:9–13.

Strickland, M. A. 1994. Harvest management of fishers and American martens. *In* Buskirk et al. (1994), 149–64.

Strickland, M. A., and C. W. Douglas. 1987. Marten. *In* Novak, Baker, et al. (1987), 530–47.

Stroganov, S. U. 1969. Carnivorous mammals of Siberia. Israel Progr. Sci. Transl., Jerusalem, x + 522 pp.

Stromberg, M. R., and M. S. Boyce. 1986. Systematics and conservation of the swift fox, *Vulpes velox,* in North America. Biol. Conserv. 35:97–110.

Stuart, C. T. 1985. The status of two endangered carnivores occurring in the Cape Province, South Africa, *Felis serval* and *Lutra maculicollis.* Biol. Conserv. 32:375–82.

Stuart, C. T., and V. J. Wilson. 1988. The cats of southern Africa. Internatl. Union Conserv. Nat. Species Survival Comm. Cat Specialist Group and Chipangali Wildl. Trust, Bulawayo, Zimbabwe, 32 pp.

Sunquist, M. E. 1974. Winter activity of striped skunks *(Mephitis mephitis)* in east-central Minnesota. Amer. Midl. Nat. 92:434–46.

———. 1981. The social organization of tigers *(Panthera tigris)* in Royal Chitawan National Park, Nepal. Smithson. Contrib. Zool., no. 336, vi + 98 pp.

———. 1982. Movements and habitat use of a sloth bear. Mammalia 46:545–47.

Sunquist, M. E., C. Leh, F. Sunquist, D. M. Hills, and R. Rajaratnam. 1994. Rediscovery of the Bornean bay cat. Oryx 28:67–70.

Svendsen, G. E. 1976. Vocalizations of the long-tailed weasel *(Mustela frenata).* J. Mamm. 57:398–99.

Swank, W. G., and J. G. Teer. 1989. Status of the jaguar—1987. Oryx 23:14–21.

T

Taberlet, P., and J. Bouvet. 1994. Mitochondrial DNA polymorphism, phylogeography, and conservation genetics of the brown bear *Ursus arctos* in Europe. Proc. Roy. Soc. London, ser. B, 255:195–200.

Tan Bangjie. 1987. Status and problems of captive tigers in China. *In* Tilson and Seal (1987), 134–48.

Tatara, M. 1994. Ecology and conservation status of the Tsushima marten. *In* Buskirk et al. (1994), 272–79.

Taylor, E. H. 1934. Philippine land mammals. Philippine Bur. Sci. Monogr., no. 30, 548 pp.

Taylor, M. E. 1972. *Ichneumia albicauda.* Mammalian Species, no. 12, 4 pp.

———. 1975. *Herpestes sanguineus.* Mammalian Species, no. 65, 5 pp.

———. 1986. Aspects on the biology of the four-toed mongoose, *Bdeogale crassicauda.* Cimbebasia, ser. A, 8:187–93.

———. 1987. *Bdeogale crassicauda.* Mammalian Species, no. 294, 4 pp.

———. 1989. New records of two species of rare viverrids from Liberia. Mammalia 53: 122–25.

———. 1992. The Liberian mongoose. Oryx 26:103–6.

Taylor, M. E., and C. A. Goldman. 1993. The taxonomic status of the African mongooses, *Herpestes sanguineus, H. nigratus, H. pulverulentus,* and *H. ochraceus* (Carnivora: Viverridae). Mammalia 57:375–91.

Taylor, P. J., and J. Meester. 1993. *Cynictis penicillata.* Mammalian Species, no. 432, 7 pp.

Tedford, R. H. 1976. Relationships of pinnipeds to other carnivores (Mammalia). Syst. Zool. 25:363–74.

Terrill, C. E. 1986. Trends of predator losses of sheep and lambs from 1940 through 1985. Proc. 12th Vert. Pest Conf., Univ. California, Davis, 347–51.

Theberge, J. B. 1991. Ecological classification, status, and management of the gray wolf, *Canis lupus,* in Canada. Can. Field-Nat. 105: 459–63.

Thenius, E. 1979. Zur systematischen und phylogenetischen Stellung des Bambusbäaren: *Ailuropoda melanoleuca* David (Carnivora, Mammalia). Z. Saugetierk. 44: 286–305.

Thiel, R. P. 1985. Relationship between road densities and wolf habitat suitability in Wisconsin. Amer. Midl. Nat. 113:404–7.

Thomas, H. H., and R. L. Dibblee. 1986. A coyote, *Canis latrans,* on Prince Edward Island. Can. Field-Nat. 100:565–67.

Thompson, I. D. 1991. Could marten become the spotted owl of eastern Canada. Forestry Chron. 67:136–40.

Thomson, P. C. 1992*a*. The behavioural ecology of dingoes in north-western Australia. II. Activity patterns, breeding season, and pup rearing. Wildl. Res. 19:519–30.

———. 1992*b*. The behavioural ecology of dingoes in north-western Australia. III. Hunting and feeding behaviour, and diet. Wildl. Res. 19:531–41.

———. 1992*c*. The behavioural ecology of dingoes in north-western Australia. IV. Social and spatial organisation, and movements. Wildl. Res. 19:543–63.

Thornback, J., and M. Jenkins. 1982. The IUCN mammal red data book. Part 1: Threatened mammalian taxa of the Americas and the Australasian zoogeographic region (excluding Cetacea). Internatl. Union Conserv. Nat., Gland, Switzerland, xl + 516 pp.

Thornton, W. A., and G. C. Creel. 1975. The taxonomic status of kit foxes. Texas J. Sci. 26:127–36.

Timmis, W. H. 1971. Observations on breeding the oriental short-clawed otter *Amblonyx cinerea* at Chester Zoo. Internatl. Zoo Yearbook 11:109–11.

Tompa, F. S. 1983. Problem wolf management in British Columbia: conflict and program evaluation. *In* Carbyn (1983), 112–19.

Trapp, G. R. 1972. Some anatomical and behavioral adaptations of ringtails *Bassariscus astutus.* J. Mamm. 53:549–57.

Trapp, G. R., and D. L. Hallberg. 1975. Ecology of the gray fox *(Urocyon cinereoargenteus):* a review. *In* Fox (1975), 164–78.

Troughton, E. Le G. 1971. The early history and relationships of the New Guinea highland dog *(Canis hallstromi).* Proc. Linnean Soc. New South Wales 96:93–98.

Tully, R. J. 1991. Results, 1991 questionnaire on damage to livestock by mountain lion. *In* Braun (1991), 68–74.

Tumanov, I. L., and E. L. Zverev. 1986. Present distribution and numbers of the European mink *(Mustela lutreola)* in the USSR. Zool. Zhur. 65:426–35.

Tumlison, R. 1987. *Felis lynx.* Mammalian Species, no. 269, 8 pp.

U

Ulmer, F. A., Jr. 1966. Hand-rearing polar bear cubs. America's First Zoo 18(1):3–5.

———. 1968. Breeding fishing cats, *Felis viverrina,* at Philadelphia Zoo. Internatl. Zoo Yearbook 8:49–55.

U.S. (United States) Fish and Wildlife Service. 1980. Administration of the Marine Mammal Protection Act of 1972, April 1, 1979 to March 31, 1980. Washington, D.C., v + 86 pp.

Urban, D. 1970. Raccoon populations, movement patterns, and predation on a managed waterfowl marsh. J. Wildl. Mgmt. 34:372–82.

Uspenski, S. M., and S. E. Belikov. 1976. Research on the polar bear in the USSR. *In* Pelton, Lentfer, and Folk (1976), 321–23.

Ustinov, S. K. 1976. The brown bear on Baikal: a few features of vital activity. *In* Pelton, Lentfer, and Folk (1976), 325–26.

V

Valtonen, M. H., E. J. Rajakoski, and J. I. Mäkelä. 1977. Reproductive features in the female raccoon dog. J. Reprod. Fert. 51:517–18.

Van Aarde, R. J., and J. D. Skinner. 1986. Pattern of space use by relocated servals, *Felis serval.* Afr. J. Ecol. 24:97–101.

Van Aarde, R. J., and A. Van Dyk. 1986. Inheritance of the king coat colour pattern in cheetahs *Acinonyx jubatus.* J. Zool. 209: 573–78.

Van Ballenberghe, V. 1983. Two litters raised in one year by a wolf pack. J. Mamm. 64:171–72.

Van Ballenberghe, V., and A. W. Erickson. 1973. A wolf pack kills another wolf. Amer. Midl. Nat. 90:490–93.

Van Ballenberghe, V., A. W. Erickson, and D. Byman. 1975. Ecology of the timber wolf in northeastern Minnesota. Wildl. Monogr., no. 43, 43 pp.

Van Ballenberghe, V., and L. D. Mech. 1975. Weights, growth, and survival of timber wolf pups in Minnesota. J. Mamm. 56:44–63.

Van Bree, P. J. H., and M. Sc. Boeadi. 1978. Notes on the Indonesian mountain weasel, *Mustela lutreolina* Robinson and Thomas, 1917. Z. Saugetierk. 43:166–71.

Van Bree, P. J. H., and M. K. B. M. Khan. 1992. On a fishing cat, *Felis (Prionailurus) viverrina* Bennett, 1833, from continental Malaysia. Z. Saugetierk. 57:179–80.

Van Camp, J., and R. Gluckie. 1979. A record long-distance movement by a wolf *(Canis lupus).* J. Mamm. 60:236–37.

Van Den Brink, F. H. 1968. A field guide to the mammals of Britain and Europe. Houghton Mifflin, Boston, 221 pp.

Van Den Brink, J. 1980. The former distribution of the Bali tiger, *Panthera tigris balica* (Schwarz, 1912). Saugetierk. Mitt. 28:286–89.

Van Gelder, R. G. 1977. Mammalian hybrids and generic limits. Amer. Mus. Novit., no. 2635, 25 pp.

———. 1978. A review of canid classification. Amer. Mus. Novit., no. 2646, 10 pp.

———. 1979. Mongooses on mainland North America. Wildl. Soc. Bull. 7:197–98.

Van Heerden, J., and F. Kuhn. 1985. Reproduction in captive hunting dogs *Lycaon pictus.* S. Afr. J. Wildl. Res. 15:80–84.

Van Lawick, H., and J. Van Lawick–Goodall. 1971. Innocent killers. Houghton Mifflin, Boston, 222 pp.

Van Rompaey, H. 1978. Longevity of a banded mongoose *(Mungos mungo gmelin)* in captivity. Internatl. Zoo News 25(3):32–33.

———. 1988. *Osbornictis piscivora.* Mammalian Species, no. 309, 4 pp.

Van Rompaey, H., and M. Colyn. 1992. *Crossarchus ansorgei.* Mammalian Species, no. 402, 3 pp.

Van Staaden, M. J. 1994. *Suricata suricatta.* Mammalian Species, no. 483, 8 pp.

Van Valkenburgh, B., F. Grady, and B. Kurten. 1990. The Plio-Pleistocene cheetah-like cat *Miracinonyx inexpectatus* of North America. J. Vert. Paleontol. 10:434–54.

Van Valkenburgh, B., and R. K. Wayne. 1994. Shape divergence associated with size convergence in sympatric East African jackals. Ecology 75:1567–81.

Van Zyll de Jong, C. G. 1972. A systematic review of the nearctic and neotropical river otters (genus *Lutra,* Mustelidae, Carnivora). Roy. Ontario Mus. Life Sci. Contrib., no. 80, 104 pp.

———. 1975. The distribution and abundance of the wolverine *(Gulo gulo)* in Canada. Can. Field-Nat. 89:431–37.

———. 1987. A phylogenetic study of the Lutrinae (Carnivora; Mustelidae) using morphological data. Can J. Zool. 65:2536–44.

———. 1992. A morphometric analysis of cranial variation in Holarctic weasels *(Mustela nivalis).* Z. Saugetierk. 57:77–93.

Vaughan, C. 1983. Coyote range expansion in Costa Rica and Panama. Brenesia 21:27–32.

Vaughn, R. 1974. Breeding the tayra *Eira barbara* at Antelope Zoo, Lincoln. Internatl. Zoo Yearbook 14:120–22.

Veitch, A. M. 1993. A unique population of black bears on the tundra. *In* Stirling (1993), 116–17.

Vereschagin, N. K. 1976. The brown bear in Eurasia, particularly the Soviet Union. *In* Pelton, Lentfer, and Folk (1976), 327–35.

Véron, G., and F. Catzeflis. 1993. Phylogenetic relationships of the endemic Malagasy carnivore *Cryptoprocta ferox* (Aeluroidea): DNA/DNA hybridization experiments. J. Mamm. Evol. 1:169–85.

Verwoerd, D. J. 1987. Observations on the food and status of the Cape clawless otter, *Aonyx capensis,* at Betty's Bay, South Africa. S. Afr. J. Zool. 22:33–39.

Vitullo, A. D., and G. A. Zuleta. 1992. Cytogenetics and fossil record: confluent evidence for speciation without chromosomal change in South American canids. Z. Saugetierk. 57:248–50.

Voigt, D. R. 1987. Red fox. *In* Novak, Baker, et al. (1987), 378–93.

Voigt, D. R., G. B. Kolenosky, and D. H. Pimlott. 1976. Changes in summer foods of wolves in central Ontario. J. Wildl. Mgmt. 40:663–68.

Voipio, P. 1990. The samson fox episode in Finland in the 1930s and 1940s, and the hypothetico-deductive method. Ann. Zool. Fennici 27:21–27.

W

Wackernagel, H. 1968. A note on breeding the serval cat, *Felis serval,* at Basle Zoo. Internatl. Zoo Yearbook 8:46–47.

Wada, L. L. L. 1995. Proposed establishment of a nonessential experimental population of black-footed ferrets in Aubrey Valley, Arizona. Federal Register 60:57387–96.

Waddell, W. 1995*a.* Red wolf *(Canis rufus).* AZA Communiqué, March, 7–8.

____. 1995*b.* Red wolf *Canis rufus gregoryi* studbook report: 1994 annual update. Point Defiance Zoo, Tacoma, 3 pp.

Wade, D. A. 1978. Coyote damage: a survey of its nature and scope, control measures and their application. *In* Bekoff (1978), 347–68.

Wade-Smith, J., and M. E. Richmond. 1975. Care, management, and biology of captive striped skunks *(Mephitis mephitis).* Lab. Anim. Sci. 25:575–84.

———. 1978. Reproduction in captive striped skunks *(Mephitis mephitis).* Amer. Midl. Nat. 100:452–55.

Wade-Smith, J., and B. J. Verts. 1982. *Mephitis mephitis.* Mammalian Species, no. 173, 7 pp.

Waechter, A. 1975. Ecologie de la fouine en Alsace. Terre Vie 29:399–457.

Wang Tsiang-Ke. 1974. On the taxonomic status of species, geological distribution, and evolutionary history of *Ailuropoda.* Acta Zool. Sinica 20:201.

Ward, G. C. 1987. India's intensifying dilemma: can tigers and people coexist? Smithsonian 18(8):52–65.

Ward, O. G., and D. H. Wurster-Hill. 1989. Ecological studies of Japanese raccoon dogs, *Nyctereutes procyonoides viverrinus.* J. Mamm. 70:330–34.

———. 1990. *Nyctereutes procyonoides.* Mammalian Species, no. 358, 5 pp.

Waser, P. M., and M. S. Waser. 1985. *Ichneumia* and the evolution of viverrid gregariousness. Z. Tierpsychol. 68:137–51.

Watson, J. P. 1990. The taxonomic status of the slender mongoose, *Galerella sanguinea* (Rüppell, 1836), in southern Africa. Navors. Nas. Mus. Bloemfontein 6:351–492.

Watson, J. P., and N. J. Dippenaar. 1987. The species limits of *Galerella sanguinea* (Rüppell, 1836), *G. pulverulenta* (Wagner, 1839), and *G. nigrita* (Thomas, 1928) in southern Africa (Carnivora: Viverridae). Navors. Nas. Mus. Bloemfontein 5:355–414.

Wayne, R. K. 1992. On the use of morphologic and molecular genetic characters to investigate species status. Conserv. Biol. 6: 590–92.

———. 1993. Molecular evolution of the dog family. Trends in Genetics 9:218–24.

Wayne, R. K., R. E. Benveniste, D. N. Janczewski, and S. J. O'Brien. 1989. Molecular and biochemical evolution of the Carnivora. *In* Gittleman (1989), 465–94.

Wayne, R. K., S. B. George, D. Gilbert, P. W. Collins, S. D. Kovach, D. Girman, and N. Lehman. 1991. A morphologic and genetic study of the island fox, *Urocyon littoralis.* Evolution 45:1849–68.

Wayne, R. K., D. A. Gilbert, N. Lehman, K. Hansen, A. Eisenhawer, D. Girman, R. O. Peterson, L. D. Mech, P. J. P. Gogan, U. S. Seal, and R. J. Krumenaker. 1991. Conservation genetics of the endangered Isle Royale gray wolf. Conserv. Biol. 5:41–51.

Wayne, R. K., and J. L. Gittleman. 1995. The problematic red wolf. Sci. Amer. 273(1):36–39.

Wayne, R. K., and S. M. Jenks. 1991. Mitochondrial DNA analysis implying extensive hybridization of the endangered red wolf, *Canis rufus.* Nature 351:565–68.

Wayne, R. K., W. S. Modi, and S. J. O'Brien. 1986. Morphological variability and asymmetry in the cheetah *(Acinonyx jubatus),* a genetically uniform species. Evolution 40: 78–85.

Wayne, R. K., and S. J. O'Brien. 1987. Allozyme divergence within the Canidae. Syst. Zool. 36:339–55.

Weaver, J. 1978. The wolves of Yellowstone. U.S. Natl. Park Serv. Nat. Res. Rept., no. 14, 38 pp.

Weeks, J. L., G. M. Tori, and M. C. Shieldcastle. 1990. Coyotes *(Canis latrans)* in Ohio. Ohio J. Sci. 90:142–45.

Weise, T. F., W. L. Robinson, R. A. Hook, and L. D. Mech. 1975. An experimental translocation of the eastern timber wolf. Audubon Conserv. Rept., no. 5, 28 pp.

Wells, D. R. 1989. Notes on the distribution and taxonomy of peninsular Malaysian mongooses *(Herpestes).* Nat. Sci. Bull. Siam Soc. 37:87–97.

Wells, D. R., and C. M. Francis. 1988. Crab-eating mongoose. *Herpestes urva,* a mammal new to peninsular Malaysia. Malayan Nat. J. 42:37–41.

Wemmer, C. M. 1977. Comparative ethology of the large-spotted genet, *Genetta tigrina,* and some related viverrids. Smithson. Contrib. Zool., no. 239, iii + 93 pp.

Wemmer, C. M., and J. Murtaugh. 1981. Copulatory behavior and reproduction in the binturong, *Arctictis binturong.* J. Mamm. 62:342–52.

Wemmer, C. M., and D. Watling. 1986. Ecology and status of the Sulawesi palm civet, *Macrogalidia musschenbroekii* Schlegel. Biol. Conserv. 35:1–17.

Wemmer, C. M., J. West, D. Watling, L. Collins, and K. Lang. 1983. External characters of the Sulawesi palm civet, *Macrogalidia musschenbroekii* Schlegel, 1879. J. Mamm. 64:133–36.

Wendell, F. E., J. A. Ames, and R. A. Hardy. 1984. Pup dependency period and length of reproductive cycle: estimates from observations of tagged sea otters, *Enhydra lutris,* in California. California Fish and Game 70: 89–100.

Wenhold, B. A., and O. A. E. Rasa. 1994. Territorial marking in the yellow mongoose *Cynictis penicillata:* sexual advertisement for subordinates? Z. Saugetierk. 59:129–38.

Werdelin, L. 1981. The evolution of lynxes. Ann. Zool. Fennici 18:37–71.

Werdelin, L., and N. Solounias. 1990. Studies of fossil hyaenids: the genus *Adcrocuta* Kretzoi and the interrelationships of some hyaenid taxa. Zool. J. Linnean Soc. 98:363–86.

———. 1991. The Hyaenidae: taxonomy, systematics, and evolution. Fossils and strata, no. 30, 104 pp.

Whitehouse, S. J. O. 1977. The diet of the dingo in Western Australia. Austral. Wildl. Res. 4:145–50.

Whitman, J. S., W. B. Ballard, and C. L. Gardner. 1986. Home range and habitat use by wolverines in southcentral Alaska. J. Wildl. Mgmt. 50:460–63.

Wiig, O., E. W. Born, and G. W. Garner, eds. 1995. Polar bears: proceedings of the eleventh working meeting of the IUCN/SSC Polar Bear Specialist Group, 25–27 January 1993, Copenhagen, Denmark. IUCN (World Conservation Union), Gland, Switzerland, v + 192 pp.

Willems, R. A. 1995. The wolf-dog hybrid: an overview of a controversial animal. Anim. Welfare Inform. Center Newsl. 5(4):3–8.

Williams, E. S., E. T. Thorne, D. R. Kwiatkowski, S. L. Anderson, and K. Lutz. 1991. Reproductive biology and management of captive black-footed ferrets *(Mustela nigripes).* Zoo Biol. 10:383–98.

Williams, T. 1986. The final ferret fiasco. Audubon 88(3):111–19.

Wilson, D(on). E. 1991. Mammals of the Tres Marías Islands. Bull. Amer. Mus. Nat. Hist. 206:214–50.

Wilson, D(on). E., M. A. Bogan, R. L. Brownell, Jr., A. M. Burdin, and M. K. Maminov. 1991. Geographic variation in sea otters, *Enhydra lutris.* J. Mamm. 72:22–36.

Wilson, D(on). E., and D. M. Reeder, eds. 1993. Mammal species of the world: a taxonomic and geographic reference. Smithsonian Inst. Press, Washington, D.C., xviii + 1206 pp.

Wirth, R., and H. Van Rompaey. 1991. The Nilgiri marten, *Martes gwatkinsii* (Horsfield, 1851). Mustelid & Viverrid Conserv. 5:6.

Wolfe, M. L., and D. L. Allen. 1973. Continued studies of the status, socialization, and relationships of Isle Royale wolves, 1967 to 1970. J. Mamm. 54:611–33.

Woodford, M. 1994. Canine distemper epidemic in lions. Species 23:29.

Wooding, J. B., J. A. Cox, and M. R. Pelton. 1994. Distribution of black bears in the southeastern coastal plain. Proc. Ann. Conf. Southeast. Assoc. Fish and Wildl. Agencies 48.

Wooding, J. B., and T. S. Hardisky. 1990. Coyote distribution in Florida. Florida Field Nat. 18:12–14.

Wozencraft, W. C. 1986. A new species of striped mongoose from Madagascar. J. Mamm. 67:561–71.

———. 1987. Emendation of species name. J. Mamm. 68:198.

———. 1989*a*. Classification of the Recent Carnivora. *In* Gittleman (1989), 569–93.

———. 1989*b*. The phylogeny of the Recent Carnivora. *In* Gittleman (1989), 495–535.

Wrigley, R. E., and D. R. M. Hatch. 1976. Arctic fox migrations in Manitoba. Arctic 29: 147–58.

Wyss, A. R., and J. J. Flynn. 1993. A phylogenetic analysis and definition of the Carnivora. *In* Szalay, Novacek, and McKenna (1993*b*), 32–52.

X

Xanten, W. A., H. Kafka, and E. Olds. 1976. Breeding the binturong *Arctictis binturong* at the National Zoological Park, Washington. Internatl. Zoo Yearbook 16:117–19.

Xiang Peilon, Tan Bangjie, and Jia Xianggang. 1987. South China tiger recovery program. *In* Tilson and Seal (1987), 323–28.

Ximenez, A. 1975. *Felis geoffroyi*. Mammalian Species, no. 54, 4 pp.

Xu Hongfa and Sheng Helin. 1994. Reproductive behaviour of the small Indian civet *(Viverricula indica)*. Small Carnivore Conserv. 11:13–15.

Y

Yalden, D. W., M. J. Largen, and D. Kock. 1980. Catalogue of the mammals of Ethiopia. 4. Carnivora. Italian J. Zool., Suppl., n.s., 13:169–272.

Yamada, J. K., and B. S. Durrant. 1989. Reproductive parameters of clouded leopards *(Neofelis nebulosa)*. Zoo Biol. 8:223–31.

Yasuma, S. 1988. Iriomote cat: king of the night. Anim. Kingdom 91(6):12–21.

Yocom, C. F. 1974. Status of marten in northern California, Oregon, and Washington. California Fish and Game 60:54–57.

Yocom, C. F., and M. T. McCollum. 1973. Status of the fisher in northern California, Oregon, and Washington. California Fish and Game 59:305–9.

Yonzon, P. B., and M. L. Hunter, Jr. 1991*a*. Cheese, tourists, and red pandas in the Nepal Himalayas. Conserv. Biol. 5:196–202.

———. 1991*b*. Conservation of the red panda *Ailurus fulgens*. Biol. Conserv. 57:1–11.

Youngman, P. M. 1975. Mammals of the Yukon Territory. Natl. Mus. Can. Publ. Zool., no. 10, 192 pp.

———. 1982. Distribution and systematics of the European mink *Mustela lutreola* Linnaeus 1761. Acta Zool. Fennica 166:1–48.

———. 1990. *Mustela lutreola*. Mammalian Species, no. 362, 3 pp.

———. 1994. Beringian ferrets: mummies, biogeography, and systematics. J. Mamm. 75:454–61.

Youngman, P. M., and F. W. Schueler. 1991. *Martes nobilis* is a synonym of *Martes americana*, not an extinct Pleistocene-Holocene species. J. Mamm. 72:567–77.

Yurco, F. J. 1990. The cat and ancient Egypt. Field Mus. Nat. Hist. Bull. 61(2):15–23.

Z

Zannier, F. 1965. Verhaltensuntersuchungen an der Zwergmanguste *Helogale undulata rufula* in Zoologischen Garten Frankfurt am Main. Z. Tierpsychol. 22:672–95.

Zheng Shengwu, Li Guihui, Song Shiying, and Han Yiping. 1988. Study on the ecology of sand badger. Acta Theriol. Sinica 8:65–72.

Zheng Yonglie and Xu Longhui. 1983. Subspecific study on the ferret badger *(Melogale moschata)* in China, with description of a new subspecies. Acta Theriol. Sinica 3:165–71.

Zimen, E. 1975. Social dynamics of the wolf pack. *In* Fox (1975), 336–62.

———. 1981. Italian wolves. Nat. Hist. 90(2):66–81.

Zumbaugh, D. M., and J. R. Choate. 1985. Historical biogeography of foxes in Kansas. Trans. Kansas Acad. Sci. 88:1–13.

Index

The scientific names of orders, families, and genera that have titled accounts in the text are in boldfaced type, as are the page numbers on which such accounts begin. Other scientific names and vernacular names appear in roman.